The Keys to Linear Algebra

Applications, Theory, and Reasoning

By Daniel Solow

Department of Operations Research
and Operations Management

Weatherhead School of Management

Case Western Reserve University

Cleveland, Ohio 44106

Dedicated to all students of mathematics

The material in Appendix A is adapted from *How to Read and Do Proofs*, second edition, by Daniel Solow, copyright 1990, John Wiley & Sons, Inc., and is reprinted with permission from John Wiley & Sons, Inc.

Order from:

BookMasters, Inc.
PO Box 388
1444 St. Route 42
Ashland, OH 44805

Fax: (419) 281-6883
Phone: (800) 247-6553

Library of Congress Catalog Card Number: 97-73562

Solow, Daniel

    The keys to linear algebra:
    applications, theory, and reasoning

ISBN 0-9644519-2-1

Printed in the United States of America

The Keys to Linear Algebra

Applications, Theory, and Reasoning

By Daniel Solow

Order from: BookMasters, Inc.   Fax: (419) 281-6883
                  PO Box 388   Phone: (800) 247-6553
                  1444 St. Route 42
                  Ashland, OH 44805

                  ISBN: 0-9644519-2-1

Complimentary desk copies and Solutions Manuals are available to instructors who adopt this book as a text by e-mailing the course title and number, enrollment, and instructor's name and address to dxs8@po.cwru.edu, by calling (216) 368-3837, or by filling out the form on the web at the following address (where comments and suggestions for improvements are also welcomed): http://weatherhead.cwru.edu/books/linalg

The related book *The Keys to Advanced Mathematics* by Daniel Solow is available from BookMasters, at the above address (ISBN: 0-9644519-0-5). *How to Read and Do Proofs*, second ed., by Daniel Solow is available from John Wiley & Sons, Inc., New York, NY (ISBN: 0-471-51004-1).

Contents

This book is designed as a text for an introductory undergraduate course in linear algebra. This preface describes the innovative ways in which applications, theory, and general mathematical reasoning are taught and how the design and pedagogical features complement the approach.

Applications

This book is driven by applications. Each chapter begins with a realistic problem to motivate the need for learning the subsequent material. Appropriate theory and technology are then applied at the end of the chapter to solve the opening problem. Related project exercises involve the student actively in technology-based problem solving—a feature students have responded to enthusiastically in class testing. These exercises do not rely on any one specific form of technology. Many other applications are drawn from physics, statistics, business, and computer science to illustrate where and how the techniques of linear algebra apply.

Theory

Almost all of the standard topics and theorems in a first course on linear algebra are included. Proofs presented in this book are based on a systematic approach in which each proof technique is given a name that is used whenever that technique arises subsequently. An explanation of how and when to use the various techniques is provided in Appendix A, to which the student is referred when necessary. The material in Appendix A is adapted from *How to Read and Do Proofs*, second edition, by Daniel Solow and is reprinted with permission from John Wiley & Sons, Inc.

Reasoning

What makes this book unique is that, in addition to teaching the standard topics of linear algebra, a primary objective is to equip students with general problem-solving skills that are needed in their subsequent mathematics and related courses. This goal is accomplished by identifying and explaining the underlying mathematical *thinking processes* that arise in linear algebra. These thinking processes—hereafter referred to collectively as the *why* mathematics—include unification, generalization, abstraction, identifying similarities and differences, converting visual images to symbolic form and vice versa, understanding and creating definitions and proofs, and developing and working with axiomatic systems. Teaching these ideas explicitly is designed to reduce the time and frustration involved in learning linear algebra and to provide the student with a deeper and more-lasting appreciation of the subject . . . and of mathematics in general.

To teach the *why* mathematics, a discussion of such concepts as unification, generalization, abstraction, and so on, is presented when those ideas arise naturally in the context of linear algebra. Each such concept is named and then referred to whenever the idea appears subsequently in the book.

To understand and to use these techniques, students need to acquire certain basic mathematical skills. These skills include doing proofs, identifying similarities and differences, converting visual images to symbolic form (and vice versa), and understanding and creating definitions.

The ability to identify similarities and differences among various mathematical concepts is essential to unification in that like properties from different problems must be isolated, identified, and then brought together in a single framework. This skill is also used in creating definitions, where it is necessary to identify a common property of all objects being defined.

Much effort is also devoted to teaching students how to convert visual images to symbolic form. When students learn mathematics, they typically develop their own ways of imagining and picturing specific concepts—such as the projection of one vector onto another. However, it is one thing to visualize such concepts; it is another thing entirely to translate such an image to symbolic form, especially when quantifiers are involved. The art of doing so is illustrated and explained with many examples. Students are also shown how to check for syntax and logic errors that can arise in the translation process. The reverse technique of converting symbolic mathematics to visual form is also taught.

One area in which the student is given special help is in understanding definitions. Although a student may be able to visualize objects with a desirable property—such as linearly independent vectors—it is quite a challenge to create (or even to understand) the symbolic definition. Merely presenting the definition is inadequate because doing so fails to explain how the definition was arrived at and why the definition is correct. The approach taken here is to teach the student to identify similarities shared by all items having the property being defined, and then to translate those observations to symbolic form. They are also taught to verify that the definition includes all objects having the desirable property while excluding all other objects.

Pedagogical Features

The pedagogical features of this book include a realistic introductory problem at the beginning of each chapter to motivate the need for learning the material presented in the chapter. That material is then applied at the end of the chapter to solve the problem.

Each mathematical thinking process is set off in the text and is easily identified by a keys icon, as in the following example:

Another important mathematical problem-solving tool is **visualization**, in which you create a visual or mental image associated with a particular concept.

These thinking processes appear in the body of the text, in the appropriate chapter summaries, and also in their entirety in Appendix B, together with a discussion of how and when to use them and their respective advantages and disadvantages.

Each chapter contains many numbered and titled examples to illustrate the topic under discussion. Definitions and theorems have special design features for easy reference. Extensive use of figures provides the student with visual images of mathematical concepts.

Each numbered section is followed by numerous exercises, some of which give the student practice in performing mechanical computations while others test their understanding of the *why* mathematics. The last numbered section in each chapter has exercises that are designed to be solved with MATLAB, Maple, Mathematica, or a graphing calculator. Solutions to exercises whose numbers are in blue are given in the back of the book. More than just answers, these fully worked-out solutions explain *how* the answer is obtained. A student can use these exercises and solutions as self-teaching problems. The instructor may therefore want to assign for homework those exercises whose numbers are in black.

The following table provides pages where samples of many of the educational features provided in this book can be found.

Samples of Pedagogical Features

Quick-Reference Table

Pedagogical Feature	Page Numbers
Applications	
Chapter Introductory Problem	
Problem Description	1
Problem Solution	51 - 54
Technology-Based Exercises	57 - 59
Miscellaneous Applications	6 - 8
Theory	
General Approach to Proofs	26 - 27
Proof Techniques	26, 55, 105, 447 - 477
Reasoning	
Generalization	2
Closed-form Solution	10
Identifying Similarities and Differences	13
Translating Visual Images to Symbolic Form	17
Numerical-method Solution	20
Unification	45
Abstraction	175 - 177
Axiomatic System	177
Creating Definitions	199 - 203

Acknowledgments

When I first approached several colleagues to provide me with comments on a preliminary version of the book, I expected them to scan through the 500-page single-spaced manuscript and, at best, to say "I like this . . ." and "I do not like that . . .." I was overwhelemed when I received extensive line-by-line comments on every chapter that ranged from suggestions for improving the exposition, to including additional examples and exercises, redrawing figures, rewording sentences, and correcting math and spelling errors. All but one gave their comments with no financial compensation. Their objective was to help me produce the best possible text. To whatever degree I succeeded is due, in large part, to the contributions of the following dedicated people:

> Patrick Driscoll, U. S. Military Academy, NY
>
> Chungsim Han, Baldwin-Wallace College, OH
>
> Bonnie Lawrence, University of South Carolina, SC
>
> Robert Stanton, St. John's University, NY

I hope they are satisfied with the final product. I am honored that they were part of this project. I also thank Bob Moore for his comments on the first four chapters. Chapter 8 benefited from a thorough review by Noah Rhee, who specializes in numerical methods.

Classroom testing of the manuscript by the following faculty provided valuable feedback from both instructor and students:

> Ted Hodgson, Montana State University, MT
>
> Richard Little, Baldwin Wallace College, OH
>
> Robert Moore, Southern Adventist University, TN
>
> Noah Rhee, University of Missouri at Kansas City, MO

With regard to the production, it was a pleasure to work with the following professionals, without whom, this book would never have come together in its current form:

> Amy Jenkins, on the design features and the cover,
>
> Thomas Engle, on preparing the figures,
>
> Arthur Ogawa, on preparing the LATEX macros.

And of course, a thanks to Donald Knuth for developing TEX.

I am also grateful for the use of the excellent computing facilities at the Weatherhead School of Management at Case Western Reserve University and to Chris Fenton, who always managed to resolve my problems and to keep the system working properly.

Chapter 1

Euclidean Vectors

This book has two objectives. One is to present the basic concepts and techniques of **linear algebra**, which is the study of ordered lists of numbers and their applications to lines, planes, and related topics. These techniques are useful in solving a variety of problems arising in mathematics, computer science, engineering, physics, chemistry, statistics, economics, business and other areas. A second objective is to describe and illustrate a number of mathematical *thinking processes* that arise not only in linear algebra, but also in many other mathematical subjects, such as discrete mathematics, abstract algebra, advanced calculus, and the like. Understanding these thinking processes greatly reduces the time and frustration involved in learning advanced mathematics. You can also use these thinking processes to solve mathematical problems in general, as you will see in this chapter for solving the following problem.

The Routing Problem of Best Paper Products

Best Paper Products supplies local-area businesses with a variety of paper products. Each day, a truck leaves the warehouse to make deliveries to a number of customers. The order in which the deliveries are made is determined manually by plotting the locations of the day's customers on a map and then rotating a horizontal line drawn through the warehouse in the counterclockwise direction 360 degrees, picking up each customer in turn. You have been asked to automate this procedure on their computer system.

1.1 Vectors in Euclidean Space and Their Applications

You already know how to work with individual numbers, each of which typically represents a physical quantity in a problem. For some problems, however, it is necessary to work with a list of two numbers. For example, you might use the following ordered list for the height (x) and weight (y) of a person:

$$(x, y) = (73, 175)$$

The number 73 represents the height of the person in inches and the number 175 represents the weight of the person in pounds.

You can extend this concept to an ordered list of three numbers, say,

$$(x, y, z) = (73, 175, 25)$$

The third number in the list, 25, represents the age of the person. In fact, you can work with a list of 4, 5, or n real numbers (where n is a positive integer), as formalized in the following definition.

DEFINITION 1.1

For a positive integer n, a **vector u** $= (u_1, \ldots, u_n)$ (also called an n-**vector**) is an ordered list of n real numbers in which n is the **dimension** of **u**. For each $i = 1, \ldots, n$, the number u_i is called **component** i of **u**.

Observe in Definition 1.1 the advantange of using *subscript notation* to represent the components of a vector rather than, say, $(x, y, z, \ldots)$. You can also use the single symbol **u** to refer to the vector as a whole, instead of writing $(u_1, \ldots, u_n)$ each time. The choice of notation is important when dealing with mathematical concepts. From here on, real numbers are denoted by italicized lowercase letters, such as a, t, and u_1. Vectors are denoted by lowercase letters in boldface—for example, **u**, **v**, and **d**. The set of all real numbers is written as R^1 or simply R. The set of all 2-vectors is R^2. More generally, the set of all n-vectors is written as R^n and is often called **Euclidean n-space**. An n-vector is written **u** $\in R^n$, where the symbol $\in$ stands for the words "is an element of."

 The progression you have just seen from an ordered list of two numbers to an ordered list of n numbers is an example of a mathematical technique called **generalization**. Generalization is the process of creating, from an original concept (problem, definition, theorem, and so on), a more general concept (problem, definition, theorem, and so on) that includes not only the original one, but many other new ones as well.

Each of the original concepts that gives rise to the generalization is called a **special case**. In the foregoing examples, the ordered lists

$$(73, 175) \quad \text{and} \quad (73, 175, 25)$$

are special cases of an n-vector **u** $= (u_1, \ldots, u_n)$. The first ordered list, (73, 175), is a special case in which $n = 2$, $u_1 = 73$, and $u_2 = 175$, so $\mathbf{u} = (73, 175) \in R^2$. The second ordered list, $(73, 175, 25)$, is a special case in which $n = 3$, $u_1 = 73$, $u_2 = 175$, and $u_3 = 25$, so $\mathbf{u} = (73, 175, 25) \in R^3$. Observe that each of the special cases is obtained from the generalization by an appropriate *substitution of values*. Each special case of an n-vector $\mathbf{u} = (u_1, \ldots, u_n)$ is obtained by substituting specific values for n and $u_1, \ldots, u_n$.

1.1.1 The Geometry of Euclidean Vectors

You have just seen the technique of generalization.

Another important mathematical problem-solving tool is **visualization**, in which you create a visual or mental image associated with a particular concept.

To illustrate visualization, consider the following example of a vector.

EXAMPLE 1.1 Representing a Complex Number as a Vector
You know from previous studies that the value $i = \sqrt{-1}$ is used in writing a complex number such as

$$3 + 2i$$

in which 3 is the real part and $2i$ is the complex part. You can represent this complex number by the vector $\mathbf{u} = (3, 2)$ in 2-space.

One advantage of representing a complex number as a vector in 2-space is that you can create an associated visual image, as you will now see.

Visualizing a Vector in 2-Space

To visualize the vector $\mathbf{u} = (u_1, u_2)$ with two components, use a horizontal line in the plane—the x-axis—to represent the first component and a vertical line—the y-axis—to represent the second component [see Figure 1.1(a)]. Collectively, these two lines are the **coordinate axes**. The point at which the two coordinate axes intersect is called the **origin** and corresponds to the values $(0, 0)$. The arrows on the coordinate axes indicate the direction of increasing values for the corresponding component. Thus, values to the right of the origin on the x-axis in Figure 1.1(a) are positive and values to the left are negative. Similarly, values above the origin on the y-axis are positive and values below are negative.

  Given the coordinate axes, there are two standard ways to picture the vector $\mathbf{u} = (u_1, u_2)$:

1. As a *point* in the plane that is located by moving from the origin u_1 units in the appropriate direction along the x-axis and then moving from there u_2 units in the appropriate direction of the y-axis. The visual image of the vector $\mathbf{u} = (3, 2)$ in Example 1.1 that corresponds to the complex number $3 + 2i$ is shown in Figure 1.1(a).

2. As an *arrow* in the plane whose **tail** is at the origin and whose **head** is at the coordinates (u_1, u_2), as seen by the vector $(3, 2)$ in Figure 1.1(b).

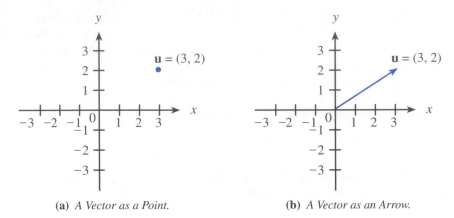

(a) *A Vector as a Point.*   (b) *A Vector as an Arrow.*

Figure 1.1 *Visualizing a Vector as a Point and as an Arrow.*

Visualizing a Vector in 3-Space

You can extend these visualizations to a vector having three components by adding a third axis—the z-axis—through the origin perpendicular to both the x-axis and the y-axis. The z-axis is used to represent the third component of the vector. For example, the vector $(2, 1, 3)$, shown as both a point and an arrow in Figure 1.2, has its tail at the origin and its head located at the point obtained by moving 2 units in the positive direction of the x-axis, then moving 1 unit in the positive direction of the y-axis, and finally moving 3 units in the positive direction of the z-axis.

You now have two different visualizations of a vector in 2-space and in 3-space—namely, as a point and as an arrow. Unfortunately, you cannot extend these visualizations to a vector having four or more components. Even though you cannot picture such vectors geometrically, you can work with them algebraically, as you will see shortly. Herein lies one of the main advantages of generalization: *the ability to create and to work with concepts that you cannot visualize*. A useful approach when working with vectors in n-space is to work first with examples in 2-space and in 3-space that you can visualize and then generalize your results algebraically to vectors in n-space.

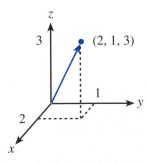

Figure 1.2 *Visualizing a Vector in 3-Space.*

1.1.2 Comparing Two Vectors

Algebra is used to work with vectors in n-space that you cannot visualize. To illustrate, observe that the components of the vectors

$$(1 + 0, 3 - 1) \quad \text{and} \quad (0 + 1, -1 + 3)$$

are written differently yet these vectors are the *same*, the algebraic meaning of which is made precise in the following definition that applies to n-vectors.

DEFINITION 1.2

The vector $\mathbf{u} = (u_1, \ldots, u_n)$ is **equal to** the vector $\mathbf{v} = (v_1, \ldots, v_n)$, written $\mathbf{u} = \mathbf{v}$, if and only if for all $i = 1, \ldots, n$, $u_i = v_i$.

This symbol "=" is used to compare two n-vectors for equality. Thus, the statement

$$\mathbf{u} = \mathbf{v}$$

means that $\mathbf{u}$ *is* equal to $\mathbf{v}$, or equivalently, the result of checking whether $\mathbf{u}$ is equal to $\mathbf{v}$ is true. Analogously, the statement

$$\mathbf{u} \neq \mathbf{v}$$

means that $\mathbf{u}$ is not equal to $\mathbf{v}$, or equivalently, the result of checking whether $\mathbf{u}$ is equal to $\mathbf{v}$ is false.

The key words *for each* (or equivalent words like *for all*, *for every*, *for any*, and so on) appearing in Definition 1.2 arise frequently in mathematical statements and are called the **universal quantifier**. A further discussion of the universal quantifier is given in Appendix A.4.

One of the advantages of creating a definition with the universal quantifier is that you can write the *negation* of the statement more easily. For example, by applying the rules for negating a statement containing the quantifier *for all* (see Appendix A.10), the statement that $\mathbf{u} \neq \mathbf{v}$ means that *there is* at least one integer i with $1 \leq i \leq n$ such that $u_i \neq v_i$. The words *there is* (or equivalent words like *there exists*, and so on) are called the **existential quantifier**. A further discussion of the existential quantifier is given in Appendix A.3.

Observe that there is a significant difference between using the symbol "=" to compare two numbers and using that same symbol to compare two n-vectors. For example, if u and v are numbers for which $u \neq v$, then you know that $u < v$ or $u > v$. This statement is not necessarily true if $\mathbf{u}$ and $\mathbf{v}$ are n-vectors. This is because

$$\mathbf{u} \neq \mathbf{v} \text{ means that there is at least one integer } i \text{ such that } u_i \neq v_i$$

However, you do not know the relationship of the other components of $\mathbf{u}$ and $\mathbf{v}$. Additional examples of using the symbols "=" and "$\neq$" to compare two n-vectors follows.

EXAMPLE 1.2 Comparing Two Vectors for Equality

The following comparisons are all true:

$$(2, 4) = (2, 4)$$

$$(2, 4) \neq (4, 2) \qquad \text{(the order is important)}$$
$$(2, -3) \neq (4, -6) \qquad \text{(note that } 2 < 4 \text{ while } -3 > -6)$$
$$(0, 1, 0) \neq (0, 0, 0)$$

Always think about what objects are being compared with the symbol "=."

1.1.3 Applications of *n*-Vectors

Various uses of n-vectors in problem solving are now presented.

Vectors as a Notational Tool

One use of vectors in problem solving is as a notational method for representing items in a problem. For example, suppose you are solving a problem that involves the polynomial

$$p(x) = 3 + 2x - x^2 + 4x^4$$

You can represent $p(x)$ as the following vector, in which each component is the coefficient of the corresponding ordered power of x in $p(x)$:

$$\mathbf{p} = (3, 2, -1, 0, 4)$$

One advantage of representing a polynomial as a vector is that you can store and manipulate this information in the memory of a computer by using a data structure called an **array**.

Another example of using vectors as a notational tool arises in the problem of Best Paper Products presented at the beginning of the chapter. To solve this problem, you need to represent the locations of the day's customers mathematically. A vector serves the purpose, but there are several ways to do so. One approach is to use a vector consisting of two components—say, (x, y)—in which the value of x is the number of miles the customer is located from the warehouse in the East-West direction. A positive x value means that the customer is located to the East of the warehouse and a negative x value means the customer is located to the West. Likewise, a positive y value indicates that the customer is located to the North of the warehouse and a negative y value means the customer is located to the South. You would then need a list of vectors—say, $(x_1, y_1), (x_2, y_2), \ldots, (x_n, y_n)$—to represent the locations of n customers.

An alternative approach is to use two n-vectors to represent the locations of the n customers in this problem. Specifically, the n-vector $\mathbf{x} = (x_1, \ldots, x_n)$ represents the list of East-West distances of the customers from the warehouse and $\mathbf{y} = (y_1, \ldots, y_n)$ represents the corresponding North-South distances. From this example, you can see that there are different ways to use vectors to represent information in solving a problem.

Vectors in the Physical Sciences

Vectors arise frequently in the study of physics, chemistry, and many other sciences. In physics, for instance, vectors often have three components—corresponding to the three dimensions of the physical world—representing properties of a particle of interest. Appropriate units of measure are chosen for the components on the basis of the quantities represented by the vector. For example, in measuring the position of a particle relative to a fixed origin using meters, the vector

$$\mathbf{r} = (3, 1, 2) \quad \text{(meters)}$$

represents a particle that is 3 meters in the positive x-coordinate direction from the origin, 1 meter in the positive y-coordinate direction, and 2 meters in the positive z-coordinate direction. The choice of units in which the distance is measured is up to you but should be specified.

Another example is a vector that describes the velocity of a particle relative to a fixed origin. For instance, if the units of measure are meters per second, then the vector

$$\mathbf{v} = (2, -1, 3) \quad \text{(meters / second)}$$

represents the fact that the particle is changing position at the rate of 2 meters per second in the direction of the positive x-axis, 1 meter per second in the direction of the negative y-axis, and 3 meters per second in the direction of the positive z-axis.

Vectors in 3-space are also used to represent such physical properties as acceleration, weight, force, torque, linear and angular momentum, electric and magnetic fields, current, and more.

Binary Vectors

Vectors are also useful in representing "yes/no" information. To illustrate, consider the following results for one student on five questions on the Scholastic Aptitude Test (SAT), in which "C" means the student selected the correct answer and "I" means the student selected an incorrect answer:

Question number	1	2	3	4	5
Result	C	I	C	C	I

One way to represent these results mathematically is by the vector

$$\mathbf{r} = (r_1, r_2, r_3, r_4, r_5) = (1, 0, 1, 1, 0)$$

The numerical value of 1 for a component of $\mathbf{r}$ means that the student selected the correct answer to the corresponding question. A value of 0 indicates that the student selected an incorrect answer to that question.

In this example, the value of each component of the vector $\mathbf{r}$ is 1 or 0 and is therefore called a **binary vector**. Binary vectors are typically used to represent "yes/no," "true/false," or similar types of information in which there are two possibilities. The choice of which possibility to represent by the value 1 and which to represent by the value 0 is up to you.

Permutation Vectors

To illustrate another use of vectors, suppose that a medical study involves the five patients whose names and ages are given in the following list:

Number	Name	Age
1	Robert Smith	25
2	Ann Jones	32
3	Ed Monroe	20
4	Betty Baker	22
5	Hector Lopez	28

One of the most important problems in computer science is to sort information in increasing or decreasing order. For instance, you might want to display the foregoing list in increasing order of age. You can easily do so in this example because there are only five patients. When there are, say, 3000 patients, vectors provide a useful method for sorting information.

To see how vectors are used in this example, consider a vector $\mathbf{a} = (25, 32, 20, 22, 28)$ whose components represent the ages of the five patients. You can sort these numbers in increasing order to obtain the vector: $(20, 22, 25, 28, 32)$. However, doing so does not easily allow you to list the patients in the correct order.

An alternative approach is to create a vector $\mathbf{p}$ whose components represent the order in which to list the patients. In this example, to list the patients in increasing order of age, use the following vector:

$$\mathbf{p} = (3, 4, 1, 5, 2) \tag{1.1}$$

The vector $\mathbf{p}$ in (1.1) indicates that patient number 3 is listed first, patient 4 is listed second, patient 1 is third, patient 5 is fourth, and patient 2 is fifth.

The vector $\mathbf{p}$ in this example is called a **permutation vector** of the set $\{1, 2, 3, 4, 5\}$, meaning a rearrangement of the five elements in this set. By using the permutation vector in (1.1), you can create the following sorted vector associated with the age vector $\mathbf{a} = (25, 32, 20, 22, 28)$:

$$\mathbf{a_p} = (a_3, a_4, a_1, a_5, a_2) = (20, 22, 25, 28, 32)$$

By listing the patients in the order indicated by the components of the permutation vector $\mathbf{p}$ in (1.1), you obtain the following list of patients in increasing order of age:

Number	Name	Age
3	Ed Monroe	20
4	Betty Baker	22
1	Robert Smith	25
5	Hector Lopez	28
2	Ann Jones	32

1.1.4 Coordinate Systems

Whenever you write the components of a vector, *you do so relative to a coordinate system that you are free to choose*. For instance, recall the example of a vector whose two components represent the height and weight of a person, respectively. If the corresponding units of measure are inches and pounds, then the vector

$$(73, 175) \qquad\qquad (1.2)$$

represents a person who is 73 inches tall and weighs 175 pounds. Alternatively, you can choose a coordinate system in which the units are meters and kilograms. In this case, the components of the vector in (1.2) are

$$(1.85, 79.55) \qquad \text{(approximately)} \qquad (1.3)$$

From this example you can see that, when choosing a coordinate system, it is important to specify the units of measurement on each coordinate axis because the visual representation of the vector can change according to the units chosen. For example, the vector in (1.3) might be drawn closer to the origin than the vector in (1.2).

Translating the Axes

Having chosen appropriate units for each of the coordinate axes, the origin is the point corresponding to all values being 0. Sometimes, however, a different choice for the location of the origin simplifies the process of finding the solution to a problem. For example, in the problem of Best Paper Products, if the locations of the customers are initially given relative to an origin located at, say, city hall, then a more suitable origin for solving this problem is the warehouse.

When changing the origin of a coordinate system in two dimensions, you have the original xy-coordinate system with its origin and a new **translated $x'y'$-coordinate system** with its origin located at the point (a, b) in the original xy-coordinate system (see Figure 1.3).

A fixed vector (point) in the plane now has two sets of components—one relative to the xy-coordinate system and one relative to the $x'y'$-coordinate system. For example, in Figure 1.3, the point corresponding to the origin of the original coordinate system has components

$$(0, 0) \quad \text{relative to the origin of the } xy\text{-coordinate system}$$

and components

$$(0 - a, 0 - b) = (-a, -b) \quad \text{relative to the origin of the } x'y'\text{-coordinate system}$$

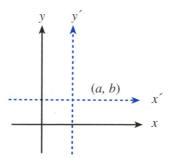

Figure 1.3 *A Translated Coordinate System.*

More generally, a given vector **u** with components

(u_1, u_2) relative to the origin of the xy-coordinate system

has components

$(u_1 - a, u_2 - b)$ relative to the origin of the $x'y'$-coordinate system

---

EXAMPLE 1.3 Expressing a Vector in Terms of a Translated Coordinate System

Suppose the components, in miles from city hall, of a customer of Best Paper Products are

$$(5, 1) \tag{1.4}$$

If the origin of that coordinate system is translated to the warehouse located at the point $(2, -3)$ (also relative to city hall), then the components of the vector in (1.4) relative to the warehouse are

$(5 - 2, 1 - (-3)) = (3, 4)$

In reverse, if the components of a vector relative to the warehouse are

$(4, -2)$

then the components of that vector relative to city hall are

$(4 + 2, -2 + (-3)) = (6, -5)$

You can generalize the translation process to an n-vector $\mathbf{u} = (u_1, \ldots, u_n)$ whose components are expressed relative to the origin of an original coordinate system. If you create a new coordinate system by translating the axes to the point $\mathbf{a} = (a_1, \ldots, a_n)$, then the components of **u** relative to the origin of the new coordinate system are

$$(u_1 - a_1, \ldots, u_n - a_n) \tag{1.5}$$

Vice versa, if the components of a vector $\mathbf{u} = (u_1, \ldots, u_n)$ are expressed relative to the origin of the new coordinate system, then the components of **u** relative to the origin of the original coordinate system are

$(u_1 + a_1, \ldots, u_n + a_n)$

Obtaining the formula in (1.5) is an example of one of the common ways in which mathematics is used to solve problems, namely, to use items—called **data**—that you know (the components of the n-vectors $u = (u_1, \ldots, u_n)$ and $a = (a_1, \ldots, a_n)$, in this case) to find items that you do not know but would like to know (the components of u relative to the origin of the translated coordinate system, in this case). For this problem, it is possible to derive a **closed-form solution**, which is a solution obtained from the problem data by a simple rule or formula, as in (1.5).

Another useful geometric property of an n-vector **v** is that this vector does not change when you move **v** parallel to itself. To see why this is so, consider moving the vector $\mathbf{v} = (v_1, v_2)$ in Figure 1.4(a) parallel to itself so that its tail is now at the point whose

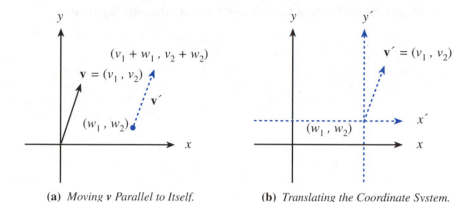

(a) *Moving v Parallel to Itself.*   (b) *Translating the Coordinate System.*

Figure 1.4 *Moving a Vector Parallel to Itself.*

coordinates are, say, (w_1, w_2). Call the resulting line segment $\mathbf{v}'$ and note that the head of $\mathbf{v}'$ is at the coordinates $(v_1 + w_1, v_2 + w_2)$. By translating the axes to the point (w_1, w_2), as shown in Figure 1.4(b), you can now see that $\mathbf{v}' = \mathbf{v}$ because the components of $\mathbf{v}'$ relative to the translated coordinate system are:

$$\mathbf{v}' = (v_1 + w_1 - w_1, v_2 + w_2 - w_2) = (v_1, v_2) = \mathbf{v}$$

Thus, $\mathbf{v}'$ and $\mathbf{v}$ are the same according to Definition 1.2, so moving a vector parallel to itself does not change the vector.

Other Coordinate Systems

Translation is not the only way to change a coordinate system. Another possibility is *rotation*. For example, keeping the origin fixed, you might rotate the axes 45 degrees in the counterclockwise direction, as in Figure 1.5. Once again, each vector in 2-space has two sets of components—one in the original coordinate system and one in the rotated system.

Up until now, the coordinate axes have been pairwise perpendicular. Such a coordinate system is called a **rectangular coordinate system**. However, in general, you can use a coordinate system that is not rectangular, such as the one shown in Figure 1.6.

Keep in mind that you are free to choose the coordinate system. Base your choice on one that facilitates obtaining the solution to your problem.

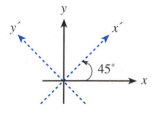

Figure 1.5 *A Rotated Coordinate System.*

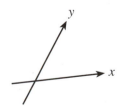

Figure 1.6 *A Coordinate System that is Not Rectangular.*

Right-handed and Left-handed Coordinate Systems

Rectangular coordinate systems in 3-space are either **right-handed** or **left-handed**. The system is right handed if, when you hold the z-axis in the palm of your right hand with the thumb pointing in the positive z direction and your other fingers pointing in the positive direction of the x-axis, you reach the positive positive y-axis by curling your fingers through a 90° counterclockwise rotation—as in Figure 1.7(a). The system is left handed if, when you hold the z-axis in the palm of your left hand with the thumb pointing in the positive z direction and your other fingers pointing in the positive direction of the x-axis, you reach the positive y-axis by curling your fingers through a 90° clockwise rotation—as in Figure 1.7(b). From here on, only right-handed coordinate systems are used.

Now that you have been introduced to various coordinate systems, you will learn about certain special vectors in n-space.

1.1.5 The Zero Vector and the Standard Unit Vectors

One special vector that arises frequently is the **zero vector**, denoted by **0**. The zero vector is the vector each of whose components is the real number 0. For example, the zero vectors in R^2 and R^3 are:

$$\mathbf{0} = (0, 0) \quad \text{(the zero vector in } R^2\text{)}$$

$$\mathbf{0} = (0, 0, 0) \quad \text{(the zero vector in } R^3\text{)}$$

The dimension of the zero vector is generally understood from the context. Geometrically, the zero vector is the point located at the origin of the coordinate system. Compare the zero vector in 2-space to other vectors in 2-space. What similarities and differences can you identify? For example, the zero vector in 2-space has two components, just like every other vector in 2-space. However, the zero vector has no length or direction, unlike other vectors in 2-space, such as the one in Figure 1.1(b).

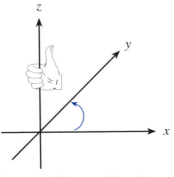

(a) *A Right-handed Coordinate System.*

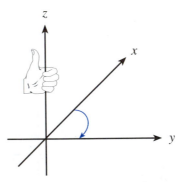

(b) *A Left-handed Coordinate System.*

Figure 1.7 *Right-handed and Left-handed Coordinate Systems.*

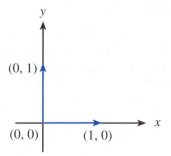

Figure 1.8 *The Standard Unit Vectors in Two Dimensions.*

Here you see another mathematical technique: **identifying similarities and differences**. This is the process of comparing and contrasting two or more mathematical concepts (problems, definitions, theorems, and so on) with the objective of gaining a deeper understanding of the relationship of these two items.

This skill is useful not only in mathematics, but in many other fields as well. For example, when a group of people suddenly develop the same illness, a doctor tries to identify what those people have in common in an attempt to discover the source of the illness. Identifying similarities and differences is used frequently throughout this book.

Returning to the topic of linear algebra, other n-vectors that arise in applications are the **standard unit vectors**, each of which is similar to the zero vector with the exception of one component, whose value is 1. The following two standard unit vectors in R^2 are illustrated in Figure 1.8:

$$(1, 0) \quad \text{and} \quad (0, 1)$$

The three standard unit vectors in R^3 are commonly denoted by $\mathbf{i}$, $\mathbf{j}$, and $\mathbf{k}$, that is,

$$\mathbf{i} = (1, 0, 0), \quad \mathbf{j} = (0, 1, 0), \quad \text{and} \quad \mathbf{k} = (0, 0, 1)$$

In n dimensions, the n standard unit vectors, denoted by $\mathbf{e}_1, \ldots, \mathbf{e}_n$, are:

$$\mathbf{e}_1 = (1, 0, 0, \ldots, 0)$$
$$\mathbf{e}_2 = (0, 1, 0, \ldots, 0)$$
$$\vdots \vdots \qquad \vdots$$
$$\mathbf{e}_n = (0, 0, \ldots, 0, 1)$$

This section has provided an introduction to vectors and their role in problem solving. You have seen the geometry of vectors and the need to choose an appropriate coordinate system in which to express the components of a vector. You have learned about the zero vector and the standard unit vectors and seen how to compare two vectors for equality and inequality. In the next section, numerous other operations on vectors are presented.

Exercises for Section 1.1

1. Write each of the following as a single vector. Indicate the units of measure for each component. What is the dimension of the vector?

   (a) A vector to represent the fact that the library is located 3 miles west and 4 miles north of city hall.

   (b) A vector whose first component is the straight-line distance from city hall to the library in part (a) and whose second component is the number of minutes needed to drive that distance to the library. Assume that the driver averages 25 miles per hour.

2. Write each of the following as a single vector. Indicate the units of measure for each component. What is the dimension of the vector?

   (a) A vector to represent the physical dimensions, the surface area, and the volume of a closed cardboard box having a height of 54 centimeters, a length of 66 centimeters, and a width of 42 centimeters.

   (b) A vector whose components are integers (whole numbers) to represent the fact that an experiment was conducted on November 25, 1996 at 10:26 P.M. (as opposed to 10:26 A.M.).

3. Write each of the following as a single vector. Indicate the units of measure for each component. What is the dimension of the vector?

   (a) Consider a cube in which each edge is 5 inches long and is drawn from the origin. Use a vector to describe the point of the cube that is diagonally opposite to, and farthest away from, the origin.

   (b) A vector to represent the list of results (heads or tails) of flipping a coin n times.

4. Write each of the following as a single vector. Indicate the units of measure for each component. What is the dimension of the vector?

   (a) A vector to represent that Projects 1 and 5 are to be funded and Projects 2, 3, 4, and 6 are not to be funded.

   (b) How do you use the vector in part (a) to determine the total number of projects that are funded?

5. (a) Three sides of a cube are colored red, two sides are colored blue, and the remaining side is white. The cube is tossed in the air and lands randomly on one of its sides. Create a vector whose components represent the probability of each color occurring on the top of the cube after a single toss.

   (b) The vector in part (a) is called a *probability vector* because each component of the vector represents the likelihood of a particular event occurring. What mathematical properties does an n-vector $\mathbf{x}$ have to satisfy to be a probability vector for n events?

6. Three jobs are to be performed on a machine, one at a time. The first job requires 50 minutes, the second job takes 30 minutes, and the third job takes 40 minutes. If the jobs are done in this order, then the first job is completed after 50 minutes, the second job is completed 30 minutes later at time 80, and the last job is completed 40 minutes after the second job, namely, at time 120. The average of these completion times is $(50 + 80 + 120)/3 = 250/3$. To maximize average customer satisfaction, the objective is to determine the order in which to perform the jobs so as to minimize the average of the times at which all three jobs are completed.

(a) Create a vector that represents the data in this problem.

(b) Define a vector whose components represent the solution to this problem. What do the components of this vector mean in the context of this problem?

(c) Solve this problem by performing the following steps:

    (i) List all permutation vectors of $\{1, 2, 3\}$.

    (ii) For each permutation in part (i), compute the average of the completion times associated with performing the jobs in the order indicated by the associated permutation vector.

    (iii) Identify the permutation in part (ii) that provides the minimum average completion time of all jobs.

7. Draw the following vectors.

    (a) The vector in Exercise 1(a).   (b) $(1, 1), (1, -1), (-1, 1), (-1, -1)$.

8. Draw the following vectors.

    (a) $(-1, 0, 1)$.   (b) $(0, 0, -1)$.

9. Draw the following vectors.

    (a) The vectors corresponding to the complex numbers $1 + 2i$ and $2 - i$. What is the geometric relation between these two vectors?

    (b) A vector that is twice as long as the vector $(-1, -2)$ and points in the opposite direction. What are the components of this vector?

10. Draw the following vectors.

    (a) The vectors $(1, 3)$ and $(3, 1)$. Explain how this shows that the order in which the components of a vector are listed is important.

    (b) A vector perpendicular to $(3, 1)$. What are the components of your vector?

    (c) The vector from the point $(1, 3)$ to the point $(3, 1)$. What are the components of this vector? (Hint: Move this vector parallel to itself so that the resulting vector is drawn from the origin.)

11. Indicate whether each of the following comparisons is true.

    (a) $(1, 0) = (0, 1)$.

    (b) $(0, 1, 0) \neq \mathbf{k}$.

    (c) $(1 - 1, 2 + (-2)) = \mathbf{0}$.

12. Indicate whether each of the following comparisons is true.

    (a) $(1, 0) \neq (0, 1)$.

    (b) $(0, 1, 0) \neq \mathbf{j}$.

    (c) $(1 - 1, 2 + (-2), 3 - 2) \neq \mathbf{0}$.

13. It is possible to compare two n-vectors with $\leq$. In particular, the n-vector $\mathbf{u} = (u_1, \ldots, u_n)$ is **less than or equal to** the n-vector $\mathbf{v} = (v_1, \ldots, v_n)$, written $\mathbf{u} \leq \mathbf{v}$, if and only if for each $i = 1, \ldots, n, u_i \leq v_i$. Also, $\mathbf{u} < \mathbf{v}$ means that for each $i = 1, \ldots, n, u_i < v_i$.

    (a) Is $(-1, 0) \leq (0, 0)$? Is $(-1, 0) < (0, 0)$?

    (b) For n-vectors $\mathbf{u}$ and $\mathbf{v}$, what is the negation of the statement $\mathbf{u} \leq \mathbf{v}$?

    (c) Suppose that $\mathbf{u} = (1, 2, -1)$ and $\mathbf{v} = (3, -1, 0)$. Is $\mathbf{u} \leq \mathbf{v}$? Is $\mathbf{v} \leq \mathbf{u}$? Why or why not? Explain.

    (d) For numbers u and v, it is true that either $u \leq v$ or $v \leq u$. Is the same true for n-vectors $\mathbf{u}$ and $\mathbf{v}$? Why or why not? Explain.

(e) Draw the set of all vectors in 2-space that are less than or equal to a given vector $\mathbf{u} = (u_1, u_2)$.

14. It is possible to compare two n-vectors with $\geq$. In particular, the n-vector $\mathbf{u} = (u_1, \ldots, u_n)$ is **greater than or equal to** the n-vector $\mathbf{v} = (v_1, \ldots, v_n)$, written $\mathbf{u} \geq \mathbf{v}$, if and only if for each $i = 1, \ldots, n$, $u_i \geq v_i$. Also, $\mathbf{u} > \mathbf{v}$ means that for each $i = 1, \ldots, n$, $u_i > v_i$.
   (a) Is $(1, 2) \geq \mathbf{0}$? Is $(1, 2) > \mathbf{0}$?
   (b) For n-vectors $\mathbf{u}$ and $\mathbf{v}$, what is the negation of the statement $\mathbf{u} \geq \mathbf{v}$?
   (c) Suppose that $\mathbf{u} = (1, 2, -1)$ and $\mathbf{v} = (3, -1, 0)$. Is $\mathbf{u} \geq \mathbf{v}$? Is $\mathbf{v} \geq \mathbf{u}$? Why or why not? Explain.
   (d) For numbers u and v, it is true that either $u < v$, $u = v$, or $u > v$. Is the same true for n-vectors $\mathbf{u}$ and $\mathbf{v}$? Why or why not? Explain.
   (e) Draw the set of all vectors in 2-space that are greater than or equal to a given vector $\mathbf{u} = (u_1, u_2)$.

15. (a) What are the components of the vector in the solution to Exercise 2(a) when the units are inches instead of centimeters? Assume that 1 inch equals 2.54 centimeters.
   (b) What are the components of the vector $(-1, 0, 3)$ when the origin of the coordinate system is moved to the point $(2, 4, 1)$?

16. Suppose that $\mathbf{u}$ is a vector whose components are (a, b), relative to the origin of the plane. Consider now a given point (c, d) in the plane and draw a horizontal and a vertical line through this point, as shown in the following figure:

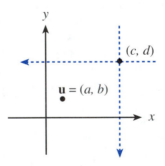

Consider a new coordinate system obtained by turning the page upside down and viewing the point (c, d) as the origin. Find a closed-form expression for the components of $\mathbf{u}$ relative to the origin of this new coordinate system in terms of $a, b, c,$ and d.

1.2 Arithmetic Operations on Vectors

In this section, various arithmetic operations on vectors used in problem solving are described. Because a vector has both a geometric representation (as a point or as an arrow) and an algebraic representation (as an ordered list of real numbers), each operation has a geometric representation with an algebraic counterpart. The advantage of the geometric version is that you can visualize the operation, at least for vectors in 2-space and 3-space. The algebraic version has two distinct advantages: (1) the algebraic operation is applicable to vectors in n-space, for any positive integer n and (2) computers are capable of performing algebraic operations but not geometric ones.

When an operation on a vector is described in an algebraic (or *symbolic*) form, use visualization to create an appropriate image of the operation.

Similarly, for each operation described in a geometric form, you should create a corresponding symbolic form by using a skill hereafter referred to as **translating visual images to symbolic form.**

To illustrate both the technique of visualization and its counterpart, translating visual images to symbolic form, some operations are described in this section first in geometric form and then translated to symbolic form. Other operations are described first in symbolic form and then visualization is used to create an appropriate geometric image.

An operation that is applied to only one vector at a time is called a **unary operation**. An operation that combines two vectors is called a **binary operation**. The unary and binary operations on vectors described in this section fall into one of the following four categories:

1. Unary operations on an n-vector that result in a real number.
2. Unary operations on an n-vector that result in an n-vector.
3. Binary operations that combine two n-vectors to yield a real number.
4. Binary operations that combine two n-vectors to produce another n-vector.

1.2.1 The Length of a Vector

As you have seen geometrically, associated with each vector is a length. You will now see how to use the individual components to compute the length of a vector.

Computing the Length of a Vector

When a vector is drawn pointing out of the origin, the length of the vector indicates how far the head is from the origin. For example, if $\mathbf{u} = (3, 4)$ represents the location of a customer of Best Paper Products relative to the warehouse, then, as seen in Figure 1.9, the straight-line distance in miles from the origin (warehouse) to $(3, 4)$ is given by the Pythagorean theorem:

$$\sqrt{(3-0)^2 + (4-0)^2} = 5 \quad \text{miles}$$

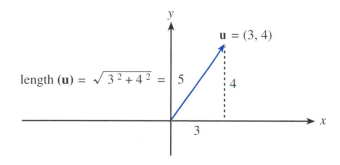

Figure 1.9 *Finding the Length of a Vector.*

For a general vector $\mathbf{u} = (u_1, u_2)$ in 2-space, the distance from (u_1, u_2) to the origin is

$$\sqrt{(u_1 - 0)^2 + (u_2 - 0)^2} = \sqrt{u_1^2 + u_2^2}$$

Generalizing to a vector $\mathbf{u} = (u_1, u_2, u_3)$ in 3-space, you have

$$\text{length}(\mathbf{u}) = \sqrt{u_1^2 + u_2^2 + u_3^2}$$

For example, if the velocity of a particle in meters per second is denoted by the vector $\mathbf{v} = (3, -1, 2)$, then the *speed* of the particle is the length of the velocity vector, that is,

$$\text{speed} = \text{length}(\mathbf{v}) = \sqrt{3^2 + (-1)^2 + 2^2} = 3.742 \quad \text{meters / second}$$

Generalizing to n dimensions results in the following closed-form expression for the length of an n-vector in terms of its components.

DEFINITION 1.3

The **length of an n-vector $\mathbf{u} = (u_1, \ldots, u_n)$** whose tail is at the origin is denoted by $\|\mathbf{u}\|$ and is computed by the following formula:

$$\|\mathbf{u}\| = \sqrt{u_1^2 + \cdots + u_n^2} \tag{1.6}$$

The length of an n-vector is also called the **norm of the vector**.

The norm is a unary operation on a vector that results in a real number. A numerical example follows.

EXAMPLE 1.4 Computing the Length of a Vector
If $\mathbf{u} = (1, -2, 3, 0)$, then

$$\|\mathbf{u}\| = \sqrt{1^2 + (-2)^2 + 3^2 + 0^2} = \sqrt{14}$$

Computing the Distance Between Two Vectors

The norm is also used to compute the *distance between two vectors*. To see how this is done, visualize two vectors $\mathbf{u} = (u_1, u_2)$ and $\mathbf{v} = (v_1, v_2)$ as two points in the plane corresponding, say, to the location of two customers of Best Paper Products. As seen in Figure 1.10, from the Pythagorean theorem, the distance from the customer located at $\mathbf{u}$ to the customer located at $\mathbf{v}$ is

$$d = \sqrt{(u_1 - v_1)^2 + (u_2 - v_2)^2}$$

Generalizing to n-vectors $\mathbf{u} = (u_1, \ldots, u_n)$ and $\mathbf{v} = (v_1, \ldots, v_n)$,

$$\text{Distance from } \mathbf{u} \text{ to } \mathbf{v} = \sqrt{(u_1 - v_1)^2 + \cdots + (u_n - v_n)^2} \tag{1.7}$$

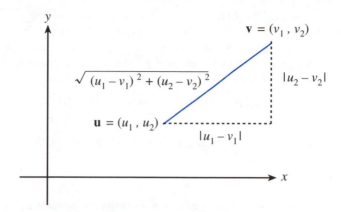

Figure 1.10 *Computing the Distance Between Two Vectors.*

1.2.2 Reversing a Vector and Multiplying a Vector by a Real Number

The operation of computing the length of a vector is a unary operation that results in a real number. The next unary operations create an n-vector from an n-vector.

Reversing a Vector

Another useful operation "reverses" an n-vector **u** to produce the new n-vector, $-\mathbf{u}$, having the same length as **u** but pointing in the opposite direction, as illustrated in Figure 1.11.

To translate the visual image in Figure 1.11 to symbolic form, compare the components of **u** to those of $-\mathbf{u}$ for numerous specific examples in 2-space. Doing so should lead you to the conclusion that when $\mathbf{u} = (u_1, u_2)$,

$$-\mathbf{u} = (-u_1, -u_2)$$

Generalizing to an n-vector $\mathbf{u} = (u_1, \ldots, u_n)$, the closed-form expression for $-\mathbf{u}$ is

$$-\mathbf{u} = (-u_1, \ldots, -u_n) \tag{1.8}$$

For example, if $\mathbf{u} = (2, -3)$, then $-\mathbf{u} = (-2, 3)$.

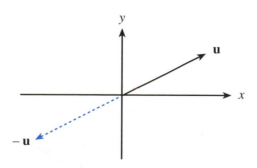

Figure 1.11 *Reversing a Vector in 2-Space.*

As an example of reversing a vector in problem solving, recall, from Section 1.1.3, the vector $\mathbf{a} = (25, 32, 20, 22, 28)$ that represents the ages of five patients. Suppose you have a method for sorting the components of a vector in increasing order, but that you now want to sort the components in decreasing order. Rather than writing a new sorting procedure, you can can use the existing one by performing the following steps:

Step 1. Reverse the vector $\mathbf{a}$ to obtain $-\mathbf{a} = (-25, -32, -20, -22, -28)$.

Step 2. Sort the vector $-\mathbf{a}$ in increasing order: $(-32, -28, -25, -22, -20)$.

Step 3. Reverse the vector in Step 2 to obtain $(32, 28, 25, 22, 20)$, which is the original vector sorted in decreasing order.

The three steps just described for sorting the components of a vector in decreasing order are an example of a **numerical-method solution**, which is a solution obtained from the problem data by performing a sequence of computations.

Comparing Numerical-method and Closed-form Solutions

To illustrate the difference between a closed-form and a numerical-method solution, consider the problem of finding the area of a circle of radius $r > 0$. The closed-form solution in terms of the known radius r is

$$\text{area} = \pi r^2$$

In contrast, an example of a numerical-method solution for this problem is to perform the following computations:

1. Divide the circle into n triangles, as in Figure 1.12, where n is a positive integer of your choosing.

2. Compute the area of each triangle.

3. Add the areas of all n triangles and use the resulting number as an approximate value for the area of the circle.

This example illustrates the advantages of a closed-form solution over a numerical-method solution. You can perform algebraic operations on a closed-form solution, which you cannot do with a numerical-method solution. A closed-form solution generally requires less computational effort to evaluate than does a numerical method (which is the case in

Figure 1.12 *A Numerical-method Solution for Finding the Area of a Circle.*

this example if you know an approximate value for π). A closed-form expression, such as πr^2, provides an exact solution to a problem (although the specific value of π can only be approximated), while a numerical method might produce only an approximate solution. For these reasons, a closed-form solution is generally preferred. When you are unable to find a closed-form solution, the next-best alternative is to develop a numerical-method solution, if possible.

Multiplying a Vector by a Real Number

A generalization of reversing a vector arises when you consider multiplying each component of a given vector $\mathbf{u} = (u_1, \ldots, u_n)$ by an arbitrary real number, t, to obtain the new vector $t\mathbf{u}$. This operation involves both a real number t and an n-vector $\mathbf{u}$ and is described symbolically as follows:

$$t\mathbf{u} = t(u_1, \ldots, u_n) = (tu_1, \ldots, tu_n) \tag{1.9}$$

Observe that $t\mathbf{u}$ is a generalization of $-\mathbf{u}$ because when t in (1.9) is replaced with the specific value -1, the result is (1.8).

One use of multiplying a vector by a real number is to change the units of the vector. For example, suppose that

$$\mathbf{a} = \left(\frac{\pi}{2}, \frac{\pi}{3}, \frac{\pi}{4} \right)$$

is a list of three angles, measured in radians. Multiplying $\mathbf{a}$ by the real number $t = 180/\pi$, you obtain the following vector $t\mathbf{a}$ whose components are those of $\mathbf{a}$ expressed in degrees:

$$t\mathbf{a} = \frac{180}{\pi} \left(\frac{\pi}{2}, \frac{\pi}{3}, \frac{\pi}{4} \right) = \left(\frac{180}{\pi} \cdot \frac{\pi}{2}, \frac{180}{\pi} \cdot \frac{\pi}{3}, \frac{180}{\pi} \cdot \frac{\pi}{4} \right) = (90, 60, 45)$$

Another example in which a vector is multiplied by a real number arises in the definition of *force* in physics. The force, $\mathbf{f}$, on a particle whose mass is $m = 3$ kilograms and whose acceleration is given by the vector $\mathbf{a} = (1, -2, 0)$ meters per second squared is the following vector:

$$\mathbf{f} = m\mathbf{a} = 3(1, -2, 0) = (3, -6, 0) \quad \text{kilogram-meters / second}^2$$

Another such example is *linear momentum* of a particle in physics. For example, the linear momentum of a particle whose mass is $m = 2$ kilograms and whose velocity is given by the vector $\mathbf{v} = (3, 1, 2)$ meters per second, as measured from a fixed origin, is described by the following vector:

$$\mathbf{p} = m\mathbf{v} = 2(3, 1, 2) = (6, 2, 4) \quad \text{kilogram-meters / second}$$

Other instances of multiplying a vector by a real number are presented in the following numerical example.

EXAMPLE 1.5 Multiplying a Vector by a Real Number

The following shows the vector $t\mathbf{u}$ for various values of the real number t and the vector $\mathbf{u}$:

t	$\mathbf{u}$	$t\mathbf{u}$
2	$(-3, 4)$	$(-6, 8)$
-2	$(-3, 4)$	$(6, -8)$
3	$(1, 0, -2)$	$(3, 0, -6)$
4	$(0, 1, 0, -1)$	$(0, 4, 0, -4)$

The next step is to create a geometric visualization of $t\mathbf{u}$. For a fixed vector $\mathbf{u}$, draw vectors $t\mathbf{u}$ in two dimensions for several specific values of t. You should conclude that the operation $t\mathbf{u}$ creates a vector whose length is $|t|$ times the length of $\mathbf{u}$. The direction of $t\mathbf{u}$ is the same as that of $\mathbf{u}$ if $t > 0$ and opposite to that of $\mathbf{u}$ if $t < 0$, as shown in Figure 1.13. If $t = 0$, then $t\mathbf{u}$ is the zero vector, that is, $0\mathbf{u} = \mathbf{0}$.

1.2.3 Adding and Subtracting Vectors

Two useful binary operations that combine two vectors to produce a new vector are addition and subtraction, the meanings of which are now made precise.

Adding Vectors

For the vectors $\mathbf{u}$ and $\mathbf{v}$ drawn from the origin in two dimensions, the operation of addition results geometrically in the new vector $\mathbf{u} + \mathbf{v}$ whose tail is at the origin and whose head is that of $\mathbf{v}$ *after* moving $\mathbf{v}$ parallel to itself until the tail of $\mathbf{v}$ coincides with the head of $\mathbf{u}$, as seen in Figure 1.14.

Look at the components of several specific examples of vectors $\mathbf{u}$, $\mathbf{v}$, and $\mathbf{u} + \mathbf{v}$ in two dimensions and then translate this geometric operation to algebraic form. Specifically, express the unknown components of $\mathbf{u} + \mathbf{v}$ in terms of the known components of $\mathbf{u}$ and $\mathbf{v}$.

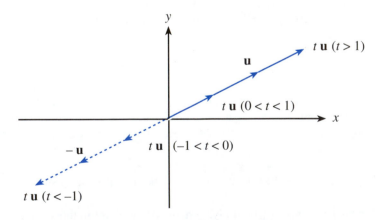

Figure 1.13 *The Vectors $t\mathbf{u}$ When $t > 0$ and When $t < 0$.*

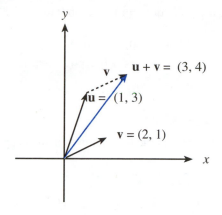

Figure 1.14 *The Geometry of Adding Two Vectors.*

You should come to the following closed-form solution:

$$\mathbf{u} + \mathbf{v} = (u_1 + v_1, u_2 + v_2)$$

Generalizing to *n*-vectors $\mathbf{u} = (u_1, \ldots, u_n)$ and $\mathbf{v} = (v_1, \ldots, v_n)$ results in

$$\mathbf{u} + \mathbf{v} = (u_1 + v_1, \ldots, u_n + v_n) \quad \text{(adding } n\text{-vectors)}$$

For example, if $\mathbf{u} = (1, -2, 2)$ and $\mathbf{v} = (0, 4, -5)$, then

$$\mathbf{u} + \mathbf{v} = (1, -2, 2) + (0, 4, -5) = (1 + 0, (-2) + 4, 2 + (-5)) = (1, 2, -3)$$

Other examples of vector addition follow.

EXAMPLE 1.6 Adding Two Vectors
The following table shows the vector $\mathbf{u} + \mathbf{v}$ for various vectors $\mathbf{u}$ and $\mathbf{v}$:

$\mathbf{u}$	$\mathbf{v}$	$\mathbf{u} + \mathbf{v}$
$(1, 2)$	$(3, -1)$	$(4, 1)$
$(1, 0, 1)$	$(0, 1, 0)$	$(1, 1, 1)$
$(2, 0, -4, 1)$	$(-1, 2, 7, 3)$	$(1, 2, 3, 4)$
$(1, -3, 2, -4)$	$(-1, 3, -2, 4)$	$(0, 0, 0, 0)$

To illustrate vector addition in problem solving, suppose that $\mathbf{u} = (-30, 45, -60)$ is a list of the degrees of the angles needed to rotate the x-axis to reach each of three points in the plane, with a negative angle representing clockwise rotation and a positive angle indicating counterclockwise rotation. You can express all angles in terms of counterclockwise rotation by adding to **u** the vector $\mathbf{v} = (360, 0, 360)$ to obtain

$$\mathbf{u} + \mathbf{v} = (-30, 45, -60) + (360, 0, 360)$$
$$= (-30 + 360, 45 + 0, -60 + 360)$$
$$= (330, 45, 300)$$

Subtracting Vectors

In a similar manner, subtracting vectors is defined algebraically as follows:

$$\mathbf{u} - \mathbf{v} = (u_1 - v_1, \ldots, u_n - v_n) \quad \text{(subtracting } n\text{-vectors)}$$

For example, if $\mathbf{u} = (1, 2)$ and $\mathbf{v} = (3, 1)$, then

$$\mathbf{u} - \mathbf{v} = (1, 2) - (3, 1) = (1 - 3, 2 - 1) = (-2, 1)$$

Other examples of subtracting vectors follow.

EXAMPLE 1.7 Subtracting Two Vectors

The following table shows the vector $\mathbf{u} - \mathbf{v}$ for various vectors **u** and **v**:

u	**v**	**u** − **v**
$(1, 0, 1)$	$(0, 1, 0)$	$(1, -1, 1)$
$(2, 0, -4, 1)$	$(-1, 2, 7, 3)$	$(3, -2, -11, -2)$
$(1, -3, 2, -4)$	$(1, -3, 2, -4)$	$(0, 0, 0, 0)$

One way to create a visual image of $\mathbf{u} - \mathbf{v}$ is first to multiply **v** by -1 to create the vector $-\mathbf{v}$, thus reversing the direction of **v**, as shown for the vectors $\mathbf{u} = (1, 2)$ and $\mathbf{v} = (3, 1)$ in Figure 1.15(a). Then add $-\mathbf{v}$ to **u** to obtain $\mathbf{u} + (-\mathbf{v}) = \mathbf{u} - \mathbf{v}$, as illustrated in Figure 1.15(b).

By, moving the vector $\mathbf{u} - \mathbf{v}$ in Figure 1.15(b) parallel to itself as shown in Figure 1.15(c), you can see that $\mathbf{u} - \mathbf{v}$ is the vector that points from the head of **v** to the head of **u**. As a result, you can now view a vector as a directed line segment that starts at any given initial point $\mathbf{v} = (v_1, \ldots, v_n)$ and ends at a given terminal point $\mathbf{u} = (u_1, \ldots, u_n)$. The components of this vector are

$$\mathbf{u} - \mathbf{v} = (u_1 - v_1, \ldots, u_n - v_n)$$

The vector with tail at **v** and head at **u** (pointing from **v** to **u**) is the same as the vector drawn at the origin whose coordinates are $(u_1 - v_1, \ldots, u_n - v_n)$.

You might also note that the formula for computing the distance from $\mathbf{u} = (u_1, \ldots, u_n)$ to $\mathbf{v} = (v_1, \ldots, v_n)$, as given in (1.7), is really the norm of $\mathbf{u} - \mathbf{v}$, that is,

Distance from **u** to $\mathbf{v} = \|\mathbf{u} - \mathbf{v}\|$

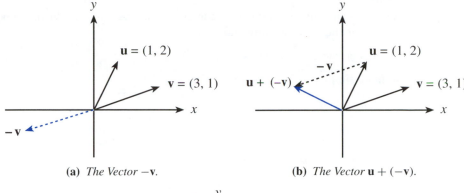

(a) *The Vector* −**v**.   **(b)** *The Vector* **u** + (−**v**).

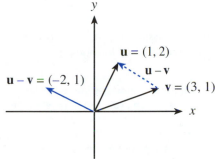

(c) *The Vector* **u** − **v** *Pointing from* **v** *to* **u**.

Figure 1.15 *The Geometry of Subtracting Two Vectors.*

Properties of Vector Addition and Subtraction

After working with vectors for some time, you will discover that the operations presented so far satisfy certain properties, some of which are summarized in the following theorem.

THEOREM 1.1

If **u**, **v**, and **w** are n-vectors and s and t are real numbers, then the following properties hold:

(a) $\mathbf{u} + \mathbf{v} = \mathbf{v} + \mathbf{u}$.

(b) $(\mathbf{u} + \mathbf{v}) + \mathbf{w} = \mathbf{u} + (\mathbf{v} + \mathbf{w})$.

(c) $\mathbf{u} + \mathbf{0} = \mathbf{0} + \mathbf{u} = \mathbf{u}$.

(d) $\mathbf{u} + (-\mathbf{u}) = \mathbf{0}$.

(e) $s(t\mathbf{u}) = (st)\mathbf{u}$.

(f) $s(\mathbf{u} + \mathbf{v}) = (s\mathbf{u}) + (s\mathbf{v})$.

(g) $(s + t)\mathbf{u} = (s\mathbf{u}) + (t\mathbf{u})$.

(h) $1\mathbf{u} = \mathbf{u}$.

(i) $0\mathbf{u} = \mathbf{0}$ (the real number 0 times the vector **u** is the vector **0**).

Proof.

A proof is an argument designed to convince someone that the following type of mathematical statement, called an **implication**, is true:

> if p is true, then q is true

or, more simply,

> if p, then q

where p is a statement referred to as the **hypothesis** and q is a statement called the **conclusion**. For example, in Theorem 1.1(a), you have the following implication with hypothesis p and conclusion q:

> if $\underbrace{\mathbf{u} \text{ and } \mathbf{v} \text{ are } n\text{-vectors}}_{p}$, then $\underbrace{\mathbf{u} + \mathbf{v} = \mathbf{v} + \mathbf{u}}_{q}$

A more detailed discussion of proofs is given in Appendix A, which you should read as needed and when indicated in this book.

> There are various proof techniques you can use to establish that the implication "if p, then q" is true. With one of them, referred to here as the **forward-backward method**, you can assume that p is true. Your objective is to use this assumption to show that q is true (see Appendix A.1 and Appendix A.2). An effective approach to doing so is to work backward from the conclusion by asking yourself the key question: "How can I show that q is true?" After asking the key question, you must provide an answer. One common way to do so is to use a definition.

To illustrate working backward with Theorem 1.1(a), in which the conclusion is $\mathbf{u} + \mathbf{v} = \mathbf{v} + \mathbf{u}$, a valid key question is "How can I show that two n-vectors (namely, $\mathbf{u} + \mathbf{v}$ and $\mathbf{v} + \mathbf{u}$) are equal?" This same question is appropriate for the conclusion in each part of Theorem 1.1.

One answer to the foregoing key question is to use Definition 1.2 to show that each component of one n-vector is equal to the corresponding component of the second n-vector. You will now see how this is done. You will also find it helpful to use visual images of vectors and their operations in two dimensions to motivate the following proofs.

(a) To see that the two vectors $\mathbf{u} + \mathbf{v}$ and $\mathbf{v} + \mathbf{u}$ are equal, use Definition 1.2 to show that each component of $\mathbf{u} + \mathbf{v}$ equals the corresponding component of $\mathbf{v} + \mathbf{u}$, that is, you must show that

> for each $i = 1, \ldots, n$, $(\mathbf{u} + \mathbf{v})_i = (\mathbf{v} + \mathbf{u})_i$

To this end, you have that

$$\mathbf{u} + \mathbf{v} = (u_1 + v_1, \ldots, u_n + v_n) \qquad \text{(definition of } \mathbf{u} + \mathbf{v}\text{)}$$
$$= (v_1 + u_1, \ldots, v_n + u_n) \qquad \text{(commutative property of addition of real numbers)}$$
$$= \mathbf{v} + \mathbf{u} \qquad \text{(definition of } \mathbf{v} + \mathbf{u}\text{)}$$

(b–d) The proofs of these properties are left to Exercise 23.

(e) Again, use Definition 1.2 to show that the vector $s(t\mathbf{u})$ is equal to the vector $(st)\mathbf{u}$. This is accomplished with the following steps, which are justified by the definition of multiplying a vector by a real number:

$$
\begin{aligned}
s(t\mathbf{u}) &= s(tu_1, \ldots, tu_n) \\
&= (s(tu_1), \ldots, s(tu_n)) \\
&= ((st)u_1, \ldots, (st)u_n) \\
&= (st)(u_1, \ldots, u_n) \\
&= (st)\mathbf{u}
\end{aligned}
$$

(f–i) The proofs of these properties are left to Exercise 23.

This completes the proof. ■

The symbol ■ indicates the end of the proof. The proofs in Theorem 1.1 illustrate the common technique of rewriting expressions until they have the new, desired form.

1.2.4 Multiplying Two Vectors: The Dot Product

Now that you know how to add and subtract two vectors, it is reasonable to ask if you can multiply two vectors. In fact, there are two useful ways to perform such a multiplication, one of which is described next.

The Algebra of the Dot Product

One way to multiply two vectors $\mathbf{u}$ and $\mathbf{v}$ results in a real number. To illustrate, consider three items whose unit prices are, say, u_1, u_2, and u_3, respectively. If you purchase v_1 units of the first item, v_2 units of the second item, and v_3 units of the third item, then the total cost is $u_1 v_1 + u_2 v_2 + u_3 v_3$. By thinking of u_1, u_2, and u_3 as the three components of a vector $\mathbf{u}$ and v_1, v_2, and v_3 as the three components of a vector $\mathbf{v}$, the total cost is obtained by combining $\mathbf{u}$ and $\mathbf{v}$ in a special way to create a single number. This particular binary operation on vectors is common in applications and is formalized in the following definition.

DEFINITION 1.4

The **dot product** of the n-vector $\mathbf{u} = (u_1, \ldots, u_n)$ and the n-vector $\mathbf{v} = (v_1, \ldots, v_n)$, denoted by $\mathbf{u} \cdot \mathbf{v}$ or $\mathbf{uv}$, is the real number computed by the following formula:

$$
\mathbf{u} \cdot \mathbf{v} = \sum_{i=1}^{n} u_i v_i = u_1 v_1 + \cdots + u_n v_n
$$

This value is also called the **inner product** of $\mathbf{u}$ and $\mathbf{v}$.

As another example of the use of the dot product, suppose that the value 1 in the binary vector $\mathbf{a} = (1, 0, 1, 1, 0)$ indicates that a student selected the correct answer on a particular question of an exam and the value 0 indicates that the student selected an incorrect answer for that question. The student's total score, that is, the total number of correct answers, is the number of 1's in the vector $\mathbf{a}$ (3, in this case). One way to obtain this total score is to add the components of the vector **a**: $1 + 0 + 1 + 1 + 0 = 3$. Equivalently, by creating the vector **u** $= (1, 1, 1, 1, 1)$, the total score is the dot product of **a** and **u** because

$$\mathbf{a} \cdot \mathbf{u} = (1, 0, 1, 1, 0) \cdot (1, 1, 1, 1, 1) = 1(1) + 0(1) + 1(1) + 1(1) + 0(1) = 3 = \text{score}$$

Another example of the use of the dot product arises in computing the *work*, w, done by exerting a force **f** $= (f_1, f_2, f_3)$ in kilogram-meters per second squared through a distance denoted by the vector **d** $= (d_1, d_2, d_3)$ meters. That is,

$$w = \mathbf{f} \cdot \mathbf{d} = f_1 d_1 + f_2 d_2 + f_3 d_3 \quad \text{kilogram-meters}^2 / \text{second}^2$$

Other numerical examples of the dot product follow.

EXAMPLE 1.8 Computing the Dot Product of Two Vectors

The following are examples of the dot product of different pairs of vectors:

u	**v**	$u_1 v_1 + \cdots + u_n v_n = \mathbf{u} \cdot \mathbf{v}$
$(1, 2)$	$(0, 3)$	$1(0) + 2(3) = 6$
$(1, 2)$	$(0, -3)$	$1(0) + 2(-3) = -6$
$(1, 2, 0)$	$(-4, 2, -2)$	$1(-4) + 2(2) + 0(-2) = 0$

The Geometry of the Dot Product

There is a useful geometric interpretation of the sign of $\mathbf{u} \cdot \mathbf{v}$. To understand this in two dimensions, consider two nonzero vectors **u** and **v** drawn at the origin. Now draw the vector **u** $-$ **v** from **v** to **u** to create a triangle and let θ be the **angle between u and v**, as shown in Figure 1.16. The next theorem establishes a relationship between θ and the sign of $\mathbf{u} \cdot \mathbf{v}$.

THEOREM 1.2

If θ is the angle between two nonzero vectors **u** $= (u_1, u_2)$ and **v** $= (v_1, v_2)$, then

(a) $\mathbf{u} \cdot \mathbf{v} = \|\mathbf{u}\| \|\mathbf{v}\| \cos(\theta)$.

(b) The angle θ satisfies one of the following properties:

θ is acute	if and only if	$\mathbf{u} \cdot \mathbf{v} > 0$
θ is obtuse	if and only if	$\mathbf{u} \cdot \mathbf{v} < 0$
$\theta = \pi/2$	if and only if	$\mathbf{u} \cdot \mathbf{v} = 0$

Proof.

Refer to Figure 1.16 throughout this proof.

(a) The Law of Cosines is used to show that $\mathbf{u} \cdot \mathbf{v} = \|\mathbf{u}\| \|\mathbf{v}\| \cos(\theta)$. Specifically, the Law of Cosines applied to Figure 1.16 yields

$$\|\mathbf{u} - \mathbf{v}\|^2 = \|\mathbf{u}\|^2 + \|\mathbf{v}\|^2 - 2\|\mathbf{u}\| \|\mathbf{v}\| \cos(\theta)$$

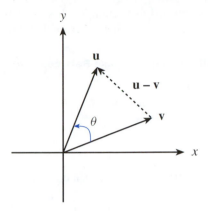

Figure 1.16 *The Angle Between Two Vectors.*

or equivalently,

$$2\|\mathbf{u}\|\|\mathbf{v}\|\cos(\theta) = \|\mathbf{u}\|^2 + \|\mathbf{v}\|^2 - \|\mathbf{u} - \mathbf{v}\|^2 \tag{1.10}$$

Substituting the relations

$$\|\mathbf{u} - \mathbf{v}\|^2 = (u_1 - v_1)^2 + (u_2 - v_2)^2$$

and

$$\|\mathbf{u}\|^2 = u_1^2 + u_2^2 \quad \text{and} \quad \|\mathbf{v}\|^2 = v_1^2 + v_2^2$$

in (1.10) and simplifying yields the desired result that

$$\|\mathbf{u}\|\|\mathbf{v}\|\cos(\theta) = u_1 v_1 + u_2 v_2 = \mathbf{u} \cdot \mathbf{v}$$

(b) The key question associated with the conclusion of this part is "How can I show that an angle (namely, θ) is acute, obtuse, or equal to $\pi/2$?" Because $0 \leq \theta \leq \pi$, one answer is to show that $\cos(\theta) > 0$, $\cos(\theta) < 0$, or $\cos(\theta) = 0$, respectively. But from part (a),

$$\cos(\theta) = \frac{\mathbf{u} \cdot \mathbf{v}}{\|\mathbf{u}\|\|\mathbf{v}\|} \tag{1.11}$$

The result now follows from (1.11) because $\|\mathbf{u}\| > 0$ and $\|\mathbf{v}\| > 0$. ■

It is possible to use Theorem 1.2(a) to define the angle between two n-vectors $\mathbf{u} = (u_1, \ldots, u_n)$ and $\mathbf{v} = (v_1, \ldots, v_n)$, as follows:

$$\theta = \arccos\left(\frac{\mathbf{u} \cdot \mathbf{v}}{\|\mathbf{u}\|\|\mathbf{v}\|}\right) \tag{1.12}$$

For the operation of the arccos in (1.12) to be defined, you need that

$$-1 \leq \frac{\mathbf{u} \cdot \mathbf{v}}{\|\mathbf{u}\|\|\mathbf{v}\|} \leq 1$$

This fact is established in Section 7.4.3, so (1.12) is a valid generalization of the angle between two n-vectors. A numerical example follows.

> **EXAMPLE 1.9 Finding the Angle Between Two n-Vectors**
> For the n-vectors
>
> $$\mathbf{u} = (-2, -4, 4) \quad \text{and} \quad \mathbf{v} = (-1, 0, 1)$$
>
> you have that
>
> $$\mathbf{u} \cdot \mathbf{v} = 6, \quad \|\mathbf{u}\| = 6, \quad \text{and} \quad \|\mathbf{v}\| = \sqrt{2}$$
>
> Now, from (1.12), the angle θ between $\mathbf{u}$ and $\mathbf{v}$ is
>
> $$\theta = \arccos\left(\frac{\mathbf{u} \cdot \mathbf{v}}{\|\mathbf{u}\|\|\mathbf{v}\|}\right) = \arccos\left(\frac{6}{6(\sqrt{2})}\right) = \arccos\left(\frac{1}{\sqrt{2}}\right) = \frac{\pi}{4}$$

From Theorem 1.2(b), you know that if $\mathbf{u} \cdot \mathbf{v} = 0$, then $\theta = \pi/2$, in other words, $\mathbf{u}$ and $\mathbf{v}$ are perpendicular. Thus, one way to show that $\mathbf{u}$ and $\mathbf{v}$ are perpendicular is to show that $\mathbf{u} \cdot \mathbf{v} = 0$. Two perpendicular vectors are also called **orthogonal vectors**. By agreement, all vectors are orthogonal to the zero vector.

1.2.5 Properties of the Norm and the Dot Product

Recall that for two n-vectors $\mathbf{u} = (u_1, \ldots, u_n)$ and $\mathbf{v} = (v_1, \ldots, v_n)$,

$$\|\mathbf{u}\| = \sqrt{u_1^2 + \cdots + u_n^2} \tag{1.13}$$

$$\mathbf{u} \cdot \mathbf{v} = u_1 v_1 + \cdots + u_n v_n \tag{1.14}$$

The next theorem establishes numerous relationships between (1.13) and (1.14) and various other properties of the dot product. Keep in mind that, geometrically, $\|\mathbf{u}\|$ is the length of $\mathbf{u}$.

> **THEOREM 1.3**
>
> Suppose that $\mathbf{u}$, $\mathbf{v}$, and $\mathbf{w}$ are n-vectors and that t is a real number. Then the following properties hold:
>
> (a) $\mathbf{u} \cdot \mathbf{u} \geq 0$.
> (b) $\mathbf{u} \cdot \mathbf{u} = 0$ if and only if $\mathbf{u} = \mathbf{0}$.
> (c) $\|t\mathbf{u}\| = |t|\|\mathbf{u}\|$.
> (d) $\mathbf{u} \cdot \mathbf{u} = \|\mathbf{u}\|^2$, that is, $\|\mathbf{u}\| = \sqrt{\mathbf{u} \cdot \mathbf{u}}$.
> (e) $\mathbf{u} \cdot \mathbf{v} = \mathbf{v} \cdot \mathbf{u}$.
> (f) $\mathbf{u} \cdot (\mathbf{v} + \mathbf{w}) = (\mathbf{u} \cdot \mathbf{v}) + (\mathbf{u} \cdot \mathbf{w})$.
> (g) $t(\mathbf{u} \cdot \mathbf{v}) = (t\mathbf{u}) \cdot \mathbf{v} = \mathbf{u} \cdot (t\mathbf{v})$.

Proof.
In all but part (a), the key question is "How can I show that two real numbers are equal?" The approach to doing so is to use (1.13) and (1.14) together with algebraic manipulations.

Throughout, let

$$\mathbf{u} = (u_1, \ldots, u_n), \quad \mathbf{v} = (v_1, \ldots, v_n), \quad \text{and} \quad \mathbf{w} = (w_1, \ldots, w_n)$$

(a) The fact that $\mathbf{u} \cdot \mathbf{u} \geq 0$ is evident from the formula in (1.14).

(b) Here you see the key words *if and only if* in the form

$$p \text{ if and only if } q$$

where p is the statement $\mathbf{u} \cdot \mathbf{u} = 0$ and q is the statement $\mathbf{u} = \mathbf{0}$. In such cases, it is necessary to prove two statements: if p, then q and the **converse**, if q, then p. So, for the current problem, suppose first that $\mathbf{u} \cdot \mathbf{u} = 0$. From (1.14), this means that

$$\mathbf{u} \cdot \mathbf{u} = u_1 u_1 + \cdots + u_n u_n = u_1^2 + \cdots + u_n^2 = 0$$

The only way this sum of nonnegative numbers can be 0 is if each number is 0, so, $u_1^2 = 0, \ldots, u_n^2 = 0$. Thus, $u_1 = 0, \ldots, u_n = 0$ and hence $\mathbf{u} = \mathbf{0}$. This completes the first proof.

To prove the converse, assume that $\mathbf{u} = \mathbf{0}$, that is, $u_1 = 0, \ldots, u_n = 0$. Then

$$\mathbf{u} \cdot \mathbf{u} = u_1 u_1 \cdots + u_n u_n = 0(0) + \cdots + 0(0) = 0$$

(c, d) The proofs of these properties are left to Exercise 24.

(e) To see that the two numbers $\mathbf{u} \cdot \mathbf{v}$ and $\mathbf{v} \cdot \mathbf{u}$ are equal, use (1.14) and rewrite $\mathbf{u} \cdot \mathbf{v}$ as follows to obtain $\mathbf{v} \cdot \mathbf{u}$:

$$
\begin{aligned}
\mathbf{u} \cdot \mathbf{v} &= u_1 v_1 + \cdots + u_n v_n && \text{(definition of } \mathbf{u} \cdot \mathbf{v}) \\
&= v_1 u_1 + \cdots + v_n u_n && \text{(commutative property of multi-} \\
& && \text{plication of real numbers)} \\
&= \mathbf{v} \cdot \mathbf{u} && \text{(definition of } \mathbf{v} \cdot \mathbf{u})
\end{aligned}
$$

(f, g) The proofs of these properties are left to Exercise 24.

This completes the proof. ■

1.2.6 Multiplying Two Vectors: The Cross Product

In physics and engineering, it is often useful to create a vector whose direction is orthogonal to two given vectors in 3-space. For example, if the position of a particle relative to a fixed origin is denoted by the vector $\mathbf{r} = (r_1, r_2, r_3)$ and a force $\mathbf{f} = (f_1, f_2, f_3)$ is applied to the particle, then the *torque* acting on the particle with respect to this origin is a vector that is perpendicular to both $\mathbf{r}$ and $\mathbf{f}$. As another example, if the position of a particle relative to a fixed origin is denoted by the vector $\mathbf{r} = (r_1, r_2, r_3)$ and the linear momentum of that particle is represented by the vector $\mathbf{p} = (p_1, p_2, p_3)$, then the *angular momentum* of that particle with respect to the origin is a vector that is orthogonal to both $\mathbf{r}$ and $\mathbf{p}$. The operation of creating this new vector is a second form of multiplying two vectors that is formalized in the following definition.

DEFINITION 1.5

Given two vectors $\mathbf{u} = (u_1, u_2, u_3)$ and $\mathbf{v} = (v_1, v_2, v_3)$ in 3-space, the **cross product**, denoted by $\mathbf{u} \times \mathbf{v}$, is the vector

$$\mathbf{u} \times \mathbf{v} = (u_2 v_3 - u_3 v_2, u_3 v_1 - u_1 v_3, u_1 v_2 - u_2 v_1)$$

EXAMPLE 1.10 The Cross Product of Two Vectors

If the position of a particle in meters from a fixed origin is $\mathbf{r} = (1, 0, -2)$ and the linear momentum of the particle in kilogram-meters per second is $\mathbf{p} = (3, 1, 2)$, then the angular momentum of the particle in kilogram-meters^2 per second is

$$\mathbf{r} \times \mathbf{p} = (0(2) - (-2)(1), (-2)(3) - 1(2), 1(1) - 0(3)) = (2, -8, 1)$$

For the standard unit vectors $\mathbf{i} = (1, 0, 0)$, $\mathbf{j} = (0, 1, 0)$, and $\mathbf{k} = (0, 0, 1)$ in 3-space, you can also verify that

$$\mathbf{i} \times \mathbf{i} = 0 \qquad \mathbf{j} \times \mathbf{j} = 0 \qquad \mathbf{k} \times \mathbf{k} = 0$$
$$\mathbf{i} \times \mathbf{j} = \mathbf{k} \qquad \mathbf{j} \times \mathbf{k} = \mathbf{i} \qquad \mathbf{k} \times \mathbf{i} = \mathbf{j}$$
$$\mathbf{j} \times \mathbf{i} = -\mathbf{k} \qquad \mathbf{k} \times \mathbf{j} = -\mathbf{i} \qquad \mathbf{i} \times \mathbf{k} = -\mathbf{j}$$

The next theorem establishes that $\mathbf{u} \times \mathbf{v}$ is orthogonal to both $\mathbf{u}$ and $\mathbf{v}$.

THEOREM 1.4

If $\mathbf{u}$ and $\mathbf{v}$ are vectors in 3-space, then

(a) $\mathbf{u} \cdot (\mathbf{u} \times \mathbf{v}) = 0$ (that is, $\mathbf{u}$ and $\mathbf{u} \times \mathbf{v}$ are orthogonal).

(b) $\mathbf{v} \cdot (\mathbf{u} \times \mathbf{v}) = 0$ (that is, $\mathbf{v}$ and $\mathbf{u} \times \mathbf{v}$ are orthogonal).

Proof.

For $\mathbf{u} = (u_1, u_2, u_3)$ and $\mathbf{v} = (v_1, v_2, v_3)$, you can establish part (a) from the definition of the dot product and the cross product, as follows:

$$
\begin{aligned}
\mathbf{u} \cdot (\mathbf{u} \times \mathbf{v}) &= (u_1, u_2, u_3) \cdot (u_2 v_3 - u_3 v_2, u_3 v_1 - u_1 v_3, u_1 v_2 - u_2 v_1) \\
&= u_1(u_2 v_3 - u_3 v_2) + u_2(u_3 v_1 - u_1 v_3) + u_3(u_1 v_2 - u_2 v_1) \\
&= 0
\end{aligned}
$$

The proof of part (b) is similar to part (a) and is omitted. ■

Additional Properties of the Cross Product

The operation of the cross product satisfies the properties given in the following theorem.

THEOREM 1.5

For any vectors **u**, **v**, **w** in 3-space and for any real number t, the following properties hold:

(a) $\mathbf{u} \times \mathbf{v} = -(\mathbf{v} \times \mathbf{u})$.

(b) $\mathbf{u} \times (\mathbf{v} + \mathbf{w}) = (\mathbf{u} \times \mathbf{v}) + (\mathbf{u} \times \mathbf{w})$.

(c) $(\mathbf{u} + \mathbf{v}) \times \mathbf{w} = (\mathbf{u} \times \mathbf{w}) + (\mathbf{v} \times \mathbf{w})$.

(d) $t(\mathbf{u} \times \mathbf{v}) = (t\mathbf{u}) \times \mathbf{v} = \mathbf{u} \times (t\mathbf{v})$.

(e) $\mathbf{u} \times \mathbf{0} = \mathbf{0} \times \mathbf{u} = \mathbf{0}$.

(f) $\mathbf{u} \times \mathbf{u} = \mathbf{0}$.

Proof.

Throughout, let $\mathbf{u} = (u_1, u_2, u_3)$ and $\mathbf{v} = (v_1, v_2, v_3)$. The key question associated with each part of this theorem is "How can I show that two vectors are equal?" The answer is to use Definition 1.2 and Definition 1.5 to show that the corresponding components of the two vectors are equal, as is now shown for part (a).

(a) You have from Definition 1.5 that

$$\begin{aligned}
\mathbf{u} \times \mathbf{v} &= (u_2 v_3 - u_3 v_2, \, u_3 v_1 - u_1 v_3, \, u_1 v_2 - u_2 v_1) \\
&= -(v_2 u_3 - v_3 u_2, \, v_3 u_1 - v_1 u_3, \, v_1 u_2 - v_2 u_1) \\
&= -(\mathbf{v} \times \mathbf{u})
\end{aligned}$$

(b–f) The proofs of these parts are left to Exercise 25.

This completes the proof.

You can gain a better understanding of how the dot product and the cross product are related by identifying similarities and differences between these two operations. For example, both are binary operations on two given vectors. However, the dot product results in a real number whereas the cross product results in a vector. Another difference is that the dot product is applicable to any two vectors of the same dimension but the cross product applies only to vectors in 3-space. The next theorem provides more insight into the relationship between these two operations.

THEOREM 1.6

For any vectors $\mathbf{u} = (u_1, u_2, u_3)$ and $\mathbf{v} = (v_1, v_2, v_3)$, the following conditions hold:

(a) $\|\mathbf{u} \times \mathbf{v}\|^2 = \|\mathbf{u}\|^2 \|\mathbf{v}\|^2 - (\mathbf{u} \cdot \mathbf{v})^2$ (Lagrange's identity).

(b) $\|\mathbf{u} \times \mathbf{v}\| = \|\mathbf{u}\| \|\mathbf{v}\| \sin(\theta)$, where θ is the angle between **u** and **v**.

Proof.

Let $\mathbf{u} = (u_1, u_2, u_3)$ and $\mathbf{v} = (v_1, v_2, v_3)$.

(a) Applying the norm to $\mathbf{u} \times \mathbf{v}$ and squaring the result yields

$$\|\mathbf{u} \times \mathbf{v}\|^2 = (u_2 v_3 - u_3 v_2)^2 + (u_3 v_1 - u_1 v_3)^2 + (u_1 v_2 - u_2 v_1)^2$$

From the definition of the norm and the dot product you obtain

$$\|\mathbf{u}\|^2\|\mathbf{v}\|^2 - (\mathbf{u}\cdot\mathbf{v})^2 = (u_1^2 + u_2^2 + u_3^2)(v_1^2 + v_2^2 + v_3^2) - (u_1v_1 + u_2v_2 + u_3v_3)^2$$

The result follows by expanding all terms and then verifying that the two foregoing equations are equal.

(b) From Theorem 1.2(a), you know that

$$\mathbf{u}\cdot\mathbf{v} = \|\mathbf{u}\|\,\|\mathbf{v}\|\cos(\theta)$$

Substituting this expression for $\mathbf{u}\cdot\mathbf{v}$ in Lagrange's identity [part (a)], you have that

$$\begin{aligned}
\|\mathbf{u}\times\mathbf{v}\|^2 &= \|\mathbf{u}\|^2\|\mathbf{v}\|^2 - [\|\mathbf{u}\|\,\|\mathbf{v}\|\cos(\theta)]^2\\
&= \|\mathbf{u}\|^2\|\mathbf{v}\|^2[1 - \cos^2(\theta)]\\
&= \|\mathbf{u}\|^2\|\mathbf{v}\|^2\sin^2(\theta)
\end{aligned}$$

The result follows by taking the positive square root of both sides of the foregoing equality.

This completes the proof.

The Geometry of the Cross Product

For two given nonzero vectors $\mathbf{u}$ and $\mathbf{v}$ in 3-space, you know from Theorem 1.4 that $\mathbf{u}\times\mathbf{v}$ is a vector that is orthogonal to both $\mathbf{u}$ and $\mathbf{v}$. However, as seen in Figure 1.17 (a) and (b), there are *two* directions that are orthogonal to both $\mathbf{u}$ and $\mathbf{v}$. Which one points in the same direction as $\mathbf{u}\times\mathbf{v}$?

Because the coordinate system is a right-handed one (see Section 1.1.4), you can use the following **right-hand rule** to determine which of these two directions corresponds to $\mathbf{u}\times\mathbf{v}$. Identify the angle θ between $\mathbf{u}$ and $\mathbf{v}$ by connecting the two points $\mathbf{u}$ and $\mathbf{v}$ to form a triangle. If you curl the fingers of your right hand so that they rotate through the angle θ from $\mathbf{u}$ toward $\mathbf{v}$, then the thumb of your right hand points approximately in the direction of $\mathbf{u}\times\mathbf{v}$, as shown in Figure 1.18. A corresponding **left-hand rule** is applicable for a left-handed coordinate system.

There is also a useful geometric interpretation for the length of $\mathbf{u}\times\mathbf{v}$ which, from

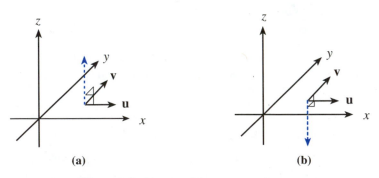

Figure 1.17 *Two Possible Directions for* $\mathbf{u}\times\mathbf{v}$.

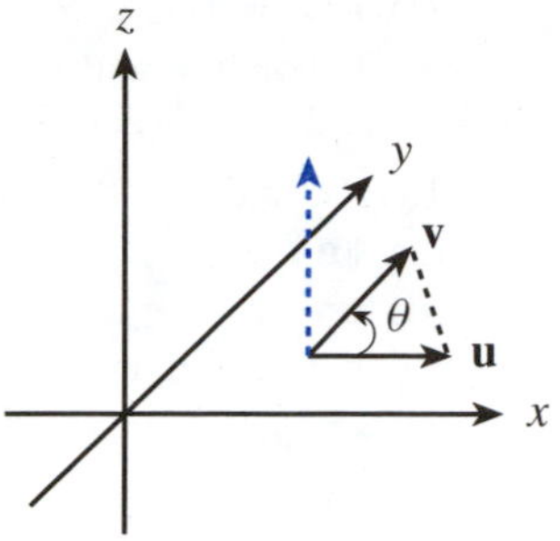

Figure 1.18 *Using the Right-hand Rule to Determine the Approximate Direction of* **u** × **v**.

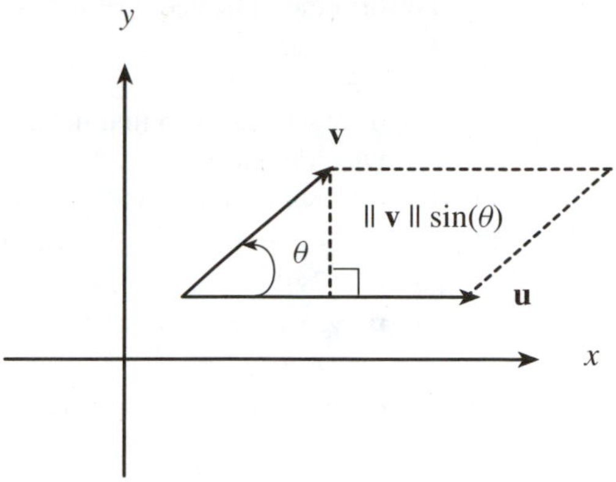

Figure 1.19 *The Geometry of the Cross Product of Two Vectors.*

Theorem 1.6(b), is described by the following expression:

$$\|\mathbf{u} \times \mathbf{v}\| = \|\mathbf{u}\|\|\mathbf{v}\| \sin(\theta)$$

As shown in Figure 1.19, $\|\mathbf{v}\| \sin(\theta)$ is the altitude of the parallelogram whose sides are **u** and **v**. You can now see that $\|\mathbf{u} \times \mathbf{v}\|$ is the area of this parallelogram because

Area = (base) (altitude) = $\|\mathbf{u}\|\|\mathbf{v}\| \sin(\theta) = \|\mathbf{u} \times \mathbf{v}\|$

In this section, various operations on vectors were described. Some of these operations result in a real number, others result in a vector. In the next section, various ways of using vectors to describe lines and planes are presented.

Exercises for Section 1.2

1. Perform the indicated operations with $\mathbf{u} = (1, 0)$ and $\mathbf{v} = (0, 2)$.
   (a) $\|\mathbf{u}\|$ and $\|\mathbf{v}\|$.
   (b) Distance from **u** to **v**.
   (c) $-\mathbf{u}$.
   (d) $2\mathbf{u} - \mathbf{v}$.
   (e) $\mathbf{u} \cdot \mathbf{v}$. Use $\mathbf{u} \cdot \mathbf{v}$ to find the angle between **u** and **v**.
2. Repeat the previous exercise with $\mathbf{u} = (1, 1)$ and $\mathbf{v} = (-1, -1)$.
3. Perform the indicated operations with $\mathbf{u} = (1, -1)$ and $\mathbf{v} = (-1, 0)$.
   (a) $\mathbf{u} + 2\mathbf{v}$.
   (b) $\|2\mathbf{u}\|$.
   (c) $\mathbf{u} \cdot \mathbf{v}$. Use $\mathbf{u} \cdot \mathbf{v}$ to find the angle between **u** and **v**.
4. Perform the indicated operations with $\mathbf{u} = (2, 0)$ and $\mathbf{v} = (1, \sqrt{3})$.
   (a) $2\mathbf{u} - 3\mathbf{v}$.
   (b) $\dfrac{1}{\|\mathbf{u}\|}\mathbf{u}$.
   (c) $\mathbf{u} \cdot \mathbf{v}$. Use $\mathbf{u} \cdot \mathbf{v}$ to find the angle between **u** and **v**.

5. Perform the indicated operations with $\mathbf{u} = (1, 0, -1)$ and $\mathbf{v} = (0, 2, 0)$.
   - (a) $\|\mathbf{u}\|$ and $\|\mathbf{v}\|$.
   - (b) Distance from $\mathbf{u}$ to $\mathbf{v}$.
   - (c) $-3\mathbf{v}$.
   - (d) $2\mathbf{u} - \mathbf{v}$.
   - (e) $\mathbf{u} \cdot \mathbf{v}$. Use $\mathbf{u} \cdot \mathbf{v}$ to find the angle between $\mathbf{u}$ and $\mathbf{v}$.
   - (f) $\mathbf{u} \times \mathbf{v}$. Use $\mathbf{u} \times \mathbf{v}$ to find the angle between $\mathbf{u}$ and $\mathbf{v}$.

6. Repeat the previous exercise with $\mathbf{u} = (1, -2, 3)$ and $\mathbf{v} = (-2, 4, -6)$.

7. Perform the indicated operations with $\mathbf{u} = (-1, 1, 0)$ and $\mathbf{v} = (0, 2, 0)$.
   - (a) $\mathbf{u} + 2\mathbf{v}$.
   - (b) $\|\mathbf{u}\|$.
   - (c) $\mathbf{u} \cdot \mathbf{v}$. Use $\mathbf{u} \cdot \mathbf{v}$ to find the angle between $\mathbf{u}$ and $\mathbf{v}$.
   - (d) $\mathbf{u} \times \mathbf{v}$. Use $\mathbf{u} \times \mathbf{v}$ to find the angle between $\mathbf{u}$ and $\mathbf{v}$.

8. Perform the indicated operations with $\mathbf{u} = (1, 0, -1)$ and $\mathbf{v} = (-\sqrt{2}, -\sqrt{2}, 0)$.
   - (a) $2\mathbf{u} - 3\mathbf{v}$.
   - (b) $\|\mathbf{v} - \mathbf{u}\|$.
   - (c) $\mathbf{u} \cdot \mathbf{v}$. Use $\mathbf{u} \cdot \mathbf{v}$ to find the angle between $\mathbf{u}$ and $\mathbf{v}$.
   - (d) $\mathbf{u} \times \mathbf{v}$. Use $\mathbf{u} \times \mathbf{v}$ to find the angle between $\mathbf{u}$ and $\mathbf{v}$.

9. Convert each of the following sets to visual form by drawing an appropriate picture. Describe the resulting figure in words.
   - (a) For a given nonzero vector $\mathbf{d} = (d_1, d_2)$, draw $\{t\mathbf{d} : t \geq 0\}$.
   - (b) For nonzero vectors $\mathbf{u} = (u_1, u_2)$ and $\mathbf{v} = (v_1, v_2)$, draw $\{s\mathbf{u} + t\mathbf{v} : s, t \geq 0\}$.

10. Convert each of the following sets to visual form by drawing an appropriate picture. Describe the resulting figure in words.
    - (a) For a given vector $\mathbf{u} = (u_1, u_2)$ and a nonzero vector $\mathbf{d} = (d_1, d_2)$, draw $\{\mathbf{u} + t\mathbf{d} : t \geq 0\}$.
    - (b) For two given nonzero vectors $\mathbf{u} = (u_1, u_2)$ and $\mathbf{v} = (v_1, v_2)$, draw $\{s\mathbf{u} + t\mathbf{v} : s, t$ are real numbers$\}$.

11. Identify as many similarities and differences as you can between a closed-form solution and a numerical-method solution.

12. Use the following numerical method to find values for the real numbers x and y that solve the problem

    $$\text{Maximize } 3x + 2y$$
    $$\text{Subject to} \quad x + 2y \leq 10 \quad (1)$$
    $$x \quad\quad \geq 0 \quad (2)$$
    $$y \geq 0 \quad (3)$$

    - (a) For each pair of numbered inequalities, replace the inequality signs with equality signs and solve the resulting two linear equations to find values for x and y.
    - (b) For each set of values for x and y found in part (a), compute the value of $3x + 2y$.
    - (c) Use your results in part (b) to identify the values for x and y that provide the largest value of $3x + 2y$.

13. Use vector operations to perfom each of the following tasks.
    - (a) The components of the vector $(183.75, 168.00, 367.50)$ represent the cost, in Japanese yen, of buying a kilogram of tomatoes, a liter of milk, and a cantaloupe, respectively. What are these costs in U.S. dollars if the current exchange rate is 105 yen to the dollar?

(b) On each of three segments of a trip, a truck driver averaged the following speeds, in miles per hour: (50, 60, 55). The amount of time, in minutes, driven on each of the segments is (240, 180, 300). What is the total distance driven?

14. Recall, from Section 1.1.3, that you can represent a polynomial $p(x)$ by a vector $\mathbf{p}$ whose components are the coefficients of the polynomial, say, in increasing order of the exponent of x.
    (a) Suppose you want to evaluate the polynomial represented by the vector $\mathbf{p}$ at a particular value of x. Use x to create an appropriate vector $\mathbf{x}$ so that the value of the polynomial is computed as $\mathbf{p} \cdot \mathbf{x}$.
    (b) Illustrate the approach in part (a) by evaluating the polynomial $x^3 - 2x + 4$ at the point $x = 2$ using $\mathbf{p}$ and $\mathbf{x}$.

15. (a) Suppose that $\mathbf{d} = (d_1, \ldots, d_n)$ is a list of measurements in inches. Use vector operations to create a list of these measurements in centimeters, assuming that 1 inch is approximately equal to 2.54 centimeters.
    (b) Apply your result in part (a) to the vector $\mathbf{d} = (2, 3, 5)$.

16. (a) Suppose that $\mathbf{f} = (f_1, \ldots, f_n)$ is a list of temperatures in degrees Fahrenheit. Use vector operations to create a list of these temperatures in degrees centigrade using the fact that $F = \frac{9}{5}C + 32$.
    (b) Apply you result in part (a) to the vector $\mathbf{f} = (0, 32, 212)$.

17. Suppose that the position of a particle in meters is denoted by the vector $\mathbf{r}$ in 3-space and the linear momentum of the particle measured in kilogram-meters per second is represented by the vector $\mathbf{p}$. Use the formula for the cross product to determine the units of the angular momentum $\mathbf{l} = \mathbf{r} \times \mathbf{p}$.

18. Suppose that the position of a particle is denoted in meters by the vector $\mathbf{r}$ in 3-space and that a force $\mathbf{f}$ measured in kilogram-meters per second^2 is applied to the particle. Use the formula for the cross product to determine the units of the torque $\mathbf{t} = \mathbf{r} \times \mathbf{f}$.

19. What is the area of the parallelogram whose sides are given by the vectors $(1, 0, -1)$ and $(0, 1, 1)$?

20. Suppose that $\mathbf{u}$ and $\mathbf{v}$ are nonzero orthogonal vectors in 3-space. Explain why the following formulas for $\|\mathbf{u} \times \mathbf{v}\|$ correctly provide the area of the rectangle whose sides are $\mathbf{u}$ and $\mathbf{v}$.
    (a) $\|\mathbf{u} \times \mathbf{v}\| = \sqrt{\|\mathbf{u}\|^2 \|\mathbf{v}\|^2 - (\mathbf{u} \cdot \mathbf{v})^2}$.
    (b) $\|\mathbf{u} \times \mathbf{v}\| = \|\mathbf{u}\| \|\mathbf{v}\| \sin(\theta)$, where θ is the angle between $\mathbf{u}$ and $\mathbf{v}$.

21. Let $\mathbf{u} = (u_1, \ldots, u_n)$ and $\mathbf{v} = (v_1, \ldots, v_n)$ be n-vectors. Classify each of the following as a unary or a binary operation. Does the operation result in a vector or a number?
    (a) dim($\mathbf{u}$) = the dimension of $\mathbf{u}$.
    (b) sum($\mathbf{u}$) = $u_1 + \cdots + u_n$.
    (c) times($\mathbf{u}, \mathbf{v}$) = $(u_1 v_1, \ldots, u_n v_n)$.

22. Let $\mathbf{u} = (u_1, \ldots, u_n)$ and $\mathbf{v} = (v_1, \ldots, v_n)$ be n-vectors. Classify each of the following as a unary or a binary operation. Does the operation result in a vector or a number?
    (a) abs($\mathbf{u}$) = $(|u_1|, \ldots, |u_n|)$.
    (b) left($\mathbf{u}$) = $(u_2, u_3, \ldots, u_n, u_1)$.
    (c) $d(\mathbf{u}, \mathbf{v})$ = $\|\mathbf{u} - \mathbf{v}\|$.

23. Suppose that **u**, **v**, and **w** are n-vectors and s and t are real numbers. Complete Theorem 1.1 in Section 1.2.3 by proving each of the following properties. Give a reason for each step in your proofs.
    (a) $(\mathbf{u} + \mathbf{v}) + \mathbf{w} = \mathbf{u} + (\mathbf{v} + \mathbf{w})$.
    (b) $\mathbf{u} + \mathbf{0} = \mathbf{0} + \mathbf{u} = \mathbf{u}$.
    (c) $\mathbf{u} + (-\mathbf{u}) = \mathbf{0}$.
    (d) $s(\mathbf{u} + \mathbf{v}) = (s\mathbf{u}) + (s\mathbf{v})$.
    (e) $(s + t)\mathbf{u} = (s\mathbf{u}) + (t\mathbf{u})$.
    (f) $1\mathbf{u} = \mathbf{u}$.
    (g) $0\mathbf{u} = \mathbf{0}$ (the real number 0 times the vector **u** equals the vector **0**).

24. Suppose that **u**, **v**, and **w** are n-vectors and t is a real number. Complete Theorem 1.3 in Section 1.2.5 by proving each of the following properties. Give a reason for each step in your proofs.
    (a) $\|t\mathbf{u}\| = |t|\|\mathbf{u}\|$.
    (b) $\mathbf{u} \cdot \mathbf{u} = \|\mathbf{u}\|^2$, that is, $\|\mathbf{u}\| = \sqrt{\mathbf{u} \cdot \mathbf{u}}$.
    (c) $\mathbf{u} \cdot (\mathbf{v} + \mathbf{w}) = (\mathbf{u} \cdot \mathbf{v}) + (\mathbf{u} \cdot \mathbf{w})$.
    (d) $t(\mathbf{u} \cdot \mathbf{v}) = (t\mathbf{u}) \cdot \mathbf{v} = \mathbf{u} \cdot (t\mathbf{v})$.

25. Suppose that **u**, **v**, and **w** are vectors in 3-space and t is a real number. Complete Theorem 1.5 in Section 1.2.6 by proving each of the following properties. Give a reason for each step in your proofs.
    (a) $\mathbf{u} \times (\mathbf{v} + \mathbf{w}) = (\mathbf{u} \times \mathbf{v}) + (\mathbf{u} \times \mathbf{w})$.
    (b) $(\mathbf{u} + \mathbf{v}) \times \mathbf{w} = (\mathbf{u} \times \mathbf{w}) + (\mathbf{v} \times \mathbf{w})$.
    (c) $t(\mathbf{u} \times \mathbf{v}) = (t\mathbf{u}) \times \mathbf{v} = \mathbf{u} \times (t\mathbf{v})$.
    (d) $\mathbf{u} \times \mathbf{0} = \mathbf{0} \times \mathbf{u} = \mathbf{0}$.
    (e) $\mathbf{u} \times \mathbf{u} = \mathbf{0}$.

1.3 Lines and Planes

In this section, vectors are used in constructing equations of lines and planes. Various symbolic ways to write the equation of a line in 2-space are given first and then generalized to 3-space and to n-space.

1.3.1 Lines in 2-Space

In this section, three different symbolic representations of a line in the plane are presented.

The Slope-Intercept Form of a Line

A line is used in problem solving to describe a particular relationship between two variables. For example, the following line relates the annual cost, y, of medical insurance for an individual to the number of years, x, since 1990:

$$y = 300x + 1500 \tag{1.15}$$

The general **slope-intercept form** of a line, such as the one in Figure 1.20, is:

$$y = mx + b, \quad \text{or equivalently,} \quad \{(x, y) : y = mx + b\} \tag{1.16}$$

where

$m =$ the slope of the line and

$b =$ the y-intercept

The slope, m, represents the change in y per unit of increase in x. The y-intercept, b, is the value of y when x is 0. For example, for the line described by (1.15), the slope, $m = 300$,

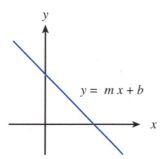

Figure 1.20 *A Line in the Plane.*

indicates that the cost of medical insurance increases at the rate of $300 for each year after 1990. The y-intercept, $b = 1500$, indicates that the annual cost of medical insurance in 1990 (corresponding to $x = 0$) was $1500.

To find specific values for m and b, you must have either of the following information:

1. The slope and the coordinates of one point on the line.

2. The coordinates of two different points on the line.

Two additional methods of representing lines in 2-space by using vectors are described next. Each is applied to the following numerical example.

EXAMPLE 1.11 The Slope-Intercept Form of a Line in the Plane

The slope and y-intercept of the line whose equation is

$$y = 4 - 2x$$

are

$$m = \text{the slope} = -2$$
$$b = y\text{-intercept} = 4$$

The Point-Normal Form of a Line

An alternate representation arises by considering a point (x_0, y_0) on the line together with a vector (a, b) whose tail is at the point (x_0, y_0) and whose direction is perpendicular to the line, as shown in Figure 1.21. You can use the point (x_0, y_0) and the direction vector (a, b) to describe the line algebraically. This is done by identifying the property a point (x, y) on the line satisfies in terms of the data (x_0, y_0) and (a, b). As seen in Figure 1.22, the vector from (x_0, y_0) to (x, y) is orthogonal to the vector (a, b).

To translate this visual image to symbolic form, first write the vector pointing from (x_0, y_0) to (x, y) which, as you learned in Section 1.2.3, is

$$(x, y) - (x_0, y_0) = (x - x_0, y - y_0) \tag{1.17}$$

Now translate the fact that the vector in (1.17) is orthogonal to the vector (a, b) to symbolic form by recalling from Theorem 1.2(b) in Section 1.2.4 that two vectors are orthogonal if

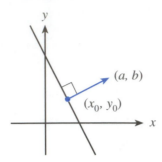

Figure 1.21 *A Line Determined by a Point and a Perpendicular Vector.*

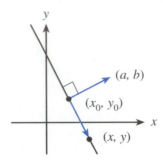

Figure 1.22 *The Relationship of Points on a Line to the Perpendicular Vector.*

and only if their dot product is zero. Thus,

$$(a, b) \cdot (x - x_0, y - y_0) = 0$$

Equivalently, any point (x, y) on the line in Figure 1.22 satisfies

$$a(x - x_0) + b(y - y_0) = 0 \qquad (1.18)$$

The representation in (1.18) is called the **point-normal form of a line**, an example of which follows.

EXAMPLE 1.12 The Point-Normal Form of a Line in the Plane

The point-normal form of the line that goes through the point $(x_0, y_0) = (1, 2)$ and is perpendicular to the vector $(a, b) = (6, 3)$ is

$$6(x - 1) + 3(y - 2) = 0$$

Solving for y in terms of x, the line described in Example 1.12 is $y = 4 - 2x$, which is the same as the line in Example 1.11.

The Parametric Representation of a Line

Yet another way to represent a line in the plane is to start once again with a point (x_0, y_0) on the line. Rather than using a vector that is perpendicular to the line, consider a vector—say, (a, b)—that points in either direction along the line, as shown in Figure 1.23, for example. The line L in Figure 1.23 consists of those points in the plane obtained by moving some amount t from (x_0, y_0) in the direction (a, b), where t is a real number whose value varies between $-\infty$ and $+\infty$. Translating this visual image to symbolic form yields:

$$
\begin{aligned}
L &= \{(x, y) : (x, y) = (x_0, y_0) + t(a, b)\} \\
&= \{(x, y) : (x, y) = (x_0, y_0) + (ta, tb)\} \\
&= \{(x, y) : (x, y) = (x_0 + ta, y_0 + tb)\}
\end{aligned}
$$

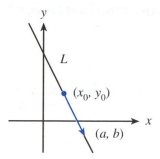

Figure 1.23 *A Line Determined by a Point and a Direction of the Line.*

Equivalently stated, points (x, y) on the line L are described by the following two **parametric equations**:

$$\left.\begin{array}{l} x = x_0 + at \\ y = y_0 + bt \end{array}\right\} \quad \text{(parametric equations for a line in the plane)} \tag{1.19}$$

where t varies between $-\infty$ and $+\infty$. From here on, it is understood that "the parameter t varies between $-\infty$ and $+\infty$" and so this phrase is omitted.

> **EXAMPLE 1.13 Parametric Equations for a Line in the Plane**
> The line that goes through the point $(x_0, y_0) = (1, 2)$ in the direction $(a, b) = (-1, 2)$ is given by the following two parametric equations:
>
> $$x = x_0 + at = 1 - t \tag{1.20}$$
> $$y = y_0 + bt = 2 + 2t \tag{1.21}$$

You can see that the line described in Example 1.13 is the same as the ones in Example 1.11 and Example 1.12 by eliminating the parameter t in (1.20) and (1.21). Specifically, using (1.20) to solve for t, you have that

$$t = 1 - x$$

Replacing this value for t in (1.21) yields

$$y = 2 + 2t = 2 + 2(1 - x) = 4 - 2x$$

which is the same as the lines described in Example 1.11 and Example 1.12.

1.3.2   Generalizations to Lines and Planes in *n*-Space

You have just seen three different symbolic representations of a line in the plane: slope-intercept form, point-normal form, and parametric-equation form. The objective now is to generalize each of these, resulting in different representations of lines and planes in *n*-space.

Generalizing the Slope-Intercept Form of a Line

One way to generalize the equation

$$y = mx + b \tag{1.22}$$

is to replace the real number x in (1.22) with an n-vector $\mathbf{x}$ to obtain

$$y = m\mathbf{x} + b$$

The foregoing equation does not make sense because the term $m\mathbf{x}$, in which you multiply the real number m by the n-vector $\mathbf{x}$, is an n-vector that cannot be added to the real number b. This is an example of a **syntax error**, which is an error in a mathematical expression that arises when you cannot perform an operation (such as adding an n-vector to a real number).

How, then, can you generalize (1.22) in a correct and meaningful way? One answer is to replace both m and x with n-vectors $\mathbf{m}$ and $\mathbf{x}$. Thus, (1.22) becomes

$$y = \mathbf{m} \cdot \mathbf{x} + b \tag{1.23}$$

Here, $\mathbf{m} \cdot \mathbf{x}$ is the dot product of the n-vectors $\mathbf{m}$ and $\mathbf{x}$ which, as you know from Section 1.2.4, results in a real number. Thus, (1.23) has no syntax error because you *can* add the real number $\mathbf{m} \cdot \mathbf{x}$ to the real number b to obtain the real number y.

Using the formula for the dot product of the n-vectors $\mathbf{m} = (m_1, \ldots, m_n)$ and $\mathbf{x} = (x_1, \ldots, x_n)$ as given in Definition 1.4 in Section 1.2.4, another way to write (1.23) is

$$y = (m_1, \ldots, m_n) \cdot (x_1, \ldots, x_n) + b = m_1 x_1 + \cdots + m_n x_n + b \tag{1.24}$$

The equations in (1.23) and (1.24) are symbolic representations of a *plane* in $(n + 1)$-space and are called the **inclination-intercept form**. A special case of an equation in this form is a plane in 3-space, which you can write as follows by substituting the symbols $y = z$, $n = 2$, $x_1 = x$, and $x_2 = y$ in (1.24):

$$z = m_1 x + m_2 y + b \tag{1.25}$$

The value m_1 in (1.25) is the change in z per unit of increase in x, assuming that the value of y remains fixed. Likewise, the value of m_2 is the change in z per unit of increase in y, assuming that the value of x remains fixed. For example, consider the following plane that describes the total monthly cost z of operating a warehouse, in which x represents the number of computers in inventory and y represents the number of cubic feet rented to a third party:

$$z = 3x - 0.25y + 300$$

The value $m_1 = 3$ means that each additional computer in inventory increases the total monthly cost by \$3, assuming the amount of rented space remains unchanged. Likewise, the value $m_2 = -0.25$ indicates that the total monthly cost decreases by 25 cents for each additional square foot of space that is rented, assuming the number of computers in inventory remains fixed. The value $b = 300$ indicates that the monthly cost is \$300 when no computers are in inventory ($x = 0$) and no space is rented ($y = 0$). The following example of a plane is used subsequently.

EXAMPLE 1.14 The Inclination-Intercept Form of a Plane in 3-Space

The equation of the plane in 3-space with inclination $(m_1, m_2) = (1, -2)$ through the intercept $b = 3$ is

$$z = m_1 x + m_2 y + b = 1x + (-2)y + 3 = x - 2y + 3$$

Generalizing the Point-Normal Form of a Line

Recall the following point-normal representation of a line in 2-space:

$$a(x - x_0) + b(y - y_0) = 0 \tag{1.26}$$

in which (x_0, y_0) is a point on the line and (a, b) is a vector that is perpendicular to the line. Generalizing this idea to 3-space results in a plane. Specifically, the **point-normal form of a plane** that contains the given point (x_0, y_0, z_0) and is perpendicular to the given **normal vector** (a, b, c), as shown in Figure 1.24, is

$$a(x - x_0) + b(y - y_0) + c(z - z_0) = 0 \tag{1.27}$$

EXAMPLE 1.15 The Point-Normal Form of a Plane in 3-Space

The point-normal form of the plane that contains the point $(x_0, y_0, z_0) = (2, 1, 3)$ and is perpendicular to the normal vector $(a, b, c) = (-1, 2, 1)$ is

$$(-1)(x - 2) + 2(y - 1) + 1(z - 3) = 0$$

Equivalently,

$$-(x - 2) + 2(y - 1) + (z - 3) = 0 \tag{1.28}$$

By eliminating the parentheses in (1.28) and then solving for z, you obtain

$$z = x - 2y + 3$$

which is the same as the equation in Example 1.14.

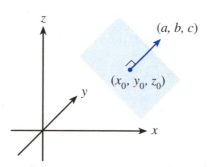

Figure 1.24 *A Plane in 3-Space Determined by a Point and a Normal Vector.*

You can generalize the point-normal form to a plane in n-space by using appropriate subscript and vector notation. Specifically, the point-normal form of a plane in n-space that contains the given point $\mathbf{x}' = (x_1', \ldots, x_n')$ and is perpendicular to the normal vector $\mathbf{a} = (a_1, \ldots, a_n)$ is

$$\mathbf{a} \cdot (\mathbf{x} - \mathbf{x}') = a_1(x_1 - x_1') + \cdots + a_n(x_n - x_n') = 0 \tag{1.29}$$

The generalization of the point-normal form from a line in 2-space, to a plane in 3-space, to a plane in n-space, is an example of the process hereafter referred to as **sequential generalization**, in which one generalization leads to further and further generalizations.

Generalizing the Parametric Equations to a Line in n-Space

Recall the following parametric equations representing a line in 2-space:

$$\left.\begin{array}{l} x = x_0 + at \\ y = y_0 + bt \end{array}\right\} \tag{1.30}$$

in which (x_0, y_0) is a known point on the line and (a, b) is a vector representing a direction of the line. It is possible to generalize these equations both to a line in n-space and a plane in n-space. Only the generalization to a line is given here.

The generalization of (1.30) to a line in 3-space through a given point (x_0, y_0, z_0) in a given direction (a, b, c) is the following parametric equations:

$$\left.\begin{array}{l} x = x_0 + at \\ y = y_0 + bt \\ z = z_0 + ct \end{array}\right\} \quad \text{(parametric equations for a line in 3-space)} \tag{1.31}$$

EXAMPLE 1.16 The Parametric Equations of a Line in 3-Space

The parametric equations that describe the line through the point $(x_0, y_0, z_0) = (1, 0, 2)$ in the direction $(a, b, c) = (2, 1, -1)$ are

$$x = x_0 + at = 1 + 2t$$
$$y = y_0 + bt = 0 + 1t$$
$$z = z_0 + ct = 2 - 1t$$

In the spirit of sequential generalization, you can extend this idea to a line in n-space. By using appropriate subscript notation, the parametric equations for the line that passes through a point $(u_1, \ldots, u_n)$ in a given direction $(d_1, \ldots, d_n)$ are

$$\left.\begin{array}{l} x_1 = u_1 + td_1 \\ x_2 = u_2 + td_2 \\ \vdots \quad \vdots \quad \vdots \\ x_n = u_n + td_n \end{array}\right\} \quad \text{(parametric equations for a line in } n\text{-space)}$$

You can take advantage of vector notation by writing these parametric equations succinctly in terms of the vectors $\mathbf{x} = (x_1, \ldots, x_n)$, $\mathbf{u} = (u_1, \ldots, u_n)$, and $\mathbf{d} = (d_1, \ldots, d_n)$, as follows:

$$\mathbf{x} = \mathbf{u} + t\mathbf{d}$$

EXAMPLE 1.17 Parametric Equations for a Line in *n*-Space

The line that passes through the point $\mathbf{u} = (3, 0, -1, 2)$ in the direction $\mathbf{d} = (-1, 1, 4, 0)$ is the set of vectors $\mathbf{x} = (x_1, x_2, x_3, x_4)$ such that

$$\mathbf{x} = \mathbf{u} + t\mathbf{d} = (3, 0, -1, 2) + t(-1, 1, 4, 0) = (3 - t, t, -1 + 4t, 2)$$

Equivalently stated, the parametric equations for this line are

$$x_1 = 3 - t, \quad x_2 = t, \quad x_3 = -1 + 4t, \quad x_4 = 2$$

1.3.3 Unifying the Equations of a Plane

In Section 1.3.2, you were introduced to the following two ways to represent a plane in *n*-space:

$$z = m_1 x_1 + \cdots + m_{n-1} x_{n-1} + b \tag{1.32}$$
$$a_1(x_1 - x_1') + \cdots + a_n(x_n - x_n') = 0 \tag{1.33}$$

It would be convenient if there were one way to represent the equation of a plane that includes both (1.32) and (1.33) as special cases.

The process of creating this new, encompassing representation is an example of another mathematical technique called **unification.** This technique involves combining two or more concepts (problems, theories, and so on) into a single framework from which you can study each of the special cases.

To illustrate, unification is used now to combine the special cases in (1.32) and (1.33) into a single equation for representing a plane in *n*-space.

Identifying Similarities and Differences

The first step in unification is to identify similarities and differences between the special cases in an attempt to find what is common. For instance, (1.32) and (1.33) exhibit the following similarities and differences.

Similarities in (1.32) and (1.33)

1. Both equations involve n variables: z and $x_1, \ldots, x_{n-1}$, in (1.32) and $x_1, \ldots, x_n$, in (1.33).
2. Each of the variables in both equations is multiplied by a known constant. For example, in (1.32), x_1 is multiplied by m_1 and in (1.33), x_1 is multiplied by a_1.
3. Both equations involve other known constants: b, in (1.32) and $a_1 x_1', \ldots, a_n x_n'$, in (1.33).

Differences in (1.32) and (1.33)

1. The left and right sides of the two equations have different forms. For example, the right side of (1.32) contains variables together with the constant b. The right side of (1.33) contains only the constant 0.
2. The specific constants that multiply the variables are different in the two equations.
3. One of the variable names in (1.32) differs from that in (1.33). Specifically, (1.32) contains the variable z and (1.33) contains x_n.

The objective now is to find ways to eliminate the differences so that you can unify (1.32) and (1.33) into a single equation.

Eliminating the Differences

You can eliminate the first difference by rewriting (1.32) and (1.33) so that all terms involving variables are on the left side of the equality signs and all terms involving constants are on the right side. For instance, moving all terms involving variables in (1.32) to the left side results in

$$-m_1 x_1 - \cdots - m_{n-1} x_{n-1} + z = b \tag{1.34}$$

Likewise, removing the parentheses in (1.33) and moving all constant terms to the right side results in

$$a_1 x_1 + \cdots + a_n x_n = a_1 x_1' + \cdots + a_n x_n' \tag{1.35}$$

Now, in both (1.34) and (1.35), all variables are on the left side of the equality sign and only known constants are on the right side.

The second difference identified above is overcome by introducing new notation, as done, for example, in the following unification of (1.34) and (1.35):

$$c_1 y_1 + \cdots + c_n y_n = d \quad \text{(unification)} \tag{1.36}$$

In the unification in (1.36), the new symbols $y_1, \ldots, y_n$ are used to represent the n variables from the special cases in (1.34) and (1.35). The symbols c_1 through c_n in the unification represent known constants that multiply the corresponding variables in the special cases. The symbol d in the unification represents a known value on the right side of the equalities in the special cases.

As you have seen, the use of appropriate notation is important. To avoid confusion, all symbols used in the unified equation (1.36) are different from those in the special cases. Often, however, mathematicians use *overlapping notation*, that is, the same symbol is used more than once, with a different meaning in each case. For example, you can write (1.36) equally well as

$$a_1 x_1 + \cdots + a_n x_n = b \tag{1.37}$$

In this case, the symbols $a_1, \ldots, a_n$ in (1.37) overlap with those same symbols in (1.33) and the symbol b in (1.37) overlaps with that same symbol in (1.32). Also, the symbols $x_1, \ldots, x_{n-1}$ in (1.37) overlap with those same symbols in (1.32) and (1.33). When this overlapping notation arises, be sure to keep the meaning of the symbols straight in your mind.

Verifying the Special Cases

After obtaining a unification, such as the one in (1.37), you should verify that each of the original items that gives rise to the unification is, in fact, a special case of the unification. This is done by showing that an appropriate substitution of the symbols in the unification results in each of the original special cases. In this example, by substituting

$$a_1 = -m_1, \ldots, a_{n-1} = -m_{n-1}, a_n = 1,$$
$$x_1 = x_1, \ldots, x_{n-1} = x_{n-1}, x_n = z, \text{ and } b = b$$

in (1.37) and rewriting, you obtain the original (1.32). Likewise, by substituting

$$a_1 = a_1, \ldots, a_n = a_n, x_1 = x_1, \ldots, x_n = x_n, \text{ and } b = a_1 x_1' + \cdots + a_n x_n'$$

in (1.37) and rewriting, you obtain the original (1.33). Thus, both (1.32) and (1.33) are special cases of the unification in (1.37).

The unified equation

$$a_1 x_1 + \cdots + a_n x_n = b \tag{1.38}$$

is called a **linear equation** in the variables $x_1, \ldots, x_n$, meaning that each variable is multiplied by a known constant, the individual terms are then added to or subtracted from each other, and the result is a known constant. The following set of examples includes some equations that are linear and others that are not linear.

EXAMPLE 1.18 Examples of Linear and Nonlinear Equations

Linear Equations	**Nonlinear Equations**
$2x + 3y = 5$	$x^2 + 3y = 5$
$x_1 = 2x_2 - 3$	$\log(x_1) = 2x_2 - 3$
$-x + 2y - 3z = 0$	$xyz = 1$
$x_1 + 2 = x_2 - x_3 + 3$	$\dfrac{1}{x_1} + 2 = x_2 - x_3 + 3$

In this section, you have seen how vectors are used to represent lines and planes. The role of vectors in problem solving is the topic of Section 1.4.

Exercises for Section 1.3

1. Interpret the slope and intercept of the line $d = 5000 + 40t$ that relates the distance d of an object in feet from a measuring device after t seconds.

2. Interpret the slope and intercept of the line $v = 16000 - 125m$ that relates the dollar value v of a car in terms of the number of thousands of miles m that the car has been driven.

3. Interpret the slope and intercept of the line $s = 88 - 7t$ that relates the speed s of a passenger car as measured in feet per second to the number t of seconds that the car has been braking.

4. Interpret the slope and intercept of the line $p = 0.567r + 5.03$ that relates the monthly mortgage payment p per thousand dollars of loan to the interest rate r, expressed as a percentage.

5. Write the equation of the line $3x + 2y = 6$ in (a) slope-intercept form, (b) point-normal form using the point $(2, 0)$ and the normal vector $(3, 2)$, and (c) parametric-equation form using the point $(2, 0)$ and the direction $(-2, 3)$.

6. Write the equation of the line $2x - 3y = 12$ in (a) slope-intercept form, (b) point-normal form using the point $(0, -4)$ and the normal vector $(2, -3)$, and (c) parametric-equation form using the point $(0, -4)$ and the direction $(3, 2)$.

7. Write the equation of the plane $x + 2y - z = 8$ in (a) inclination-intercept form and (b) point-normal form using the point $(2, 3, 0)$ and the normal vector $(1, 2, -1)$.

8. Write the equation of the plane $6x - 3y - 9z = 36$ in (a) inclination-intercept form and (b) point-normal form using the point $(6, 0, 0)$ and the normal vector $(6, -3, -9)$.

9. Indicate an appropriate substitution to verify that each of the following equations is a special case of the general linear equation $a_1 x_1 + \cdots + a_n x_n = b$.
   (a) $2x + 3y = 3$.
   (b) $x_1 = 2x_2 - 3$.

10. Indicate an appropriate substitution to verify that each of the following equations is a special case of the general linear equation $a_1 x_1 + \cdots + a_n x_n = b$.
    (a) $-x + 2y - 3z = 0$.
    (b) $x_1 + 2 = x_2 - x_3 + 3$.

11. For each of the following problems, use appropriate notation to identify the data. Then find a closed-form solution in terms of the given data.
    (a) Find the slope-intercept equation for a line in the plane using one point on the line and the slope.
    (b) Find the slope-intercept equation for a line in the plane using two different points on the line.

12. For each of the following problems, use appropriate notation to identify the data. Then find a closed-form solution in terms of the given data.
    (a) Find the slope-intercept equation for a line in the plane using the parametric equations of the line.
    (b) Find the slope-intercept equation for a line in the plane using the point-normal form of the line.

13. For each of the following problems, use appropriate notation to identify the data. Then find a closed-form solution in terms of the given data.
    (a) Find parametric equations for a line in the plane using the slope-intercept equation of the line.
    (b) Find parametric equations for a line in the plane using the point-normal equation of the line.

14. For each of the following problems, use appropriate notation to identify the data. Then find a closed-form solution in terms of the given data.
    (a) Find the point-normal equation for a line in the plane using the slope-intercept equation of the line.
    (b) Find the point-normal equation for a line in the plane using the parametric equations of the line.

15. Suppose that $\mathbf{u} = (u_1, u_2, u_3)$ and $\mathbf{v} = (v_1, v_2, v_3)$ are two vectors drawn at the origin and that both lie in a given plane P that also goes through the origin.
    (a) Write the point-normal equation of the plane P using $\mathbf{u} \times \mathbf{v}$ as the normal vector.
    (b) Apply the result in part (a) to find the point-normal equation of the plane that goes through the three points $(0, 0, 0)$, $(0, 2, 0)$, and $(1, 1, 1)$.

16. (a) Identify similarities and differences between the inclination-intercept form of a plane and the point-normal form of a plane in 3-space. (Hint: Consider a plane that is parallel to the z-axis.)
    (b) In view of part (a), show how and when you can use the point-normal equation of a plane in 3-space to write the point-inclination form.

17. Consider the following two ways to measure the distance between two vectors $\mathbf{u} = (u_1, u_2)$ and $\mathbf{v} = (v_1, v_2)$ in the plane:

$$|u_1 - v_1| + |u_2 - v_2| \quad \text{and} \quad \sqrt{(u_1 - v_1)^2 + (u_2 - v_2)^2}$$

    (a) Create a visual image of these two expressions on a single graph.
    (b) Unify the foregoing two measures into a single expression and show that, by an appropriate substitution, each of these two measures is a special case of your unified expression.

18. Unify the following three problems into a single problem and show that, by an appropriate substitution, the unified problem includes each one as a special case.
    (i) Each gallon of premium gasoline produced requires 1 hour of labor in the Purification Department and 2 hours in the Blending Department. Each gallon of regular gasoline requires 2 hours of labor in the Purification Department and 1 hour in the Blending Department. Given that there are a total of 240 hours of labor available in the Purification Department and 280 hours of labor in the Blending Department this week, determine how many gallons of each type of gasoline can be produced with this amount of labor.
    (ii) Find a point where two lines in the plane intersect.
    (iii) Find values for the variables x and y so that

$$2x = 2 + y$$
$$x = -2y$$

19. For n-vectors $\mathbf{u} = (u_1, \ldots, u_n)$ and $\mathbf{v} = (v_1, \ldots, v_n)$, consider the following unary and binary operations:

    $$\text{dim}(\mathbf{u}) = \text{dimension of } \mathbf{u}$$
    $$\text{sum}(\mathbf{u}) = u_1 + \cdots + u_n$$
    $$\text{times}(\mathbf{u}, \mathbf{v}) = (u_1 v_1, \ldots, u_n v_n)$$

    If possible, use the components of $\mathbf{u}$ and $\mathbf{v}$ to indicate the result of performing the following expressions. If you are unable to do so, then identify all syntax errors that arise.
    (a) times(2$\mathbf{u}$).   (b) sum(times($\mathbf{u}$, $\mathbf{v}$)).
    (c) times($\mathbf{u}$, sum($\mathbf{v}$)).   (d) dim(sum($\mathbf{v}$)).

20. For n-vectors $\mathbf{u} = (u_1, \ldots, u_n)$ and $\mathbf{v} = (v_1, \ldots, v_n)$, consider the following unary and binary operations:

    $$\text{abs}(\mathbf{u}) = (|u_1|, \ldots, |u_n|)$$
    $$\text{left}(\mathbf{u}) = (u_2, u_3, \ldots, u_n, u_1)$$
    $$d(\mathbf{u}, \mathbf{v}) = \|\mathbf{u} - \mathbf{v}\|$$

    If possible, use the components of $\mathbf{u}$ and $\mathbf{v}$ to indicate the result of performing the following expressions. If you are unable to do so, then identify all syntax errors that arise.
    (a) abs(left($\mathbf{u}$)).   (b) left(abs($\mathbf{u}$)).
    (c) d(abs($\mathbf{u}$), $\mathbf{u} \cdot \mathbf{v}$).   (d) abs($d$($\mathbf{u}$, $\mathbf{v}$)).

21. Suppose that t, u, v, and w are real numbers with $t \geq 0$ that satisfy the following condition:

    $$\left| \frac{u}{v} - w \right| \leq t \tag{1.39}$$

    (a) Identify all syntax errors that arise in the following generalization of (1.39) when $\mathbf{v}$ is a n-vector:

    $$|u - w\mathbf{v}| \leq t|\mathbf{v}|$$

    (b) Using the expression in part (a) as a guide, generalize (1.39) when both $\mathbf{v}$ and $\mathbf{w}$ are n-vectors. Be sure your resulting expression contains no syntax error. (Hint: Consider the norm of a vector.)

22. Suppose that k is a positive integer and u is a real number.
    (a) How can you generalize the expression u^k to the case where u is an n-vector, say, $\mathbf{u} = (u_1, \ldots, u_n)$?
    (b) Can you generalize the expression u^k to the case where k is an n-vector, say, $\mathbf{k} = (k_1, \ldots, k_n)$? If not, explain what difficulties there are.

23. Suppose that $\mathbf{u} = (u_1, u_2)$ is a given nonzero vector in two dimensions drawn at the origin. Now draw the line through the origin that is perpendicular to $\mathbf{u}$, as in the following figure:

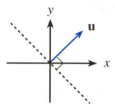

Observe that this line divides the plane into two parts (each called a *half space*). That is, the line and all points on one side of the line constitute a half space. Translate these two half-spaces to symbolic form using the components (u_1, u_2) of the given vector **u**.

24. Generalize your result in the previous exercise to a given vector **u** in *n*-space. That is, write a symbolic representation of the *plane* that goes through the origin and is perpendicular to **u** together with the two associated half spaces created by this plane.

1.4 Problem Solving with Vectors

The material in this chapter is brought together by solving problems involving vectors and lines.

1.4.1 Practical Problem Solving: The Problem of Best Paper Products

The problem of Best Paper Products presented at the beginning of this chapter is designed to show you how vector operations are used to obtain the solution. You will also see how technology is used to perform these operations.

The Problem of Best Paper Products

Recall the following problem of Best Paper Products that supplies local-area businesses with a variety of paper products. Each day, a truck leaves the warehouse to make deliveries to a number of customers. The order in which the deliveries are made is determined manually by plotting the locations of the day's customers on a map and then rotating a horizontal line drawn through the warehouse in the counterclockwise direction 360 degrees, picking up each customer, in turn. You have been asked to automate this procedure.

The first step in solving this, and virtually all mathematical problems, is to identify the given data and the desired output. Here, the data are the known locations of the customers. The desired output is the order in which to visit those customers. From the description, you know how to solve this problem visually by plotting the customer locations on the map and rotating a horizontal line counterclockwise through the customers. The goal is to translate this visual solution to a numerical method in algebraic form—using vectors and their operations—so that the solution can be obtained using the company's computer system. The specific implementation of the steps described in what follows depends on whether the company uses a graphing calculator, a software package, such as MATLAB, Maple, and Mathematica, or a programming language, such as C++.

Solving the Problem

Solving this problem involves representing the data as vectors, performing operations on those vectors to obtain the solution, and displaying the results in an easy-to-understand manner.

Representing the Data. The first issue is how to represent the data corresponding to the locations of the customers on the day's delivery route. One approach is to use two numbers for each location. The first number represents the East-West distance in miles from the warehouse to the customer, in which a positive value means that the customer is to the East of the warehouse and a negative value means that the customer is to the West. The second number represents the North-South distance in miles, with a positive value meaning that the customer is to the North of the warehouse and a negative value to the South. For example, the following data represent the locations of six customers on today's delivery schedule:

Customer number	1	2	3	4	5	6
East-West distance	1.2	−2.2	4.1	−3.5	4.0	−1.0
North-South distance	−2.8	−2.5	3.0	1.4	−1.0	2.5

These six locations are plotted in Figure 1.25, with the warehouse located at the origin. As described in Section 1.1.3, you can represent these six locations by the six vectors $(1.2, -2.8)$, $(-2.2, -2.5)$, $(4.1, 3.0)$, $(-3.5, 1.4)$, $(4.0, -1.0)$, and $(-1.0, 2.5)$. An alternative—the one used from here on—is to represent these data by two vectors: one for the six East-West distances and one for the six North-South distances, as follows:

$$\mathbf{x} = (1.2, -2.2, 4.1, -3.5, 4.0, -1.0) \tag{1.40}$$
$$\mathbf{y} = (-2.8, -2.5, 3.0, 1.4, -1.0, 2.5) \tag{1.41}$$

Obtaining and Displaying the Solution. As indicated in the problem description, the order in which the customers are visited is determined by rotating the x-axis counterclockwise through 360 degrees. Doing so in Figure 1.25 leads to the order indicated by the following customer numbers: 3, 6, 4, 2, 1, 5. It is now necessary to translate this visual solution to symbolic form in terms of operations on the vectors $\mathbf{x}$ in (1.40) and $\mathbf{y}$ in (1.41). Doing so involves the following steps.

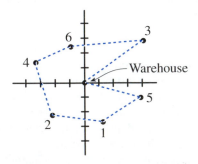

Figure 1.25 *The Locations of the Customers of Best Paper Products.*

A Numerical-method Solution for Best Paper Products

Step 1. Use the vectors **x** and **y** to compute, for each customer, an angle between 0 and 360 degrees associated with rotating the x-axis in the counterclockwise direction until that axis reaches the location of that customer. In this example, that list of angles in degrees is:

Customer	1	2	3	4	5	6
Angle	292.96	228.79	36.10	158.14	346.25	111.73

Step 2. Sort the list of angles obtained in Step 1 in increasing order and record the sequence of the customer numbers that produces this ordering. In this example, you obtain the following:

Customer	3	6	4	2	1	5
Angle	36.10	111.73	158.14	228.79	292.96	346.25

Step 3. Draw a graph indicating the routing from the warehouse through all of the customers in the order obtained in Step 2. For this example, that graph is shown in Figure 1.25.

Implementing the Solution Using Technology

Now it is necessary to express these steps in terms of the given vectors **x** and **y** using vector operations. A general discussion of how to do so follows but the specific method depends on whether you are using a graphing calculator, a software package, or a programming language.

Finding the Angles. To perform Step 1 of the solution procedure, it is necessary to use the vectors **x** and **y** to determine the angles from the x-axis to the locations of the customers. Graphing calculators, computer software packages, and programming languages all provide built-in functions such as *arccos*, *arcsin*, and *arctan* for doing so. Care is needed when using these functions to ensure that the angle obtained represents a rotation in the counterclockwise direction.

To illustrate, suppose that the function *atan2* in MATLAB is used. This function returns the arctangent, in radians between $-\pi$ and π, of a point located in a quadrant of the plane. (This may not be the case for *arctan* functions in graphing calculators and programming languages.) In this numerical example, *atan2*(**y**, **x**) applied to the vectors **y** in (1.41) and **x** in (1.40) yields the following vector of angles (rounded to two decimal places):

$$\mathbf{a} = (-1.17, -2.29, 0.63, 2.76, -0.24, 1.95)$$

For ease of understanding, you can convert these angles from radians to degrees by multiplying **a** by $180/\pi$:

$$\mathbf{b} = \left(\frac{180}{\pi}\right)\mathbf{a} = (-67.04, -131.21, 36.10, 158.14, -13.75, 111.73)$$

It is now necessary to make all of these angles positive so that they reflect rotation of the x-axis in the counterclockwise direction. This is accomplished by adding 360 degrees to those angles in **b** whose values are negative. To do so using vector operations, first create a vector **c** whose components are 0, if the corresponding angle is positive and 360, if the corresponding angle is negative. In this case, the vector **c** is:

$$\mathbf{c} = (360, 360, 0, 0, 360, 0)$$

The result of adding **c** to **b** is the following list of angles, in degrees, as measured from the x-axis in the counterclockwise direction:

$$\mathbf{d} = \mathbf{b} + \mathbf{c} = (292.96, 228.79, 36.10, 158.14, 346.25, 111.73)$$

Sorting the Angles. Now you can sort these angles in increasing order to obtain

$$\mathbf{e} = (36.10, 111.73, 158.14, 228.79, 292.96, 346.25)$$

However, what you need is the *order* of the customers that leads to the vector **e**. That is, you need a permutation vector of $\{1, 2, 3, 4, 5, 6\}$, as described in Section 1.1.3. In this case, that permutation vector is

$$\mathbf{p} = (3, 6, 4, 2, 1, 5)$$

Visiting the customers in the sequence indicated by **p** corresponds to the list of angles in increasing order, that is,

$$\mathbf{e} = (d_{p_1}, d_{p_2}, d_{p_3}, d_{p_4}, d_{p_5}, d_{p_6}) = (d_3, d_6, d_4, d_2, d_1, d_5)$$

Displaying the Final Results. Having found the permutation vector, all that remains is to display the final sequence—as in Figure 1.25—that starts at the warehouse located at the origin and connects the locations of the customers in the order indicated by the permutation vector **p**.

1.4.2 Conceptual Problem Solving: Orthogonal Projections in 2-space

The solution to the following problem again illustrates the use of vector operations and translating visual images to symbolic form.

Problem Description

Suppose that a force, in the form of a vector **f** in 2-space, is applied to an object moving in a nonzero direction represented by the vector **d** drawn at the origin, as shown in Figure 1.26(a). The objective is to determine the component **v** of the force **f** acting in the direction **d** and the component of **f** acting orthogonal to **d**. The vector **v** shown in Figure 1.26(b) is obtained by drawing the line through the origin in the direction of **d** and dropping a perpendicular from the head of **f** to the line. The component of **f** orthogonal to **d** is then $\mathbf{f} - \mathbf{v}$.

Solving the Problem

The first step is to identify the given data and the desired output. The data are the known vectors **f** and **d**. The desired output is a closed-form expression for the vectors **v** and $\mathbf{f} - \mathbf{v}$ in terms of **f** and **d**.

To find **v**, observe in Figure 1.26(b) that "**v** points in the same direction as **d** but has a different length." Now translate the expression in quotation marks to symbolic form.

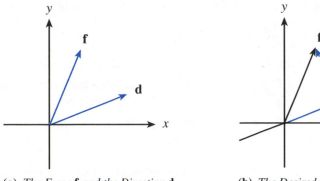

(a) *The Force* **f** *and the Direction* **d**.   **(b)** *The Desired Vectors* **v** *and* **f** − **v**.

Figure 1.26 *The Given and Desired Vectors.*

Recalling from Section 1.2.2 the operation of multiplying a vector by a real number, you can express **v** as some real number t times **d**. Using the existential quantifier *there is*, you can state that

there is a real number t such that $\mathbf{v} = t\mathbf{d}$ (1.42)

The only question is, What is the value for t?

Whenever you want to show that "there is an object with a certain property such that something happens," you should use a technique referred to here as the **construction method**. With this technique, you must somehow produce the object, by guessing, by trial-and-error, by devising an algorithm, or by some other means. After doing so, be sure to show that the object you construct is the right one—that is, that your object satisfies the certain property and the something that happens. (See Appendix A.3 for a more detailed description of this technique.)

For the current problem, you can construct the value for t in (1.42) by using the fact that **f** − **v** is perpendicular to **d**. Translate this visual image to symbolic form by writing that the dot product of the two vectors is 0:

$$(\mathbf{f} - \mathbf{v}) \cdot \mathbf{d} = 0 \tag{1.43}$$

From the laws of vector operations you learned in Section 1.2.3 and from (1.42), you can rewrite (1.43) as follows:

$$(\mathbf{f} - \mathbf{v}) \cdot \mathbf{d} = (\mathbf{f} \cdot \mathbf{d}) - (\mathbf{v} \cdot \mathbf{d}) = (\mathbf{f} \cdot \mathbf{d}) - t(\mathbf{d} \cdot \mathbf{d}) = 0 \tag{1.44}$$

Noting that $\mathbf{d} \cdot \mathbf{d} \neq 0$ because $\mathbf{d} \neq \mathbf{0}$, you can solve (1.44) for t to obtain

$$t = \frac{\mathbf{f} \cdot \mathbf{d}}{\mathbf{d} \cdot \mathbf{d}} \tag{1.45}$$

Observe that no syntax error is present in (1.45) because both the numerator $\mathbf{f} \cdot \mathbf{d}$ and the denominator $\mathbf{d} \cdot \mathbf{d}$ are real numbers.

Having found the value for t, you can now use (1.42) to find $\mathbf{v}$, namely,

$$\mathbf{v} = t\mathbf{d} = \frac{\mathbf{f} \cdot \mathbf{d}}{\mathbf{d} \cdot \mathbf{d}}\mathbf{d} \tag{1.46}$$

Also, the desired vector $\mathbf{f} - \mathbf{v}$ is

$$\mathbf{f} - \mathbf{v} = \mathbf{f} - \frac{\mathbf{f} \cdot \mathbf{d}}{\mathbf{d} \cdot \mathbf{d}}\mathbf{d} \tag{1.47}$$

The vector $\mathbf{v}$ is called the **orthogonal projection of f onto d** and is often written

$$\text{proj}_\mathbf{d}\mathbf{f}$$

The vector $\mathbf{f} - \mathbf{v}$ is called the **vector component of f orthogonal to d**. You now have the following closed-form solutions for two given vectors $\mathbf{f}$ and $\mathbf{d}$ in 2-space with $\mathbf{d} \neq \mathbf{0}$:

$$\text{proj}_\mathbf{d}\mathbf{f} = \frac{\mathbf{f} \cdot \mathbf{d}}{\mathbf{d} \cdot \mathbf{d}}\mathbf{d} \tag{1.48}$$

$$\mathbf{f} - \text{proj}_\mathbf{d}\mathbf{f} = \mathbf{f} - \frac{\mathbf{f} \cdot \mathbf{d}}{\mathbf{d} \cdot \mathbf{d}}\mathbf{d} \tag{1.49}$$

Here you see an additional benefit of using vector notation. Because the solutions in (1.48) and (1.49) are written in vector notation, those solutions apply to vectors in n-space, not just to vectors in 2-space. An application of these solutions to vectors in 4-space follows.

EXAMPLE 1.19 Orthogonal Projections

For vectors

$$\mathbf{f} = (4, -3, 0, -8) \quad \text{and} \quad \mathbf{d} = (1, 0, -2, -1)$$

the formulas in (1.48) and (1.49) result in

$$\begin{aligned}
\text{proj}_\mathbf{d}\mathbf{f} &= \frac{\mathbf{f} \cdot \mathbf{d}}{\mathbf{d} \cdot \mathbf{d}}\mathbf{d} \\[2mm]
&= \frac{(4, -3, 0, -8) \cdot (1, 0, -2, -1)}{(1, 0, -2, -1) \cdot (1, 0, -2, -1)}(1, 0, -2, -1) \\[2mm]
&= \frac{12}{6}(1, 0, -2, -1) = (2, 0, -4, -2)
\end{aligned}$$

$$\mathbf{f} - \text{proj}_\mathbf{d}\mathbf{f} = (4, -3, 0, -8) - (2, 0, -4, -2) = (2, -3, 4, -6)$$

In this section, you have seen how vectors are used in problem solving to represent data and how to perform operations on those vectors to obtain the solution. For example, you have learned to find the orthogonal projection of one vector onto another vector and the associated vector component. This was accomplished by translating a visual image to symbolic form using vectors and their operations. In Chapter 2, linear algebra is used to solve another important class of problems.

Exercises for Section 1.4

1. Find the projection of the vector $(1, 3)$ onto the vector $(4, 2)$ and the component of $(1, 3)$ orthogonal to $(4, 2)$. Draw an appropriate picture.

2. Find the projection of the vector $(2, 1)$ onto the vector $(-3, -2)$ and the component of $(2, 1)$ orthogonal to $(-3, -2)$. Draw an appropriate picture.

3. Suppose that **u** and **w** are n-vectors, neither of which is the zero vector.
   (a) Translate the statement that "the projection of **u** onto **w** is **0**" to visual form. What is the geometric relation of **u** to **w**?
   (b) What algebraic property holds for **u** and **w** if the projection of **u** onto **w** is **0**?

PROJECT 1.1: The Routing Problem of Best Paper Products

Use a graphing calculator, MATLAB, or other software package, as indicated by your instructor, to perform the instructions in Exercises 4 through 10 for the problem of Best Paper Products described in Section 1.4.1 with these locations for the six customers:

Customer number	1	2	3	4	5	6
East-West distance	1.2	−2.2	4.1	−3.5	4.0	−1.0
North-South distance	−2.8	−2.5	3.0	1.4	−1.0	2.5

Whenever possible, use vector operations available on your system. Write the answer and the sequence of operations you performed to obtain the answer.

4. Store the six East-West distances of the customers in a vector **x** and the six North-South distances in a vector **y**.

5. (a) Find the straight-line distance from the warehouse to each of the customers. Use a vector dot product to find the sum of these distances.
   (b) Suppose the truck driver averages 30 miles per hour. For each customer, determine how long it would take (to the nearest minute) to drive from the warehouse to that one customer and return to the warehouse. Use a vector dot product to find the sum of these individual driving times.

6. Repeat the previous exercise assuming a rectangular road system.

7. (a) For each customer, find the angle, in radians between 0 and 2π, through which a horizontal line must be rotated in the counterclockwise direction until passing through the location of that customer.
   (b) Express the angles in part (a) in degrees between 0 and 360.

8. Write instructions that use the vector of angles obtained in the previous exercise to create a vector of the locations of the customers in order of increasing angle. (Note: The instructions you use depend on the specific form of technology you are working with. If you are unable to write such instructions, then create the desired vector of locations manually.)

9. Use your technology to draw a graph indicating the routing of the truck from the warehouse to the six customers in the order obtained in the previous exercise.

10. What is the total straight-line distance the truck travels in visiting the customers in the order determined in Exercise 8?

PROJECT 1.2: The Prediction Problem of Measuring Devices
Measuring Devices is developing a device for estimating the speed and initial distance of
an object that moves in a straight line at a constant, but unknown, speed. If the object is
initially at a distance of p feet from the device and travels at a constant speed of s feet per
second, then, at any time $t \geq 0$, the distance, d, of the object is given by the linear relation

$$d = p + st$$

The objective is to estimate values for s and p on the basis of data recorded by the device.
Specifically, the device measures and records the distance of the object from the device at
various times, such as the following data pertaining to the location of a power boat:

Time (in seconds)	0	2	4	7	10	12
Distance (in feet)	5000	5075	5175	5300	5400	5500

Use a graphing calculator, MATLAB, or other software package, as indicated by your
instructor, to perform the instructions in Exercises 11 through 16. Whenever possible, use
vector operations available on your system. Write the answer and the sequence of operations
you performed to obtain the answer.

11. Create a vector **t** to represent the six times at which the device recorded measurements.
    Create a vector **d** to represent the six recorded distances.

12. Use your technology to plot the six pairs of recorded values on a graph whose x-
    axis represents time and whose y-axis represents distance. Due to inaccuracies in the
    measurements, observe that these points do not lie precisely on a straight line.

13. Recall that the objective is to find values for p and s so that the line

    $$d = p + st \qquad (1.50)$$

    best fits the data. For specific values of p and s, a measure is needed to determine
    how well the line in (1.50) fits the data. One such measure is obtained by comparing
    how close the recorded distance at a given time is to the distance predicted by (1.50).
    For example, suppose that $p = 5000$ and $s = 40$. Then, at time $t = 2$, the distance
    predicted by (1.50) is $d = p + st = 5000 + 40(2) = 5080$ while the recorded distance
    is 5075. The difference between the predicted and recorded values is $5080 - 5075 = 5$.
    To obtain an overall measure of how close the line $d = 5000 + 40t$ fits the given data,
    perform the following computations:
    (a) Use (1.50) with $p = 5000$ and $s = 40$ to create a vector of predicted distance at
        each of the six times.
    (b) Use your results in part (a) to create a vector of the differences between the predicted
        and the recorded distances.
    (c) Create a vector whose components are the squares of the components of the vector
        in part (b).
    (d) Obtain a final measure of how well the line $d = 5000 + 40t$ fits the data—called
        the *sum of squared errors*—by adding up the components of the vector in part (c).

14. (a) Create a vector of five slopes in which component i of this vector is the slope of
        the line that goes through the first data point and the data point $i + 1$.

(b) Repeat Exercise 13 five times, each time setting $p = 5000$ and s equal to one of the five components of the vector from part (a). Report the sum of squared errors for each slope. Which of these slopes provides the least sum of squared errors?

15. (a) For the slope producing the smallest sum of squared errors in Exercise 14(b), find the intercept of the line that goes through the point obtained by computing the average of the times and the average of the distances in the recorded data.

(b) Compute the sum of squared errors for the line whose slope and intercept you found in part (a). Does this line fit the data better than the line in Exercise 13? Why or why not? Explain.

(c) Draw on a graph the best line you have found together with the original data points.

16. Use the best line you have found so far to predict how far the boat is, in miles, at each of the following times: (a) 15 minutes, (b) 30 minutes, (c) 1 hour.

Note: In Project 1.2, trial-and-error is used to find a line that fits the data well. This approach is time consuming and there is no guarantee that you find the best line. In Chapter 7, you will learn a systematic method for determining the best line.

Chapter Summary

In this chapter, you have learned that an n-vector $\mathbf{u} = (u_1, \ldots, u_n)$ is an ordered list of n real numbers that has two visualizations when $n = 2$ or 3: (1) as a point and (2) as an arrow connecting two points. The components of a vector are always determined relative to a fixed coordinate system whose units of measure you can choose. All vectors other than the zero vector have a positive length and a direction.

Operations on Vectors

The algebraic (or symbolic) form and the corresponding geometric visualization of various operations on vectors were presented. Some of these operations result in a real number and others result in a vector. The algebraic formulas for these various operations on n-vectors $\mathbf{u} = (u_1, \ldots, u_n)$ and $\mathbf{v} = (v_1, \ldots, v_n)$ are summarized in Table 1.1 and satisfy the properties in Theorems 1.1 - 1.6 in Section 1.2.

Operation	Symbol	Closed-form Expression
A real number t times a vector	$t\mathbf{u}$	$(tu_1, \ldots, tu_n)$
Length (or norm) of a vector	$\|\mathbf{u}\|$	$\sqrt{u_1^2 + \cdots + u_n^2}$
Adding two vectors	$\mathbf{u} + \mathbf{v}$	$(u_1 + v_1, \ldots, u_n + v_n)$
Subtracting two vectors	$\mathbf{u} - \mathbf{v}$	$(u_1 - v_1, \ldots, u_n - v_n)$
Distance between two vectors	$\|\mathbf{u} - \mathbf{v}\|$	$\sqrt{(u_1 - v_1)^2 + \cdots + (u_n - v_n)^2}$
The dot product of two vectors	$\mathbf{u} \cdot \mathbf{v}$	$u_1 v_1 + \cdots + u_n v_n = \|\mathbf{u}\| \|\mathbf{v}\| \cos(\theta)$
The cross product of two vectors	$\mathbf{u} \times \mathbf{v}$	$(u_2 v_3 - u_3 v_2, u_3 v_1 - u_1 v_3, u_1 v_2 - v_1 u_2)$

Table 1.1 *Algebraic Operations on Vectors.*

Uses of Vectors

Vectors are used as a notational tool in solving problems that involve lines and planes. You can obtain the solution to such problems by performing appropriate operations on the vectors that represent the problem data. The solution thus obtained applies not only to the problem in 2-space that you can visualize, but also to the same problem in n-space that you cannot visualize. Various ways of writing the equations of lines and planes are summarized in Table 1.2.

General Mathematical Concepts

In addition to the specific topics of vectors, you have seen that a mathematical problem consists of some known information, in the form of data, that is used to obtain some unknown but desired information. You would ideally like to find a closed-form solution, which is a solution obtained from the problem data by a simple rule or formula. When you cannot obtain a closed-form solution, the next best approach is to develop a numerical method, which is a solution obtained from the problem data by performing a sequence of computations. You have also learned the following mathematical thinking processes:

1. **Generalization**   The process of creating, from an original mathematical concept (problem, definition, theorem, and so on) a more general concept (problem, definition, theorem, and so on), that includes not only the original one, but many other new and perhaps different ones as special cases. Each special case is obtained from the generalization by an appropriate substitution of symbols.

2. **Identifying similarities and differences**   The process of comparing and contrasting two or more mathematical concepts (problems, definitions, theorems, and so on) with the objective of gaining a deeper understanding of the relationship of these two items.

Representation	Lines	Planes
Slope-intercept	$y = mx + b$, where m = slope and b = y-intercept	$y = \mathbf{m} \cdot \mathbf{x} + b$, where $\mathbf{m}$ = inclination and b = y-intercept
Point-normal	$a(x - x_0) + b(y - y_0) = 0$, where (x_0, y_0) = a point on the line and (a, b) = a vector perpendicular to the line	$\mathbf{a} \cdot (\mathbf{x} - \mathbf{x}') = 0$, where $\mathbf{x}'$ = a point on the plane and $\mathbf{a}$ = a vector perpendicular to the plane
General form	$a_1 x_1 + a_2 x_2 = b$, where (a_1, a_2) = a vector perpendicular to the line and b = the dot product of (a_1, a_2) with any one point on the line	$\mathbf{a} \cdot \mathbf{x} = b$, where $\mathbf{a}$ = a vector perpendicular to the plane and b = the dot product of $\mathbf{a}$ with any one point in the plane
Parametric equations in 2 dimensions	$(x_0, y_0) + t(a, b)$, where (x_0, y_0) = a point on the line and (a, b) = a direction of the line	Not applicable here
Parametric equations in n dimensions	$\mathbf{x} = \mathbf{u} + t\mathbf{d}$, where $\mathbf{u}$ = a point on the line and $\mathbf{d}$ = a direction of the line	Not applicable here

Table 1.2 *Equations of Lines and Planes.*

3. Visualization   The process of creating an image that captures the essential features of a mathematical concept so that you can think about and work with that concept more easily.

4. Translating visual images to symbolic form   The process of converting a visual image of a mathematical concept to a formal written form in terms of the vocabulary and syntax of the language of mathematics.

5. Unification   The process of combining two or more problems (ideas, concepts, theories, and so on) into a single framework from which to study each of the special cases.

6. Proof   An argument for convincing someone that a statement of the form "if p, then q" is true, where the hypothesis p and the conclusion q are given statements. Various techniques for doing so are described in Appendix A.

Chapter 2

Using Matrices to Solve ($m \times n$) Linear Equations

In Chapter 1, you learned that an n-vector is an ordered list of real numbers and saw how vectors are used in problem solving. In this chapter, you will learn about another type of ordered list used for solving linear equations. You will also see how unification, generalization, identifying similarities and differences, and translating visual images to symbolic form arise in the study of these lists and in solving problems such as the following one.

The Investment Problem of Portfolio Planners

The managers of Portfolio Planners want to invest 3 million dollars in five mutual funds identified by the Research Department as meeting the minimum requirements of risk and reward. The risk category and estimated annual rates of return are given in the following table:

Fund number	1	2	3	4	5
Rate of return	0.25	0.15	0.12	0.08	0.05
Risk category	high	high	medium	medium	low

To increase the chance of making money while keeping the risk of losing money as low as possible, management has decided to diversify their investments in the five funds as follows. The ratio of the amounts invested in Funds 1 and 2 will be 1 to 3. The ratio of the amounts invested in Funds 3 and 4 will be 1 to 2. Also, the total amount invested in the medium-risk Funds 3 and 4 should be twice as much as the total amount invested in the high-risk Funds 1 and 2. Management knows intuitively that the more money invested in the low-risk Fund 5, the lower the return on their total investment. They have asked you to quantify this relationship so they can decide how much to invest in each of the funds.

2.1 Applications of Solving Linear Equations

In this section, various examples of fundamental problems arising in physics, chemistry, engineering, computer science, statistics, economics, business, and many other areas are presented. The art of formulating these problems requires identifying variables, also called **unknowns**, whose values, once determined, constitute a solution to the problem. In each case, these values must satisfy a **system of linear equations**, which is a finite number of linear equations, as you will now see.

2.1.1 Leontief Economic Models

One example in which the need to solve a system of linear equations arises is the study of an economic system divided into a number of *sectors*. Each sector produces goods, some of which are used internally by that sector. The rest are sold to the remaining sectors.

Problem Description

To illustrate, consider a hypothetical world economy made up of three sectors: Germany, Japan, and the United States. The fraction of each country's annual output that is consumed internally and sold to the other countries is given in the following table:

	From		
To	**Germany**	**Japan**	**U.S.**
Germany	0.5	0.2	0.2
Japan	0.1	0.4	0.1
U.S.	0.4	0.4	0.7

For example, from the first column, you can see that Germany consumes 50% of its annual production and sells 10% to Japan and the remaining 40% to the U.S.

The objective is to use the data in the foregoing table to determine a dollar value for the annual production of each country such that the total amount spent by each country on consumption equals the dollar value of its annual production. Such a balance between the two sets of values is called an *equilibrium point*. (Wassily Leontief proved that an equilibrium point must exist.)

Problem Formulation

To find this equilibrium point, first identify variables whose values you can control. In this problem the variables are the values of the annual production of each of the three countries. To maintain consistency of units, the annual production of each country should be measured in one currency, say, dollars, thus giving rise to the following three variables:

G = the annual production of Germany, in dollars

J = the annual production of Japan, in dollars

U = the annual production of the U.S., in dollars

You want to find values for these variables so that the amount of each country's annual production is equal to the amount that country consumes. For example, for Germany,

$$\begin{pmatrix} \text{Value of} \\ \text{German} \\ \text{production} \end{pmatrix} = \begin{pmatrix} \text{Purchases} \\ \text{from} \\ \text{Germany} \end{pmatrix} + \begin{pmatrix} \text{Purchases} \\ \text{from} \\ \text{Japan} \end{pmatrix} + \begin{pmatrix} \text{Purchases} \\ \text{from} \\ \text{U. S.} \end{pmatrix}$$

In terms of the variables and the problem data from the first row of the table, you have

$$G = 0.5G + 0.2J + 0.2U \quad \text{(balance for Germany)} \tag{2.1}$$

Likewise, using the second row of the table, the balance equation for Japan is

$$J = 0.1G + 0.4J + 0.1U \quad \text{(balance for Japan)} \tag{2.2}$$

The balance equation for the U. S. from the third row of the table is

$$U = 0.4G + 0.4J + 0.7U \quad \text{(balance for the U.S.)} \tag{2.3}$$

Moving all variables to the left side of the equalities in (2.1), (2.2), and (2.3), your goal is to find values for the variables that satisfy the following system of linear equations:

$$0.5G - 0.2J - 0.2U = 0$$
$$-0.1G + 0.6J - 0.1U = 0$$
$$-0.4G - 0.4J + 0.3U = 0$$

Problem Solution

The technique you will learn in Section 2.3 for solving this problem leads to the conclusion that there are an infinite number of values for the variables that simultaneously satisfy the foregoing equations. Given any value for the variable U, the values of the remaining variables are

$$G = 0.50U \quad \text{and} \quad J = 0.25U$$

For example, if the value of the U. S. annual production is $U = 10$ billion dollars, then $G = 0.5(10) = 5$ billion dollars and $J = 0.25(10) = 2.5$ billion dollars constitute equilibrium values satisfying all three balance equations simultaneously.

2.1.2 Linear Equations in Network-Flow Problems

Another class of problems in which it is necessary to solve linear equations arises when material flows through a network system, as is now illustrated.

Problem Description

The California Gas Company sends 80 thousand gallons of gasoline per day from its refinery in Oakland to a distributor in Sacramento through a pipeline network with pumping stations at Concord, Antioch, Livermore, and Stockton, as represented by the diagram in Figure 2.1. The number of thousands of gallons shipped per day in the following segments is fixed:

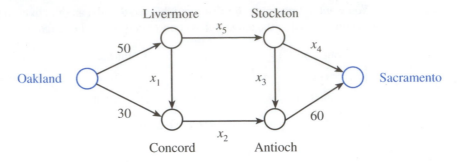

Figure 2.1 *The Pipeline Network of the California Gas Company.*

From	To	Amount
Oakland	Concord	30
Oakland	Livermore	50
Antioch	Sacramento	60

The amounts shipped in the remaining segments can vary according to daily demands. Management wants to know the least and most amount of gasoline per day that can be shipped from Livermore to Stockton and the impact these values will have on the amounts shipped in the remaining pipeline segments.

Problem Formulation

Figure 2.1 is a visual representation of a **network**. The finite collection of circles, called **nodes**, represent physical items in a problem (pumping stations, in this example). Certain pairs of these nodes are connected by a line with an arrow, called an **arc**, representing a relationship between the pair of connected nodes. In this problem, an arc represents the ability to ship gasoline in the direction of the arrow between the locations of the two connected nodes. Gasoline cannot flow in the opposite direction.

A network problem often has data associated with the nodes and/or arcs. In the current problem, the amounts to ship along three segments are known. These values are written next to the corresponding arcs in Figure 2.1. To determine the amount of flow in the remaining five arcs, over which you have some control, introduce the five variables $x_1, \ldots, x_5$. These variables are written next to the associated arcs in Figure 2.1. The values of these variables represent the number of thousands of gallons of gasoline per day to ship along the associated pipeline segment.

The mathematical representation of the problem, then, is to use the given data to determine a range of allowable values for x_5 and to evaluate the impact of the value of x_5 on the remaining variables. A system of four linear equations is used to accomplish this goal. These equations arise from the requirement that the total amount of gasoline per day entering a node, other than the nodes corresponding to Oakland and Sacramento, must equal the total amount shipped out. These equations are expressed in terms of the data and variables in the following table (see Figure 2.1):

Node	Flow in = Flow out
Concord	$30 + x_1 = x_2$
Livermore	$50 = x_1 + x_5$
Antioch	$x_2 + x_3 = 60$
Stockton	$x_5 = x_3 + x_4$

By rearranging terms so that all variables are on one side of the equality and all numbers are on the other side, you obtain the following system of four linear equations in five unknowns:

$$
\begin{aligned}
x_2 - x_1 &= 30 \\
x_1 + x_5 &= 50 \\
x_2 + x_3 &= 60 \\
x_5 - x_3 - x_4 &= 0
\end{aligned}
$$

(2.4)

The values of the five variables must be nonnegative because of the assumption that gasoline can flow only in the direction indicated by the arrows.

Problem Solution

Solving this problem leads again to the conclusion that there are an infinite number of values for the variables that simultaneously satisfy (2.4). Given a value for x_5, the values of the remaining variables are determined as follows:

$$x_1 = 50 - x_5 \tag{2.5}$$
$$x_2 = 80 - x_5 \tag{2.6}$$
$$x_3 = -20 + x_5 \tag{2.7}$$
$$x_4 = 20 \tag{2.8}$$

Recalling that all variables must be nonnegative, you can now answer both questions posed by management. For example, from (2.7), you can say that x_5 must be at least 20 so that $x_3 \geq 0$. Likewise, from (2.5), x_5 cannot exceed 50, otherwise, $x_1 < 0$, and from (2.6), $x_5 \leq 80$. Putting the pieces together, to keep all variables nonnegative, you can ship between 20 and 50 thousand gallons per day from Livermore to Stockton.

2.1.3 Linear Programming Problems

In the preceding two examples, you sought values for variables that satisfy certain linear equations. When there are many values for the variables that satisfy the equations, it is often desirable to determine values that minimize or maximize a linear function, called the *objective function*. A problem consisting of real-valued variables whose values must satisfy a system of linear equations and nonnegativity constraints while simultaneously minimizing or maximizing a linear objective function is called a **linear programming problem**. Consider the following problem of Fiber Optics.

Problem Description

Fiber Optics produces both a low-density and a high-density fiber optic cable. The company has just received a one-time order for 2000 feet of the low-density cable and 4500 feet of the high-density cable for the following week. Each foot of the low-density cable costs $4 to produce and requires half a minute of machine time. Each foot of the high-density cable costs $6 to produce and requires three-fourths of a minute on the same machine. Because only 40 hours of machine time are available next week, Fiber Optics cannot meet these demands. However, the company can make up the difference by buying some amount of these cables from a Japanese company at a cost of $7 per foot for the low-density cable and $8 per foot for the high-density cable. You have been asked to find a production/purchase plan for the company that minimizes the cost of meeting the demand while not exceeding the available machine time.

Problem Formulation

The first step is to use the problem data to develop a mathematical model consisting of variables, an objective function, and constraint equations (or inequalities).

Identifying the Variables. Begin by identifying variables whose values you can control. In this case, you are free to determine how much of each cable to produce and how much to purchase from the Japanese company. This gives rise to the following four variables:

LP = the number of feet of low-density cable to produce

HP = the number of feet of high-density cable to produce

LJ = the number of feet of low-density cable to purchase

HJ = the number of feet of high-density cable to purchase

Identifying the Objective Function. You want to choose values for the variables that minimize the total cost of meeting the demand. The total cost is equal to the cost of producing the cables plus the cost of purchasing the cables from the Japanese company. As stated in the problem description, it costs $4 to produce each foot of low-density cable, so, producing LP feet costs $4LP$ dollars. Likewise, each foot of high-density cable costs $6 to produce, so HP feet costs $6HP$ dollars. Combining the production costs with the costs of purchasing LJ feet of low-density cable at $7 per foot and HJ feet of high-density cable at $8 per foot, the objective is to

$$\text{minimize} \quad \underbrace{4LP + 6HP}_{\text{production cost}} + \underbrace{7LJ + 8HJ}_{\text{purchase cost}} \qquad (2.9)$$

Identifying the Constraints. The values of these variables must satisfy certain constraints. For example, the plan must meet the demand of 2000 feet for the low-density cable. So, the amount of low-density cable produced plus the amount purchased from the Japanese company must total 2000 feet. In terms of the variables, this constraint is:

$$LP + LJ = 2000 \quad \text{(demand for low-density cable)} \qquad (2.10)$$

To meet the demand for high-density cable, the following constraint must be satisfied:

$$HP + HJ = 4500 \quad \text{(demand for high-density cable)} \tag{2.11}$$

Another constraint is needed to ensure that the 40 hours of available machine time is not exceeded. The total machine time used is the amount used to produce low-density cable plus the amount used to produce high-density cable. As stated in the problem description, each foot of low-density cable requires 0.5 minutes of machine time, so LP feet of low-density cable uses $0.5LP$ minutes. Likewise, each foot of high-density cable requires 0.75 minutes of machine time, so HP feet of high-density cable uses $0.75HP$ minutes. Thus, in terms of the variables and problem data, you might be tempted write the following *incorrect* machine-time constraint:

$$0.5LP + 0.75HP \leq 40 \quad \text{(an incorrect constraint)}$$

This constraint is incorrect because the unit of measure on the left side of the inequality is minutes and that on the right side is hours. One way to correct the problem is to convert the 40 hours of available machine time to $40(60) = 2400$ minutes. Also, because it is less expensive to produce the cables than it is to purchase them, the company should use all the available machine time. Thus, the machine-time constraint becomes

$$0.5LP + 0.75HP = 2400 \quad \text{(machine time in minutes)} \tag{2.12}$$

In addition, the value of each variable must be nonnegative, so you should include the following constraints:

$$LP \geq 0, \quad HP \geq 0, \quad LJ \geq 0, \quad HJ \geq 0 \tag{2.13}$$

Putting together the results in (2.9) through (2.13), your goal is to find values for the variables LP, HP, LJ, and HJ so as to solve the following linear programming problem:

Objective Function

$$\text{minimize} \quad 4LP + 6HP + 7LJ + 8HJ \quad \text{(total cost)}$$

Constraints

$$
\begin{array}{llll}
\text{subject to} \quad & LP & + LJ & = 2000 \quad \text{(low-density demand)} \\
& HP & + HJ = 4500 \quad \text{(high-density demand)} \\
& 0.5LP + 0.75HP & = 2400 \quad \text{(machine time)} \\
& LP, \quad HP, \quad LJ, \quad HJ \geq & 0
\end{array}
$$

Ignoring the objective function and the nonegativity constraints, you have a system of 3 linear equations in the 4 unknowns LP, HP, LJ, and HJ. The technique you will learn in Section 2.3 leads to the conclusion that there are an infinite number of values for the variables that satisfy these linear equations. Among all of these solutions, you want to find one, called an *optimal solution*, that satisfies all of the constraints and simultaneously provides the smallest value of the objective function. A systematic method called the *simplex algorithm* exists for doing so, but the details are beyond the scope of this book. In the exercises in Section 2.5, a trial-and-error approach is proposed for finding an optimal solution.

In this section, you have learned how to formulate a system of linear equations to solve a problem. Doing so requires identifying variables whose values you can control and creating a system of linear equations—expressed in terms of the variables and problem data—that the values of the variables must satisfy. In the rest of this chapter, you will learn several systematic methods for finding values of the variables that simultaneously satisfy all of the linear equations. In Section 2.2, a formal mathematical statement of the problem of solving a system of linear equations is given.

Exercises for Section 2.1

1. Suppose the U.S. economy is divided in two sectors: goods and services.
   (a) Write the balance equations for an economy in which the goods sector uses 70% of its annual output and purchases 40% of the annual output of the services sector.
   (b) Find an equilibrium point for your model in part (a) by expressing the annual output of goods in terms of the annual output of services.
   (c) What is the equilibrium point if the annual production of services is $30 billion?

2. Write the balance equations for a North American economy in which the fractions of each region's annual production used internally and sold to the other regions are given in the columns of the following table:

	From		
To	Canada	Mexico	U.S.
Canada	0.4	0.1	0.3
Mexico	0.4	0.5	0.1
U.S.	0.2	0.4	0.6

3. Repeat the previous exercise for the following world economy:

	From			
To	Asia	Europe	North America	South America
Asia	0.4	0.1	0.2	0.1
Europe	0.2	0.6	0.2	0.1
North America	0.3	0.2	0.4	0.3
South America	0.1	0.1	0.2	0.5

4. The following table shows the fraction of people who move between rural living, urban living, and suburban living over a one-year period:

	From		
To	Rural	Suburban	Urban
Rural	0.90	0.01	0.02
Suburban	0.03	0.95	0.08
Urban	0.07	0.04	0.90

For example, from the first column, of all the people living in a rural environment, 90% stay, 3% move to suburban living, and 7% move to urban living. Write appropriate equations to determine how many people would have to be living in each group (rural, suburban, and urban) so that, after one year, the number of people in each group remains unchanged.

5. Recall the problem of the California Gas Company in Section 2.1.2. Suppose that the company can control the amount of gasoline shipped in each segment of the pipeline in Figure 2.1. Formulate a linear programming problem to determine the maximum amount of gasoline that can be shipped each day from Oakland through the network to Sacramento.

6. Microtech has 1500 computers in a warehouse in Bakersfield and another 2000 computers in a warehouse in Fresno. The company needs to send 1300 computers to a retail store in San Francisco, 1700 computers to a store in Los Angeles, and 500 computers to a customer in San Luis Obispo. The mileages between these locations are given in the following table:

	To		
From	**San Francisco**	**Los Angeles**	**San Luis Obispo**
Bakersfield	284	103	122
Fresno	176	211	136

(a) Draw a network diagram for this problem. Indicate the meaning of the nodes and arcs. Include all of the given data in the diagram.

(b) Formulate a mathematical model consisting of variables, an objective function, and constraints to determine a shipping plan from the warehouses to the customers that minimizes the total number of "computer-miles" while satisfying the customer demands and not exceeding the available warehouse supplies.

7. Texas Bus Company is a new company that will provided service among Austin (A), Corpus Christi (C), Dallas (D), El Paso (E), Houston (H), Lubbock (L), and San Antonio (S). One-way fares between cities having nonstop service are given in the following table:

	To						
From	**A**	**C**	**D**	**E**	**H**	**L**	**S**
A	–	–	45	–	40	–	–
C	–	–	–	95	65	–	40
D	45	–	–	–	60	–	–
E	–	95	–	–	–	75	–
H	40	65	60	–	–	–	–
L	–	–	–	75	–	–	85
S	–	40	–	–	–	85	–

(a) Draw a network diagram to illustrate which pairs of cities have nonstop bus service. Indicate the meaning of the nodes and arcs. Include the given cost data in the diagram.

(b) Formulate a mathematical model consisting of variables, an objective function, and constraints to determine the least-cost fare between Dallas and El Paso. The least-cost fare is the one that provides the smallest value of the sum of the nonstop fares along any route from Dallas to El Paso through connecting cities. (Hint: In constructing the least-cost route, you can choose to use or not to use each arc.)

8. An oil refinery buys two types of oil: light crude oil at a cost of $40 per barrel and heavy crude oil at $34 per barrel. The following quantities (in barrels) of gasoline, kerosene, and jet fuel are produced from each barrel of each type of crude oil:

	Gasoline	Kerosene	Jet Fuel
Light crude oil	0.45	0.18	0.32
Heavy crude oil	0.34	0.36	0.22

Management has contracted to deliver 1,200,000 barrels of gasoline, 500,000 barrels of kerosene, and 300,000 barrels of jet fuel. Formulate a linear programming problem to determine how much of each type of crude oil to purchase so as to minimize the cost while at least meeting the demand.

9. Each gallon of milk, pound of cheese, and pound of apples produces a known number of milligrams of protein and vitamins A, B, and C, as given in the following table. The weekly requirements of the nutritional ingredients and the costs of the foods are also supplied.

	Milk (gals.)	Cheese (lb.)	Apples (lb.)	Weekly Requirement
Protein (mg.)	40	20	10	80
Vitamin A (mg.)	5	40	30	60
Vitamin B (mg.)	20	30	40	50
Vitamin C (mg.)	30	50	60	30
Unit cost ($)	2	3.5	0.9	

Formulate a linear programming problem to determine how much of each food to consume to meet or exceed the weekly requirements of nutrients while minimizing the total cost.

10. Clear Chemicals needs to prepare a combined total of 5000 liters of two of their products, H_2SO_4 and H-CL, which are made from concentrated acids, as follows. Each liter of concentrated sulfuric acid cost $12 and is diluted with 20 liters of distilled water to produce 21 liters of H_2SO_4 that then sells for $1 per liter. Each liter of concentrated hydrochloric acid cost $18 and is diluted with 30 liters of distilled water to produce 31 liters of H-CL that then sells for $1 per liter. Formulate an appropriate linear programming problem assuming that distilled water costs $0.15 per liter and that 200 liters of concentrated sulfuric acid and 150 liters of concentrated hydrochloric acid are available.

2.2 The Problem of Solving Linear Equations

In Section 2.1, various examples and applications of the problem of solving a system of linear equations were presented. In this section, a mathematical representation of that problem is developed through a sequential generalization of the following problem of solving one linear equation in one unknown.

PROBLEM 2.1 Solving One Linear Equation in One Unknown
Given values for the real numbers a and b, find a value for the variable x that satisfies the following linear equation:

$$ax = b \tag{2.14}$$

2.2.1 The Problem of Solving Two Linear Equations in Two Unknowns

One way to generalize Problem 2.1 arises when the system has additional variables and/or equations. For example, consider the following problem of solving two linear equations in the two unknowns x and y:

$$2x + 3y = 8$$
$$6x - 2y = 2$$

More generally, the the problem of solving a (2×2) (pronounced "two by two") system of linear equations, that is, two linear equations in two unknowns, is stated as follows.

PROBLEM 2.2 Solving a (2 x 2) System of Linear Equations
Given values for the real numbers p, q, r, s, t, and u, find values for the variables x and y so that

$$px + qy = t \tag{2.15}$$
$$rx + sy = u \tag{2.16}$$

Note that Problem 2.2 has more data than the original problem of solving one linear equation in one unknown. In particular, Problem 2.2 has six known data items (p, q, r, s, t, and u) whereas the single linear equation has only two known data items (a and b).

There are various ways to verify that Problem 2.2 includes Problem 2.1 as a special case. For example, you can use the values of a and b from the linear equation to create the following data for the (2×2) system:

$$p = a, \quad q = 0, \quad t = b \tag{2.17}$$

$$r = 0, \quad s = 0, \quad u = 0 \tag{2.18}$$

Substituting these values in (2.15) and (2.16) yields

$$px + qy = t \quad \text{or} \quad ax + 0y = b \quad \text{or} \quad ax = b \tag{2.19}$$

$$rx + sy = u \quad \text{or} \quad 0x + 0y = 0 \quad \text{or} \quad 0 = 0 \tag{2.20}$$

Ignoring equation (2.20), you have in (2.19) the linear equation as a special case of two linear equations in two unknowns.

Another way to verify that the linear equation is a special case of Problem 2.2 is to substitute

$$p = a, \quad q = 0, \quad t = b \tag{2.21}$$

$$r = 0, \quad s = 1, \quad u = 0 \tag{2.22}$$

in (2.15) and (2.16) to obtain

$$px + qy = t \quad \text{or} \quad ax + 0y = b \quad \text{or} \quad ax = b \tag{2.23}$$

$$rx + sy = u \quad \text{or} \quad 0x + 1y = 0 \quad \text{or} \quad y = 0 \tag{2.24}$$

Once again, ignoring the bottom equation (2.24), you obtain in (2.23) the linear equation as a special case of two linear equations in two unknowns.

2.2.2 The Problem of Solving *n* Linear Equations in *n* Unknowns

A further generalization arises when you consider the problem of solving 3 linear equations in 3 unknowns, as illustrated in the following example.

EXAMPLE 2.1 A (3 x 3) System of Linear Equations
The following is an example of 3 linear equations in the 3 unknowns x, y, and z:

$$
\begin{aligned}
2x + 4y - z &= -1 \\
-x + 2y + 2z &= 2 \\
x - y + z &= 0
\end{aligned}
$$

More generally, by using appropriate subscript notation, the problem of solving n linear equations in n unknowns follows.

PROBLEM 2.3 Solving an (*n* x *n*) System of Linear Equations
Given a positive integer n and values for the real numbers a_{ij}, for each $i, j = 1, \ldots, n$, and also for the real numbers $b_1, \ldots, b_n$, find values for the variables $x_1, \ldots, x_n$ so that

the following system is satisfied:

$$
\begin{aligned}
a_{11}x_1 + a_{12}x_2 + \cdots + a_{1n}x_n &= b_1 \\
a_{21}x_1 + a_{22}x_2 + \cdots + a_{2n}x_n &= b_2 \\
&\vdots \\
a_{n1}x_1 + a_{n2}x_2 + \cdots + a_{nn}x_n &= b_n
\end{aligned}
\tag{2.25}
$$

2.2.3 The Problem of Solving *m* Linear Equations in *n* Unknowns

The final generalization considered here arises when the number of equations is permitted to be different from the number of unknowns, as illustrated in the following example.

EXAMPLE 2.2 An (*m* x *n*) System of Linear Equations
The following is a group of $m = 3$ linear equations in the $n = 4$ unknowns x_1, x_2, x_3, x_4:

$$
\begin{aligned}
2x_1 + 4x_2 - x_3 + 4x_4 &= -1 \\
-x_1 + 2x_2 + 2x_3 - x_4 &= 1 \\
x_1 - x_2 + x_3 &= 3
\end{aligned}
$$

More generally, by using subscript notation, the problem of solving *m* linear equations in *n* unknowns is written as follows.

PROBLEM 2.4 Solving an (*m* x *n*) System of Linear Equations
Given positive integers m and n and values for the real numbers a_{ij}, for each $i = 1, \ldots, m$ and $j = 1, \ldots, n$, and also for the real numbers $b_1, \ldots, b_m$, find values for the variables $x_1, \ldots, x_n$ so that the following system is satisfied:

$$
\begin{aligned}
a_{11}x_1 + a_{12}x_2 + \cdots + a_{1n}x_n &= b_1 \\
a_{21}x_1 + a_{22}x_2 + \cdots + a_{2n}x_n &= b_2 \\
&\vdots \\
a_{m1}x_1 + a_{m2}x_2 + \cdots + a_{mn}x_n &= b_m
\end{aligned}
\tag{2.26}
$$

There are different ways to solve the system in (2.26). To understand these methods, you must first understand precisely what it means to "solve" a system of linear equations.

2.2.4 What it Means to "Solve" a System of Linear Equations

You know that the unique value of x that satisfies the equation $2x = 8$ is $x = 4$. More generally, a set of values $s_1, \ldots, s_n$ is a **solution to the system of linear equations** in (2.26) if all m equations are satisfied simultaneously when the values $x_1 = s_1, \ldots, x_n = s_n$ are substituted. For example, $x_1 = 2, x_2 = 0, x_3 = 1, x_4 = -1$ is a solution to the system in Example 2.2 because all three equations are satisfied simultaneously for these values of the variables. In contrast, $x_1 = 0, x_2 = 2, x_3 = 1, x_4 = -2$ is not a solution to this system because only the first equation is satisfied.

Solving a system of linear equations in general requires care, even when the system consists of only one equation and one unknown. One reason for this is the possibility that the system might be **inconsistent**, that is, the system might have no solution, such as the following equation:

$$0x = 1 \quad \text{(no value of } x \text{ satisfies this equation)}$$

A system of linear equations that has a solution is said to be **consistent**. The system in Example 2.2 is consistent, as is the following system.

EXAMPLE 2.3 The Homogeneous System of Linear Equations
The following system of linear equations is consistent because setting the value of each variable to 0 satisfies all of the equations:

$$a_{11}x_1 + a_{12}x_2 + \cdots + a_{1n}x_n = 0$$
$$a_{21}x_1 + a_{22}x_2 + \cdots + a_{2n}x_n = 0$$
$$\vdots$$
$$a_{m1}x_1 + a_{m2}x_2 + \cdots + a_{mn}x_n = 0$$

This particular system of equations is called **homogeneous**, meaning that the constant values on the right side of the equalities are all 0.

It is also possible for a system to have an infinite number of solutions, as in the equation

$$0x = 0 \quad \text{(any value of } x \text{ satisfies this equation)}$$

You cannot list all such solutions explicitly, so an alternative approach is to describe the set of solutions in a form called the **general solution**, which is presented in the following example.

EXAMPLE 2.4 The General Solution to a System of Linear Equations Having an Infinite Number of Solutions
To find all the solutions to the linear equation

$$2x + 3y = 6 \tag{2.27}$$

solve (2.27) for either of the variables in terms of the remaining one. For example, solving (2.27) for y yields

$$y = 2 - \frac{2}{3}x \tag{2.28}$$

By assigning an arbitrary value, say, t, to x and substituting this value in (2.28), you obtain the following general solution to (2.27):

$$x = t \quad \text{and} \quad y = 2 - \frac{2}{3}t$$

For example, when $t = 3$, $x = 3$ and $y = 0$. Because there are an infinite number of values for t, there are an infinite number of solutions to (2.27). Alternatively, you can solve (2.27) for x to obtain

$$x = 3 - \frac{3}{2}y \tag{2.29}$$

By assigning the arbitrary value of t to y and substituting this value in (2.29), you obtain the following form of the general solution to (2.27):

$$x = 3 - \frac{3}{2}t \quad \text{and} \quad y = t$$

For example, when $t = 4$, $x = -3$ and $y = 4$.

From these various foregoing examples, when solving a system of linear equations, you can identify the following three possible outcomes, the proof of which is given in Theorem 2.5 in Section 2.4.6:

Three Possibilities When Solving a System of Linear Equations

1. The system has no solution.
2. The system has exactly one solution.
3. The system has infinitely many solutions.

Visualizing the Possible Solutions to a System of Linear Equations

A visualization of these three possibilities is created by considering the following system of two linear equations in the two unknowns x and y:

$$ax + by = e \tag{2.30}$$
$$cx + dy = f \tag{2.31}$$

Each of these equations corresponds geometrically to a line in the plane, as described in Section 1.3. The algebraic operation of solving these two equations translates to the visual image of finding a point where the two corresponding lines intersect. By testing different values for a, b, c, d, e, and f, you will discover that one of the following three conditions occurs: (1) the two lines do not intersect at all, as in Figure 2.2(a), which means that the system has no solution, or (2) the two lines intersect in exactly one point, as in Figure 2.2(b), which means that the system has exactly one solution, or (3) the two lines lie on top of each other, as in Figure 2.2(c), which means that the system has infinitely many solutions.

As a result of the foregoing discussion, from here on, **solving a system of linear equations** will mean to do one of the following:

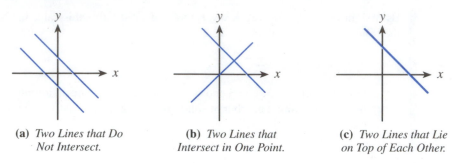

(a) *Two Lines that Do Not Intersect.*   **(b)** *Two Lines that Intersect in One Point.*   **(c)** *Two Lines that Lie on Top of Each Other.*

Figure 2.2 *Three Possible Relations Between Two Lines in the Plane.*

1. Determine that the system has no solution.
2. Determine that the system has exactly one solution and then find values for the variables $x_1, \ldots, x_n$ that simultaneously satisfy all m equations in (2.26).
3. Determine that the system has infinitely many solutions and then describe the general solution in an algebraic form similar to the one in Example 2.4.

2.2.5   Some Closed-form Solutions

In view of the preceding discussion, the following constitutes a closed-form solution to one linear equation in one unknown.

SOLUTION to Problem 2.1

The solution to Problem 2.1 of finding a value for x such that

$$ax = b$$

is described as follows:

$$\begin{cases} \text{no solution,} & \text{if } a = 0 \text{ and } b \neq 0 \\ x = b/a, & \text{if } a \neq 0 \\ x = \text{any value } t, & \text{if } a = 0 \text{ and } b = 0 \end{cases} \qquad (2.32)$$

The foregoing solution is in closed-form because the solution is expressed as a simple formula in terms of the data a and b. It is also possible to develop a closed-form solution to the problem of solving a (2×2) system of linear equations when the system has a unique solution, as illustrated next.

A Closed-form Solution for a (2 x 2) System of Linear Equations

One way to solve the following system of two linear equations:

$$px + qy = t \qquad (2.33)$$
$$rx + sy = u \qquad (2.34)$$

for x and y is to multiply (2.33) through by s and (2.34) through by q:

$$psx + qsy = st \tag{2.35}$$
$$qrx + qsy = qu \tag{2.36}$$

Subtracting (2.36) from (2.35) and dividing both sides of the resulting equation by $ps - qr$, assuming that this number is not 0, yields the following value for x:

$$x = \frac{st - qu}{ps - qr} \tag{2.37}$$

The value for y is then found by substituting (2.37) in (2.33), resulting in the following:

SOLUTION to Problem 2.2
The closed-form solution to the two linear equations in (2.33) and (2.34) in terms of the data p, q, r, s, t, and u, assuming that $ps - qr \neq 0$, is:

$$x = \frac{st - qu}{ps - qr} \tag{2.38}$$

$$y = \frac{pu - rt}{ps - qr} \tag{2.39}$$

Whenever you develop a solution to a more general problem (such as two linear equations in two unknowns), it is important to make sure that you can use that method to solve the special cases (the single linear equation, in this case). You will then have some assurance that the general solution is correct, at least for the special cases. In this example, the data for the (2×2) problem associated with the single linear equation $ax = b$ are given in (2.17) and (2.18), or, alternatively, in (2.21) and (2.22) in Section 2.2.1. You cannot substitute the values $p = a, q = 0, t = b, r = 0, s = 0$, and $u = 0$ from (2.17) and (2.18) in (2.38) and (2.39) because, in that case,

$$ps - qr = a(0) - 0(0) = 0$$

However, you *can* use the values $p = a, q = 0, t = b, r = 0, s = 1$, and $u = 0$ from (2.21) and (2.22). Substituting these values in (2.38) and (2.39) yields the following solution:

$$x = \frac{st - qu}{ps - qr} = \frac{1(b) - 0(0)}{a(1) - 0(0)} = \frac{b}{a}$$

$$y = \frac{pu - rt}{ps - qr} = \frac{a(0) - 0(b)}{a(1) - 0(0)} = 0$$

Looking at the value for x, note that you have obtained the solution to the linear equation as a special case of solving a (2×2) system of linear equations. You might also notice that *using the solution for the more general problem to solve the special case is less efficient computationally than solving the special case directly*. This is one of the disadvantages of generalization.

There is no known closed-form expression for solving a general system of m linear equations in n unknowns. The next best approach is to develop a numerical-method solution. Such a method is presented in Section 2.3.

Exercises for Section 2.2

1. Show, by an appropriate substitution, that Problem 2.2 in Section 2.2.1 is a special case of Problem 2.3 in Section 2.2.2.

2. Show, by an appropriate substitution, that the system of linear equations in Example 2.1 in Section 2.2.2 is a special case of Problem 2.3.

3. After considering the values $x_1 = 2$, $x_2 = -1$, and $x_3 = 0$, a student said that the following system of equations is inconsistent:

$$2x_1 - x_2 + 3x_3 = 5$$
$$-x_1 + 3x_2 - 4x_3 = -4$$
$$x_1 + x_2 + x_3 = 1$$

Do you agree? Why or why not? Explain.

4. Recall that a homogeneous system of equations is one in which the constant values on the right side of all equalities are 0.
   (a) Give a specific numerical example of a homogeneous system of two equations in two unknowns whose only solution is one in which the two variables have value 0.
   (b) Give a specific numerical example of a homogeneous system of two equations in two unknowns that has an infinite number of solutions.

5. Determine geometrically whether each of the following systems of linear equations has no solution; exactly one solution, in which case, find that solution; or infinitely many solutions, in which case, write the general solution.

   (a) $\begin{aligned} x - 2y &= 2 \\ -2x + 4y &= -4 \end{aligned}$
   (b) $\begin{aligned} x - 2y &= 2 \\ 2x + 2y &= 4 \end{aligned}$
   (c) $\begin{aligned} x - 2y &= 2 \\ -3x + 6y &= -6 \end{aligned}$

6. Repeat the previous exercise for the following systems of equations.

   (a) $\begin{aligned} 2x + y &= 4 \\ x - y &= 0 \end{aligned}$
   (b) $\begin{aligned} 2x + y &= 4 \\ -4x - 2y &= -8 \end{aligned}$
   (c) $\begin{aligned} 2x + y &= 4 \\ -3x - \tfrac{3}{2}y &= 0 \end{aligned}$

7. For given real numbers a, b, c, d, e, and f, consider the problem of finding values for x and y so that

$$ax + by = e$$
$$cx + dy = f$$

   (a) What condition on the data indicates that this system has no solution?
   (b) What condition on the data indicates that this system has an infinite number of solutions?

8. Write the general solution to each of the following systems of linear equations in as many ways as possible by assigning, each time, an arbitrary value to a different variable. For each system, find two different numerical solutions.

   (a) $2x - 4y = 6$
   (b) $\begin{aligned} 2x - 4y &= 6 \\ 3x + 4y + 2z &= 9 \end{aligned}$

9. Given real numbers a, b, c, and d, none of which is 0, consider the linear equation:

$$ax + by + cz = d$$

(a) Find a closed-form expression for the general solution by solving for x and then assigning arbitrary, but different, values to y and z.

(b) Find a closed-form expression for the general solution by solving for y and then assigning arbitrary, but different, values to x and z.

(c) Find a closed-form expression for the general solution by solving for z and then assigning arbitrary, but different, values to x and y.

10. Apply the results of the previous exercise to the linear equation $3x - y + 2z = 6$ by writing the general solution three different ways, each time assigning arbitrary values to two of the three variables.

2.3 Solving Linear Equations by Row Operations

There are many numerical methods for solving a system of m linear equations in n unknowns. These methods are composed of the following two phases:

1. **Preparation Phase**, in which operations are performed to prepare the system of equations for finding a subsequent solution.

2. **Solution Phase**, in which additional operations are performed to provide a solution to the system of equations.

Two methods that work in this way are presented in this section.

2.3.1 Gauss-Jordan Elimination

For the numerical method described in this section, virtually all of the effort is devoted to preparing the system of equations so that, with relatively little effort, you can then obtain the desired solution. More specifically, the method called **Gauss-Jordan elimination** is used to "reduce" (that is, transform) the original system of equations to an "equivalent" system that has an obvious solution. *Equivalence* in this context means that the set of solutions to the reduced system is the same as the set of solutions to the original system. Thus, if the reduced system has no solution, then neither does the original system. If the reduced system has exactly one solution, then that same solution solves the original system. If the reduced system has an infinite number of solutions, then each of those solutions is also a solution to the original system. For example, as you will see, Gauss-Jordan elimination reduces the original system

$$\begin{aligned} y - 2z &= 2 \\ 2x + 4y - 8z &= 6 \\ -x - 2y + 5z &= -2 \end{aligned} \tag{2.40}$$

to the following equivalent system, whose solution is obvious:

$$
\begin{aligned}
x & = -1 \\
y & = 4 \\
z & = 1
\end{aligned}
\tag{2.41}
$$

You can verify by substitution that (2.41) is the solution to the original system in (2.40).

Elementary Row Operations

The next issue is how, exactly, Gauss-Jordan elimination is used to reduce the original system. The approach is a generalization of the one used in Section 2.2.5 for solving a system of two linear equations in two unknowns and is based on the following three **elementary row operations**:

Elementary Row Operations

1. Multiply an equation through by a nonzero number.
2. Interchange two equations.
3. Add a multiple of one equation to another equation.

It is tedious to rewrite a system of equations each time one of these row operations is applied, so mathematicians have devised a more concise way to represent a system of linear equations. The idea is to use a *matrix*, which is a rectangular table of numbers organized in rows and columns.

Associated with each system of *m* linear equations in *n* unknowns is an **augmented matrix** consisting of *m* rows and *n* + 1 columns. This augmented matrix is obtained by first aligning the system of equations so that the terms in a single column pertain to the same variable, as in (2.40). The variable names are then dropped, leaving only the coefficients with their signs. Next, all of the equality signs are replaced with a single vertical bar. The last column in the augmented matrix then contains the values on the right side of the equalities. Such an augmented matrix is presented in the following example.

EXAMPLE 2.5 Representing a System of Linear Equations by an Augmented Matrix
The system in (2.40) and its associated augmented matrix are:

System of Equations $\qquad\qquad$ Augmented Matrix

$$
\begin{aligned}
y - 2z & = 2 \\
2x + 4y - 8z & = 6 \\
-x - 2y + 5z & = -2
\end{aligned}
\qquad
\left[\begin{array}{ccc|c}
0 & 1 & -2 & 2 \\
2 & 4 & -8 & 6 \\
-1 & -2 & 5 & -2
\end{array}\right]
$$

Because each augmented matrix corresponds to a system of equations, and vice versa, you can perform the elementary row operations equally well on the augmented matrix as on the linear equations, as shown in the following example.

EXAMPLE 2.6 Elementary Row Operations

The following is a sequence of elementary row operations that reduces the augmented matrix associated with the system in (2.40) to the augmented matrix associated with the equivalent system in (2.41) (each operation is performed on the matrix in the previous step):

Step 0. The original augmented matrix:

$$\left[\begin{array}{ccc|c} 0 & 1 & -2 & 2 \\ 2 & 4 & -8 & 6 \\ -1 & -2 & 5 & -2 \end{array}\right]$$

Step 1. Interchange row 1 and row 2:

$$\left[\begin{array}{ccc|c} 2 & 4 & -8 & 6 \\ 0 & 1 & -2 & 2 \\ -1 & -2 & 5 & -2 \end{array}\right]$$

Step 2. Multiply row 1 by 1/2:

$$\left[\begin{array}{ccc|c} 1 & 2 & -4 & 3 \\ 0 & 1 & -2 & 2 \\ -1 & -2 & 5 & -2 \end{array}\right]$$

Step 3. Add 1 times row 1 to row 3:

$$\left[\begin{array}{ccc|c} 1 & 2 & -4 & 3 \\ 0 & 1 & -2 & 2 \\ 0 & 0 & 1 & 1 \end{array}\right]$$

Step 4. Add -2 times row 2 to row 1:

$$\left[\begin{array}{ccc|c} 1 & 0 & 0 & -1 \\ 0 & 1 & -2 & 2 \\ 0 & 0 & 1 & 1 \end{array}\right]$$

Step 5. Add 2 times row 3 to row 2:

$$\left[\begin{array}{ccc|c} 1 & 0 & 0 & -1 \\ 0 & 1 & 0 & 4 \\ 0 & 0 & 1 & 1 \end{array}\right]$$

Observe that the solution $x_1 = -1$, $x_2 = 4$, and $x_3 = 1$ to the final system (and to the original system) is contained in the last column of the final augmented matrix in Step 5.

Reduced Row-Echelon Form

The objective now is to develop a systematic sequence of elementary row operations for reducing a given system of linear equations to a final system whose solution is easy to recognize. The final system has an easily recognizable solution when the associated augmented matrix has the properties described in the following definition.

DEFINITION 2.1

A matrix is in **reduced row-echelon form** if and only if the matrix has all of the following properties:

(1) All rows consisting entirely of zeros, if any, occur together at the bottom of the matrix.

(2) The first nonzero number in a row, if any, is a 1 (called a **leading 1**).

(3) For any two consecutive rows that do not consist entirely of zeros, the leading 1 in the lower row occurs further to the right than the leading 1 in the higher row.

(4) Each column that has a leading 1 contains zeros above and below the leading 1.

EXAMPLE 2.7 Matrices in Reduced Row-Echelon Form

All of the following augmented matrices are in reduced row-echelon form:

$$(a) \begin{bmatrix} 1 & 0 & 0 & 2 \\ 0 & 1 & 0 & -1 \\ 0 & 0 & 1 & 3 \end{bmatrix} \qquad (b) \begin{bmatrix} 1 & 0 & 0 & -1 & 2 \\ 0 & 1 & 0 & 1 & -3 \\ 0 & 0 & 1 & 2 & 1 \end{bmatrix}$$

$$(c) \begin{bmatrix} 1 & 0 & 0 & -1 \\ 0 & 1 & 0 & 4 \\ 0 & 0 & 0 & 1 \end{bmatrix} \qquad (d) \begin{bmatrix} 1 & 2 & 0 & 0 & -1 & 5 \\ 0 & 0 & 1 & 0 & 2 & -1 \\ 0 & 0 & 0 & 1 & -3 & 4 \\ 0 & 0 & 0 & 0 & 0 & 0 \end{bmatrix}$$

As mentioned previously, the advantage of an augmented matrix in reduced row-echelon form is that the solution to the associated system of linear equations is easily identified, as illustrated now for the reduced row-echelon matrices in Example 2.7.

The system of equations associated with the augmented matrix in Example 2.7(a) is

$$\begin{aligned} x_1 \quad\quad &= \quad 2 \\ x_2 \quad &= -1 \\ x_3 &= \quad 3 \end{aligned}$$

By inspection, the solution is $x_1 = 2, x_2 = -1, x_3 = 3$.

The system of equations associated with the augmented matrix in Example 2.7(b) is

$$\begin{aligned} x_1 \quad\quad - \ x_4 &= \quad 2 \\ x_2 \ + \ x_4 &= -3 \\ x_3 + 2x_4 &= \quad 1 \end{aligned}$$

The variables x_1, x_2, and x_3 corresponding to the leading 1's are called the **leading variables**. The remaining variables (x_4, in this case) are called **free variables**. When a system of equations has free variables, there are infinitely many solutions that you can represent algebraically by expressing the leading variables in terms of the free variables. In this case,

$$\begin{aligned} x_1 &= \quad 2 + \ x_4 \\ x_2 &= -3 - \ x_4 \\ x_3 &= \quad 1 - 2x_4 \end{aligned}$$

By assigning an arbitrary value of t to the free variable x_4, the general solution to this system is

$$x_1 = 2 + t, \quad x_2 = -3 - t, \quad x_3 = 1 - 2t, \quad x_4 = t$$

For example, the solution corresponding to $t = 1$ is $x_1 = 3, x_2 = -4, x_3 = -1, x_4 = 1$.

Turning to Example 2.7(c), the last equation associated with the augmented matrix is

$$0x_1 + 0x_2 + 0x_3 = 1$$

No values of the variables satisfy this equation, so this system is inconsistent.

The system of equations associated with the augmented matrix in Example 2.7(d) is

$$
\begin{aligned}
x_1 + 2x_2 \quad\quad\quad - \ x_5 &= 5 \\
x_3 \quad\ + 2x_5 &= -1 \\
x_4 - 3x_5 &= 4 \\
0x_1 + 0x_2 + 0x_3 + 0x_4 + 0x_5 &= 0
\end{aligned}
$$

You can ignore the bottom equation containing all zeros because any values chosen for the variables satisfy this equation. Expressing the leading variables x_1, x_3, and x_4 in terms of the free variables x_2 and x_5 yields

$$
\begin{aligned}
x_1 &= 5 - 2x_2 + \ x_5 \\
x_3 &= -1 \quad\quad\ - 2x_5 \\
x_4 &= \quad 4 \quad\quad\ + 3x_5
\end{aligned}
$$

Assigning arbitrary values to the free variables, say, s for x_2 and t for x_5, the general solution to this system of equations is

$$x_1 = 5 - 2s + t, \quad x_2 = s, \quad x_3 = -1 - 2t, \quad x_4 = 4 + 3t, \quad x_5 = t$$

The Steps of Gauss-Jordan Elimination

What remains is to specify rules for performing a sequence of elementary row operations on the augmented matrix associated with a given system of linear equations so that the final matrix is in reduced row-echelon form. The steps of Gauss-Jordan elimination are given below and are illustrated on the augmented matrix associated with the system in (2.40).

Preparation Phase—Prepare the system of equations by performing the following steps on the rows of the augmented matrix, starting at the top and progressing row by row toward the bottom:

Step 1. Try to create a leading 1 in the top row of the matrix by locating the first column from the left that does not consist entirely of zeros. If there is no such column, go to Step 6. In this example, the first nonzero column from the left is column 1:

$$
\left[
\begin{array}{ccc|c}
0 & 1 & -2 & 2 \\
2 & 4 & -8 & 6 \\
-1 & -2 & 5 & -2
\end{array}
\right]
$$

↑ (the first nonzero column from the left)

Step 2. If necessary, interchange the top row with any other row to obtain a nonzero number, say a, at the top of the column identified in Step 1. In this example, interchange the top row of the matrix in Step 1 with row 2 to obtain $a = 2$ at the top of column 1:

$$\begin{bmatrix} 2 & 4 & -8 & | & 6 \\ 0 & 1 & -2 & | & 2 \\ -1 & -2 & 5 & | & -2 \end{bmatrix}$$

Step 3. Create a leading 1 in the top row by multiplying the top row through by $1/a$. In this example, multiply the top row of the matrix in Step 2 by $1/a = 1/2$ to obtain the following:

$$\begin{bmatrix} 1 & 2 & -4 & | & 3 \\ 0 & 1 & -2 & | & 2 \\ -1 & -2 & 5 & | & -2 \end{bmatrix}$$

Step 4. Add appropriate multiples of the top row to each row below the top row so that all entries below the leading 1 are reduced to 0. In this example, add 1 times the top row of the matrix in Step 3 to row 3 to obtain the following:

$$\begin{bmatrix} 1 & 2 & -4 & | & 3 \\ 0 & 1 & -2 & | & 2 \\ 0 & 0 & 1 & | & 1 \end{bmatrix}$$

Step 5. If there is another row below the top row, repeat Steps 1 through 4 on the matrix obtained by covering up the current top row. Otherwise, go to Step 6. In this example, when row 1 of the matrix in Step 4 is covered up, performing Steps 1, 2, 3, and 4 causes no change in the matrix. Then, when row 2 is covered up, performing Steps 1, 2, 3, and 4 again causes no change in the matrix. Because there are no more rows below row 3, proceed to Step 6.

Step 6. Starting with the bottom-most row that has a leading 1 and progressing row by row toward the top of the matrix, add appropriate multiples of the current row to each row *above* the current row so that all entries above the leading 1 are reduced to 0. When the procedure is completed, the matrix is in reduced row-echelon form.

Using row 3 of the matrix in Step 4, add 2 times row 3 to row 2 and add 4 times row 3 to row 1:

Using row 2 of the foregoing matrix, add −2 times row 2 to row 1:

$$\begin{bmatrix} 1 & 2 & 0 & | & 7 \\ 0 & 1 & 0 & | & 4 \\ 0 & 0 & 1 & | & 1 \end{bmatrix} \qquad \begin{bmatrix} 1 & 0 & 0 & | & -1 \\ 0 & 1 & 0 & | & 4 \\ 0 & 0 & 1 & | & 1 \end{bmatrix}$$

Solution Phase—Solve the system of equations by performing the following steps:

Step 1. If any of the rows in the reduced row-echelon matrix contains all zeros except for a nonzero value in the last column, then stop. The system of linear equations has no solution. Otherwise, go to Step 2. No such row occurs in the final matrix in Step 6, so go to Step 2.

Step 2. Write the system of linear equations corresponding to the final augmented matrix obtained in Step 6, ignoring any rows consisting entirely of zeros. Performing this step in the example yields the following:

$$
\begin{aligned}
x &= -1 \\
y &= 4 \\
z &= 1
\end{aligned}
$$

Step 3. Solve the equations in Step 2 for the leading variables in terms of the free variables, if any. In this example, there are no free variables, so go to Step 4.

Step 4. Identify the general solution by assigning an arbitrary value to each of the free variables. In this example, there are no free variables, so you can identify the solution directly from Step 2:

$$
\begin{aligned}
x &= -1 \\
y &= 4 \\
z &= 1
\end{aligned}
$$

The two phases of Gauss-Jordan elimination are illustrated again in the following numerical examples.

EXAMPLE 2.8 The Preparation Phase of Gauss-Jordan Elimination

The preparation phase of Gauss-Jordan elimination is illustrated on the following system and augmented matrix:

System of Equations   Augmented Matrix

$$
\begin{aligned}
3x + 3y + 3z &= 9 \\
-2x - 2y - 2z &= -6 \\
x + y - z &= 5
\end{aligned}
\qquad
\left[\begin{array}{ccc|c}
3 & 3 & 3 & 9 \\
-2 & -2 & -2 & -6 \\
1 & 1 & -1 & 5
\end{array}\right]
$$

Now perform the following steps.

Step 1. Identify column 1 as the first nonzero column from the left.

Step 2. No interchange of rows is necessary because the nonzero value of $a = 3$ appears at the top of column 1.

Step 3. Create a leading 1 in row 1 by multiplying row 1 by $1/a = 1/3$:

$$
\left[\begin{array}{ccc|c}
1 & 1 & 1 & 3 \\
-2 & -2 & -2 & -6 \\
1 & 1 & -1 & 5
\end{array}\right]
$$

Step 4. Add 2 times row 1 to row 2 and add -1 times row 1 to row 3:

$$
\left[\begin{array}{ccc|c}
1 & 1 & 1 & 3 \\
0 & 0 & 0 & 0 \\
0 & 0 & -2 & 2
\end{array}\right]
$$

Step 1 (again). Cover up row 1 and identify column 3 as the first nonzero column from the left:

$$\left[\begin{array}{ccc|c} 1 & 1 & 1 & 3 \\ 0 & 0 & 0 & 0 \\ 0 & 0 & -2 & 2 \end{array}\right]$$
$\uparrow$

Step 2 (again). Interchange row 2 with row 3 to obtain $a = -2$ at the top of column 3:

$$\left[\begin{array}{ccc|c} 1 & 1 & 1 & 3 \\ 0 & 0 & -2 & 2 \\ 0 & 0 & 0 & 0 \end{array}\right]$$

Step 3 (again). Create a leading 1 in row 2 by multiplying row 2 by $1/a = -1/2$:

$$\left[\begin{array}{ccc|c} 1 & 1 & 1 & 3 \\ 0 & 0 & 1 & -1 \\ 0 & 0 & 0 & 0 \end{array}\right]$$

Step 4 (again). This step is not necessary because a 0 already occurs below the leading 1 in row 2. This also completes Step 5.

Step 6. Add -1 times row 2 to row 1:

$$\left[\begin{array}{ccc|c} 1 & 1 & 0 & 4 \\ 0 & 0 & 1 & -1 \\ 0 & 0 & 0 & 0 \end{array}\right]$$

The matrix is now in reduced row-echelon form.

EXAMPLE 2.9 The Solution Phase for Example 2.8

Step 1. This step of the solution phase of Gauss-Jordan elimination does not apply to the final augmented matrix in Example 2.8. The remaining steps follow.

Step 2. The system of linear equations corresponding to the final augmented matrix in Example 2.8, ignoring the bottom row consisting entirely of zeros, is:

$$x + y = 4$$
$$z = -1$$

Step 3. Solving for the leading variables x and z in terms of the free variable y yields

$$x = 4 - y$$
$$z = -1$$

Step 4. Assigning an arbitrary value of t to the free variable y yields the general solution to the system of equations:

$$x = 4 - t, \quad y = t, \quad z = -1$$

EXAMPLE 2.10 The Preparation Phase of Gauss-Jordan Elimination

The preparation phase of Gauss-Jordan elimination is illustrated on the following system and augmented matrix:

System of Equations

$$2x + 2y + 4z = 6$$
$$y + 2z = 4$$
$$x + 2y + 4z = 8$$

Augmented Matrix

$$\begin{bmatrix} 2 & 2 & 4 & | & 6 \\ 0 & 1 & 2 & | & 4 \\ 1 & 2 & 4 & | & 8 \end{bmatrix}$$

Step 1. Identify column 1 as the first nonzero column from the left.

Step 2. No interchange of rows is necessary because the nonzero value of $a = 2$ appears at the top of column 1.

Step 3. Multiply row 1 by $1/a = 1/2$:

$$\begin{bmatrix} 1 & 1 & 2 & | & 3 \\ 0 & 1 & 2 & | & 4 \\ 1 & 2 & 4 & | & 8 \end{bmatrix}$$

Step 4. Add -1 times row 1 to row 3:

$$\begin{bmatrix} 1 & 1 & 2 & | & 3 \\ 0 & 1 & 2 & | & 4 \\ 0 & 1 & 2 & | & 5 \end{bmatrix}$$

Step 1 (again). Cover up row 1 and identify column 2 as the first nonzero column from the left:

$$\begin{bmatrix} 1 & 1 & 2 & | & 3 \\ 0 & 1 & 2 & | & 4 \\ 0 & 1 & 2 & | & 5 \end{bmatrix}$$
$$\uparrow$$

Step 2 (again). No interchange of rows is necessary because $a = 1$ is at the top of column 2.

Step 3 (again). A leading 1 is already in row 2, so you need not perform this step.

Step 4 (again). Add -1 times row 2 to row 3 to obtain the following (which also completes Step 5):

$$\begin{bmatrix} 1 & 1 & 2 & | & 3 \\ 0 & 1 & 2 & | & 4 \\ 0 & 0 & 0 & | & 1 \end{bmatrix}$$

Step 6 (using row 3). Add -4 times row 3 to row 2 and -3 times row 3 to row 1:

$$\begin{bmatrix} 1 & 1 & 2 & | & 0 \\ 0 & 1 & 2 & | & 0 \\ 0 & 0 & 0 & | & 1 \end{bmatrix}$$

Step 6 (using row 2). Add -1 times row 2 to row 1:

$$\begin{bmatrix} 1 & 0 & 0 & | & 0 \\ 0 & 1 & 2 & | & 0 \\ 0 & 0 & 0 & | & 1 \end{bmatrix}$$

The matrix is now in reduced row-echelon form.

EXAMPLE 2.11 The Solution Phase for Example 2.10

Step 1 of the solution phase of Gauss-Jordan elimination applies to the final augmented matrix in Step 6 of Example 2.10 because row 3 of that matrix contains all zeros, except for a 1 in the last column. You can therefore conclude that the original system has no solution.

2.3.2 Gaussian Elimination

Another numerical method for solving linear equations, called **Gaussian elimination**, involves less computational effort in the preparation phase and more effort in the solution phase. These two phases are described now.

The Preparation Phase of Gaussian Elimination

The preparation phase of Gaussian elimination differs from that of Gauss-Jordan elimination in that you need not obtain the reduced row-echelon form of the augmented matrix, but rather, a matrix in the form described in the following definition:

DEFINITION 2.2

A matrix is in **row-echelon form** if and only if

(1) All rows consisting entirely of zeros, if any, occur together at the bottom of the matrix.

(2) The first nonzero number in a row, if any, is called a **leading value** and can be any positive or negative number.

(3) For any two consecutive rows that do not consist entirely of zeros, the leading value in the lower row occurs further to the right than the leading value in the higher row.

(4) Each column that has a leading value contains zeros below the leading value.

Any matrix in reduced row-echelon form is in row-echelon form, as are each of the matrices in the following example.

EXAMPLE 2.12 Matrices in Row-Echelon Form

Each of the following matrices is in row-echelon form:

$$(a) \begin{bmatrix} -1 & 2 & 3 & -4 \\ 0 & -2 & -8 & -4 \\ 0 & 0 & 3 & 3 \end{bmatrix} \qquad (b) \begin{bmatrix} 3 & -9 & 0 & 6 \\ 0 & 1 & -1 & -4 \\ 0 & 0 & 0 & 1 \end{bmatrix}$$

$$(c) \begin{bmatrix} 2 & -2 & -6 & -8 & 4 \\ 0 & 0 & -1 & 2 & 5 \end{bmatrix} \qquad (d) \begin{bmatrix} 0 & 1 & -2 & -4 & -2 & -1 \\ 0 & 0 & 0 & 1 & 3 & 5 \end{bmatrix}$$

Comparing Definition 2.2 and Definition 2.1, you can see that the preparation phase of Gaussian elimination requires less computational effort than with Gauss-Jordan elimination because (a) no row operation is needed to obtain a leading 1 in each row, so Step 3 of the preparation phase of Gauss-Jordan elimination is not needed, and (b) for a given nonzero column and the row containing a leading value, row operations are needed to obtain zeros below, but not above, the leading value. The steps of the preparation phase of Gaussian elimination applied to the system in Example 2.8 are illustrated now.

EXAMPLE 2.13 The Preparation Phase of Gaussian Elimination

The preparation phase of Gaussian elimination is illustrated on the following system and augmented matrix:

System of Equations	Augmented Matrix

$$
\begin{aligned}
3x + 3y + 3z &= 9 \\
-2x - 2y - 2z &= -6 \\
x + y - z &= 5
\end{aligned}
\qquad
\left[\begin{array}{rrr|r}
3 & 3 & 3 & 9 \\
-2 & -2 & -2 & -6 \\
1 & 1 & -1 & 5
\end{array}\right]
$$

Step 1. Identify column 1 as the first nonzero column from the left.

Step 2. No interchange of rows is necessary because the nonzero value of $a = 3$ appears at the top of column 1.

Step 3. The leading value $a = 3$ in row 1 is acceptable in a row-echelon form of a matrix, so it is not necessary to perform a row operation to obtain a leading 1.

Step 4. To obtain zeros in column 1 below the leading value, add 2/3 times row 1 to row 2 and add $-1/3$ times row 1 to row 3:

$$
\left[\begin{array}{rrr|r}
3 & 3 & 3 & 9 \\
0 & 0 & 0 & 0 \\
0 & 0 & -2 & 2
\end{array}\right]
$$

Step 1 (again). Cover up row 1 and identify column 3 as the first nonzero column from the left:

$$
\left[\begin{array}{rrr|r}
3 & 3 & 3 & 9 \\
0 & 0 & 0 & 0 \\
0 & 0 & -2 & 2
\end{array}\right]
$$
$$\uparrow$$

The matrix is now in row-echelon form.

Step 2 (again). Interchange row 2 with row 3 to obtain $a = -2$ at the top of column 3:

$$
\left[\begin{array}{rrr|r}
3 & 3 & 3 & 9 \\
0 & 0 & -2 & 2 \\
0 & 0 & 0 & 0
\end{array}\right]
$$

The Solution Phase of Gaussian Elimination

After using Gaussian elimination to prepare the system and obtaining an augmented matrix in row-echelon form, it is then necessary to solve the system of linear equations. To illustrate how this is done, consider the following system associated with the row-echelon matrix in

Example 2.12(a):

$$-x_1 + 2x_2 + 3x_3 = -4$$
$$- 2x_2 - 8x_3 = -4$$
$$3x_3 = 3$$

First solve each equation for its leading variable in terms of the other variables:

$$x_1 = 4 + 2x_2 + 3x_3$$
$$x_2 = 2 - 4x_3 \quad\quad\quad (2.42)$$
$$x_3 = 1$$

From the bottom equation in (2.42), you know that $x_3 = 1$. You can therefore substitute $x_3 = 1$ in the other two equations:

$$x_1 = 7 + 2x_2$$
$$x_2 = -2 \quad\quad\quad (2.43)$$

From the bottom equation in (2.43), you know that $x_2 = -2$. You can therefore substitute $x_2 = -2$ in the first equation:

$$x_1 = 3$$

Having found the values of all variables, the solution to the system of equations in Example 2.12(a) is

$$x_1 = 3, \quad x_2 = -2, \quad x_3 = 1$$

In summary, after using Gaussian elimination to prepare the system and obtaining an augmented matrix in row-echelon form, solve the system using the following process called **back substitution**. The steps for this process are illustrated again with the row-echelon matrix obtained at the end of Example 2.13.

The Solution Phase of Gaussian Elimination

Step 1. If any row in the row-echelon matrix contains all zeros except for a nonzero value in the last column, then stop—the system of linear equations has no solution. In this example, this step is not applicable because the row-echelon matrix in Example 2.13 is the following:

$$\begin{bmatrix} 3 & 3 & 3 & | & 9 \\ 0 & 0 & -2 & | & 2 \\ 0 & 0 & 0 & | & 0 \end{bmatrix}$$

Step 2. Write the system of linear equations corresponding to the row-echelon matrix, ignoring any rows that consist entirely of zeros. In this example, row 3 of the matrix in Step 1 is ignored, so the remaining equations are as follows:

$$3x_1 + 3x_2 + 3x_3 = 9$$
$$- 2x_3 = 2$$

Step 3. Solve the equations for the leading variables in terms of all other variables. In this example, solve the first equation for x_1 and the second equation for x_3:

$$x_1 = 3 - x_2 - x_3$$
$$x_3 = -1$$

Step 4. Starting with the bottom equation, sequentially substitute the value of the variable on the right side of the equality in each equation above the current one. In this example, the value $x_3 = -1$ is substituted in the first equation:

$$x_1 = 4 - x_2$$
$$x_3 = -1$$

Step 5. Identify the general solution by assigning an arbitrary value to each free variable, if any. In this example, assign the arbitrary value t to the free variable x_2, so the general solution is described as follows:

$$x_1 = 4 - t, \quad x_2 = t, \quad x_3 = -1$$

In this section, you have seen two numerical methods for solving a system of m linear equations in n unknowns: Gauss-Jordan and Gaussian elimination. In Section 2.4, the role of matrices is developed in greater detail.

Exercises for Section 2.3

1. Write the augmented matrix associated with each of the following systems of linear equations.

(a)
$$2x_1 \quad + \quad x_3 - 3x_4 - 4 = 0$$
$$-x_2 + 2x_3 + \quad x_4 + 2 = 0$$

(b)
$$2x + 3y - \quad z = -1$$
$$-y + 4z - 3x = \quad 0$$
$$5z + 6x + 7y = \quad 1$$

2. Write the augmented matrix associated with each of the following systems of linear equations.

(a)
$$-x \quad + 2z = -3$$
$$3x + \quad y - \quad z = \quad 4$$
$$2y + 3z = \quad 0$$

(b)
$$2x_1 \quad + \quad x_3 - 2x_4 = -2$$
$$\tfrac{1}{2}x_2 + 2x_3 - \quad x_4 = \quad 0$$
$$- \quad x_3 + 4x_4 = \quad 1$$

3. Write the system of linear equations corresponding to each of the following augmented matrices.

(a)
$$\begin{bmatrix} 1 & 2 & -4 & 3 \\ 0 & 1 & -2 & 2 \\ -1 & -2 & 5 & -2 \end{bmatrix}$$

(b)
$$\begin{bmatrix} 3 & -1 & -4 & 3 & 0 \\ 2 & 1 & 3 & -2 & -1 \\ -1 & 1 & -1 & 1 & 2 \end{bmatrix}$$

4. Write the system of linear equations corresponding to each of the following augmented matrices.

(a) $\begin{bmatrix} 1 & 2 & -4 & | & 3 \\ 0 & 1 & -2 & | & 2 \\ 0 & 0 & 1 & | & 1 \end{bmatrix}$ (b) $\begin{bmatrix} -1 & 0 & 2 & -1 & 0 & | & 2 \\ 0 & -1 & 0 & 2 & 0 & | & -3 \\ 1 & 1 & 1 & 1 & 1 & | & 1 \end{bmatrix}$

5. Show the result of performing each of the following elementary row operations on the given system of equations. Write the resulting system of equations and the associated augmented matrix.
   (a) Multiply the first equation in Exercise 2(a) by -1.
   (b) Add -3 times the first equation in part (a) to the second equation.
   (c) Interchange the second and third equations in part (b).

6. Show the result of performing the following elementary row operations on the given system of equations. Write the resulting system of equations and the associated augmented matrix.
   (a) Multiply the second equation in Exercise 2(b) by 2.
   (b) Add 1 times the third equation in part (a) to the first equation.
   (c) Interchange the first and third equations in part (b).

7. For each of the following augmented matrices that is in reduced row-echelon form, write the general solution to the associated system of linear equations. For those that are not, explain why not.

(a) $\begin{bmatrix} 1 & 2 & | & 4 \\ 0 & 0 & | & 0 \end{bmatrix}$ (b) $\begin{bmatrix} 1 & 0 & 1 & -1 & | & 2 \\ 0 & 1 & 0 & 1 & | & -3 \\ 0 & 0 & 1 & 2 & | & 1 \end{bmatrix}$

(c) $\begin{bmatrix} 1 & 0 & 0 & -1 & | & 2 \\ 0 & 1 & 0 & 1 & | & -3 \\ 0 & 0 & 2 & 2 & | & 1 \end{bmatrix}$ (d) $\begin{bmatrix} 1 & -1 & 0 & 1 & 0 & | & 1 \\ 0 & 0 & 1 & -2 & 0 & | & 0 \\ 0 & 0 & 0 & 0 & 1 & | & -2 \\ 0 & 0 & 0 & 0 & 0 & | & 0 \end{bmatrix}$

8. For each of the following augmented matrices that are in reduced row-echelon form, write the general solution to the associated system of linear equations. For those that are not, explain why not.

(a) $\begin{bmatrix} 1 & 0 & 0 & -1 & | & -2 \\ 0 & 0 & 0 & 0 & | & 0 \\ 0 & 0 & 1 & 2 & | & 1 \end{bmatrix}$ (b) $\begin{bmatrix} 1 & -1 & 0 & 0 & 2 & | & -1 \\ 0 & 0 & 1 & 0 & 4 & | & 0 \\ 0 & 0 & 0 & 1 & -3 & | & 3 \\ 0 & 0 & 0 & 0 & 0 & | & 1 \end{bmatrix}$

(c) $\begin{bmatrix} 1 & 0 & 0 & 0 & -1 & | & 2 \\ 0 & 1 & 1 & 0 & 2 & | & -3 \\ 0 & 0 & 0 & 1 & 0 & | & 4 \\ 0 & 0 & 0 & 0 & 0 & | & 0 \end{bmatrix}$ (d) $\begin{bmatrix} 1 & 2 & 0 & 0 & -1 & | & 5 \\ 0 & 0 & 1 & 0 & 2 & | & -1 \\ 0 & 0 & 0 & 1 & -3 & | & 4 \\ 0 & 0 & 0 & 0 & 1 & | & 0 \end{bmatrix}$

9. Indicate which of the following augmented matrices are in row-echelon form. For those that are, use back substitution to find the general solution to the associated system of linear equations.

(a) $\begin{bmatrix} 1 & -1 & -3 & -4 & | & 2 \\ 0 & 0 & 1 & -2 & | & -5 \end{bmatrix}$ (b) $\begin{bmatrix} 0 & 1 & -2 & -4 & -2 & | & -1 \\ 1 & 0 & 0 & 1 & 3 & | & 5 \end{bmatrix}$

(c) $\begin{bmatrix} 1 & -3 & 0 & | & 1 \\ 0 & 1 & -1 & | & -2 \\ 0 & 0 & 1 & | & 3 \end{bmatrix}$ (d) $\begin{bmatrix} 1 & -2 & -3 & | & 5 \\ 0 & 3 & 6 & | & 0 \\ 0 & 0 & 2 & | & 4 \end{bmatrix}$

10. Indicate which of the following augmented matrices are in row-echelon form. For those that are, use back substitution to find the general solution to the associated system of linear equations.

(a) $\begin{bmatrix} 1 & -1 & -3 & -4 & | & 2 \\ 0 & 1 & 1 & -2 & | & -5 \end{bmatrix}$ (b) $\begin{bmatrix} 0 & 1 & -2 & -4 & -2 & | & -1 \\ 0 & 0 & 0 & 0 & 0 & | & 5 \end{bmatrix}$

(c) $\begin{bmatrix} 1 & -2 & -3 & | & 4 \\ 0 & 1 & 4 & | & 2 \\ 0 & 1 & 0 & | & 1 \end{bmatrix}$ (d) $\begin{bmatrix} 1 & -3 & 0 & | & 1 \\ 0 & 1 & -1 & | & -2 \\ 0 & 0 & 0 & | & 0 \end{bmatrix}$

11. (a) Use Gauss-Jordan elimination to solve the following system of linear equations for x and y, assuming that $a \neq 0$ and $ad - bc \neq 0$:

$$ax + by = 1$$
$$cx + dy = 0$$

(b) Use Gauss-Jordan elimination to solve the following system of linear equations for x and y, assuming that $a \neq 0$ and $ad - bc \neq 0$:

$$ax + by = 0$$
$$cx + dy = 1$$

(c) Solve both systems in parts (a) and (b) simultaneously by applying Gauss-Jordan elimination to the following augmented matrix, assuming that both $a \neq 0$ and $ad - bc \neq 0$. How does your solution compare with those in parts (a) and (b)?

$$\begin{bmatrix} a & b & | & 1 & 0 \\ c & d & | & 0 & 1 \end{bmatrix}$$

(d) Explain the advantage of the approach in part (c) over those in parts (a) and (b).

12. (a) Apply Gauss-Jordan elimination by hand to the following augmented matrix:

$$\begin{bmatrix} 2 & -2 & 0 & | & 1 & 0 & 0 \\ 4 & -4 & 1 & | & 0 & 1 & 0 \\ -6 & 5 & -3 & | & 0 & 0 & 1 \end{bmatrix}$$

(b) Verify your answer in part (a) using technology indicated by your instructor. Write the sequence of instructions you perform.

13. (a) Use Gauss-Jordan elimination by hand to solve the following system of equations:

$$
\begin{aligned}
-2y + z &= -6 \\
x \quad\quad - 3z &= 1 \\
-x + y + 2z &= 2
\end{aligned}
$$

(b) Verify your answer in part (a) using technology indicated by your instructor. Write the sequence of instructions you perform.

14. (a) Use Gauss-Jordan elimination by hand to solve the following system of equations:

$$
\begin{aligned}
-2x_3 + x_4 &= -6 \\
-x_1 + x_2 + x_3 + 2x_4 &= 2 \\
x_1 - x_2 \quad\quad - \tfrac{5}{2}x_4 &= 1
\end{aligned}
$$

(b) Verify your answer in part (a) using technology indicated by your instructor. Write the sequence of instructions you perform.

15. Repeat Exercise 13 using Gaussian elimination.

16. Repeat Exercise 14 using Gaussian elimination.

17. Is reduced row-echelon form a special case or a generalization of row-echelon form? Explain.

18. Show that for a given system of linear equations, it is possible to use row operations to create two different matrices in row-echelon form. Specifically, create a numerical example of a (3 × 3) system of linear equations whose associated augmented matrix is already in row-echelon form. Then apply a single row operation to produce another augmented matrix that is also in row-echelon form.

19. Explain why you can stop Gauss-Jordan elimination as soon as you encounter a row consisting of all zeros, execept for a nonzero value in the last column.

2.4 Matrices and Their Operations

In Section 2.3, the concept of a matrix was introduced and you saw how an augmented matrix is used to solve a system of linear equations. A matrix is an important concept in its own right. In this section, you will learn more about matrices and the various operations you can perform on them.

2.4.1 Matrices and Their Representations

You already know generally what a matrix is. A formal definition follows.

DEFINITION 2.3

An ($m \times n$) **matrix** A (read as "an m by n matrix A") is a rectangular table of real numbers organized in m rows and n columns. The values of m and n constitute the **dimension** or **size** of A. The number in row i ($1 \leq i \leq m$) and column j ($1 \leq j \leq n$) of A is denoted by A_{ij} and is called an **element** or **entry** of the matrix.

In contrast to vectors—which are denoted by lowercase boldface letters such as $\mathbf{d}$, $\mathbf{x}$, $\mathbf{y}$—matrices are denoted in this book by uppercase italicized letters—for example, A, B, C, X. The first subscript of a matrix refers to the row number and the second subscript refers to the column number. Thus, A_{ij} is the number in row i and column j of A. The set of all ($m \times n$) matrices is denoted by $R^{m \times n}$. An ($m \times n$) matrix is often described by the notation $A \in R^{m \times n}$. The following is an example of a matrix.

EXAMPLE 2.14 An Example of a Matrix
The following (4×7) matrix represents a calendar for the month of February:

$$A = \begin{bmatrix} 1 & 2 & 3 & 4 & 5 & 6 & 7 \\ 8 & 9 & 10 & 11 & 12 & 13 & 14 \\ 15 & 16 & 17 & 18 & 19 & 20 & 21 \\ 22 & 23 & 24 & 25 & 26 & 27 & 28 \end{bmatrix}$$

From Example 2.14, you can see that a matrix is a generalization of a vector in the sense that a matrix consists of an ordered list of vectors. In fact, there are two different ways to visualize a matrix in terms of vectors.

Row Representation of a Matrix

Realizing that each row of a matrix $A \in R^{m \times n}$ is an n-vector, you can think of A as consisting of its m rows, ordered from top to bottom:

$$A = \begin{bmatrix} A_{11} & A_{12} & \cdots & A_{1n} \\ A_{21} & A_{22} & \cdots & A_{2n} \\ \vdots & \vdots & \vdots & \vdots \\ A_{m1} & A_{m2} & \cdots & A_{mn} \end{bmatrix} \begin{matrix} \leftarrow \text{row } 1 \\ \leftarrow \text{row } 2 \\ \vdots \\ \leftarrow \text{row } m \end{matrix}$$

For each $i = 1, \ldots, m$, row i of A is a ($1 \times n$) matrix denoted in this book by A_{i*}. For instance, for the matrix in Example 2.14,

$$A_{1*} = [1 \quad 2 \quad 3 \quad 4 \quad 5 \quad 6 \quad 7]$$
$$A_{2*} = [8 \quad 9 \quad 10 \quad 11 \quad 12 \quad 13 \quad 14]$$
$$A_{3*} = [15 \quad 16 \quad 17 \quad 18 \quad 19 \quad 20 \quad 21]$$
$$A_{4*} = [22 \quad 23 \quad 24 \quad 25 \quad 26 \quad 27 \quad 28]$$

In this example, A_{i*} represents the dates of the seven days in week i of February. For instance, A_{3*} represents the dates of the seven days in the third week of February.

Column Representation of a Matrix

Alternatively, realizing that each column of a matrix $A \in R^{m \times n}$ is an m-vector, you can think of A as consisting of n columns, ordered from left to right:

$$
\begin{array}{cccc}
\text{Column} & 1 & 2 & \cdots & n \\
 & \downarrow & \downarrow & \cdots & \downarrow
\end{array}
$$

$$
A = \begin{bmatrix}
A_{11} & A_{12} & \cdots & A_{1n} \\
A_{21} & A_{22} & \cdots & A_{2n} \\
\vdots & \vdots & \vdots & \vdots \\
A_{m1} & A_{m2} & \cdots & A_{mn}
\end{bmatrix}
$$

For each $j = 1, \ldots, n$, column j of A is an $(n \times 1)$ matrix denoted in this book by A_{*j}. For instance, for the matrix in Example 2.14,

$$
A_{*1} = \begin{bmatrix} 1 \\ 8 \\ 15 \\ 22 \end{bmatrix}, \quad A_{*2} = \begin{bmatrix} 2 \\ 9 \\ 16 \\ 23 \end{bmatrix}, \quad \ldots, \quad A_{*7} = \begin{bmatrix} 7 \\ 14 \\ 21 \\ 28 \end{bmatrix}
$$

In this example, A_{*j} represents those dates in February that fall on the same day of the week. For instance, the dates in A_{*1} might all be Sundays.

Now you will learn about certain special matrices and the various comparisons and operations that you can perform on matrices in general.

2.4.2 Special Matrices

After working with matrices in various applications, you will find that certain matrices arise frequently and play a special role. Some of these matrices are described now.

The Zero Matrix

One special matrix is the **zero matrix**, denoted by 0, in which each element of the matrix is the number 0. For example, the (2×3) zero matrix is

$$
0 = \begin{bmatrix} 0 & 0 & 0 \\ 0 & 0 & 0 \end{bmatrix}
$$

The dimension of the zero matrix is generally understood from the context.

Square Matrices

Another special type of matrix is one in which the number of rows equals the number of columns. Such a matrix with n rows and n columns is a **square matrix of order** n.

The elements $A_{11}, A_{22}, \ldots, A_{nn}$ of a square matrix A of order n are called the **diagonal elements**. Several examples of square matrices are described next.

Diagonal Matrices. A **diagonal matrix** is a square matrix of order n in which all elements, except possibly the diagonal elements, are 0. The $(n \times n)$ zero matrix is a diagonal matrix, as are the following matrices:

$$A = \begin{bmatrix} 1 & 0 \\ 0 & -2 \end{bmatrix} \quad \text{and} \quad B = \begin{bmatrix} 4 & 0 & 0 \\ 0 & 0 & 0 \\ 0 & 0 & 3 \end{bmatrix}$$

The Identity Matrix. One particular diagonal matrix that arises when solving a system of linear equations by the method described in Chapter 3 is the $(n \times n)$ **identity matrix**, denoted by I_n, in which the diagonal elements are all 1:

$$I_n = \begin{bmatrix} 1 & 0 & \cdots & 0 \\ 0 & 1 & \cdots & 0 \\ \vdots & \vdots & \ddots & \vdots \\ 0 & 0 & \cdots & 1 \end{bmatrix}$$

When the dimension is clear, the identity matrix is denoted by I.

Elementary Matrices. An **elementary matrix** is a matrix obtained by applying one elementary row operation to the identity matrix. The identity matrix itself is an elementary matrix, as are the following matrices:

$$A = \begin{bmatrix} 1 & 0 \\ 0 & -2 \end{bmatrix} \qquad \text{(multiply row 2 of } I_2 \text{ by } -2)$$

$$B = \begin{bmatrix} 0 & 0 & 1 \\ 0 & 1 & 0 \\ 1 & 0 & 0 \end{bmatrix} \qquad \text{(interchange row 1 and row 3 of } I_3)$$

$$C = \begin{bmatrix} 1 & 0 & 0 & 0 \\ 0 & 1 & 0 & 0 \\ 0 & 0 & 1 & 0 \\ 0 & 0 & -3 & 1 \end{bmatrix} \qquad \text{(add } -3 \text{ times row 3 of } I_4 \text{ to row 4 of } I_4)$$

2.4.3 Comparing Two Matrices

The primary comparison between two matrices is to determine whether they are the same, the precise meaning of which is made clear in the following definition.

DEFINITION 2.4

Two ($m \times n$) matrices A and B are **equal**, written $A = B$, if and only if for each $i = 1, \ldots, m$ and for each $j = 1, \ldots, n$, $A_{ij} = B_{ij}$.

Alternatively, thinking of a matrix in terms of its rows, $A = B$ if and only if for each row $i = 1, \ldots, m$, $A_{i*} = B_{i*}$. Equivalently, in terms of the columns, $A = B$ if and only if for each column $j = 1, \ldots, n$, $A_{*j} = B_{*j}$.

Because the symbol "=" is used to compare real numbers, vectors, and matrices, always be careful to determine what objects are being compared. For example, the first occurrence of the symbol "=" in Definition 2.4 denotes a comparison of the two matrices A and B but the last occurrence of "=" denotes a comparison of the two real numbers A_{ij} and B_{ij}.

When the symbol "=" is used to compare two matrices of the same dimension for equality, the statement

$$A = B$$

means that A *is* equal to B, or equivalently, that the result of checking whether A is equal to B is true. Analogously, the statement

$$A \neq B$$

means that A is not equal to B, or equivalently, that the result of checking whether A is equal to B is false. By using the rules for negating a statement that contains the quantifier *for all*, observe that $A \neq B$ means that there exist $1 \leq i \leq m$ and $1 \leq j \leq n$ such that $A_{ij} \neq B_{ij}$. The use of these symbols for comparing matrices is illustrated in the next example.

EXAMPLE 2.15 Comparing Two Matrices for Equality

It is true that

$$\begin{bmatrix} -1 & 0 & 2 \\ 0 & 3 & -2 \end{bmatrix} = \begin{bmatrix} 1-2 & 1-1 & 4-2 \\ 3-3 & 3-0 & 2-4 \end{bmatrix}$$

Also, for the ($n \times n$) identity matrix I_n and the ($n \times n$) zero matrix 0, it is true that $I_n \neq 0$.

2.4.4 Operations on Matrices

Several operations on matrices that result in a matrix are described in this section.

Multiplying a Matrix by a Real Number

One useful operation is to multiply a matrix $A \in R^{m \times n}$ by a real number t to obtain an ($m \times n$) matrix, denoted by tA. The elements of tA are those of A multiplied by t, that is:

$$(tA)_{ij} = tA_{ij}, \quad \text{for each } i = 1, \ldots, m \text{ and for each } j = 1, \ldots, n$$

EXAMPLE 2.16 Multiplying a Matrix by a Real Number

If

$$t = 2 \quad \text{and} \quad A = \begin{bmatrix} -1 & 0 & 2 \\ 0 & 3 & -2 \end{bmatrix}$$

then

$$tA = 2 \begin{bmatrix} -1 & 0 & 2 \\ 0 & 3 & -2 \end{bmatrix} = \begin{bmatrix} -2 & 0 & 4 \\ 0 & 6 & -4 \end{bmatrix}$$

When $t = 0$, tA is the zero matrix, that is, $0A = 0$. Also, for the special case of $t = 1$, $1A = A$.

The Transpose of a Matrix

Another operation is to **transpose** a matrix $A \in R^{m \times n}$, in which you create a new matrix, denoted by A^t, that has n rows and m columns. Specifically, if the elements of A are A_{ij}, then the elements of A^t are A_{ji}, that is,

$$(A^t)_{ji} = A_{ij}, \quad \text{for each } i = 1, \ldots, m \text{ and for each } j = 1, \ldots, n$$

Equivalently stated, row i of A becomes column i of A^t.

EXAMPLE 2.17 The Transpose of a Matrix

If $A = \begin{bmatrix} -1 & 0 & 2 \\ 0 & 3 & -2 \end{bmatrix}$, then $A^t = \begin{bmatrix} -1 & 0 \\ 0 & 3 \\ 2 & -2 \end{bmatrix}$

The remaining arithmetic operations of addition, subtraction, and multiplication of matrices constitute binary operations that combine two matrices to produce another matrix.

Adding and Subtracting Matrices

You can add two matrices of the same dimension to create a new matrix of that dimension. Specifically, **matrix addition** is defined as follows: If $A, B \in R^{m \times n}$, then $A + B$ is the $(m \times n)$ matrix obtained by adding the corresponding elements of A and B, that is:

$$(A + B)_{ij} = A_{ij} + B_{ij}, \quad \text{for each } i = 1, \ldots, m \text{ and } j = 1, \ldots, n \tag{2.44}$$

Once again, be careful to interpret symbols correctly. For example, the first occurrence of the symbol "+" in (2.44) denotes the addition of the two matrices A and B but the second occurrence of "+" denotes the addition of the two real numbers A_{ij} and B_{ij}. An example of matrix addition follows.

EXAMPLE 2.18 Adding Two Matrices

If

$$A = \begin{bmatrix} -1 & 0 & 2 \\ 0 & 3 & -2 \end{bmatrix} \quad \text{and} \quad B = \begin{bmatrix} 2 & 3 & -2 \\ -1 & 0 & 4 \end{bmatrix}$$

then

$$A + B = \begin{bmatrix} -1 & 0 & 2 \\ 0 & 3 & -2 \end{bmatrix} + \begin{bmatrix} 2 & 3 & -2 \\ -1 & 0 & 4 \end{bmatrix}$$

$$= \begin{bmatrix} -1+2 & 0+3 & 2+(-2) \\ 0+(-1) & 3+0 & -2+4 \end{bmatrix}$$

$$= \begin{bmatrix} 1 & 3 & 0 \\ -1 & 3 & 2 \end{bmatrix}$$

Matrix subtraction is defined analogously. If $A, B \in R^{m \times n}$, then $A - B$ is the $(m \times n)$ matrix in which

$$(A - B)_{ij} = A_{ij} - B_{ij}, \quad \text{for each } i = 1, \ldots, m \text{ and } j = 1, \ldots, n$$

Multiplying Matrices

The operation of multiplying two matrices also results in a matrix. One way to multiply two matrices $A, B \in R^{m \times n}$ is similar to addition and subtraction in that you multiply the corresponding elements of A and B. However, much experience in applications has led to a different and more useful method for multiplying two matrices.

The more practical method of multiplying the matrix A by the matrix B results in a new matrix, denoted by AB, in which the element in row i and column j of AB is the dot product of A_{i*} and B_{*j}, when viewed as vectors, that is,

$$(AB)_{ij} = A_{i*} \cdot B_{*j} \tag{2.45}$$

Observe that to compute the dot product in (2.45), the vectors A_{i*} and B_{*j} must have the same dimension. In terms of the matrices A and B, this means that the number of columns of A must be the same as the number of rows of B. In other words, using the method in (2.45), it is only possible to multiply a matrix $A \in R^{m \times p}$ by a matrix B if $B \in R^{p \times n}$. That is, if A has p columns, then B must have p rows. In this case, **matrix multiplication** is defined as follows. If $A \in R^{m \times p}$ and $B \in R^{p \times n}$, then AB is the $(m \times n)$ matrix in which

$$(AB)_{ij} = A_{i*} \cdot B_{*j}, \quad \text{for each } i = 1, \ldots, m \text{ and } j = 1, \ldots, n \tag{2.46}$$

Alternatively, using the definition of the dot product, you can rewrite (2.46) as follows:

$$(AB)_{ij} = A_{i*} \cdot B_{*j} = \sum_{k=1}^{p} A_{ik} B_{kj} \tag{2.47}$$

EXAMPLE 2.19 Multiplying Two Matrices

If

$$A = \begin{bmatrix} 1 & 2 & 0 \\ 0 & -1 & 3 \end{bmatrix} \quad \text{and} \quad B = \begin{bmatrix} 1 & 0 & -1 & 2 \\ 2 & -1 & 3 & 0 \\ 1 & -1 & 0 & 4 \end{bmatrix}$$

then $A \in R^{2 \times 3}$ and $B \in R^{3 \times 4}$, so $AB \in R^{2 \times 4}$ and, writing A_{i*} and B_{*j} as vectors, you have

$$(AB)_{11} = A_{1*} \cdot B_{*1} = (1, 2, 0) \cdot (1, 2, 1) \quad = \quad 5$$

$$(AB)_{12} = A_{1*} \cdot B_{*2} = (1, 2, 0) \cdot (0, -1, -1) = -2$$

$$(AB)_{13} = A_{1*} \cdot B_{*3} = (1, 2, 0) \cdot (-1, 3, 0) \quad = \quad 5$$

$$(AB)_{14} = A_{1*} \cdot B_{*4} = (1, 2, 0) \cdot (2, 0, 4) \quad = \quad 2$$

$$(AB)_{21} = A_{2*} \cdot B_{*1} = (0, -1, 3) \cdot (1, 2, 1) \quad = \quad 1$$

$$(AB)_{22} = A_{2*} \cdot B_{*2} = (0, -1, 3) \cdot (0, -1, -1) = -2$$

$$(AB)_{23} = A_{2*} \cdot B_{*3} = (0, -1, 3) \cdot (-1, 3, 0) \quad = -3$$

$$(AB)_{24} = A_{2*} \cdot B_{*4} = (0, -1, 3) \cdot (2, 0, 4) \quad = \quad 12$$

By inserting each of the foregoing numbers in its correct position in the matrix AB, you obtain the following:

$$AB = \begin{bmatrix} 5 & -2 & 5 & 2 \\ 1 & -2 & -3 & 12 \end{bmatrix}$$

When you want to work with only one row, say, i, of the matrix AB, you do not have to perform the entire matrix multiplication. Rather, as you are asked to show in Exercise 17, you need only compute:

$$(AB)_{i*} = A_{i*}B \tag{2.48}$$

Similarly, to find column j of AB, you need only compute

$$(AB)_{*j} = AB_{*j} \tag{2.49}$$

As you have just seen, **another use of mathematics is to reduce the computational effort needed to solve a problem.** For example, it is more efficient to find column *j* of *AB* by using (2.49) than it is to compute *AB* and then write column *j* of *AB*.

You can use matrix multiplication to perform an elementary row operation on an $(m \times n)$ matrix A. This is accomplished by multiplying A on the left by the corresponding $(m \times m)$ elementary matrix, E, as shown in the next example.

EXAMPLE 2.20 Performing an Elementary Row Operation by Using an Elementary Matrix

To interchange rows 2 and 3 of the matrix

$$A = \begin{bmatrix} 1 & -1 & 0 \\ 0 & 0 & 1 \\ 0 & -1 & 3 \end{bmatrix}$$

construct the following elementary matrix E by interchanging rows 2 and 3 of I_3:

$$E = \begin{bmatrix} 1 & 0 & 0 \\ 0 & 0 & 1 \\ 0 & 1 & 0 \end{bmatrix}$$

Multiplying A on the left by E interchanges rows 2 and 3 of A:

$$EA = \begin{bmatrix} 1 & 0 & 0 \\ 0 & 0 & 1 \\ 0 & 1 & 0 \end{bmatrix} \begin{bmatrix} 1 & -1 & 0 \\ 0 & 0 & 1 \\ 0 & -1 & 3 \end{bmatrix} = \begin{bmatrix} 1 & -1 & 0 \\ 0 & -1 & 3 \\ 0 & 0 & 1 \end{bmatrix}$$

2.4.5 Properties of Matrices and Their Operations

After working with matrices for some time, you will discover that the various operations satisfy the properties in the following theorems.

THEOREM 2.1

If $A, B, C, 0 \in R^{m \times n}$, then

(a) $A + B = B + A$.

(b) $(A + B) + C = A + (B + C)$.

(c) $A + 0 = 0 + A = A$.

(d) $A - A = 0$.

(e) $0 - A = -A$.

Proof.

The key question associated with each part of this theorem is, "How can I show that two matrices are equal?" Definition 2.3 states that you must show that all corresponding elements of the matrices are equal.

Whenever you need to show that

> **for every "object" with a "certain property", "something happens,"**

you could make a list of all those objects with the certain property. For each object on the list, you would have to show that the something happens. When the list is too long (or even infinite), you can accomplish the task by establishing that the something happens for one generic object on the list. This argument could then, in theory, be repeated for every object on the list. This technique is referred to in this book as the choose method and is described in detail in Appendix A.4.

The use of the choose method is now demonstrated.

(a) To show that the two matrices $A + B$ and $B + A$ are equal, it is necessary to verify that each element of $A + B$ is equal to the corresponding element of $B + A$, that is, that for each $i = 1, \ldots, m$ and for each $j = 1, \ldots, n$, $(A + B)_{ij} = (B + A)_{ij}$. Applying the choose method, choose integers i and j with $1 \leq i \leq m$ and $1 \leq j \leq n$. You must show that $(A + B)_{ij} = (B + A)_{ij}$. This goal is achieved by transforming $(A + B)_{ij}$ into $(B + A)_{ij}$, as follows:

$$
\begin{aligned}
(A + B)_{ij} &= A_{ij} + B_{ij} && \text{(definition of matrix addition)} \\
&= B_{ij} + A_{ij} && \text{(commutative property of adding real numbers)} \\
&= (B + A)_{ij} && \text{(definition of matrix addition)}
\end{aligned}
$$

(b–e) These proofs are similar to part (a) and are left to Exercise 19.

The proof is now complete.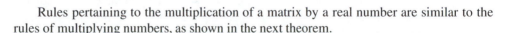

Rules pertaining to the multiplication of a matrix by a real number are similar to the rules of multiplying numbers, as shown in the next theorem.

THEOREM 2.2

If $A, B, 0 \in R^{m \times n}$ and s and t are real numbers, then

(a) $s(A + B) = sA + sB$.

(b) $s(A - B) = sA - sB$.

(c) $(s + t)A = sA + tA$.

(d) $(s - t)A = sA - tA$.

(e) $s(tA) = (st)A$.

(f) $0A = 0$ (the number 0 times the matrix A = the zero matrix).

(g) $s0 = 0$ (the number s times the zero matrix = the zero matrix).

Proof.

Each part of this theorem requires showing that two matrices are equal. Definition 2.3 leads you to show that all corresponding elements of the matrices are equal. The choose method is used to accomplish this task.

(a) To see that $s(A + B) = sA + sB$, let i and j be integers with $1 \le i \le m$ and $1 \le j \le n$. (The word *let* in a proof often indicates the use of the choose method.) You must show that $[s(A + B)]_{ij} = (sA + sB)_{ij}$, but

$$
\begin{aligned}
[s(A + B)]_{ij} &= s[(A + B)_{ij}] \\
&= s(A_{ij} + B_{ij}) \\
&= sA_{ij} + sB_{ij} \\
&= (sA)_{ij} + (sB)_{ij} \\
&= (sA + sB)_{ij}
\end{aligned}
$$

(b–g) These proofs are similar to part (a) and are left to Exercise 20.

The proof is now complete. ■

THEOREM 2.3

If s is a real number and A and B are matrices for which the following operations can be performed, then

(a) $(A^t)^t = A$.

(b) $(sA)^t = sA^t$.

(c) $(A + B)^t = A^t + B^t$.

(d) $(AB)^t = B^t A^t$.

Proof.

Each part of this theorem requires you to show that two matrices are equal. Definition 2.3 leads you to show that all corresponding elements of the matrices are equal. The choose method is used to do so.

(a) Observe that $(A^t)^t$ and A are both, say, $(m \times n)$ matrices, although their elements are computed in different ways. To show that these two matrices are equal, let i and j be integers with $1 \le i \le m$ and $1 \le j \le n$. You must show that $[(A^t)^t]_{ij} = A_{ij}$. This goal is achieved by transforming $[(A^t)^t]_{ij}$ into A_{ij}, as follows:

$$
\begin{aligned}
[(A^t)^t]_{ij} &= (A^t)_{ji} \qquad \text{(definition of matrix transpose)} \\
&= A_{ij} \qquad\ \ \text{(definition of matrix transpose)}
\end{aligned}
$$

(b, c) These proofs are similar to part (a) and are left to Exercise 21.

(d) To be able to compute $(AB)^t$ and $B^t A^t$, the dimensions of A and B must be such that $A \in R^{n \times p}$ and $B \in R^{p \times m}$. To show that $(AB)^t = B^t A^t$, use the choose method to choose integers i and j with $1 \le i \le m$ and $1 \le j \le n$. You must show that $[(AB)^t]_{ij} = (B^t A^t)_{ij}$. This goal is achieved by transforming $[(AB)^t]_{ij}$ into $(B^t A^t)_{ij}$, as follows:

$$[(AB)^t]_{ij} = (AB)_{ji} \qquad \text{(definition of matrix transpose)}$$
$$= A_{j*} \cdot B_{*i} \qquad \text{(definition of matrix multiplication)}$$
$$= (A^t)_{*j} \cdot (B^t)_{i*} \qquad \text{(definition of matrix transpose)}$$
$$= (B^t)_{i*} \cdot (A^t)_{*j} \qquad \text{(property of the dot product)}$$
$$= (B^t A^t)_{ij} \qquad \text{(definition of matrix multiplication)}$$

The proof is now complete. ■

Similarities and Differences Between Multiplying Two Real Numbers and Multiplying Two Matrices

The operation of multiplying two matrices differs in several important ways from that same operation applied to real numbers. For example, if a and b are real numbers, then you know that $ab = ba$. However, if A and B are matrices, then it is not necessarily true that $AB = BA$. One reason is that you may not be able to perform these computations because the dimensions of A and B are not appropriate. For example, if A is a (2×3) matrix and B is a (3×4) matrix, then AB is a (2×4) matrix. However, you cannot compute BA because the number of columns of B (4, in this case) is not equal to the number of rows of A (2, in this case). Even when you *can* compute both AB and BA, the resulting matrices might not be equal, as shown in the next example.

EXAMPLE 2.21 Two Matrices A and B for Which AB ≠ BA
If

$$A = \begin{bmatrix} 1 & 2 \\ -1 & -2 \end{bmatrix} \quad \text{and} \quad B = \begin{bmatrix} 2 & 4 \\ -1 & -2 \end{bmatrix}$$

then $AB \neq BA$ because

$$AB = \begin{bmatrix} 1 & 2 \\ -1 & -2 \end{bmatrix} \begin{bmatrix} 2 & 4 \\ -1 & -2 \end{bmatrix} = \begin{bmatrix} 0 & 0 \\ 0 & 0 \end{bmatrix}$$

and

$$BA = \begin{bmatrix} 2 & 4 \\ -1 & -2 \end{bmatrix} \begin{bmatrix} 1 & 2 \\ -1 & -2 \end{bmatrix} = \begin{bmatrix} -2 & -4 \\ 1 & 2 \end{bmatrix}$$

Another difference between the product of two real numbers and the product of two matrices is the property that, for real numbers a and b, if $ab = 0$, then $a = 0$ or $b = 0$. This need not be the case for matrices. For example, the matrices A and B in Example 2.21 satisfy $AB = 0$ yet $A \neq 0$ and $B \neq 0$. In contrast, certain properties of matrix multiplication are the same as those of multiplication of real numbers, as summarized in the next theorem.

> **THEOREM 2.4**
>
> If s is a real number and A, B, C, 0, and I (the identity matrix) are matrices for which you can perform the following operations on the basis of their dimensions, then
>
> (a) $A(B + C) = AB + AC$.
>
> (b) $A(B - C) = AB - AC$.
>
> (c) $(A + B)C = AC + BC$.
>
> (d) $(A - B)C = AC - BC$.
>
> (e) $(AB)C = A(BC)$.
>
> (f) $s(AB) = (sA)B = A(sB)$.
>
> (g) $IA = A$ and $BI = B$.
>
> (h) $0A = 0$ and $B0 = 0$.

Proof.

Each part of this theorem requires showing that two matrices are equal. Definition 2.3 leads you to show that all corresponding elements of the matrices are equal. The choose method is used to accomplish this task.

(a) The fact that you can compute $A(B + C)$, AB, and AC means that the dimensions of A, B, and C must be such that $A \in R^{m \times p}$ and $B, C \in R^{p \times n}$. So now, let i and j be integers with $1 \leq i \leq m$ and $1 \leq j \leq n$. You must show that $[A(B + C)]_{ij} = (AB + AC)_{ij}$, but

$$
\begin{aligned}
[A(B + C)]_{ij} &= A_{i*} \cdot (B + C)_{*j} \\
&= A_{i*} \cdot (B_{*j} + C_{*j}) \\
&= (A_{i*} \cdot B_{*j}) + (A_{i*} \cdot C_{*j}) \\
&= (AB)_{ij} + (AC)_{ij} \\
&= (AB + AC)_{ij}
\end{aligned}
$$

(b–h) These proofs are similar to part (a) and are left to Exercise 22.

The proof is now complete. ■

Part (g) of Theorem 2.4 is particularly important because it states that the identity matrix plays the same role for matrices as the number 1 does for real numbers—when you multiply any number by 1 (or any matrix by the identity matrix), the result is whatever you started with. Similarly, part (h) of Theorem 2.4 indicates that for matrices, the zero matrix behaves like the number 0 for real numbers—zero times anything is zero.

2.4.6 Matrix-Vector Notation

One use of vectors and matrices is as a notational tool. Several examples are presented in this section.

Vectors as a Special Case of Matrices

A matrix is a collection of its row (or column) vectors. Analogously, a given n-vector $\mathbf{x} = (x_1, \ldots, x_n)$ is a special type of matrix that has two possible representations, depending on whether you think of $\mathbf{x}$ as a column or a row.

Column Vectors. You can think of the n-vector $\mathbf{x} = (x_1, \ldots, x_n)$ as an $(n \times 1)$ matrix whose one column consists of the entries in $\mathbf{x}$:

$$\mathbf{x} = \begin{bmatrix} x_1 \\ x_2 \\ \vdots \\ x_n \end{bmatrix}$$

In this case, $\mathbf{x}$ is referred to as a **column vector**. Keep in mind that a column vector $\mathbf{x}$ is a *matrix*, even though $\mathbf{x}$ is written in boldface.

Row Vectors. You can also think of the n-vector $\mathbf{x} = (x_1, \ldots, x_n)$ as a $(1 \times n)$ matrix whose one row consists of the entries in $\mathbf{x}$:

$$\mathbf{x} = [x_1, x_2, \ldots, x_n]$$

In this case, $\mathbf{x}$ is referred to as a **row vector**. Once again, a row vector $\mathbf{x}$ is a matrix.

To make the distinction between an n-vector, a row vector, and a column vector clear, an n-vector $\mathbf{x} = (x_1, \ldots, x_n)$, when thought of as a row vector, is written as "the row vector $\mathbf{x} \in R^{1 \times n}$." The corresponding column vector is written as "the column vector $\mathbf{x} \in R^{n \times 1}$." For example, recall the standard unit vector $\mathbf{e}_i$ in which each component is 0, except for component i whose value is 1 (see Section 1.1.4). The corresponding row vector $\mathbf{e}_i \in R^{1 \times n}$ satisfies $\mathbf{e}_i = I_{i*}$, where I is the $(n \times n)$ identity matrix. In contrast, the column vector $\mathbf{e}_i \in R^{n \times 1}$ satisfies $\mathbf{e}_i = I_{*i}$.

Representing Systems of Linear Equations

To illustrate another notational use of vectors and matrices, consider again the problem of finding values for the variables $x_1, \ldots, x_n$ that satisfy the following m linear equations:

$$\begin{aligned}
a_{11}x_1 + a_{12}x_2 + \cdots + a_{1n}x_n &= b_1 \\
a_{21}x_1 + a_{22}x_2 + \cdots + a_{2n}x_n &= b_2 \\
&\vdots \\
a_{m1}x_1 + a_{m2}x_2 + \cdots + a_{mn}x_n &= b_m
\end{aligned} \tag{2.50}$$

You can use vectors and matrices to write (2.50) in a simpler form that is easier to work with. To see how, think of the n unknown variables $x_1, \ldots, x_n$ collectively as a column vector $\mathbf{x} \in R^{n \times 1}$. Likewise, view the known values of $b_1, \ldots, b_m$ on the right-hand side of (2.50) as the components of a column vector, say, $\mathbf{b} \in R^{m \times 1}$. The remaining values a_{ij} in (2.50) constitute a rectangular table of numbers that you can represent by an $(m \times n)$ matrix

A. You therefore have the following:

$$
A = \begin{bmatrix} a_{11} & a_{12} & \cdots & a_{1n} \\ a_{21} & a_{22} & \cdots & a_{2n} \\ \vdots & \vdots & \vdots & \vdots \\ a_{m1} & a_{m2} & \cdots & a_{mn} \end{bmatrix}, \quad \mathbf{x} = \begin{bmatrix} x_1 \\ \vdots \\ x_n \end{bmatrix}, \quad \mathbf{b} = \begin{bmatrix} b_1 \\ \vdots \\ b_m \end{bmatrix}
$$

Using the foregoing column vectors **x** and **b** and the (*m* × *n*) matrix *A*, you can restate (2.50) in terms of matrix multiplication. The result is the following more concise description of the problem of solving *m* linear equations in *n* unknowns.

> **PROBLEM 2.5 Solving *m* Linear Equations in *n* Unknowns in Matrix-Vector Notation**
>
> Given positive integers *m* and *n*, values for the elements of an (*m* × *n*) matrix *A*, and a column vector $\mathbf{b} \in R^{m \times 1}$, find values for the column vector of variables $\mathbf{x} \in R^{n \times 1}$ so that
>
> $$A\mathbf{x} = \mathbf{b} \tag{2.51}$$

The advantage of matrix-vector notation is shown in the proof of the following theorem.

> **THEOREM 2.5**
>
> A system of *m* linear equations in *n* unknowns has no solution, exactly one solution, or infinitely many solutions.

Proof.

A system of linear equations has either no solution, one solution, or more than one solution. In the first two cases, the theorem is true, so suppose that the system has more than one solution. It remains to show that, in this case, the system has infinitely many solutions.

To produce infinitely many solutions, consider the notational representation of the system in Problem 2.5, namely,

$$A\mathbf{x} = \mathbf{b} \tag{2.52}$$

where *A* is a known (*m* × *n*) matrix, $\mathbf{x} \in R^{n \times 1}$ is a column vector of unknowns, and $\mathbf{b} \in R^{m \times 1}$ is a known column vector.

Because this system is assumed to have more than one solution, let $\mathbf{x} = \mathbf{u}$ and $\mathbf{x} = \mathbf{v}$ be two different solutions to (2.52), so

$$A\mathbf{u} = \mathbf{b} \quad \text{and} \quad A\mathbf{v} = \mathbf{b} \tag{2.53}$$

The idea is to use **u** and **v** to create an infinite number of solutions to (2.52). In fact, all points on the line through **u** and **v** satisfy (2.52) (see Figure 2.3).

To see why this is so, consider the parametric representation of the line in Figure 2.3, described as follows by using the point **u** on the line and the direction **v** − **u** pointing from

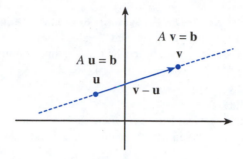

Figure 2.3 *A Line Each of Whose Points is a Solution to* $A\mathbf{x} = \mathbf{b}$.

u to **v** along the line:

$$\mathbf{u} + t(\mathbf{v} - \mathbf{u})$$

For any point **w** on this line, there is a real number t such that

$$\mathbf{w} = \mathbf{u} + t(\mathbf{v} - \mathbf{u})$$

Each such value of **w** is a solution to (2.52) because, from (2.53) and the rules of algebra,

$$A\mathbf{w} = A[\mathbf{u} + t(\mathbf{v} - \mathbf{u})] = A\mathbf{u} + tA\mathbf{v} - tA\mathbf{u} = \mathbf{b} + t\mathbf{b} - t\mathbf{b} = \mathbf{b}$$

The proof is now complete. ■

A consistent system has an infinite number of solutions when there are strictly fewer equations than variables, as proved in the next theorem.

THEOREM 2.6

If $A\mathbf{x} = \mathbf{b}$ is a consistent system of m linear equations in n unknowns in which $m < n$, then this system has an infinite number of solutions.

Proof.

Apply the preparation phase of Gauss-Jordan elimination to the augmented matrix $[A \mid \mathbf{b}]$ to obtain the reduced row-echelon matrix R having at most m leading ones. The hypothesis that $m < n$ then ensures that there is at least one free variable.

Turning to the solution phase, the hypothesis that the system is consistent ensures that R does not have a row with all zeros except for a nonzero value in the last column. The free variable then provides an infinite number of solutions to the system of equations, thus completing the proof. ■

The power of matrix-vector notation is shown again in the following generalization of Problem 2.5.

PROBLEM 2.6 Solving *k* Systems of *m* Linear Equations in *n* Unknowns

Given positive integers k, m, and n, values for the elements of an $(m \times n)$ matrix A, and column vectors $\mathbf{b}_1, \ldots, \mathbf{b}_k$, each in $R^{m \times 1}$, find k column vectors of variables $\mathbf{x}_1, \ldots, \mathbf{x}_k$, each in $R^{n \times 1}$, so that for each $j = 1, \ldots, k$,

$$A\mathbf{x}_j = \mathbf{b}_j \tag{2.54}$$

Equivalently stated, letting B be the $(m \times k)$ matrix in which column j is $\mathbf{b}_j$ and X be the $(n \times k)$ matrix in which column j is $\mathbf{x}_j$, find values for the elements of X so that

$$AX = B$$

You will appreciate the value of matrix-vector notation when you try writing Problem 2.6 without using matrices and vectors.

In any event, one way to solve Problem 2.6 is to apply Gauss-Jordan (or Gaussian) elimination k different times, each time using a different value for the right-hand side of the equations. As metioned previously, one use of mathematics is to develop more efficient solution procedures, such as the following one for Problem 2.6.

SOLUTION to Problem 2.6

To solve Problem 2.6, apply Gauss-Jordan elimination to the augmented matrix $[A \mid B]$.

An example of solving such a collection of systems of equations is shown in Chapter 3.

In this section, you have learned about matrices and their role as a notational tool. You have also seen how to perform various matrix operations and learned some of the properties these operations satisfy. This material is brought together in Section 2.5 by solving the investment problem of Portfolio Planners introduced at the beginning of this chapter.

Exercises for Section 2.4

1. Identify the matrix A, the column vector $\mathbf{x}$ of variables, and the column vector $\mathbf{b}$ for the system of linear equations $A\mathbf{x} = \mathbf{b}$ associated with the Leontief economic model in Section 2.1.1. Ignore the nonnegativity constraints.

2. Suppose that P is an $(n \times n)$ matrix in which P_{ij} represents the fraction of goods purchased by sector i from sector j in a Leontief economic model. Let $\mathbf{x} \in R^{n \times 1}$ be the column vector of unknowns whose components are the dollar value of each sector's output. Use matrix-vector notation in terms of P to write the system of linear equations for finding the value of $\mathbf{x}$ that constitutes an equilibrium point. (Hint: Use the $(n \times n)$ identity matrix, I.)

3. Identify the matrix A, the column vector $\mathbf{x}$ of variables, and the column vector $\mathbf{b}$ for the system of linear equations $A\mathbf{x} = \mathbf{b}$ associated with the network-flow problem in Section 2.1.2. Ignore the nonnegativity constraints.

4. Identify the matrix A, the column vector $\mathbf{x}$ of variables, and the column vector $\mathbf{b}$ for the system of linear equations $A\mathbf{x} = \mathbf{b}$ associated with the make-or-buy problem in Section 2.1.3. Ignore the objective function and the nonnegativity constraints.

5. Write each of the following using matrix-vector notation.
   (a) A homogeneous system of m linear equations in n unknowns.
   (b) A homogeneous system of n linear equations in n unknowns.

6. Rewrite Problem 2.6 in Section 2.4.6 without using matrix-vector notation.

7. Perform each of the indicated operations using the following matrices or explain why you cannot do so:

$$A = \begin{bmatrix} -1 & 2 & 0 \\ 0 & -3 & 1 \end{bmatrix}, \quad B = \begin{bmatrix} 1 & -4 & -1 \\ 1 & 0 & 3 \end{bmatrix}, \quad C = \begin{bmatrix} 1 \\ 0 \\ -2 \end{bmatrix}$$

   (a) $A - B$. (b) $2A + B$. (c) AC. (d) CB. (e) $C^t B$.

8. Perform each of the indicated operations using the following matrices or explain why you cannot do so:

$$A = \begin{bmatrix} -1 & 2 \\ 0 & -3 \\ 1 & 0 \end{bmatrix}, \quad B = \begin{bmatrix} 0 & 3 & -1 \\ 1 & -2 & 0 \end{bmatrix}, \quad C = \begin{bmatrix} 2 \\ -1 \end{bmatrix}$$

   (a) $A + B^t$. (b) $C^t A$. (c) AC. (d) AB.

9. Perform each of the indicated operations using the following matrices or explain why you cannot do so:

$$A = \begin{bmatrix} 2 & 1 \\ 0 & -1 \end{bmatrix}, \quad B = \begin{bmatrix} 1 & 3 \\ -1 & 2 \end{bmatrix}, \quad C = \begin{bmatrix} 1 & 0 & 0 \\ 0 & 2 & -1 \end{bmatrix}$$

   (a) C^t. (b) $A + 2B$. (c) AB. (d) AC. (e) CA.

10. Perform each of the indicated operations using the following matrices or explain why you cannot do so:

$$A = \begin{bmatrix} 1 & 0 \\ 0 & 2 \\ 0 & -1 \end{bmatrix}, \quad B = \begin{bmatrix} -1 & 0 \\ 1 & 2 \end{bmatrix}, \quad C = \begin{bmatrix} -1 & 1 \\ 2 & 0 \\ 1 & -3 \end{bmatrix}$$

   (a) $2A - B$. (b) AB. (c) BA. (d) $C^t C$.

11. For each of the given elementary matrices, E, describe the elementary row operation that results by computing EA for the following matrix A:

$$A = \begin{bmatrix} -1 & 0 & -3 \\ 4 & -2 & 6 \\ 3 & 1 & 5 \end{bmatrix}$$

(a) $E = \begin{bmatrix} 1 & 0 & 0 \\ 0 & 0 & 1 \\ 0 & 1 & 0 \end{bmatrix}$ (b) $E = \begin{bmatrix} -1 & 0 & 0 \\ 0 & 1 & 0 \\ 0 & 0 & 1 \end{bmatrix}$ (c) $E = \begin{bmatrix} 1 & 0 & 0 \\ 0 & 1 & 0 \\ 0 & 1/2 & 1 \end{bmatrix}$

12. Write an appropriate elementary matrix E to perform the indicated operation on the matrix A in the previous exercise and verify your result by computing EA.
    (a) Interchange row 1 and row 3 of A. (b) Multiply row 2 of A by $-1/2$.
    (c) Add 4 times row 1 of A to row 2. (d) Add 3 times row 1 of A to row 3.

13. Suppose that A is an $(m \times n)$ matrix and $\mathbf{x}$ is a vector. To multiply A on the right by $\mathbf{x}$, think of the vector $\mathbf{x}$ as a matrix.
    (a) What must be the dimensions of the matrix $\mathbf{x}$ so that you can compute $A\mathbf{x}$? What are the dimensions of the resulting matrix $A\mathbf{x}$?
    (b) For the matrices

$$A = \begin{bmatrix} 1 & -1 & 0 \\ 2 & 0 & -2 \\ -3 & 1 & 3 \end{bmatrix} \quad \text{and} \quad \mathbf{x} = \begin{bmatrix} -1 \\ 1 \\ 2 \end{bmatrix}$$

   compute the value of $A\mathbf{x}$ by using the rows of A according to the following formula:

$$(A\mathbf{x})_i = A_{i*} \cdot \mathbf{x}, \quad \text{for } i = 1, 2, 3$$

    (c) For the matrices A and $\mathbf{x}$ in part (b), compute $A\mathbf{x}$ by using the columns of A according to the following formula:

$$A\mathbf{x} = A_{*1}x_1 + A_{*2}x_2 + A_{*3}x_3$$

14. Suppose that A is an $(m \times n)$ matrix and $\mathbf{x}$ is a vector. To multiply A on the left by $\mathbf{x}$, think of the vector $\mathbf{x}$ as a matrix.
    (a) What must be the dimensions of the matrix $\mathbf{x}$ so that you can compute $\mathbf{x}A$? What are the dimensions of the resulting matrix $\mathbf{x}A$?
    (b) For the values of A and $\mathbf{x}$ in the previous exercise, compute the value of $\mathbf{x}A$ by using the rows of A according to the following formula:

$$\mathbf{x}A = x_1 A_{1*} + x_2 A_{2*} + x_3 A_{3*}$$

    (c) For the values of A and $\mathbf{x}$ in the previous exercise, compute $\mathbf{x}A$ by using the columns of A according to the following formula:

$$(\mathbf{x}A)_j = \mathbf{x} \cdot A_{*j}, \quad \text{for } j = 1, 2, 3$$

15. Identify as many similarities and differences as you can between the following operations on vectors and on matrices.
    (a) Multiplying a vector by a real number and multiplying a matrix by a real number.
    (b) Adding two vectors and adding two matrices.
    (c) Multiplying two vectors by the dot product and multiplying two matrices.

16. For an $(n \times n)$ matrix A, the **trace of** A, denoted by $\text{tr}(A)$, is the sum of the diagonal elements of A.

(a) Find the trace of each of the following matrices:

$$(i) \quad \begin{bmatrix} a_{11} & a_{12} & a_{13} \\ a_{21} & a_{22} & a_{23} \\ a_{31} & a_{32} & a_{33} \end{bmatrix} \qquad (ii) \quad \begin{bmatrix} -1 & 0 & 2 & -4 \\ 2 & 0 & -1 & 3 \\ 0 & -1 & 5 & 0 \\ 0 & 0 & 8 & -2 \end{bmatrix}$$

(b) Compare and contrast this operation with the operation of transposing a matrix. What similarities and differences do you see?

17. Suppose that $A \in R^{m \times n}$, $B \in R^{n \times p}$, and i and j are integers with $1 \leq i \leq m$ and $1 \leq j \leq p$.
    (a) Prove that $(AB)_{i*} = A_{i*}B$ and $(AB)_{*j} = AB_{*j}$.
    (b) How many fewer multiplications are required in performing the operations in part (a) compared to computing AB? Express your answer in terms of m, n, and p.

18. Suppose that $A \in R^{m \times n}$, $B \in R^{n \times p}$, $C \in R^{p \times q}$. From the properties of matrix multiplication, you know that

    $$A(BC) = (AB)C$$

    (a) In terms of m, n, p, and q, how many multiplications are needed to compute $A(BC)$?
    (b) In terms of m, n, p, and q, how many multiplications are needed to compute $(AB)C$?
    (c) For the special case in which $m = 1$, $n = 100$, $p = 99$, and $q = 500$, which of these two alternatives is more efficient?

19. Complete Theorem 2.1 in Section 2.4.5 by proving each of the following statements, in which $A, B, C, 0 \in R^{m \times n}$.
    (a) $(A + B) + C = A + (B + C)$.   (b) $A + 0 = 0 + A = A$.
    (c) $A - A = 0$.   (d) $0 - A = -A$.

20. Complete Theorem 2.2 in Section 2.4.5 by proving each of the following statements, in which $A, B, 0 \in R^{m \times n}$ and s and t are real numbers.
    (a) $s(A - B) = sA - sB$.   (b) $(s + t)A = sA + tA$.
    (c) $(s - t)A = sA - tA$.   (d) $s(tA) = (st)A$.
    (e) $0A = 0$ (the number 0 times the (f) $s0 = 0$ (the number s times the zero
        matrix A = the zero matrix).   matrix = the zero matrix).

21. Complete Theorem 2.3 in Section 2.4.5 by proving each of the following statements, in which s is a real number and A and B are $(m \times n)$ matrices.
    (a) $(sA)^t = sA^t$.   (b) $(A + B)^t = A^t + B^t$.

22. Complete Theorem 2.4 in Section 2.4.5 by proving each of the following statements, in which s is a real number and $A, B, C, 0$, and I (the identity matrix) are matrices for which you can perform the operations.
    (a) $A(B - C) = AB - AC$.   (b) $(A + B)C = AC + BC$.
    (c) $(A - B)C = AC - BC$.   (d) $(AB)C = A(BC)$.
    (e) $s(AB) = (sA)B = A(sB)$.   (f) $IA = A$ and $BI = B$.
    (g) $0A = 0$ and $B0 = 0$.

2.5 Problem Solving with Linear Equations

The material in this chapter is now used to solve the problem of Portfolio Planners presented at the beginning of the chapter. Doing so involves formulating a system of linear equations and then using Gauss-Jordan elimination to obtain the solution.

2.5.1 The Problem of Portfolio Planners

Recall the problem of Portfolio Planners. The managers of Portfolio Planners want to invest 3 million dollars in a number of mutual funds. The Research Department has identified five funds that meet the minimum requirements of risk and reward. The risk category and the estimated annual rate of return for each fund are given in the following table:

Fund number	1	2	3	4	5
Rate of return	0.25	0.15	0.12	0.08	0.05
Risk category	high	high	medium	medium	low

To control the risk, management has decided that the amounts invested in Funds 1 and 2 will be in the ratio of 1 to 3 and the amounts invested in Funds 3 and 4 will be in the ratio of 1 to 2. Also, the total amount invested in the medium-risk Funds 3 and 4 should be twice as much as the total amount invested in the high-risk Funds 1 and 2.

Management knows intuitively that the more money invested in the low-risk Fund 5, the lower the return on their total investment. They have asked you to quantify this relationship so that they can decide how much to invest in each of the funds.

The data in this problem are the amount of the portfolio (3 million dollars), the rates of returns of the five funds, and the information pertaining to the investment ratios and guidelines. The objective is to use these data to determine how the rate of return for the portfolio varies as a function of the amount invested in Fund 5. You will now see how solving a system of linear equations is used to provide the answer.

2.5.2 Formulating a System of Linear Equations

To solve this problem, first note that your goal is to determine how much to invest in each of the five funds, thus giving rise to five variables. One approach is to define variables that represent the number of dollars to invest in each fund. You are asked to pursue this approach in Exercise 4. An alternative—the one used here—is to define the following five variables:

f_1 = the fraction of the portfolio to invest in Fund 1

f_2 = the fraction of the portfolio to invest in Fund 2

f_3 = the fraction of the portfolio to invest in Fund 3

f_4 = the fraction of the portfolio to invest in Fund 4

f_5 = the fraction of the portfolio to invest in Fund 5

Once the values of these variables are found, you can multiply each one by the portfolio value of 3 million dollars to determine the dollar amount to invest in each fund.

The values of these variables must satisfy the investment guidelines specified in the problem description. These guidelines give rise to a system of linear equations. For example, the requirement that the ratio of the dollar amounts invested in Fund 1 and Fund 2 be 1 to 3 translates to the following expression:

$$\frac{f_1}{f_2} = \frac{1}{3}$$

The foregoing expression is not a linear equation because of the division of the two variables. However, you can create a linear equation by multiplying through by $3f_2$ and then subtracting f_2 from both sides to obtain

$$3f_1 - f_2 = 0 \qquad (2.55)$$

Likewise, the requirement that the ratio of the dollar amounts invested in Fund 3 and Fund 4 be 1 to 2 translates to the following expression:

$$\frac{f_3}{f_4} = \frac{1}{2}$$

Multiplying through by $2f_4$ and then subtracting f_4 from both sides results in the following linear equation:

$$2f_3 - f_4 = 0 \qquad (2.56)$$

The final investment guideline is that the total amount invested in Funds 3 and 4 must be twice as much as the total amount invested in Funds 1 and 2. The corresponding mathematical expression is

$$f_3 + f_4 = 2(f_1 + f_2)$$

Removing the parentheses and moving all variables to the left side results in the following linear equation:

$$-2f_1 - 2f_2 + f_3 + f_4 = 0 \qquad (2.57)$$

It is important to note that additional constraints are needed to ensure that the values of the variables in fact represent fractions. In particular, all variables must be nonnegative and sum to 1, which gives rise to the following linear equation:

$$f_1 + f_2 + f_3 + f_4 + f_5 = 1 \qquad (2.58)$$

Putting together the pieces from (2.55), (2.56), (2.57), and (2.58), you want to find nonnegative values for the variables f_1, f_2, f_3, f_4, and f_5 that satisfy the following system of linear equations:

$$
\begin{aligned}
3f_1 - f_2 &&&&&= 0 \\
2f_3 - f_4 &&&= 0 \\
-2f_1 - 2f_2 + f_3 + f_4 &&&= 0 \\
f_1 + f_2 + f_3 + f_4 + f_5 &= 1
\end{aligned}
\qquad (2.59)
$$

2.5.3 Using Gauss-Jordan Elimination to Obtain the Solution

The next step is to apply Gauss-Jordan elimination to solve the system in (2.59). To do so, create the associated augmented matrix:

$$\left[\begin{array}{ccccc|c} 3 & -1 & 0 & 0 & 0 & 0 \\ 0 & 0 & 2 & -1 & 0 & 0 \\ -2 & -2 & 1 & 1 & 0 & 0 \\ 1 & 1 & 1 & 1 & 1 & 1 \end{array}\right]$$

You can perform the steps of Gauss-Jordan elimination on this augmented matrix by hand, by using a software package such as MATLAB, Maple, and Mathematica, by using a graphing calculator, or by writing a computer program. Doing so leads to the following matrix in reduced row-echelon form:

$$\left[\begin{array}{ccccc|c} 1 & 0 & 0 & 0 & 0.0833 & 0.0833 \\ 0 & 1 & 0 & 0 & 0.2500 & 0.2500 \\ 0 & 0 & 1 & 0 & 0.2222 & 0.2222 \\ 0 & 0 & 0 & 1 & 0.4444 & 0.4444 \end{array}\right]$$

The corresponding system of linear equations is as follows:

$$\begin{aligned} f_1 \quad\quad\quad\quad + 0.0833 f_5 &= 0.0833 \\ f_2 \quad\quad\quad + 0.2500 f_5 &= 0.2500 \\ f_3 \quad + 0.2222 f_5 &= 0.2222 \\ f_4 + 0.4444 f_5 &= 0.4444 \end{aligned} \tag{2.60}$$

From (2.60), you can see that f_5 is a free variable and so, from the results in Section 2.3, there are an infinite number of solutions. That is, given a value for f_5 (between 0 and 1), the values of the remaining variables are:

$$\begin{aligned} f_1 &= 0.0833 - 0.0833 f_5 \\ f_2 &= 0.2500 - 0.2500 f_5 \\ f_3 &= 0.2222 - 0.2222 f_5 \\ f_4 &= 0.4444 - 0.4444 f_5 \end{aligned} \tag{2.61}$$

You can now provide the desired information to management, who asked how the rate of return varies with f_5. In terms of the original variables and the problem data, the rate of return for the portfolio is computed as follows:

$$0.25 f_1 + 0.15 f_2 + 0.12 f_3 + 0.08 f_4 + 0.05 f_5$$

You can now use (2.61) to substitute the values of f_1, f_2, f_3, and f_4 in the foregoing expression to obtain the following rate of return in terms of f_5:

$$\text{rate of return on the portfolio} = 0.1206 - 0.0706 f_5 \tag{2.62}$$

The following table provides the values of the variables and the rate of return on the portfolio when $f_5 = 0.10, 0.20$, and 0.30:

	Value of f_5		
	0.1	0.2	0.3
$f_1 = 0.0833 - 0.0833 f_5$	0.0750	0.0667	0.0583
$f_2 = 0.2500 - 0.2500 f_5$	0.2250	0.2000	0.1750
$f_3 = 0.2222 - 0.2222 f_5$	0.2000	0.1778	0.1556
$f_4 = 0.4444 - 0.4444 f_5$	0.4000	0.3556	0.3111
return $= 0.1206 - 0.0706 f_5$	0.1135	0.1064	0.0994

2.5.4 Further Analysis

The foregoing information confirms management's intuition that, as the amount invested in Fund 5 increases, the rate of return on the portfolio decreases. After some discussion, management decided that the portfolio should return a rate of 11%. They have asked you to determine how much of the 3 million dollars to invest in each of the funds so as to achieve this return while still meeting all of the previous investment guidelines.

You can do so by using (2.62) to find the value of f_5 that provides a return on the portfolio of 0.11 and then substitute that value for f_5 in (2.61) to obtain the values of f_1, f_2, f_3, and f_4. So, solving (2.62) for f_5, you have:

$$f_5 = \frac{0.1206 - \text{return}}{0.0706} = \frac{0.1206 - 0.11}{0.0706} = 0.1496$$

Substituting $f_5 = 0.1496$ in (2.61) yields the following values for the remaining variables:

$$f_1 = .0709, \quad f_2 = 0.2126, \quad f_3 = 0.1890, \quad f_4 = 0.3780$$

These values represent the fraction of the 3 million dollars to invest in each of the five funds. The dollar amounts are obtained by multiplying each of these fraction by $3,000,000, resulting in the following amounts to invest in the five funds:

Fund	1	2	3	4	5
Invest	212,600	637,800	566,900	1,133,900	448,800

In this section, you have seen how to solve a problem by formulating a system of linear equations for which you can use Gauss-Jordan elimination to find the values of the variables. In Chapter 3, you will learn yet another method for solving a system of linear equations in which the number of equations is equal to the number of unknowns.

Exercises for Section 2.5

1. Use technology indicated by your instructor to find the solution to the system of linear equations for the Leontief economic model in Section 2.1.1. Write the sequence of instructions you perform.

2. Use technology indicated by your instructor to find the solution to the system of linear equations associated with the network-flow problem in Section 2.1.2. Write the sequence of instructions you perform.

3. Use technology indicated by your instructor to find the solution to the system of linear equations associated with the make-or-buy problem in Section 2.1.3. Write the sequence of instructions you perform.

PROJECT 2.1: The Investment Problem of Portfolio Planners

Recall, from Section 2.5, the problem of Portfolio Planners, in which the managers want to invest 3 million dollars in a number of mutual funds. The Research Department has identified five funds that meet the minimum requirements of risk and reward. The risk category and estimated annual rate of return for each fund are given in the following table:

Fund number	1	2	3	4	5
Rate of return	0.25	0.15	0.12	0.08	0.05
Risk category	high	high	medium	medium	low

To control the risk, management has decided that the amounts invested in Funds 1 and 2 will be in the ratio of 1 to 3 and the amounts invested in Funds 3 and 4 will be in the ratio of 1 to 2. Also, the total amount invested in the medium-risk Funds 3 and 4 should be twice as much as the total amount invested in the high-risk Funds 1 and 2.

Management knows intuitively that the more money invested in the low-risk Fund 5, the lower the return on their total investment. They have asked you to quantify this relationship so that they can decide how much to invest in each of the funds.

Use a graphing calculator, MATLAB, or other software package, as indicated by your instructor, to perform the operations in Exercise 4 through Exercise 7. Whenever possible, use vector and matrix operations available on your system. Write the answer and the associated sequence of operations you perform. (**Note: Be careful to display the answers with enough digits of accuracy.**)

4. Suppose you define the following five variables to solve this problem:

f_1 = the number of millions of dollars to invest in Fund 1

f_2 = the number of millions of dollars to invest in Fund 2

f_3 = the number of millions of dollars to invest in Fund 3

f_4 = the number of millions of dollars to invest in Fund 4

f_5 = the number of millions of dollars to invest in Fund 5

Briefly explain why the only change to the system of linear equations (2.59) in Section 2.5.2 for solving this problem is to replace the equation

$$f_1 + f_2 + f_3 + f_4 + f_5 = 1.0 \quad \text{with} \quad f_1 + f_2 + f_3 + f_4 + f_5 = 3.0$$

5. In view of the previous exercise, find the reduced row-echelon form of the augmented matrix associated with the following system of linear equations and show how to find

the values of the variables for a given value of f_5:

$$3f_1 - f_2 \qquad\qquad\qquad = 0$$
$$2f_3 - f_4 \qquad = 0$$
$$-2f_1 - 2f_2 + f_3 + f_4 \qquad = 0$$
$$f_1 + f_2 + f_3 + f_4 + f_5 = 3$$

6. Find the values of the variables that satisfy the system of linear equations in the previous exercise and the associated rate of return for the portfolio when $300,000 are invested in Fund 5. Do this by adding the new equation $f_5 = 0.3$ and solving the new system of equations. Repeat this process when $600,000 are invested in Fund 5 and when $900,000 are invested in Fund 5. Prepare a table that summarizes your results.

7. Formulate and solve a system of 5 linear equations in 5 unknowns to determine how much to invest in each of the five funds to achieve a return of 11% for the portfolio.

PROJECT 2.2: The Make-or-Buy Problem of Fiber Optics

Recall the following linear programming problem for solving the make-or-buy problem of Fiber Optics, described in Section 2.1.3, in which you want to find values for the variables LP, HP, LJ, and HJ so as to

$$
\begin{array}{llll}
\text{minimize} & 4LP + & 6HP + 7LJ + 8HJ \\
\text{subject to} & LP & + LJ & = 2000 & \text{(low-density demand)} \\
& HP & + HJ = 4500 & \text{(high-density demand)} \\
& 0.5LP + 0.75HP & = 2400 & \text{(machine time)} \\
& LP, & HP, LJ, HJ \geq & 0
\end{array}
$$

Mathematicians have proved that if the constraints of a linear programming problem that has an optimal solution are described by a system of m linear equations in n unknowns (with $n \geq m$), then there is an optimal solution in which at least $n - m$ of the variables are 0. In the linear program of Fiber Optics, $n = 4$ and $m = 3$, so there is an optimal solution in which at least $n - m = 1$ variable is 0. Use this knowledge and a graphing calculator, MATLAB, or other software package, as indicated by your instructor, to solve this problem by performing the operations in Exercise 8 through Exercise 10. Whenever possible, use vector and matrix operations available on your system. Write the answer and the associated sequence of operations you perform. (**Note: Be careful to display the answers with enough digits of accuracy.**)

8. Include with the three linear equations in the constraints of the linear programming problem of Fiber Optics a fourth equation that sets the value of LP equal to 0 and solve the resulting system of four linear equations in four unknowns. Use the resulting values of the variables to compute the cost, $4LP + 6HP + 7LJ + 8HJ$.

9. Repeat the previous exercise four times, each time setting the value of a different variable to 0 in the fourth equation. Report your results by filling in the following table:

LP	HP	LJ	HJ	Cost
0				
	0			
		0		
			0	

10. Use the results of the previous exercise to identify the solution to the linear programming problem of Fiber Optics. Explain why your answer is not on the line of the table in which the cost is the smallest.

Note: The technique used in Exercises 8, 9, and 10 requires listing all combinations of $n - m$ variables and, for each combination, setting the values of the $n - m$ variables to 0 and solving the resulting system of m linear equations in m unknowns. In the problem of Fiber Optics, it is computationally feasible to list the four combinations and solve the associated systems of equations. However, for linear programs in which n and m are large, this method of listing combinations is not computationally feasible. In 1951, George Dantzing developed a systematic approach for solving all linear programming problems. The method moves from one combination to another in such a way that the nonnegativity constraints are always satisfied and so that the objective function value improves each time a new combination is tried. This approach leads to the solution of the linear programming problem and, in general, requires examining only a very small fraction of the total number of combinations.

Chapter Summary

In this chapter, you have seen how to formulate a system of linear equations to solve a problem. Doing so first requires identifying variables whose values you can control. Requirements on the variables are then translated to a system of linear equations expressed in terms of the variables and other problem data. By solving the resulting system, you obtain values for the variables that constitute the solution to the original problem.

You have also seen the role of a matrix in solving a system of linear equations. A matrix is a rectangular table of numbers arranged in m rows and n columns and is therefore a generalization of a vector. Several important matrices include the zero matrix, square matrices in which the number of rows is the same as the number of columns, and the $(n \times n)$ identity matrix.

Various operations on matrices were presented, including multiplying a matrix by a real number, transposing a matrix, and adding, subtracting, and multiplying matrices of appropriate sizes. Many of these operations share properties of the corresponding operations on real numbers and vectors. However, the operation of multiplying matrices requires special care because it is only possible to perform the multiplication AB when the number of columns of A is the same as the number of rows of B. Even when this is the case, it is not necessarily true that $AB = BA$. Also, it may be that $AB = 0$ yet $A \neq 0$ and $B \neq 0$.

Matrices are useful both as a notational and as a problem-solving tool. One important example is solving the following system of m linear equations in the n unknowns $x_1, \ldots, x_n$:

$$
\begin{aligned}
a_{11}x_1 + a_{12}x_2 + \cdots + a_{1n}x_n &= b_1 \\
a_{21}x_1 + a_{22}x_2 + \cdots + a_{2n}x_n &= b_2 \\
&\vdots \\
a_{m1}x_1 + a_{m2}x_2 + \cdots + a_{mn}x_n &= b_m
\end{aligned}
\tag{2.63}
$$

in which for each $i = 1, \ldots, m$ and $j = 1, \ldots, n$, a_{ij} and b_i are known data. By defining A to be the $(m \times n)$ matrix in which $A_{ij} = a_{ij}$, $b_1, \ldots, b_m$ to be the elements of a column vector $\mathbf{b} \in R^{m \times 1}$, and $x_1, \ldots, x_n$ to be the elements of a column vector $\mathbf{x} \in R^{n \times 1}$, you can write (2.63) more concisely in matrix-vector notation, as follows:

$$
A\mathbf{x} = \mathbf{b}
\tag{2.64}
$$

Various methods for solving a system of linear equations—that is, for determining if the system has no solution, exactly one solution, or infinitely many solutions—were presented. In some cases, it is possible to develop a closed-form solution expressed as a simple formula in terms of the given data. When no closed-form solution is possible, or when the closed-form solution is not easy to evaluate, a numerical method in the form of a sequence of computational steps is used to obtain a solution.

Numerical methods for solving a system of linear equations work in two phases: a preparation phase and a solution phase. With Gauss-Jordan elimination, the system is prepared by performing a sequence of elementary row operations that reduces the augmented matrix $[A \mid \mathbf{b}]$ associated with (2.64) to reduced row-echelon form. Once in this form, the solution to the system is easy to obtain. These row operations on a matrix consist of

1. Multiplying a row by a nonzero constant.

2. Interchanging two rows.

3. Adding a multiple of one row to another row.

Gaussian elimination is a variation of Gauss-Jordan elimination that involves less work in the preparation phase to obtain an augmented matrix in row-echelon form, but more work in the the solution phase to perform back substitution.

Chapter 3

Using Matrices to Solve (*n* × *n*) Linear Equations

From Chapter 2, you know how Gauss-Jordan and Gaussian elimination are used to solve a system of *m* linear equations in *n* unknowns. In this chapter, a new approach is presented for solving a special case of this problem in which $m = n$, that is, when the number of equations is equal to the number of unknowns. The need to do so arises in the following problem faced by Solar Technologies.

The Production-Planning Problem of Solar Technologies

Solar Technologies produces low-grade, medium-grade, and high-grade solar cells, each made of either silicon or geranium. As Manager of Production, you have been asked to determine a daily production plan that satisfies the following requirements:

1. The total number of solar cells of all grades made of silicon should be three times the number made of geranium because the available supply of silicon is three times as much as that of geranium.

2. The total demand for low-grade and medium grade cells combined is twice that for high-grade cells.

3. The company wants to achieve a net profit of $9,000 per day, based on known profitability figures for each type of cell.

4. For each grade you can produce either silicon or geranium cells, but not both.

3.1 Applications of (*n* x *n*) Linear Equations

In this section, you will see several examples of problems whose solutions are found by solving a system of n linear equations in n unknowns. The first step in each application is to identify variables whose values, once determined, constitute the solution to the problem. The problem data are then used to formulate an appropriate system of linear equations whose solution provides the desired values for the variables.

3.1.1 Leontief Input-Output Models

In Section 2.1.1, you learned about a particular type of economic model. A different economic model is presented here.

Problem Description

As in the problem in Section 2.1.1, consider an economy that consists of a number of *sectors*. A portion of each sector's *output* (that is, annual production), is used internally, some is consumed by the other sectors, and the remaining portion is left over for use by what is called the *open sector*. The excess amount for the open sector might represent an amount to be stored in inventory, to be exported, or to be used to meet some other external demand. The amount of output from all of the sectors used by one particular sector is called that sector's *input*. All input and output amounts are measured in, say, billions of dollars.

A basic assumption of the problem is that for each sector, there is a known *unit consumption vector* whose components represent the amount of input from each sector needed to produce one dollar of that sector's output. The objective is to determine the annual production of each sector needed to meet each sector's input needs while leaving a surplus of a specified dollar amount for the open sector. This problem, called an *input-output model*, was studied by Wassily Leontief.

As an example, suppose that an economy has a goods, an energy, and an agriculture sector. The unit consumption vector for each sector is given in the columns of the following table:

	Goods	Energy	Agriculture
Goods	0.5	0.2	0.2
Energy	0.3	0.4	0.2
Agriculture	0.1	0.1	0.4

For example, from the first column of the foregoing table, you can see that to produce each dollar of goods requires 0.5 dollars of goods, 0.3 dollars of energy, and 0.1 dollars of agriculture. What amount of annual production is needed to ensure that, in addition to meeting the input demand of each sector, the open sector receives 1 billion dollars of goods, 3 billion dollars of energy, and 4 billion dollars of agricultural products?

Formulating a System of Linear Equations

To formulate a mathematical model of this problem, begin by identifying variables whose values you can control. In this case, you can control the amount of output from each sector, thus giving rise to the following three variables:

G = the annual production of goods, in billions of dollars

E = the annual production of energy, in billions of dollars

A = the annual production of agriculture, in billions of dollars

The objective is to find values for the variables that satisfy a certain condition—in the form of a linear equation—for each sector. For example, for goods,

$$\left(\begin{array}{c} \text{Amount of} \\ \text{goods produced} \end{array} \right) = \left(\begin{array}{c} \text{Amount needed for} \\ \text{input to all sectors} \end{array} \right) + \left(\begin{array}{c} \text{Amount needed for} \\ \text{the open sector} \end{array} \right)$$

The amount of goods produced is precisely the variable G. The amount of goods needed for input to all sectors is the amount used by goods plus the amount used by energy plus the amount used by agriculture. Using the data from the first *row* of the given table together with the variables G, E, and A, you have that:

Amount of goods needed for input to all sectors $= 0.5G + 0.2E + 0.2A$

The amount of goods needed for the open sector, as specified in the problem description, is 1 billion dollars. Thus, the linear equation for the goods sector is

$$G = 0.5G + 0.2E + 0.2A + 1$$

Equivalently, by moving all variables to the left,

$$0.5G - 0.2E - 0.2A = 1$$

Similar equations are needed for the energy and agriculture sectors. The data in the corresponding rows of the table together with the amounts needed for the open sector give rise to the following system of three linear equations in three unknowns:

$$\begin{array}{ll} 0.5G - 0.2E - 0.2A = 1 & \text{(goods)} \\ -0.3G + 0.6E - 0.2A = 3 & \text{(energy)} \\ -0.1G - 0.1E + 0.6A = 4 & \text{(agriculture)} \end{array} \qquad (3.1)$$

Of course the value of each variable must also be nonnegative.

A General Input-Output Model

You can write the system of equations for a general input-output model concisely in matrix-vector notation. To illustrate, suppose that there are n sectors whose consumption vectors are given in the columns of an $(n \times n)$ matrix C and that the known quantities needed by the open sector are represented by the elements of a column vector $\mathbf{d} \in R^{n \times 1}$. For the special case of the numerical example above, $n = 3$ and

$$C = \begin{bmatrix} 0.5 & 0.2 & 0.2 \\ 0.3 & 0.4 & 0.2 \\ 0.1 & 0.1 & 0.4 \end{bmatrix} \quad \text{and} \quad \mathbf{d} = \begin{bmatrix} 1 \\ 3 \\ 4 \end{bmatrix}$$

Letting $\mathbf{x} \in R^{n \times 1}$ be the column vector of variables representing the annual production in the n sectors, the input-output model is to find values for the elements of $\mathbf{x}$ so that

$$\mathbf{x} = C\mathbf{x} + \mathbf{d}$$

By writing the identity matrix in front of $\mathbf{x}$ on the left side and then subtracting $C\mathbf{x}$ from both sides, the foregoing system becomes:

$$(I - C)\mathbf{x} = \mathbf{d}$$

Problem Solution

Applying Gauss-Jordan elimination, or the techniques you will learn in this chapter, yields the following solution to the system in (3.1):

$$G = 12.5, \quad E = 15, \quad \text{and} \quad A = 11.25$$

In other words, if 12.5 billion dollars of goods, 15 billion dollars of energy, and 11.25 billion dollars of agriculture are produced, there will be enough to meet each sectors input requirements while leaving 1 billion dollars of goods, 3 billion dollars of energy, and 4 billion dollars of agricultural products for the open sector.

3.1.2 Electrical Networks

Another area in which a system of n linear equations in n unknowns arises is in the study of electrical networks. A simple network consists of a voltage source, such as a battery, and one or more resistors, such as a heater, light bulb, motor, and so on. An example of such a network is shown in Figure 3.1.

The battery is a voltage source that causes a current of electrons to flow through the network. The direction of current flow is arbitrary but is typically shown moving away from the longer (positive) side of the battery. Each resistor impedes the current and consumes some of the voltage causing a *voltage drop* across the resistor. Ohm's law states that the voltage drop V across the resistor, measured in *volts*, equals the resistance R of the resistor, measured in *ohms*, multiplied by the current flow I, measured in *amps*:

$$\text{Voltage drop} = V = RI \qquad \text{(Ohm's Law)} \tag{3.2}$$

Knowing any two of the quantities in (3.2) allows you to find the third value. Usually, the battery voltage and the resistance are known and the current is unknown. For example, if the battery in Figure 3.1 generates 18 volts, then the voltage drop across a 6-ohm resistor means that the current is $I = V/R = 18/6 = 3$ amps.

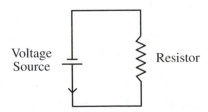

Figure 3.1 *A Simple Electrical Network.*

Much of the study of electrical networks deals with *loops*, which are closed paths that traverse *branches* (containing voltage sources, resistors, and so on) that begin and end at the same point, called a *node*. The network in Figure 3.2 has two nodes, labeled A and B, and three loops: the left loop begins and ends at the node labeled A and contains the 50-volt battery and the 4-ohm and 10-ohm resistors; the right loop begins and ends at Node A and contains the 100-volt battery and the 10-ohm and 15-ohm resistors; the outside loop begins and ends at Node A and contains both batteries and the 4-ohm and 15-ohm resistors.

The flow in these networks obey the following two laws of Kirchoff:

1. **(Current Law)** The current flowing into a node is equal to the current flowing out of that node.

2. **(Voltage Law)** The sum of the voltage drops in a given direction around a loop is equal to the sum of the voltage sources in the same direction around that loop.

These two laws applied to the loops give rise to a system of linear equations whose solution provides the amount of current in the various branches of the network. For example, in Figure 3.2, the three unknowns are the branch currents I_1, I_2, and I_3. Kirchoff's current law provides one equation for each node:

Node	Current In	=	Current Out
A	I_1	=	$I_2 + I_3$
B	$I_2 + I_3$	=	I_1

The second equation is the same as the first and is therefore omitted from further consideration.

Kirchoff's voltage law provides one equation for each loop. For example, the counterclockwise direction for the left loop and the corresponding directions of the currents I_1 and I_2 in Figure 3.2 give rise to the following equation:

Sum of Voltage Drops	=	Sum of Voltage Sources	
$4I_1 + 10I_2$	=	50	(left loop)

For the right loop, the counterclockwise direction is opposite to the direction of I_2, so the

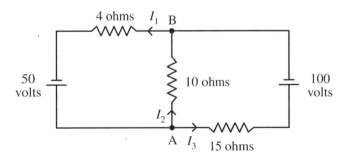

Figure 3.2 *An Electrical Network with Loops.*

equation is

Sum of Voltage Drops	=	Sum of Voltage Sources	
$-10I_2 + 15I_3$	=	100	(right loop)

The outside loop also gives rise to an equation. Using the counterclockwise direction:

Sum of Voltage Drops	=	Sum of Voltage Sources	
$4I_1 + 15I_3$	=	$50 + 100$	(outside loop)

However, this equation for the outside loop is the sum of the equations for the two inside loops. Providing no additional information, this equation is also omitted.

Putting together the equation for Node A and the two voltage equations for the left and right loops in such a way that all variables appear on the left of the equality sign and all constants appear on the right gives rise to the following system of three linear equations in the three unknowns I_1, I_2, and I_3:

$$
\begin{aligned}
I_1 \ - \ \ I_2 \ - \ \ I_3 &= \ \ \ 0 \quad \text{(current law for Node A)} \\
4I_1 \ + \ 10I_2 \ \ \ \ \ \ &= \ \ 50 \quad \text{(voltage law for left loop)} \\
- \ 10I_2 \ + \ 15I_3 &= 100 \quad \text{(voltage law for right loop)}
\end{aligned}
$$

Applying Gauss-Jordan elimination, or the techniques you will learn in this chapter, results in the following solution to this system and hence the branch currents:

$$I_1 = 9 \text{ amps}, \quad I_2 = 1.4 \text{ amps}, \quad \text{and} \quad I_3 = 7.6 \text{ amps}$$

You have seen various problems whose solution requires solving a system of n linear equations in n unknowns. Another method for obtaining the solution to such a system is presented in the rest of this chapter.

Exercises for Section 3.1

1. Consider a model of the U.S. economy consisting of a goods sector and a services sector. The goods sector uses 30% of its annual output and needs 10% of the annual output from the services sector. Services uses 40% of its own output and needs 20% of the output from goods.
   (a) Formulate a system of two linear equations in two unknowns to ensure that the output of goods is adequate to meet the needs of the two sectors and have an additional $3 billion left for consumers and that the output of services is able to meet the needs of the two sectors and have $5 billion left for consumers.
   (b) Solve the system in part (a) to find the annual production of each sector.

2. Germany uses 30% of its own annual production and imports 20% of Japan's annual production and 10% of America's annual production. Japan uses 25% of its own annual production, 15% of America's and 20% of Germany's annual production. The U.S. uses 40% of its own annual production, 15% of Germany's and 25% of Japan's annual production. Write a system of linear equations to determine the annual production of each country to ensure that, in addition to supplying itself and the other two countries,

Germany has $20 billion left for other exports, Japan has $30 billion left for other exports, and the U.S. has $50 billion for other exports.

3. Use Kirchoff's laws and a counterclockwise direction to write and solve a system of linear equations for finding the branch current in the following network:

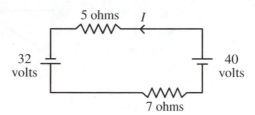

4. Use Kirchoff's laws and a counterclockwise direction to write and solve a system of linear equations for finding the branch current in the following network:

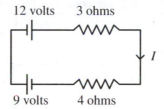

5. Use Kirchoff's laws and a counterclockwise direction to write and solve a system of linear equations for finding the branch currents in the following network:

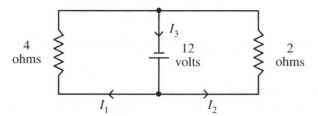

6. Use Kirchoff's laws and a counterclockwise direction to write and solve a system of linear equations for finding the branch currents in the following network:

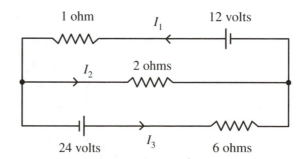

7. Use Kirchoff's current and voltage laws and a counterclockwise direction to write and solve a system of linear equations for finding the branch currents in the following network:

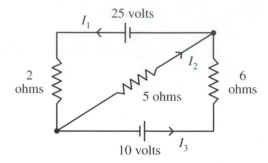

8. Use Kirchoff's current and voltage laws and a counterclockwise direction to write and solve a system of linear equations for finding the branch currents in the following network:

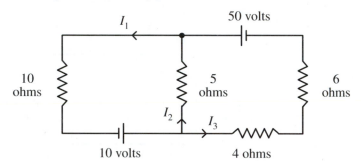

9. Consider approximating a real-valued function, f, of one variable by a quadratic function $q(x) = ax^2 + bx + c$ and let x, y, and z be three given real numbers. Write a system of 3 linear equations in 3 unknowns for determining values for a, b, and c in such a way that the graph of the quadratic function goes through the three points $(x, f(x))$, $(y, f(y))$, and $(z, f(z))$, as shown in the following figure:

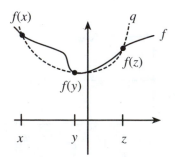

10. Suppose you want to approximate a real-valued function, f, of one variable by a polynomial of degree n, say, $p(x) = p_0 + p_1 x + p_2 x^2 + \cdots + p_n x^n$. Generalize the results in the previous exercise by showing that, if you choose $n + 1$ real numbers, say, $x_0 < x_1 < \cdots < x_n$, then you can solve a system of $n + 1$ linear equations in $n + 1$ unknowns to determine values for $p_0, p_1, \ldots, p_n$ so that the graph of the polynomial goes through the points $(x_0, f(x_0)), (x_1, f(x_1)), \ldots, (x_n, f(x_n))$.

3.2 Using the Inverse of a Matrix to Solve (*n* × *n*) Linear Equations

As you learned in Section 2.4.6, you can represent the problem of solving a system of n linear equations in n unknowns in matrix-vector notation, as follows:

$$A\mathbf{x} = \mathbf{b} \tag{3.3}$$

where A is a known $(n \times n)$ matrix, $\mathbf{b} \in R^{n \times 1}$ is a known column vector, and $\mathbf{x} \in R^{n \times 1}$ is a column vector of unknowns. (Throughout this chapter, A, $\mathbf{x}$, and $\mathbf{b}$ have this meaning.)

The new approach used to solve (3.3) is a generalization of the method for solving the single linear equation

$$ax = b$$

whose solution is

$$x = \frac{b}{a} \quad \text{(assuming that } a \neq 0) \tag{3.4}$$

One approach to generalizing (3.4) that enables you to solve the system in (3.3) is to replace, in (3.4), the real number x with the column vector $\mathbf{x}$, the real number a with the matrix A, and the real number b with the column vector $\mathbf{b}$, to obtain

$$\mathbf{x} = \frac{\mathbf{b}}{A}$$

The foregoing expression contains a syntax error because you cannot divide the column vector $\mathbf{b}$ by the matrix A. Another approach is needed.

There are several alternatives, one of which is to rewrite (3.4) so that generalization by substitution *is* possible. Doing so requires eliminating the operation of division. For example, you can rewrite (3.4) as follows:

$$x = a^{-1}b \tag{3.5}$$

In (3.5), the operation of multiplication appears instead of division. Now, generalizing (3.5) to matrices and vectors results in the following:

$$\mathbf{x} = A^{-1}\mathbf{b} \tag{3.6}$$

If you assume that A^{-1} is an $(n \times n)$ matrix, then no syntax error arises because you can perform the matrix multiplication in (3.6). The remaining question is, What, precisely, is A^{-1}? The answer is provided in what follows.

3.2.1 The Inverse of a Matrix

To determine what A^{-1} in (3.6) is, consider the corresponding value of a^{-1} in (3.5), which is a number that has certain properties in relation to a. In particular,

$$aa^{-1} = a^{-1}a = 1 \tag{3.7}$$

Analogously, in (3.6), A^{-1} is an $(n \times n)$ matrix with certain properties in relation to A. To determine what those properties are, try to generalize (3.7). If you replace the symbol a everywhere with A, you obtain

$$AA^{-1} = A^{-1}A = 1$$

The foregoing expression has a syntax error because it is not possible to use the last equality sign to compare the matrix $A^{-1}A$ with the number 1.

One way to resolve this error is to replace the number 1 with an appropriate $(n \times n)$ matrix. To determine what the elements of that matrix are, observe that the number 1 has the property that when multiplied by any other number, the result is that number. Correspondingly, you need a matrix that, when multiplied with any other matrix, results in that other matrix. From Theorem 2.4(g) in Section 2.4.5, that desired matrix is the $(n \times n)$ identity matrix, denoted in the rest of this chapter by I. Thus, the generalization of (3.7) requires finding a matrix A^{-1} with the property that

$$AA^{-1} = A^{-1}A = I$$

This type of matrix is formalized in the following definition.

DEFINITION 3.1

Supppose that $A \in R^{n \times n}$. A matrix $A^{-1} \in R^{n \times n}$ is an **inverse** of A if and only if

$$AA^{-1} = I \quad \text{and} \quad A^{-1}A = I$$

The inverse of an $(n \times n)$ matrix A is important because, if you can find A^{-1}, then the solution to the system of linear equations in (3.3) is

$$\mathbf{x} = A^{-1}\mathbf{b} \tag{3.8}$$

since

$$A\mathbf{x} = A(A^{-1}\mathbf{b}) = (AA^{-1})\mathbf{b} = I\mathbf{b} = \mathbf{b}$$

EXAMPLE 3.1 Using the Inverse to Solve a System of Equations

Consider the following system of linear equations:

$$2x_1 - x_2 = -3$$
$$-x_1 + x_2 = 4$$

in which

$$A = \begin{bmatrix} 2 & -1 \\ -1 & 1 \end{bmatrix}, \quad \mathbf{x} = \begin{bmatrix} x_1 \\ x_2 \end{bmatrix}, \quad \text{and} \quad \mathbf{b} = \begin{bmatrix} -3 \\ 4 \end{bmatrix}$$

The matrix

$$A^{-1} = \begin{bmatrix} 1 & 1 \\ 1 & 2 \end{bmatrix}$$

is the inverse of A because

$$AA^{-1} = \begin{bmatrix} 2 & -1 \\ -1 & 1 \end{bmatrix} \begin{bmatrix} 1 & 1 \\ 1 & 2 \end{bmatrix} = \begin{bmatrix} 1 & 0 \\ 0 & 1 \end{bmatrix} = I$$

and

$$A^{-1}A = \begin{bmatrix} 1 & 1 \\ 1 & 2 \end{bmatrix} \begin{bmatrix} 2 & -1 \\ -1 & 1 \end{bmatrix} = \begin{bmatrix} 1 & 0 \\ 0 & 1 \end{bmatrix} = I$$

As a result, from (3.8), the solution to $A\mathbf{x} = \mathbf{b}$ is

$$\mathbf{x} = A^{-1}\mathbf{b} = \begin{bmatrix} 1 & 1 \\ 1 & 2 \end{bmatrix} \begin{bmatrix} -3 \\ 4 \end{bmatrix} = \begin{bmatrix} 1 \\ 5 \end{bmatrix} = \begin{bmatrix} x_1 \\ x_2 \end{bmatrix}$$

That is, $x_1 = 1$ and $x_2 = 5$ solves the original system of equations.

In summary, the new approach for solving the $(n \times n)$ system of linear equations $A\mathbf{x} = \mathbf{b}$ consists of a preparation phase to find A^{-1} and a solution phase to compute $\mathbf{x} = A^{-1}\mathbf{b}$. The issue now is how to find A^{-1}.

3.2.2 Closed-form Inverses of Certain Matrices

In this section, closed-form expressions for A^{-1} in terms of the elements of A are given for various matrices A.

The Inverse of a (2 × 2) Matrix

For the following (2×2) matrix in which $ad - bc \neq 0$,

$$A = \begin{bmatrix} a & b \\ c & d \end{bmatrix} \quad \text{you have} \quad A^{-1} = \frac{1}{ad-bc} \begin{bmatrix} d & -b \\ -c & \cdot a \end{bmatrix} = \begin{bmatrix} \dfrac{d}{ad-bc} & \dfrac{-b}{ad-bc} \\ \dfrac{-c}{ad-bc} & \dfrac{a}{ad-bc} \end{bmatrix}$$

This is because

$$AA^{-1} = \frac{1}{ad-bc} \begin{bmatrix} a & b \\ c & d \end{bmatrix} \begin{bmatrix} d & -b \\ -c & a \end{bmatrix} = \frac{1}{ad-bc} \begin{bmatrix} ad-bc & 0 \\ 0 & ad-bc \end{bmatrix} = I$$

You can also verify that $A^{-1}A = I$.

The Inverse of a Diagonal Matrix

Recall, from Section 2.4.2, that a diagonal matrix is a square matrix that has the form:

$$
A = \begin{bmatrix} A_{11} & 0 & \cdots & 0 \\ 0 & A_{22} & \cdots & 0 \\ \vdots & \vdots & \ddots & \vdots \\ 0 & 0 & \cdots & A_{nn} \end{bmatrix}
$$

In the event that none of the diagonal elements $A_{11}, A_{22}, \ldots, A_{nn}$ is zero, you can compute the following inverse of A:

$$
A^{-1} = \begin{bmatrix} \dfrac{1}{A_{11}} & 0 & \cdots & 0 \\ 0 & \dfrac{1}{A_{22}} & \cdots & 0 \\ \vdots & \vdots & \ddots & \vdots \\ 0 & 0 & \cdots & \dfrac{1}{A_{nn}} \end{bmatrix}
$$

This is because

$$
AA^{-1} = \begin{bmatrix} A_{11} & 0 & \cdots & 0 \\ 0 & A_{22} & \cdots & 0 \\ \vdots & \vdots & \ddots & \vdots \\ 0 & 0 & \cdots & A_{nn} \end{bmatrix} \begin{bmatrix} \dfrac{1}{A_{11}} & 0 & \cdots & 0 \\ 0 & \dfrac{1}{A_{22}} & \cdots & 0 \\ \vdots & \vdots & \ddots & \vdots \\ 0 & 0 & \cdots & \dfrac{1}{A_{nn}} \end{bmatrix}
$$

$$
= \begin{bmatrix} A_{11}\dfrac{1}{A_{11}} & 0 & \cdots & 0 \\ 0 & A_{22}\dfrac{1}{A_{22}} & \cdots & 0 \\ \vdots & \vdots & \ddots & \vdots \\ 0 & 0 & \cdots & A_{nn}\dfrac{1}{A_{nn}} \end{bmatrix}
$$

$$
= I
$$

A similar computation shows that $A^{-1}A = I$.

Because the $(n \times n)$ identity matrix I is a special case of a diagonal matrix in which the diagonal elements are each 1, you now know by the foregoing result that $I^{-1} = I$.

The Inverse of an Elementary Matrix

Recall, from Section 2.4.2, that an elementary matrix E is obtained by performing an elementary row operation on the identity matrix I. An inverse of E must satisfy

$$E^{-1}E = I \quad \text{and} \quad EE^{-1} = I$$

For example, consider the elementary matrix E obtained by multiplying, say, the first row of the identity matrix by the nonzero constant c:

$$E = \begin{bmatrix} c & 0 & \cdots & 0 \\ 0 & 1 & \cdots & 0 \\ \vdots & \vdots & \ddots & \vdots \\ 0 & 0 & \cdots & 1 \end{bmatrix}$$

Because E is a diagonal matrix none of whose diagonal elements is zero, an inverse matrix of E is

$$E^{-1} = \begin{bmatrix} 1/c & 0 & \cdots & 0 \\ 0 & 1 & \cdots & 0 \\ \vdots & \vdots & \ddots & \vdots \\ 0 & 0 & \cdots & 1 \end{bmatrix}$$

Notice that E^{-1} is an elementary matrix that corresponds to the inverse operation of multiplying the first row of I by $1/c$. In Exercise 6 and Exercise 7, you are asked to verify that the following inverse operations provide an inverse of each of the associated elementary matrices:

Elementary Row Operation	Inverse Row Operation
Multiply row i by a constant $c \neq 0$.	Multiply row i by $1/c$.
Interchange row i and row j.	Interchange row i and row j.
Add c times row i to row j.	Add $-c$ times row i to row j.

In conclusion, you now know the following result.

THEOREM 3.1

An elementary matrix always has an inverse matrix.

From the examples presented so far, you see that some matrices have an inverse and others do not. Even the (1×1) matrix $[a]$ has no inverse when $a = 0$. This observation gives rise to the following definition.

DEFINITION 3.2

An (*n* × *n*) matrix *A* is **invertible** (or **nonsingular**) if and only if *A* has an inverse matrix, that is, if and only if there is an (*n* × *n*) matrix *B* such that

$$AB = I \quad \text{and} \quad BA = I$$

In view of Definition 3.2, you will only be able to find an inverse of a matrix *A* when *A* is invertible. In Section 3.3.3, a closed-form expression for finding an inverse of an invertible (*n* × *n*) matrix is given. Because that closed-form expression is not computationally easy to evaluate, a numerical method is used to find an inverse. This method is based on the properties that an invertible matrix satisfies, some of which are presented next.

3.2.3 Properties of Invertible Matrices

If a matrix *A* is invertible, then, by definition, there is an (*n* × *n*) matrix *B* such that

$$AB = I \quad \text{and} \quad BA = I$$

In this case, *B* is an inverse of *A*. The next theorem establishes that *B* is the only inverse of an invertible matrix *A*.

THEOREM 3.2

If the matrix *A* is invertible, then there is a unique matrix *B* such that

$$AB = I \quad \text{and} \quad BA = I$$

Proof.

Because *A* is invertible, by Definition 3.2, there is a matrix *B* such that

$$AB = I \quad \text{and} \quad BA = I \tag{3.9}$$

To show that *B* is the only matrix with this property, use the following general proof technique.

Whenever you want to prove that

there is a unique "object" with a "certain property" such that "something happens,"

first use the construction method to show that there is an object, say, *X*, with the certain property and for which the something happens. Then use the **direct uniqueness method** of assuming that there is another object, say, *Y*, with the certain property and for which the something also happens. To establish that *X* is the unique such object, you must show that *X* = *Y* (see Appendix A.12).

To see how this uniqueness method is used in the current proof, assume that there is another matrix, say, *C*, such that

$$AC = I \quad \text{and} \quad CA = I \tag{3.10}$$

If B is supposed to be unique, then it had better be the case that $C = B$. In other words, you must now show that $C = B$.

This is accomplished by using (3.9), (3.10), and the properties of matrix operations to perform the following steps:

$$CA = I \qquad \text{[from (3.10)]}$$
$$(CA)B = IB \qquad \text{(multiply both sides on the right by } B\text{)}$$
$$(CA)B = B \qquad \text{(property of } I\text{)}$$
$$C(AB) = B \qquad \text{(associative property of matrix multiplication)}$$
$$C(I) = B \qquad \text{[(from (3.9)]}$$
$$C = B \qquad \text{(property of } I\text{)}$$

This uniqueness method shows that B is the only matrix for which $AB = I$ and $BA = I$ and so the proof is complete. ■

In view of Theorem 3.2, an invertible matrix A has a unique inverse matrix, hereafter denoted A^{-1}. One advantage of knowing that the inverse of an invertible matrix A is unique is that, if you ever have a matrix B such that $AB = I$ and $BA = I$, then you can conclude that $B = A^{-1}$, as illustrated in the proof of the next theorem.

THEOREM 3.3

If A and B are invertible $(n \times n)$ matrices, then

(a) A^{-1} is invertible and $(A^{-1})^{-1} = A$.

(b) For any real number $t \neq 0$, the matrix tA is invertible and $(tA)^{-1} = \dfrac{1}{t}A^{-1}$.

(c) AB is invertible and $(AB)^{-1} = B^{-1}A^{-1}$.

Proof.
A and B are invertible so, by Definition 3.2, there are matrices A^{-1} and B^{-1} such that

$$AA^{-1} = I, \quad A^{-1}A = I, \quad BB^{-1} = I, \quad \text{and} \quad B^{-1}B = I$$

These inverse matrices are used in the subsequent proof.

(a) To show that A^{-1} is invertible, by Definition 3.2, it is necessary to produce a matrix C such that $CA^{-1} = I$ and $A^{-1}C = I$. Such a matrix is $C = A$ because $AA^{-1} = I$ and $A^{-1}A = I$. As per the discussion preceding this theorem, having produced the matrix A for which $A^{-1}A = I$ and $AA^{-1} = I$, it follows that $A = (A^{-1})^{-1}$.

(b) As in the proof of part (a), it is necessary to produce a matrix C such that $(tA)C = I$ and $C(tA) = I$. Now, $t \neq 0$, so $C = (1/t)A^{-1}$ is such a matrix because

$$(tA)C = (tA)\left(\frac{1}{t}A^{-1}\right) = t\left(\frac{1}{t}\right)(AA^{-1}) = (1)I = I$$

Likewise, $C(tA) = I$, and so tA is invertible and $(tA)^{-1} = \dfrac{1}{t}A^{-1}$.

(c) As in the proofs of parts (a) and (b), it is necessary to produce a matrix C such that $(AB)C = I$ and $C(AB) = I$. Now $C = B^{-1}A^{-1}$ is such a matrix because

$$(AB)C = (AB)(B^{-1}A^{-1}) = A(BB^{-1})A^{-1} = A(I)A^{-1} = AA^{-1} = I$$

Likewise, $C(AB) = I$ and so AB is invertible and $(AB)^{-1} = B^{-1}A^{-1}$.

This completes the proof. ■

The result in Theorem 3.3(c) generalizes to any product of invertible matrices. That is, the product of a finite number of invertible matrices is invertible and the inverse of the product is the product of the inverses of the individual matrices, in reverse order. This fact is used in the proof of the next theorem, which is one of the main results in the theory of solving linear equations.

THEOREM 3.4

The following are equivalent for an ($n \times n$) matrix A:

(a) A is invertible.

(b) For every column vector **b**, there is a unique column vector **x** such that $A\mathbf{x} = \mathbf{b}$.

(c) The unique solution to the homogeneous system $A\mathbf{x} = \mathbf{0}$ is $\mathbf{x} = \mathbf{0}$, where $\mathbf{0}$ is the zero column vector of dimension n.

(d) The reduced row-echelon form of A is I.

(e) A can be expressed as a product of elementary matrices.

Proof.
To show that these five statements are equivalent, you could show that (a) is true if and only if (b) is true, (b) is true if and only if (c) is true, (c) is true if and only if (d) is true, (d) is true if and only if (e) is true, and (e) is true if and only if (a) is true. Doing so requires ten proofs. An equally valid but more efficient approach is to prove that (a) implies (b), (b) implies (c), (c) implies (d), (d) implies (e), and (e) implies (a), which requires only five proofs.

(a) implies (b). Assume that A is invertible. Then by Definition 3.2, there is a matrix A^{-1} such that $AA^{-1} = I$ and $A^{-1}A = I$. To prove that (b) is true, because of the quantifier *for every*, use the choose method to choose a column vector **b** for which you must show that there is a unique column vector **x** such that $A\mathbf{x} = \mathbf{b}$. First construct the column vector **x**. In particular, using the matrix A^{-1}, construct

$$\mathbf{x} = A^{-1}\mathbf{b}$$

It follows that $A\mathbf{x} = \mathbf{b}$ because

$$A\mathbf{x} = A(A^{-1}\mathbf{b}) = (AA^{-1})\mathbf{b} = I\mathbf{b} = \mathbf{b}$$

It remains to show the uniqueness, that is, that $\mathbf{x} = A^{-1}\mathbf{b}$ is the only solution to $A\mathbf{x} = \mathbf{b}$. According to the uniqueness method used in the proof of Theorem 3.2, you should assume that **y** also satisfies $A\mathbf{y} = \mathbf{b}$ and show that $\mathbf{y} = \mathbf{x}$. This follows from the fact that $A\mathbf{x} = \mathbf{b}$

and $A\mathbf{y} = \mathbf{b}$ because

$$A\mathbf{x} = A\mathbf{y} \qquad \text{(because both } A\mathbf{x} \text{ and } A\mathbf{y} \text{ are equal to } \mathbf{b})$$
$$A^{-1}(A\mathbf{x}) = A^{-1}(A\mathbf{y}) \quad \text{(multiply both sides on the left by } A^{-1})$$
$$(A^{-1}A)\mathbf{x} = (A^{-1}A)\mathbf{y} \quad \text{(associativity of matrix multiplication)}$$
$$I\mathbf{x} = I\mathbf{y} \qquad \text{(property of } A^{-1})$$
$$\mathbf{x} = \mathbf{y} \qquad \text{(property of } I)$$

This establishes the uniqueness and completes the proof of (a) implies (b).

(b) implies (c). Observe that statement (b) contains the universal quantifier *for every*.

> **Whenever a statement you are assuming is true has the general form:**
>
> **For every "object" with a "certain property," "something happens,"**
>
> **you can use the technique of specialization.** (Do not confuse specialization with the choose method, which you use when you want to show that for every object with a certain property, something happens.) To use specialization, look for one particular object of interest. Once you verify that this object has the certain property, you can conclude by specialization that the something happens for that particular object. Specialization is described in greater detail in Appendix A.6.

Returning to the current proof, you are assuming that the following statement (b) is true:

For every column vector $\mathbf{b}$, there is a unique column vector $\mathbf{x}$ such that $A\mathbf{x} = \mathbf{b}$.

In particular, for $\mathbf{b} = \mathbf{0}$, you can conclude by specialization that there is a unique value of $\mathbf{x}$ such that $A\mathbf{x} = \mathbf{0}$. Evidently, that unique value is $\mathbf{x} = \mathbf{0}$. It has thus been shown that $\mathbf{x} = \mathbf{0}$ is the only solution to $A\mathbf{x} = \mathbf{0}$, as desired, and so (b) implies (c).

(c) implies (d). To prove that the reduced row-echelon form of A is I, apply Gauss-Jordan elimination to the augmented matrix $[A \mid \mathbf{0}]$ corresponding to the homogeneous system:

$$A\mathbf{x} = \mathbf{0} \tag{3.11}$$

In so doing, you obtain a finite number of elementary matrices, say, $E_1, E_2, \ldots, E_p$, with the property that the matrix

$$R = E_p \cdots E_2 E_1 [A \mid \mathbf{0}] \tag{3.12}$$

is in reduced row-echelon form. The last column of R is $\mathbf{0}$ because when you apply any row operation to (3.11), the right side remains $\mathbf{0}$.

Observe that R in (3.12) has no row of zeros, for otherwise, removing that unnecessary equation from (3.11) would leave a system with fewer equations than variables. Hence, by Theorem 2.6 in Section 2.4.6, (3.11) would have an infinite number of solutions which, according to statement (c), is not the case. It now follows that each of the n rows of R has a leading 1. In other words, $R = [I \mid \mathbf{0}]$. Dropping the zero column vector from R, you can see that the reduced row-echelon form of A is I, so (c) implies (d).

(d) implies (e). To show that A can be expressed as a product of elementary matrices, use the assumption that the reduced row-echelon form of A is I. Specifically, let E_1, E_2, $\ldots$, E_p be elementary matrices for which

$$E_p \cdots E_2 E_1 A = I \tag{3.13}$$

By Theorem 3.1, each elementary matrix in (3.13) is invertible. The desired result that A is a product of elementary matrices now follows by multiplying both sides of (3.13) on the left sequentially by the elementary matrices $E_p^{-1}, \ldots, E_1^{-1}$ to obtain

$$A = E_1^{-1} E_2^{-1} \cdots E_p^{-1} I \tag{3.14}$$

(e) implies (a). If A is a product of elementary matrices, each of which is invertible by Theorem 3.1, then, by the comment immediately following Theorem 3.3, the product of these matrices, and hence A, is invertible, as desired.

The proof is now complete because (a) implies (b), (b) implies (c), (c) implies (d), (d) implies (e), and (e) implies (a). ■

As you know, one of the uses of mathematics is to reduce the amount of effort needed to solve a problem. This use is illustrated in the proof of Theorem 3.4, where five proofs instead of ten are used to establish the theorem. As another example of this use of mathematics, observe that to show a specific matrix B is the inverse of a given matrix A, by Definition 3.2, you must show that both $AB = I$ and $BA = I$. The final theorem in this section shows that you can accomplish the same objective by verifying only one of these two statements.

THEOREM 3.5

Suppose that A and B are $(n \times n)$ matrices.

(a) If $AB = I$, then $B = A^{-1}$.

(b) If $BA = I$, then $B = A^{-1}$.

Proof.

(a) The conclusion that $B = A^{-1}$ will follow if you can show that B is invertible, for then, multiplying both sides of $AB = I$ on the right by B^{-1} results in

$$(AB)B^{-1} = IB^{-1} \quad \text{or} \quad A(BB^{-1}) = B^{-1} \quad \text{or} \quad AI = B^{-1} \quad \text{or} \quad A = B^{-1}$$

Then, from Theorem 3.3(a), B^{-1} (and hence A) is invertible and

$$B = (B^{-1})^{-1} = A^{-1}$$

To show that B is invertible, by Theorem 3.4(c) it suffices to show that the unique solution to $B\mathbf{x} = \mathbf{0}$ is $\mathbf{x} = \mathbf{0}$. To do so by the uniqueness method, suppose that $\mathbf{x}$ satisfies $B\mathbf{x} = \mathbf{0}$ and conclude that $\mathbf{x} = \mathbf{0}$. Multiplying both sides of $B\mathbf{x} = \mathbf{0}$ on the left by A and using the hypothesis that $AB = I$, you have that

$$A(B\mathbf{x}) = A\mathbf{0} \quad \text{or} \quad (AB)\mathbf{x} = \mathbf{0} \quad \text{or} \quad I\mathbf{x} = \mathbf{0} \quad \text{or} \quad \mathbf{x} = \mathbf{0}$$

This shows that B is invertible and hence that $B = A^{-1}$.

(b) The proof of this part is left to Exercise 10.

The proof is now complete.

Now that you know the basic properties of an invertible matrix, a numerical method for finding the inverse of such a matrix is presented.

3.2.4 A Numerical Method for Finding the Inverse of an Invertible Matrix

For a general $(n \times n)$ matrix A, a numerical method for finding the inverse must produce an $(n \times n)$ matrix B for which

$$BA = I \quad \text{and} \quad AB = I$$

However, by Theorem 3.5(a), it is sufficient to show that B satisfies only

$$AB = I \tag{3.15}$$

Because you do not know B, think of B as an $(n \times n)$ matrix of variables. For example, to find the inverse B of the matrix

$$A = \begin{bmatrix} 2 & -1 \\ -1 & 1 \end{bmatrix}$$

think of B as the following matrix of unknowns:

$$B = \begin{bmatrix} a & b \\ c & d \end{bmatrix}$$

From (3.15), the elements of B must satisfy the following:

$$AB = \begin{bmatrix} 2 & -1 \\ -1 & 1 \end{bmatrix} \begin{bmatrix} a & b \\ c & d \end{bmatrix} = \begin{bmatrix} 1 & 0 \\ 0 & 1 \end{bmatrix}$$

Equivalently stated, you want solve the following two systems of linear equations:

$$\begin{array}{ccc} 2a - c = 1 & & 2b - d = 0 \\ & \text{and} & \\ -a + c = 0 & & -b + d = 1 \end{array}$$

The solutions to these two systems yield $a = 1$, $b = 1$, $c = 1$, and $d = 2$, so you obtain the following inverse of A:

$$B = \begin{bmatrix} a & b \\ c & d \end{bmatrix} = \begin{bmatrix} 1 & 1 \\ 1 & 2 \end{bmatrix}$$

To find the elements of the inverse, B, of an $(n \times n)$ matrix A, you want to solve the systems of linear equations in (3.15). You might recognize that (3.15) is a special case of Problem 2.6 in Section 2.4.6. You can therefore apply Gauss-Jordan elimination to the augmented matrix whose first n columns are those of A and whose subsequent n columns are those of I, as shown in the following example.

EXAMPLE 3.2 Using Gauss-Jordan Elimination to Find the Inverse of a Matrix

To find the inverse of the (3×3) matrix

$$A = \begin{bmatrix} 0 & 1 & -2 \\ 2 & 4 & -8 \\ -1 & -2 & 5 \end{bmatrix}$$

create the following augmented matrix by appending the columns of I_3 to A:

$$\begin{bmatrix} 0 & 1 & -2 & | & 1 & 0 & 0 \\ 2 & 4 & -8 & | & 0 & 1 & 0 \\ -1 & -2 & 5 & | & 0 & 0 & 1 \end{bmatrix}$$

Applying Gauss-Jordan elimination to this augmented matrix yields

$$\begin{bmatrix} 1 & 0 & 0 & | & -2 & \frac{1}{2} & 0 \\ 0 & 1 & 0 & | & 1 & 1 & 2 \\ 0 & 0 & 1 & | & 0 & \frac{1}{2} & 1 \end{bmatrix}$$

The columns of A^{-1} are the final 3 columns of the foregoing matrix:

$$A^{-1} = \begin{bmatrix} -2 & \frac{1}{2} & 0 \\ 1 & 1 & 2 \\ 0 & \frac{1}{2} & 1 \end{bmatrix}$$

You have seen how the inverse of an invertible matrix is used to solve a system of n linear equations in n unknowns. When you cannot find the inverse in closed form, a numerical method is used. Another approach to solving such systems is presented in Section 3.3.

Exercises for Section 3.2

1. Consider the following system of linear equations:

$$2x - 3y = 4$$
$$-x + 2y = -2$$

(a) Verify that the following is the inverse of the (2×2) matrix associated with this system of linear equations:

$$\begin{bmatrix} 2 & 3 \\ 1 & 2 \end{bmatrix}$$

(b) Use the inverse matrix in part (a) to solve the system of equations.

2. Consider the following system of linear equations:

$$
\begin{aligned}
x - y \quad\;\; &= 3 \\
5y + 2z &= 1 \\
x + y + z &= 4
\end{aligned}
$$

(a) Verify that the following is the inverse of the (3×3) matrix associated with this system of linear equations:

$$
\begin{bmatrix}
3 & 1 & -2 \\
2 & 1 & -2 \\
-5 & -2 & 5
\end{bmatrix}
$$

(b) Use the inverse matrix in part (a) to solve the system of equations.

3. Suppose that a, b, c, d, e, and f are real numbers with $ad - bc \neq 0$. Use the closed-form expression for the inverse of a (2×2) matrix in Section 3.2.2 to find the solution to the following system of equations for x and y:

$$
\begin{aligned}
ax + by &= e \\
cx + dy &= f
\end{aligned}
$$

4. In developing the solution to the $(n \times n)$ system $A\mathbf{x} = \mathbf{b}$, the solution to one equation in one unknown, $ax = b$, is written as $x = a^{-1}b$. Thus, by generalization, $\mathbf{x} = A^{-1}\mathbf{b}$ is the solution to $A\mathbf{x} = \mathbf{b}$. What problems arise if you were to write the solution to $ax = b$ as $x = ba^{-1}$ and then generalize to obtain $\mathbf{x} = \mathbf{b}A^{-1}$? Explain.

5. Find the inverse of each of the following matrices or explain why you cannot do so.

(a) $\begin{bmatrix} 2 & 0 & 0 \\ 0 & -1 & 0 \\ 0 & 0 & 0 \end{bmatrix}$
(b) $\begin{bmatrix} 3 & 0 & 0 \\ 0 & 1 & 1 \end{bmatrix}$
(c) $\begin{bmatrix} 1 & 0 & 0 & 0 \\ 0 & -3 & 0 & 0 \\ 0 & 0 & 1 & 0 \\ 0 & 0 & 0 & 1 \end{bmatrix}$

6. Prove that the inverse of the elementary matrix E obtained by interchanging two rows of I is E.

7. Find the inverse of the elementary matrix obtained by adding c times row i of I to row j of I and prove that your answer is correct. (Hint: Recall from Section 3.2.2 that the inverse operation is to add $-c$ times row i of I to row j.)

8. Prove that if A is an invertible $(n \times n)$ matrix, then A^t is invertible and $(A^t)^{-1} = (A^{-1})^t$.

9. Provide a numerical example of (2×2) matrices A and B for which A and B are invertible but $A + B$ is not invertible.

10. Complete Theorem 3.5 in Section 3.2.3 by proving that if A and B are $(n \times n)$ matrices for which $BA = I$, then $B = A^{-1}$.

11. Use Gauss-Jordan elimination to find the inverse of

$$\begin{bmatrix} -1 & 1 \\ 2 & -3 \end{bmatrix}$$

12. Use Gauss-Jordan elimination to find the inverse of

$$\begin{bmatrix} 2 & 3 \\ 1 & 2 \end{bmatrix}$$

13. Use Gauss-Jordan elimination to find the inverse of

$$\begin{bmatrix} -1 & 3 & -2 \\ 0 & 1 & -3 \\ 0 & 0 & -1 \end{bmatrix}$$

14. Use Gauss-Jordan elimination to find the inverse of

$$\begin{bmatrix} -1 & 0 & 0 \\ 3 & 1 & 0 \\ -2 & -3 & -1 \end{bmatrix}$$

3.3 Using Determinants to Solve Linear Equations

In this section, a closed-form solution to the problem of solving n linear equations in n unknowns is presented. Because the resulting expression is not easy to evaluate computationally, the numerical methods you learned in previous sections are generally used to find the solution. Nevertheless, the closed-form solution presented here is useful for analyzing certain systems of equations without having to resort to a computer.

3.3.1 The Determinant of a Matrix

One of the uses of matrices is to represent a system of linear equations by the notation

$$A\mathbf{x} = \mathbf{b} \tag{3.16}$$

where A is a known $(n \times n)$ matrix and $\mathbf{b}$ is a known column vector. Associated with each $(n \times n)$ matrix A, is a "critical" number—critical in the sense that, if the number is not 0, then there is a unique column vector $\mathbf{x}$ that satisfies (3.16). This critical number, when nonzero, means that A is invertible, as you will see in Section 3.3.2. This critical number is also used in developing a closed-form solution to (3.16).

To illustrate, consider the problem of solving a single linear equation:

$$ax = b \quad \text{(where } a \text{ and } b \text{ are given real numbers)} \tag{3.17}$$

In this case, the critical number is a itself because, if $a \neq 0$, then (3.17) has a unique solution.

Proceeding to (2×2) linear equations, consider the problem of finding values for the variables x and y that satisfy

$$px + qy = t$$
$$rx + sy = u$$

(3.18)

The critical number associated with the (2×2) matrix

$$\begin{bmatrix} p & q \\ r & s \end{bmatrix}$$

is $ps - qr$. If $ps - qr \neq 0$, then (3.18) has a unique solution, as shown in Section 2.2.5.

Mathematicians have generalized the idea of a critical number to an arbitrary $(n \times n)$ matrix A. The following material explains how the entries of A are used to compute the critical number associated with A.

One of the key ideas in computing this critical number involves *permutations* of the set $\{1, \ldots, n\}$ which, as you will recall from Section 1.1.3, are orderings of all n of these numbers with no repetitions. For example, one permutation of $\{1, 2, 3, 4, 5\}$ is $(3, 1, 2, 5, 4)$.

There are $n! = n(n-1) \cdots (2)(1)$ (pronounced n factorial) permutations of $\{1, \ldots, n\}$. A systematic way to list the $n!$ permutations of $\{1, \ldots, n\}$ is to begin by listing first all permutations that begin with 1, then listing all permutations that begin with 2, and so on, that is

Permutations that Begin with 1	**Permutations that Begin with 2**	...	**Permutations that Begin with n**
$(1, ?, \ldots, ?)$	$(2, ?, \ldots, ?)$	...	$(n, ?, \ldots, ?)$
$\vdots$	$\vdots$	$\vdots$	$\vdots$

To list all permutations that begin with 1, consider the second position of the permutation. That number can be any integer from $\{1, \ldots, n\}$ other than 1, so first list all permutations of the form $(1, 2, ?, \ldots, ?)$, then list all permutations of the form $(1, 3, ?, \ldots, ?)$ and so on. Likewise, to list all permutations that begin with 2, the second position can be any integer from $\{1, \ldots, n\}$ other than 2, so first list all permutations of the form $(2, 1, ?, \ldots, ?)$, then list all permutations of the form $(2, 3, ?, \ldots, ?)$, and so on. Each time you repeat this procedure, one more number in each list is filled in.

To illustrate how this process is done on $\{1, 2, 3\}$, first list all permutations that begin with 1, then all permutations that begin with 2, and finally, all permutations that begin with 3. To find all permutations that begin with 1, first list all permutations of the form $(1, 2, ?)$—of which there is only $(1, 2, 3)$—and then list all permutations of the form $(1, 3, ?)$—of which there is only $(1, 3, 2)$. Likewise, to find all permutations that begin with 2, first list all permutations of the form $(2, 1, ?)$—of which there is only $(2, 1, 3)$—and then list all permutations of the form $(2, 3, ?)$—of which there is only $(2, 3, 1)$. Finally, to find all permutations that begin with 3, first list all permutations of the form $(3, 1, ?)$—of which there is only $(3, 1, 2)$—and then list all permutations of the form $(3, 2, ?)$—of which there is only (3, 2, 1). In total, you have the following six permutations of {1, 2, 3}:

Permutations that Begin with 1	Permutations that Begin with 2	Permutations that Begin with 3
(1, 2, 3)	(2, 1, 3)	(3, 1, 2)
(1, 3, 2)	(2, 3, 1)	(3, 2, 1)

A general permutation of $\{1, \ldots, n\}$ is written $(j_1, j_2, \ldots, j_n)$, where j_1 is the first integer of the permutation, j_2 is the second integer of the permutation, and so on. An **inversion** occurs in a permutation whenever a smaller number appears anywhere to the right of a larger number. For instance, an inversion occurs in the permutation (2, 1, 3) because the smaller number 1 is to the right of the larger number 2. The permutation (2, 3, 1) has two inversions: one because 1 is to the right of 2 and one because 1 is to the right of 3. A systematic way to determine the total number of inversions in the permutation $(j_1, j_2, \ldots, j_n)$ is to add the number of integers that follow j_1 and that are less than j_1 to the number of integers that follow j_2 and are less than j_2 and so on, up to j_{n-1}. The permutation is said to be **odd** or **even**, depending on whether the total number of inversions is odd or even, respectively. These concepts are illustrated in the following example.

EXAMPLE 3.3 Inversions in a Permutation

The following table shows the number of inversions in each permutation of {1, 2, 3}:

Permutation	Inversions for j_1	Inversions for j_2	Total Inversions	Odd/Even
(1, 2, 3)	0	0	0	even
(1, 3, 2)	0	1	1	odd
(2, 1, 3)	1	0	1	odd
(2, 3, 1)	1	1	2	even
(3, 1, 2)	2	0	2	even
(3, 2, 1)	2	1	3	odd

Elementary Products of a Matrix

An **elementary product** of an $(n \times n)$ matrix A is a product of one element from each row of A, no two of which are in the same column. For example, one elementary product of the (2×2) matrix

$$A = \begin{bmatrix} a & b \\ c & d \end{bmatrix}$$

is ad, which is a product of the element a in row 1 with the element d in row 2. Also note that a and d are in different columns. Another elementary product from this matrix is bc. In contrast, ac is not an elementary product because both a and c are in column 1.

There are $n!$ elementary products of an $(n \times n)$ matrix A, one for each permutation of

$\{1, \ldots, n\}$. A systematic method for listing all $n!$ elementary products from A is to list all permutations of $\{1, \ldots, n\}$ and, next to each one, write an associated elementary product, as follows:

Permutation	Elementary Product
$(j_1, j_2, \ldots, j_n)$	$A_{1j_1} A_{2j_2} \cdots A_{nj_n}$

Recall that a permutation is odd or even, depending on the total number of inversions. Correspondingly, a **signed elementary product** of A is an elementary product multiplied by $+1$ if the associated permutation is even and by -1 if the permutation is odd. An example of listing all signed elementary products of a (2×2) and a (3×3) matrix follows.

EXAMPLE 3.4 Listing All Signed Elementray Products of a Matrix

The signed elementary products of a (2×2) matrix A are as follows:

Permutation	Odd/Even	Signed Elementary Product
$(1, 2)$	even	$+A_{11}A_{22}$
$(2, 1)$	odd	$-A_{12}A_{21}$

The signed elementary products of a (3×3) matrix A are as follows:

Permutation	Odd/Even	Signed Elementary Product
$(1, 2, 3)$	even	$+A_{11}A_{22}A_{33}$
$(1, 3, 2)$	odd	$-A_{11}A_{23}A_{32}$
$(2, 1, 3)$	odd	$-A_{12}A_{21}A_{33}$
$(2, 3, 1)$	even	$+A_{12}A_{23}A_{31}$
$(3, 1, 2)$	even	$+A_{13}A_{21}A_{32}$
$(3, 2, 1)$	odd	$-A_{13}A_{22}A_{31}$

Closed-form Expressions for the Determinant of a Matrix

The critical number associated with an $(n \times n)$ matrix A involves all the signed elementary products of A, as presented in the following definition.

DEFINITION 3.3

The **determinant** of an $(n \times n)$ matrix A, denoted det A, det(A), or $|A|$, is the sum of all signed elementary products of A.

EXAMPLE 3.5 The Determinant of a (2 x 2) Matrix

From Example 3.4, the following is the determinant of a (2×2) matrix A:

$$\det \begin{bmatrix} A_{11} & A_{12} \\ A_{21} & A_{22} \end{bmatrix} = A_{11}A_{22} - A_{12}A_{21}$$

For instance,

$$\det \begin{bmatrix} 2 & -1 \\ -3 & 4 \end{bmatrix} = 2(4) - (-1)(-3) = 5$$

EXAMPLE 3.6 The Determinant of a (3 x 3) Matrix

From Example 3.4, the following is the determinant of a (3×3) matrix A:

$$\det \begin{bmatrix} A_{11} & A_{12} & A_{13} \\ A_{21} & A_{22} & A_{23} \\ A_{31} & A_{32} & A_{33} \end{bmatrix} = \begin{array}{l} +A_{11}A_{22}A_{33} - A_{11}A_{23}A_{32} - A_{12}A_{21}A_{33} \\ +A_{12}A_{23}A_{31} + A_{13}A_{21}A_{32} - A_{13}A_{22}A_{31} \end{array}$$

For instance,

$$\det \begin{bmatrix} 2 & -3 & 0 \\ -1 & 4 & -2 \\ 0 & 3 & 1 \end{bmatrix} = \begin{array}{l} 2(4)(1) - 2(-2)(3) - (-3)(-1)(1) \\ +(-3)(-2)(0) + 0(-1)(3) - 0(4)(0) \end{array} = 17$$

Another symbolic way to represent the determinant of A is

$$\det(A) = \sum \pm A_{1j_1} A_{2j_2} \cdots A_{nj_n} \tag{3.19}$$

where the sum is taken over all permutations $(j_1, \ldots, j_n)$ of $\{1, \ldots, n\}$ and the $\pm$ sign indicates that the appropriate sign, either $+$ or $-$, precedes each term.

It is possible to develop a closed-form expression for the determinant of certain matrices, such as (2×2) and (3×3) matrices. Also, the determinant of the zero matrix 0 is 0. This is because each term in the sum in (3.19) is zero. In fact, if any row of a matrix A consists entirely of zeros, then $\det(A) = 0$, for the same reason.

Look again at (3.19). One special type of matrix for which computing the determinant is relatively easy is when only one term in the summation is nonzero. The matrices described in the following definition have such a property.

DEFINITION 3.4

An $(n \times n)$ matrix A is **lower triangular** if and only if all elements above the diagonal are 0. An $(n \times n)$ matrix A is **upper triangular** if and only if all elements below the diagonal are 0.

EXAMPLE 3.7 Lower and Upper Triangular Matrices

The following are examples of lower and upper triangular matrices:

Lower Triangular $\qquad\qquad$ Upper Triangular

$$
\begin{bmatrix}
-3 & 0 & 0 & 0 \\
5 & 1 & 0 & 0 \\
4 & -5 & 2 & 0 \\
0 & 6 & -4 & -1
\end{bmatrix}
\qquad
\begin{bmatrix}
-2 & -1 & 0 & 4 \\
0 & 2 & -3 & 6 \\
0 & 0 & 3 & -5 \\
0 & 0 & 0 & 1
\end{bmatrix}
$$

To find the determinant of the lower triangular matrix in Example 3.7, consider any elementary product. Such a product contains an element in row 1 of the matrix. The only nonzero such element is $A_{11} = -3$. This product also contains an element in row 2, from a clomun other than column 1. The only nonzero such element is $A_{22} = 1$. The product also contains an element in row 3, from a column other than columns 1 and 2. The only nonzero such element is $A_{33} = 2$. Finally, the product contains an element in row 4, from a column other than columns 1, 2, and 3. The only such element is $A_{44} = -1$. Thus, the determinant of the lower triangular matrix in Example 3.7 is the product of its diagonal elements, namely, $(-3)(1)(2)(-1) = 6$. Applying this result to the identity matrix, you obtain that $\det(I) = 1$.

A similar argument, starting with row 4 and working toward row 1, leads to the conclusion that the determinant of the upper triangular matrix in Example 3.7 is also the product of its diagonal elements, namely, $(-2)(2)(3)(1) = -12$. These observations for lower and upper triangular matrices are summarized in the following theorem.

THEOREM 3.6

The determinant of a lower or upper triangular $(n \times n)$ matrix A is the product of the diagonal elements, that is,

$$
\det(A) = A_{11} A_{22} \cdots A_{nn}
$$

The formula in (3.19) provides a closed-form expression for the determinant of a matrix. However, as the value of n increases, the amount of computational effort involved becomes prohibitive because doing so requires listing all $n!$ signed elementary products from A. For instance, for $n = 9$, there are 362,880 such products. In the spirit of using mathematics to solve problems more efficiently, mathematicians have developed a numerical method for reducing the amount of computational effort needed to find the determinant of a matrix, as presented next.

A Numerical Method for Computing the Determinant

One approach to evaluating the determinant of a matrix A is to use Gaussian elimination (see Section 2.3.2) to reduce A to a matrix R in row-echelon form. Such a matrix R is upper triangular so, by Theorem 3.6, the determinant of R is the product of its diagonal elements. The question is, "How is the determinant of R related to the determinant of A?" Because R is obtained from A by a sequence of elementary row operations, an equivalent question is, "How does an elementary row operation affect the determinant of a matrix?" More specifically, if B is obtained by performing an elementary row operation on A, then how do you use det(B) to compute det(A)? The answer is provided in the following theorem, whose use is illustrated in the subsequent example.

THEOREM 3.7

For any $(n \times n)$ matrix A,

(a) If B is the matrix obtained by multiplying one row of A by the real number $c \neq 0$, then $\det(A) = (1/c)\det(B)$.

(b) If B is the matrix obtained by interchanging two rows of A, then $\det(A) = -\det(B)$.

(c) If B is the matrix obtained by adding a multiple of one row of A to another row of A, then $\det(A) = \det(B)$.

Proof.

These proofs follow from Definition 3.3 and are left to Exercise 14. ■

EXAMPLE 3.8 Computing the Determinant by Row Reduction

By applying the elementary row operations of Gaussian elimination to the matrix

$$A = \begin{bmatrix} 0 & 1 & -2 \\ 2 & 4 & -8 \\ -1 & -2 & 5 \end{bmatrix}$$

and using Theorem 3.7, you have that

$$\begin{vmatrix} 0 & 1 & -2 \\ 2 & 4 & -8 \\ -1 & -2 & 5 \end{vmatrix} = - \begin{vmatrix} 2 & 4 & -8 \\ 0 & 1 & -2 \\ -1 & -2 & 5 \end{vmatrix}$$

By Theorem 3.7(b), because rows 1 and 2 of A are interchanged.

$$= - \begin{vmatrix} 2 & 4 & -8 \\ 0 & 1 & -2 \\ 0 & 0 & 1 \end{vmatrix}$$

By Theorem 3.7(c), because 1/2 of row 1 of the preceding matrix is added to row 3.

$$= -2(1)(1) = -2$$

Because the preceding matrix is upper triangular.

In summary, this numerical method for finding the determinant of an $(n \times n)$ matrix A utilizes Gaussian elimination to reduce the matrix to a row-echelon form. As each step is performed, Theorem 3.7 is applied and the determinant of the new matrix is multiplied by an appropriate value, depending on the specific row operation performed.

3.3.2 Properties of the Determinant

The determinant of a matrix satisfies numerous properties, two of which follow.

THEOREM 3.8

For an $(n \times n)$ matrix A,

(a) $\det(A) = \det(A')$.

(b) For any real number c, $\det(cA) = c^n \det(A)$.

Proof.

(a) To show that $\det(A) = \det(A')$, it suffices to show that A and A' have the same set of signed elementary products. In particular, an elementary product from A is a product that contains one element from each row and column, so A and A' have the same set of elementary products. The details that the signs of these elementary products are the same is omitted.

(b) Because of the universal quantifier *for any*, choose an arbitrary real number c and show that $\det(cA) = c^n \det(A)$. To that end, consider an elementary product from A, say,

$$A_{1j_1} A_{2j_2} \cdots A_{nj_n} \tag{3.20}$$

Because each element of cA is c times the corresponding element of A, the elementary product of cA corresponding to (3.20) is

$$(cA_{1j_1})(cA_{2j_2}) \cdots (cA_{nj_n}) = c^n A_{1j_1} A_{2j_2} \cdots A_{nj_n}$$

By factoring out c^n from each signed elementary product of cA, it follows that

$$\det(cA) = c^n \sum \pm A_{1j_1} A_{2j_2} \cdots A_{nj_n} = c^n \det(A)$$

This completes the proof.

The following is an example of matrices A and B for which $\det(A + B) \neq \det(A) + \det(B)$.

EXAMPLE 3.9 Two Matrices for Which det(A + B) $\neq$ det(A) + det(B)

For

$$A = \begin{bmatrix} -2 & -1 \\ 3 & 0 \end{bmatrix}, \quad B = \begin{bmatrix} 4 & -3 \\ -2 & 1 \end{bmatrix}, \quad \text{and} \quad A + B = \begin{bmatrix} 2 & -4 \\ 1 & 1 \end{bmatrix}$$

you can see that $\det(A + B) \neq \det(A) + \det(B)$ because

$$\det(A) = 3, \quad \det(B) = -2, \quad \text{and} \quad \det(A + B) = 6$$

It is perhaps surprising that the determinant of the product of two matrices is equal to the product of their determinants, as stated in the following theorem.

THEOREM 3.9

If A and B are $(n \times n)$ matrices, then $\det(AB) = \det(A)\det(B)$.

Proof.

The proof is given only for (2×2) matrices, so let

$$A = \begin{bmatrix} a & b \\ c & d \end{bmatrix}, \quad B = \begin{bmatrix} e & f \\ g & h \end{bmatrix}, \quad \text{and so} \quad AB = \begin{bmatrix} ae + bg & af + bh \\ ce + dg & cf + dh \end{bmatrix}$$

Now $\det(A) = ad - bc$ and $\det(B) = eh - fg$, so you can compute:

$$\det(A)\det(B) = (ad - bc)(eh - fg) = adeh - bceh - adfg + bcfg$$

and

$$\begin{aligned} \det(AB) &= (ae + bg)(cf + dh) - (af + bh)(ce + dg) \\ &= acef + bcfg + adeh + bdgh - acef - adfg - bceh - bdgh \\ &= bcfg + adeh - adfg - bceh \end{aligned}$$

Comparing the expressions for $\det(A)\det(B)$ and $\det(AB)$, you see that $\det(A)\det(B) = \det(AB)$, thus completing the proof for (2×2) matrices. ■

EXAMPLE 3.10 An Example for Which det(A) det(B) = det(AB)

For

$$A = \begin{bmatrix} 1 & 6 \\ 0 & 4 \end{bmatrix}, \quad B = \begin{bmatrix} -1 & 2 \\ 1 & 0 \end{bmatrix}, \quad \text{and} \quad AB = \begin{bmatrix} 5 & 2 \\ 4 & 0 \end{bmatrix}$$

you can see that $\det(A)\det(B) = \det(AB)$ because

$$\det(A) = 4, \quad \det(B) = -2, \quad \text{and} \quad \det(AB) = -8$$

In Theorem 3.4 in Section 3.2.3, four equivalent concepts of an invertible matrix were presented. Yet another alternative is available from Theorem 3.9, as shown next.

THEOREM 3.10

An $(n \times n)$ matrix A is invertible if and only if $\det(A) \neq 0$.

Proof.

The words *if and only if* require two proofs. So assume first that A is invertible. By definition, this means that there is a matrix A^{-1} such that $AA^{-1} = I$. Taking the determinant of both sides and using Theorem 3.9, it follows that $\det(A)\det(A^{-1}) = \det(I) = 1$. It therefore must be that $\det(A) \neq 0$.

For the converse, assume that $\det(A) \neq 0$. To show that A is invertible, by Theorem 3.4(c) in Section 3.2.3, it is sufficient to show that the only solution to $A\mathbf{x} = \mathbf{0}$ is $\mathbf{x} = \mathbf{0}$. To that end, let $[R \mid \mathbf{0}]$ be the reduced row-echelon matrix obtained by applying Gauss-Jordan elimination to $[A \mid \mathbf{0}]$, so

$$[R \mid \mathbf{0}] = E_p \cdots E_1 [A \mid \mathbf{0}] \quad \text{(where } E_1, \ldots, E_p \text{ are elementary matrices)}$$

Dropping the last column from both sides, you have that

$$R = E_p \cdots E_1 A$$

Now each elementary matrix is invertible (see Theorem 3.1) and therefore, by the proof of the first part of this theorem, the determinant of each matrix is nonzero, as is $\det(A)$, by assumption. Thus, $\det(R) \neq 0$, so R contains no zero rows and hence $R = I$. It then follows that $\mathbf{x} = \mathbf{0}$ is the unique solution to the system $A\mathbf{x} = \mathbf{0}$. Thus, from Theorem 3.4(c), A is invertible and so the proof is complete. ■

The following theorem is a result of Theorem 3.10 and is called a **corollary**, meaning a result that is obtained as an immediate consequence of a previously proved theorem.

THEOREM 3.11

If A is an invertible $(n \times n)$ matrix, then $\det(A^{-1}) = 1/\det(A)$.

Proof.
From Theorem 3.10, because A is invertible, $\det(A) \neq 0$. Then, because $AA^{-1} = I$, from Theorem 3.9 it follows that

$$\det(A)\det(A^{-1}) = \det(I) = 1$$

or equivalently,

$$\det(A^{-1}) = \frac{1}{\det(A)}$$

Thus the proof is complete. ■

Now that you know the basic properties, the determinant is used to develop a closed-form solution to an $(n \times n)$ system of linear equations.

3.3.3 A Closed-form Solution to an ($n \times n$) System of Linear Equations

In Section 3.2.1, you learned that the solution to the $(n \times n)$ system

$$A\mathbf{x} = \mathbf{b} \tag{3.21}$$

when A has an inverse is

$$\mathbf{x} = A^{-1}\mathbf{b} \tag{3.22}$$

It follows that if you can find a closed-form expression for A^{-1}, then, substituting that expression in (3.22) would yield a closed-form solution to (3.21). Finding this closed-form expression is the goal of this section.

A Closed-form Expression for the Inverse of a Matrix

Mathematicians have discovered a way to express the inverse of a matrix in terms of the determinants of the $((n-1) \times (n-1))$ matrices described in the following definition.

DEFINITION 3.5

The **minor** of entry A_{ij} of an $(n \times n)$ matrix A is the determinant of the $((n-1) \times (n-1))$ matrix obtained from A by deleting row i and column j and is denoted by M_{ij}. The **cofactor** of entry A_{ij} is $C_{ij} = (-1)^{i+j} M_{ij}$.

EXAMPLE 3.11 Minors and Cofactors of a Matrix

The minor of element A_{11} of the matrix

$$A = \begin{bmatrix} 0 & 1 & -2 \\ 2 & 4 & -8 \\ -1 & -2 & 5 \end{bmatrix}$$

is the determinant of the matrix obtained by crossing off row 1 and column 1 of A, so

$$M_{11} = \begin{vmatrix} 4 & -8 \\ -2 & 5 \end{vmatrix} = 4(5) - (-2)(-8) = 4$$

The cofactor of A_{11} is

$$C_{11} = (-1)^{1+1} M_{11} = M_{11} = 4$$

The minor of A_{12} is the determinant of the matrix obtained by crossing off row 1 and column 2 of A, so

$$M_{12} = \begin{vmatrix} 2 & -8 \\ -1 & 5 \end{vmatrix} = 2(5) - (-1)(-8) = 2$$

The cofactor of A_{12} is

$$C_{12} = (-1)^{1+2} M_{12} = -M_{12} = -2$$

The cofactors provide a method for computing the determinant of the matrix. To illustrate how, recall, from Example 3.6, that the determinant of a (3×3) matrix A is

$$\det(A) = A_{11}A_{22}A_{33} - A_{11}A_{23}A_{32} - A_{12}A_{21}A_{33} +$$
$$A_{12}A_{23}A_{31} + A_{13}A_{21}A_{32} - A_{13}A_{22}A_{31}$$

Consider any row of A, say, row 1. Using the three elements A_{11}, A_{12}, and A_{13} in row 1, you can rewrite the determinant of A as follows:

$$\det(A) = A_{11}(A_{22}A_{33} - A_{23}A_{32}) + A_{12}(A_{23}A_{31} - A_{21}A_{33}) +$$
$$A_{13}(A_{21}A_{32} - A_{22}A_{31})$$

The expressions in parentheses are the cofactors C_{11}, C_{12}, and C_{13} of A, so,

$$\det(A) = A_{11}C_{11} + A_{12}C_{12} + A_{13}C_{13} \qquad (3.23)$$

The method for evaluating $\det(A)$ in (3.23) is called the **cofactor expansion** along row 1.

In fact, you can find the determinant of an $(n \times n)$ matrix A by a cofactor expansion along any one row i or column j of A, as summarized in the following closed-form expression for finding the determinant of A:

Cofactor Expansion for det(A)

Along row i: $\quad A_{i1}C_{i1} + A_{i2}C_{i2} + \cdots + A_{in}C_{in}$

Along column j: $\quad A_{1j}C_{1j} + A_{2j}C_{2j} + \cdots + A_{nj}C_{nj}$

EXAMPLE 3.12 Using Cofactors to Compute the Determinant
To create the cofactor expansion of the matrix in Example 3.11 along row 1, you need the cofactors C_{11}, C_{12}, and C_{13}. From Example 3.11, $C_{11} = 4$ and $C_{12} = -2$. You can also verify that $C_{13} = 0$, so

$$\det(A) = \begin{vmatrix} 0 & 1 & -2 \\ 2 & 4 & -8 \\ -1 & -2 & 5 \end{vmatrix} = A_{11}C_{11} + A_{12}C_{12} + A_{13}C_{13}$$
$$= 0(4) + 1(-2) + (-2)(0)$$
$$= -2$$

The cofactors are also used to construct a closed-form expression for the inverse of a matrix. The formula requires the matrix described in the following definition.

DEFINITION 3.6

For an $(n \times n)$ matrix A in which C_{ij} is the cofactor of A_{ij}, the matrix

$$\begin{bmatrix} C_{11} & C_{12} & \cdots & C_{1n} \\ C_{21} & C_{22} & \cdots & C_{2n} \\ \vdots & \vdots & \vdots & \vdots \\ C_{n1} & C_{n2} & \cdots & C_{nn} \end{bmatrix}$$

is called the **matrix of cofactors from** A. The transpose of the matrix of cofactors is called the **adjoint of** A and is denoted adj(A).

EXAMPLE 3.13 Cofactors and Adjoint of a Matrix

For the matrix A in Example 3.11, you have

$$
\begin{array}{cc}
A & \text{Cofactors} \\
\begin{bmatrix} 0 & 1 & -2 \\ 2 & 4 & -8 \\ -1 & -2 & 5 \end{bmatrix} &
\begin{array}{lll}
C_{11} = & 4, & C_{12} = -2, & C_{13} = & 0 \\
C_{21} = & -1, & C_{22} = -2, & C_{23} = -1 \\
C_{31} = & 0, & C_{32} = -4, & C_{33} = -2
\end{array}
\end{array}
$$

so

$$
\begin{array}{c}
\text{matrix of} \\
\text{cofactors of } A
\end{array}
=
\begin{bmatrix} 4 & -2 & 0 \\ -1 & -2 & -1 \\ 0 & -4 & -2 \end{bmatrix}
\quad \text{and} \quad
\text{adj}(A) =
\begin{bmatrix} 4 & -1 & 0 \\ -2 & -2 & -4 \\ 0 & -1 & -2 \end{bmatrix}
$$

The following theorem provides a closed-form expression for finding the inverse of a matrix A, provided that $\det(A) \neq 0$.

THEOREM 3.12

If A is an $(n \times n)$ invertible matrix, then

$$
A^{-1} = \frac{1}{\det(A)} \text{adj}(A)
$$

Proof.

Because A is invertible, $\det(A) \neq 0$. To obtain the conclusion, by Theorem 3.5 in Section 3.2.3, it suffices to show that

$$
A \left(\frac{1}{\det(A)} \text{adj}(A) \right) = \frac{1}{\det(A)} A \, \text{adj}(A) = I
$$

or equivalently, that

$$
A \, \text{adj}(A) = \det(A)I
$$

But

$$
A \, \text{adj}(A) =
\begin{bmatrix}
A_{11} & A_{12} & \cdots & A_{1n} \\
\vdots & \vdots & & \vdots \\
A_{i1} & A_{i2} & \cdots & A_{in} \\
\vdots & \vdots & & \vdots \\
A_{n1} & A_{n2} & \cdots & A_{nn}
\end{bmatrix}
\begin{bmatrix}
C_{11} & \cdots & C_{j1} & \cdots & C_{n1} \\
C_{12} & \cdots & C_{j2} & \cdots & C_{n2} \\
\vdots & & \vdots & & \vdots \\
C_{1n} & \cdots & C_{jn} & \cdots & C_{nn}
\end{bmatrix}
\tag{3.24}
$$

The element in row i and column j of $A \, \text{adj}(A)$ in (3.24) is

$$
A_{i1}C_{j1} + A_{i2}C_{j2} + \cdots + A_{in}C_{jn}
\tag{3.25}
$$

If $i \neq j$, then the elements A_{ik} and C_{jk} come from different rows of A. In this case, (3.25) is 0 because, for any term in $A_{ik}C_{jk}$, there is an identical term, with opposite sign, in $A_{ip}C_{jp}$, for some p. To illustrate, consider the (2×2) matrix

$$A = \begin{bmatrix} A_{11} & A_{12} \\ A_{21} & A_{22} \end{bmatrix}$$

When $i = 1$ and $j = 2$, for example, the term $-A_{11}A_{12}$ in $A_{11}C_{21}$ is cancelled by that same term, with opposite sign, in $A_{12}C_{22}$. That is,

$$A_{11}C_{21} + A_{12}C_{22} = -A_{11}A_{12} + A_{12}A_{11} = 0$$

Alternatively, if $i = j$, then (3.25) is the cofactor expansion of A along row i, so (3.25) is $\det(A)$. It therefore follows that (3.24) becomes

$$A \, \text{adj}(A) = \det(A)I$$

and so

$$A^{-1} = \frac{1}{\det(A)} \, \text{adj}(A)$$

The proof is now complete. ■

EXAMPLE 3.14 Using the Adjoint to Compute the Inverse of a Matrix

The adjoint of the matrix

$$A = \begin{bmatrix} 0 & 1 & -2 \\ 2 & 4 & -8 \\ -1 & -2 & 5 \end{bmatrix}$$

is given in Example 3.13 and, from Example 3.12, $\det(A) = -2$, so

$$A^{-1} = \frac{1}{\det(A)} \, \text{adj}(A) = \frac{1}{-2} \begin{bmatrix} 4 & -1 & 0 \\ -2 & -2 & -4 \\ 0 & -1 & -2 \end{bmatrix} = \begin{bmatrix} -2 & \frac{1}{2} & 0 \\ 1 & 1 & 2 \\ 0 & \frac{1}{2} & 1 \end{bmatrix}$$

Closed-form Solution to a System of Linear Equations: Cramer's Rule

It is now possible to derive a closed-form expression for the solution to

$$A\mathbf{x} = \mathbf{b}$$

under suitable circumstances. Specifically, when A is invertible, the solution to this system of equations is

$$\mathbf{x} = A^{-1}\mathbf{b} \tag{3.26}$$

Then, because $\det(A) \neq 0$, you can use the closed-form expression for A^{-1} from Theorem 3.12 to write (3.26) as

$$\mathbf{x} = \frac{1}{\det(A)} \text{adj}(A)\mathbf{b} \tag{3.27}$$

It is also possible to rewrite (3.27) without using $\text{adj}(A)$. To see how, multiply the matrix $\text{adj}(A)$ by the column vector $\mathbf{b}$ to obtain

$$\mathbf{x} = \frac{1}{\det(A)} \begin{bmatrix} C_{11} & C_{21} & \cdots & C_{n1} \\ C_{12} & C_{22} & \cdots & C_{n2} \\ \vdots & \vdots & \vdots & \vdots \\ C_{1n} & C_{2n} & \cdots & C_{nn} \end{bmatrix} \begin{bmatrix} b_1 \\ b_2 \\ \vdots \\ b_n \end{bmatrix}$$

$$= \frac{1}{\det(A)} \begin{bmatrix} C_{11}b_1 + C_{21}b_2 + \cdots + C_{n1}b_n \\ C_{12}b_1 + C_{22}b_2 + \cdots + C_{n2}b_n \\ \vdots \\ C_{1n}b_1 + C_{2n}b_2 + \cdots + C_{nn}b_n \end{bmatrix}$$

Thus, the value of component j of the solution vector $\mathbf{x}$ is

$$x_j = \frac{C_{1j}b_1 + C_{2j}b_2 + \cdots + C_{nj}b_n}{\det(A)} \tag{3.28}$$

The numerator of (3.28) is the cofactor expansion of the determinant of the matrix A_j obtained by replacing column j of A with the vector $\mathbf{b}$. To see that this is so, consider the matrix A_j:

Column   1 $\cdots$ $j-1$ j $j+1$ $\cdots$ n

$$A_j = \begin{bmatrix} A_{11} & \cdots & A_{1\,j-1} & b_1 & A_{1\,j+1} & \cdots & A_{1n} \\ A_{21} & \cdots & A_{2\,j-1} & b_2 & A_{2\,j+1} & \cdots & A_{2n} \\ \vdots & & \vdots & \vdots & \vdots & & \vdots \\ A_{n1} & \cdots & A_{n\,j-1} & b_n & A_{n\,j+1} & \cdots & A_{nn} \end{bmatrix}$$

The cofactor of A_j associated with each entry in column j is the same as the corresponding cofactor of A. Therefore, if you compute the determinant of A_j by cofactor expansion along column j, you have that

$$\det(A_j) = C_{1j}b_1 + C_{2j}b_2 + \cdots + C_{nj}b_n$$

In other words, the numerator in (3.28) is $\det(A_j)$ and so the value of component j of the solution vector $\mathbf{x}$ to the system of linear equations is

$$x_j = \frac{\det(A_j)}{\det(A)}$$

This closed-form solution to a system of linear equations is called **Cramer's Rule** and is summarized in the following theorem.

THEOREM 3.13

If A is an $(n \times n)$ matrix with $\det(A) \neq 0$ and $\mathbf{b}$ is a given column vector, then a closed-form solution to the system of equations

$$A\mathbf{x} = \mathbf{b}$$

is

$$x_1 = \frac{\det(A_1)}{\det(A)}, \quad x_2 = \frac{\det(A_2)}{\det(A)}, \quad \ldots, \quad x_n = \frac{\det(A_n)}{\det(A)}$$

where A_j is the matrix obtained by replacing column j of A with $\mathbf{b}$.

EXAMPLE 3.15 Using Cramer's Rule to Solve a System of Linear Equations

To solve the system of equations $A\mathbf{x} = \mathbf{b}$, in which

$$A = \begin{bmatrix} 0 & 1 & -2 \\ 2 & 4 & -8 \\ -1 & -2 & 5 \end{bmatrix} \quad \text{and} \quad \mathbf{b} = \begin{bmatrix} 2 \\ 6 \\ -2 \end{bmatrix}$$

by Cramer's Rule, compute the determinants of the matrices A, A_1, A_2, and A_3, which are

$$\det(A) = \begin{vmatrix} 0 & 1 & -2 \\ 2 & 4 & -8 \\ -1 & -2 & 5 \end{vmatrix} = -2, \qquad \det(A_1) = \begin{vmatrix} 2 & 1 & -2 \\ 6 & 4 & -8 \\ -2 & -2 & 5 \end{vmatrix} = 2$$

$$\det(A_2) = \begin{vmatrix} 0 & 2 & -2 \\ 2 & 6 & -8 \\ -1 & -2 & 5 \end{vmatrix} = -8, \qquad \det(A_3) = \begin{vmatrix} 2 & 1 & 2 \\ 6 & 4 & 6 \\ -2 & -2 & -2 \end{vmatrix} = -2$$

According to Theorem 3.13, the solution to the system of equations is

$$x_1 = \frac{\det(A_1)}{\det(A)} = -1, \quad x_2 = \frac{\det(A_2)}{\det(A)} = 4, \quad x_3 = \frac{\det(A_3)}{\det(A)} = 1$$

Solving a system by Cramer's Rule requires evaluating $n + 1$ determinants of $(n \times n)$ matrices. When $n > 3$, it is more efficient to solve the system by reducing a single $(n \times (n + 1))$ matrix using Gauss-Jordan elimination. Nevertheless, Cramer's Rule does provide a closed-form solution, whereas Gauss-Jordan elimination is a numerical method.

Exercises for Section 3.3

1. List all permutations of $\{1, 2, 3, 4\}$.

2. How many permutations of $\{1, 2, 3, 4, 5\}$ are there?

3. Find the number of inversions in each of the following permutations of $\{1, 2, 3, 4\}$. Is the given permutation odd or even?
   (a) $\{3, 1, 4, 2\}$.   (b) $\{4, 2, 1, 3\}$.

4. Find the number of inversions in each of the following permutations of $\{1, 2, \ldots, n\}$. Is the given permutation odd or even?
   (a) $\{1, 2, \ldots, n\}$.   (b) $\{n, n - 1, \ldots, 1\}$.

5. Write the signed elementary product of a (4×4) matrix A that is associated with each permutation in Exercise 3.

6. Write the signed elementary product of an $(n \times n)$ matrix A that is associated with each permutation in Exercise 4.

7. Find the determinant of each of the following matrices:

   (a) $\begin{bmatrix} 1 & -2 \\ -3 & 4 \end{bmatrix}$ (b) $\begin{bmatrix} 1 & -2 & 0 \\ -3 & 4 & 1 \\ 0 & 3 & -1 \end{bmatrix}$

8. Find the determinant of each of the following matrices:

   (a) $\begin{bmatrix} 2 & -1 \\ -4 & 2 \end{bmatrix}$ (b) $\begin{bmatrix} 1 & 0 & 0 & 0 \\ -3 & 4 & 0 & 0 \\ 5 & 6 & -2 & 0 \\ 7 & -1 & 8 & 3 \end{bmatrix}$

9. Suppose that A and B are (3×3) matrices and that $\det(A) = 5$ and $\det(B) = -2$. Use these values to compute each of the following.
   (a) $\det(A^{-1})$.   (b) $\det(AB)$.   (c) $\det(B')$.   (d) $\det(2A)$.

10. Suppose that A and B are (4×4) matrices and that $\det(A) = -3$ and $\det(B) = 4$. Use these values to compute each of the following.
    (a) $\det(A^2)$.   (b) $\det(AB^{-1})$.   (c) $\det(B^4)$.   (d) $\det(5B)$.

11. Recall that $\mathbf{i}, \mathbf{j}$, and $\mathbf{k}$ are the standard unit vectors in 3-space. Show that the determinant of the following matrix is equal to the cross product of the vectors $\mathbf{u} = (u_1, u_2, u_3)$ and $\mathbf{v} = (v_1, v_2, v_3)$ (see Section 1.2.6):

$$\begin{bmatrix} \mathbf{i} & \mathbf{j} & \mathbf{k} \\ u_1 & u_2 & u_3 \\ v_1 & v_2 & v_3 \end{bmatrix}$$

12. A *permutation matrix* is an $(n \times n)$ matrix, P, in which each row and each column has exactly one 1 and all other entries are 0. Use (3.19) in Section 3.3.1 to prove that $\det(P) = \pm 1$.

13. Use (3.19) in Section 3.3.1 to prove that if every element of an $(n \times n)$ matrix A is $-1, 0,$ or $+1$, then $|\det(A)| \leq n!$.

14. Complete Theorem 3.7 in Section 3.3.1 by proving each of the following statements for an $(n \times n)$ matrix A:
    (a) If B is the matrix obtained by multiplying one row of A by the real number $c \neq 0$, then $\det(A) = (1/c)\det(B)$.
    (b) If B is the matrix obtained by interchanging two different rows of A, then $\det(A) = -\det(B)$.
    (c) If B is the matrix obtained by adding a multiple of one row of A to another row of A, then $\det(A) = \det(B)$.

15. (a) Find the minor and cofactor of each entry of the following matrix:

$$\begin{bmatrix} a & b \\ c & d \end{bmatrix}$$

    (b) Find the matrix of cofactors and the adjoint of the matrix in part (a).

16. On the basis of your results in Exercise 15,
    (a) Find the inverse of the matrix in Exercise 15.
    (b) Use Cramer's rule to find the solution to the following system of linear equations:

$$ax + by = e$$
$$cx + dy = f$$

17. Use Gaussian elimination and Theorem 3.7 in Section 3.3.1 to find the determinant of the following matrix:

$$\begin{bmatrix} -1 & -2 & 5 \\ 0 & 1 & -2 \\ -2 & 0 & 3 \end{bmatrix}$$

18. Use Gaussian elimination and Theorem 3.7 in Section 3.3.1 to find the determinant of the following matrix:

$$\begin{bmatrix} 1 & -1 & 2 \\ 0 & 3 & -4 \\ -2 & 1 & -3 \end{bmatrix}$$

19. Find the minor and cofactor of the entries (a) A_{11}, (b) A_{12}, and (c) A_{13} of the matrix in Exercise 17.

20. Find the minor and cofactor of the entries (a) A_{11}, (b) A_{12}, and (c) A_{13} of the matrix in Exercise 18.

21. Find the minor and cofactor of the entries (a) A_{21}, (b) A_{22}, and (c) A_{23} of the matrix in Exercise 17.

22. Find the minor and cofactor of the entries (a) A_{21}, (b) A_{22}, and (c) A_{23} of the matrix in Exercise 18.

23. Find the minor and cofactor of the entries (a) A_{31}, (b) A_{32}, and (c) A_{33} of the matrix in Exercise 17.

24. Find the minor and cofactor of the entries (a) A_{31}, (b) A_{32}, and (c) A_{33} of the matrix in Exercise 18.

25. Use the results in Exercise 19 to find the determinant of the matrix in Exercise 17.

26. Use the results in Exercise 20 to find the determinant of the matrix in Exercise 18.

27. Use the results in Exercises 19, 21, and 23 to find the matrix of cofactors and the adjoint of the matrix in Exercise 17.

28. Use the results in Exercises 20, 22, and 24 to find the matrix of cofactors and the adjoint of the matrix in Exercise 18.

29. Use the results in Exercise 25 and Exercise 27 to find the inverse of the matrix in Exercise 17.

30. Use the results in Exercise 26 and Exercise 28 to find the inverse of the matrix in Exercise 18.

31. Use Cramer's rule and the results in Exercise 25 to solve the following system of linear equations:

$$-x - 2y + 5z = 6$$
$$y - 2z = -4$$
$$-2x + 3z = -3$$

32. Use Cramer's rule and the results in Exercise 26 to solve the following system of linear equations:

$$x - y + 2z = -5$$
$$3y - 4z = 6$$
$$-2x + y - 3z = 8$$

3.4 Problem Solving with Linear Equations

The material in this chapter is brought together by solving the production-planning problem of Solar Technologies posed at the beginning of this chapter.

3.4.1 The Problem of Solar Technologies

Solar Technologies produces low-grade, medium-grade, and high-grade solar cells, each made of either silicon or geranium. As Manager of Production, you have been asked to determine a daily production plan that satisfies the following requirements:

1. The total number of solar cells of all grades made of silicon should be three times the number made of geranium because the available supply of silicon is three times as much as that of geranium.

2. The total demand for low-grade and medium grade cells combined is twice that for high-grade cells.

3. The company wants to achieve a net profit of $9,000 per day, based on the following profitability figures for each type of cell:

	Low-Grade		Medium-Grade		High-Grade	
	Sil.	Ger.	Sil.	Ger.	Sil.	Ger.
Profit ($/cell)	0.25	0.50	0.50	0.75	1.00	1.50

4. For each grade you can produce either silicon or geranium cells, but not both.

3.4.2 Formulating a Mathematical Model

The data in this problem are the supply and demand information and the profitability figures. The desired output is a daily production plan. How can you use variables to represent a production plan?

Identifying the Variables

Asking yourself what you are free to control should lead you to identify the following six variables:

LS = the number of low-grade silicon cells to produce

LG = the number of low-grade geranium cells to produce

MS = the number of medium-grade silicon cells to produce

MG = the number of medium-grade geranium cells to produce

HS = the number of high-grade silicon cells to produce

HG = the number of high-grade geranium cells to produce

Identifying the Constraints

The values of these variables, once determined, constitute the daily production plan. Those values must satisfy the restrictions given in the problem statement, which must be specified in a mathematical form. From the problem description, the first constraint in verbal form is:

"Total number of silicon cells produced is 3 times that of geranium."

In terms of the variables, this constraint is described as follows:

$$LS + MS + HS = 3(LG + MG + HG)$$

By moving all variables to the left side, you have:

$$LS - 3LG + MS - 3MG + HS - 3HG = 0 \tag{3.29}$$

The second constraint in verbal form is:

"Total demand for low-grade and medium-grade cells combined is twice that for the high-grade cells."

In terms of the variables, this constraint is described as follows:

$$LS + LG + MS + MG = 2(HS + HG)$$

By moving all variables to the left side, you have:

$$LS + LG + MS + MG - 2HS - 2HG = 0 \tag{3.30}$$

The third constraint—pertaining to achieving a desired level of profit—in verbal form is the statement:

"Daily profit should be $9000."

In terms of the variables and the profitability figures given in the problem description, this constraint is described as follows:

$$0.25LS + 0.50LG + 0.50MS + 0.75MG + HS + 1.50HG = 9000 \tag{3.31}$$

The fourth group of constraints in verbal form is:

"For each grade, you can produce either silicon or geranium cells, but not both."

This group gives rise to three separate constraints, one for each of the three grades of cells. For low-grade cells, for example, you have:

"Produce either silicon or geranium low-grade cells, but not both."

You need to translate this verbal constraint to a mathematical expression in terms of the decision variables. One approach is the following:

$$LS(LG) = 0 \quad \text{(restriction for low-grade cells)} \tag{3.32}$$

Similar constraints are needed for medium-grade and high-grade cells:

$$MS(MG) = 0 \quad \text{(restriction for medium-grade cells)} \tag{3.33}$$
$$HS(HG) = 0 \quad \text{(restriction for high-grade cells)} \tag{3.34}$$

This completes the constraints stated explicitly in the problem, however, there are several *implicit* constraints. For example, the values of the variables must be nonnegative integers because you cannot produce a fraction of a solar cell. By putting together the pieces and the constraints in (3.29) through (3.34), you want to find integer values of the variables LS, LG, MS, MG, HS, and HG that satisfy the following constraints:

Complete Mathematical Model for Solar Technologies

$$
\begin{aligned}
LS - \quad 3LG + \quad MS - \quad 3MG + HS - \quad 3HG &= \quad 0 \text{ (Supply)} \\
LS + \quad LG + \quad MS + \quad MG - 2HS - \quad 2HG &= \quad 0 \text{ (Demand)} \\
0.25LS + 0.50LG + 0.50MS + 0.75MG + \quad HS + 1.50HG &= 9000 \text{ (Profit)} \\
LS , \quad LG , \quad MS , \quad MG , \quad HS , \quad HG &\geq \quad 0 \text{ (Nonneg.)} \\
LS(LG) = 0 , \quad MS(MG) = 0 , \quad HS(HG) = 0 \quad &\quad \text{(Either/or)}
\end{aligned}
$$

3.4.3 Solving the Problem

The first three constraints—corresponding to the supply, demand, and profit—give rise to a system of 3 linear equations in 6 unknowns. However, this model also involves (1) integer restrictions on the variables, (2) nonnegativity constraints, which are inequalities, and (3) either/or constraints, which are nonlinear. Creativity is needed to obtain a solution that satisfies all of these conditions.

You can ignore temporarily the integer and nonnegativity constraints and see if, by good fortune, the values you obtain for the variables are nonnegative integers. If so, you have a solution to the problem, otherwise, another approach is needed.

Turning to the either/or constraints, the only way $LS(LG) = 0$ is if $LS = 0$ or $LG = 0$. So why not assume, temporarily, that $LS = 0$ and see what happens. Should doing so not produce acceptable values for the remaining variables, then you can try setting $LG = 0$. Likewise, because $MS(MG) = 0$, either $MS = 0$ or $MG = 0$, so assume temporarily that $MS = 0$. Similarly, from $HS(HG) = 0$, assume temporarily that $HS = 0$.

By assuming that $LS = 0$, $MS = 0$, and $HS = 0$, you can eliminate these three variables from the supply, demand, and profit constraints, thus leaving you with the following system of 3 linear equations in 3 unknowns:

$$-3LG - 3MG - 3HG = 0$$
$$LG + MG - 2HG = 0$$
$$0.50LG + 0.75MG + 1.50HG = 9000$$

Equivalently, in matrix-vector notation, you have

$$\begin{bmatrix} -3 & -3 & -3 \\ 1 & 1 & -2 \\ 0.50 & 0.75 & 1.50 \end{bmatrix} \begin{bmatrix} LG \\ MG \\ HG \end{bmatrix} = \begin{bmatrix} 0 \\ 0 \\ 9000 \end{bmatrix}$$

You can solve this system using the techniques of this chapter by multiplying the vector on the right side of the equality sign by the inverse of the (3×3) matrix on the left side. Using appropriate technology to do so leads to the following values of the variables:

$$LG = -36000, \quad MG = 36000, \quad HG = 0$$

The value of $LG = -36000$ violates the nonnegativitiy constraint, so these values do not constitute an acceptable production plan.

The foregoing values of the variables LG, MG, and HG are based on the assumption that $LS = 0$, $MS = 0$, and $HS = 0$. You now know that this assumption does not produce acceptable values for the variables, so you should try another set of assumptions, such as $LS = 0$, $MS = 0$, and $HG = 0$ and see what happens. That is, set the values of these variables to 0 and solve the resulting system of three linear equations to obtain the values of the remaining variables LG, MG, and HS. Should these values be acceptable (nonnegative integers), you have a solution to the problem. If not, you must try another combination. There are eight possible combinations of three of the original six variables that you can set to 0, as listed in the eight rows of Table 3.1. This table also includes the values of the remaining variables obtained by solving the resulting system of three linear

Variables Set to 0	Values of the Remaining Variables		
LS, MS, HS	LG = −36000	MG = 36000	HG = 0
LS, MS, HG	LG = −36000	MG = 36000	HS = 0
LS, MG, HS	LG = −900	MS = 8100	HG = 3600
LS, MG, HG	LG = 3375	MS = 5625	HS = 4500
LG, MS, HS	LS = 10800	MG = −1200	HG = 4800
LG, MS, HG	LS = 6000	MG = 3600	HS = 4800
LG, MG, HS	LS = −36000	MS = 366000	HG = 0
LG, MG, HG	LS = −36000	MS = 36000	HS = 0

Table 3.1 *The Eight Possible Solutions to the Problem of Solar Technologies.*

equations in three unknowns. From the results in Table 3.1, you can see that the following two combinations in rows 4 and 6 result in nonnegative integer values for all variables:

$$LS = 0, \quad MG = 0, \quad HG = 0, \quad LG = 3375, \quad MS = 5625, \quad HS = 4500 \ \text{(row 4)}$$
$$LG = 0, \quad MS = 0, \quad HG = 0, \quad LS = 6000, \quad MG = 3600, \quad HS = 4800 \ \text{(row 6)}$$

You should thus recommend that the company produce either (a) 3375 low-grade geranium cells, 5625 medium-grade silicon cells, and 4500 high-grade silicon cells or (b) 6000 low-grade silicon cells, 3600 medium-grade geranium cells, and 4800 high-grade silicon cells.

In this chapter, you have learned how to use the inverse of an invertible matrix to solve a system of linear equations. Between this and the previous chapter, you have seen how matrices are used in solving linear equations. You might have noticed that some operations on matrices, such as addition and multiplication, satisfy many of the same properties as the corresponding operations on *n*-vectors. In Chapter 4, you will see how matrices, vectors, and their operations are brought together in a single framework through the use of several new and powerful mathematical techniques.

Exercises for Section 3.4

PROJECT 3.1: The Production-Planning Problem of Solar Technologies

Use a graphing calculator, MATLAB, or other software package, as indicated by your instructor, to perform the instructions in Exercises 1, 2, and 3 for the following mathematical model of the production-planning problem of Solar Technologies described in Section 3.4.1:

$$
\begin{aligned}
LS - \ \ 3LG + \ \ \ MS - \ \ 3MG + \ HS - \ \ 3HG &= \ \ 0 \ \text{(Supply)} \\
LS + \ \ \ LG + \ \ \ MS + \ \ \ MG - 2HS - \ \ 2HG &= \ \ 0 \ \text{(Demand)} \\
0.25LS + 0.50LG + 0.50MS + 0.75MG + \ HS + 1.50HG &= 9000 \ \text{(Profit)} \\
LS , \ \ \ \ LG , \ \ \ \ MS , \ \ \ \ MG , \ \ HS , \ \ \ \ HG &\geq \ \ 0 \ \text{(Nonneg.)} \\
LS(LG) = 0 , \ \ \ \ \ MS(MG) = 0 , \ \ \ \ HS(HG) &= 0 \ \ \ \ \ \ \text{(Either/or)}
\end{aligned}
$$

Whenever possible, use vector operations available on your system. Write the answer and the sequence of operations you performed to obtain the answer.

1. Obtain the solutions in Table 3.1 by solving the eight systems of three equations in three unknowns, in which each system has (a) $LS = 0$ or $LG = 0$, (b) $MS = 0$ or $MG = 0$, and (c) $HS = 0$ or $HG = 0$. Try to avoid entering eight different (3×3) systems of linear equations. Rather, enter the (3×6) matrix of data in the foregoing linear equations and use your technology to set three of the appropriate variables to 0 in a systematic way, if possible. Write the instructions you perform.

   When presenting your results, the Chief Executive Officer (CEO) noticed that in one of the two feasible plans, the total number of solar cells produced is 14,400 and in the other, 13,500. He said that the plant could produce up to 15,000 solar cells per day. He therefore asked you to determine a production plan that satisfies the original supply, demand, and either/or constraints, results in a total of 15,000 solar cells, and yields the largest possible profit. Do so in each of the two ways described in Exercise 2 and 3. In each case, write the instructions you perform and indicate the production plan you present to the CEO.

2. Replace the profit constraint in the original problem with a constraint to ensure that the total number of solar cells produced is 15,000. Now run the eight combinations as you did in the previous exercise. Each time you obtain a solution, compute the profit and fill in the following table:

Variables Set to 0			Values of the Other Variables						Profit
LS,	MS,	HS	LG	=	MG	=	HG	=	
LS,	MS,	HG	LG	=	MG	=	HS	=	
LS,	MG,	HS	LG	=	MS	=	HG	=	
LS,	MG,	HG	LG	=	MS	=	HS	=	
LG,	MS,	HS	LS	=	MG	=	HG	=	
LG,	MS,	HG	LS	=	MG	=	HS	=	
LG,	MG,	HS	LS	=	MS	=	HG	=	
LG,	MG,	HG	LS	=	MS	=	HS	=	

3. Add to the original (3×6) matrix a new constraint to ensure that the total number of solar cells produced is 15,000. Then replace the figure of 9000 on the right side of the profit constraint with a variable, say, P, whose value is equal to the expression on the left side of the profit equation. Now run all eight combinations as you did in the previous exercise and report the results in a similar table.

PROJECT 3.2: The Problem of Technology Investments

Obtain the solution to the following problem of Technology Investments by performing the instructions in Exercise 4 and Exercise 5. The Research Department has identified three projects in which to consider investing: Bio-Med, Tele-Com, and Laser-Tech. The number of thousands of dollars that must be invested in each of the next three years, the amount of money Technology Investments has to invest in each year, and the expected total return, in today's dollars, for investing in each of these projects is given in the following table:

	Bio-Med	Tele-Com	Laser-Tech	Available Funds
Year 1	50	30	12	75
Year 2	15	30	40	50
Year 3	15	30	15	40
Total return	200	350	350	

Any funds not invested in a given year earn 10% interest per year, in today's dollars.

4. Use the following variables to formulate a model that maximizes the total amount of money the company will have after three years so as to satisfy the budget constraints:

$$B = \begin{cases} 1, & \text{if the company invests in Bio-Med} \\ 0, & \text{if the company does not invest in Bio-Med} \end{cases}$$

$$T = \begin{cases} 1, & \text{if the company invests in Tele-Com} \\ 0, & \text{if the company does not invest in Tele-Com} \end{cases}$$

$$L = \begin{cases} 1, & \text{if the company invests in Laser-Tech} \\ 0, & \text{if the company does not invest in Laser-Tech} \end{cases}$$

$U1 = $ the number of thousands of dollars not invested in year 1

$U2 = $ the number of thousands of dollars not invested in year 2

$U3 = $ the number of thousands of dollars not invested in year 3

5. Use a graphing calculator, MATLAB, or other software package, as indicated by your instructor, to solve the problem. Do this by solving all eight possible strategies of investing or not investing in each of the three projects and fill in the following table:

B	*T*	*L*	*U1*	*U2*	*U3*	Total Return
0	0	0				
0	0	1				
0	1	0				
0	1	1				
1	0	0				
1	0	1				
1	1	0				
1	1	1				

Try to avoid entering eight different systems of linear equations. Rather, enter the matrix of data in the linear equations of the model in the previous exercise and use your technology to set three of the appropriate variables to 0 or 1 in a systematic way, if possible. Write the instructions you perform.

Chapter Summary

In this chapter, a new approach was presented for solving the following system of n linear equations in n unknowns:

$$A\mathbf{x} = \mathbf{b} \tag{3.35}$$

in which A is a known $(n \times n)$ matrix and $\mathbf{b}$ is a known column vector. The need to solve such a system arises in the study of economic models and electrical networks, for example.

In the event that A is invertible, there is an inverse matrix A^{-1} that satisfies

$$AA^{-1} = A^{-1}A = I$$

In this case, the solution to (3.35) is

$$\mathbf{x} = A^{-1}\mathbf{b}$$

Thus, another approach to solving a system of linear equations involves a preparation phase to find A^{-1} and a solution phase to compute $\mathbf{x} = A^{-1}\mathbf{b}$. For certain matrices, it is possible to find a simple closed-form expression for the inverse of A in terms of the elements of A. As an alternative, you can apply Gauss-Jordan elimination to the matrix $[A \mid I]$ to find A^{-1}.

Associated with each $(n \times n)$ matrix A is a number called the determinant of A. This number is important because, if $\det(A) \neq 0$, then the system of linear equations in (3.35) has a closed-form solution expressed by Cramer's Rule. For certain matrices, it is possible to find a simple closed-form expression for the determinant. Alternatively, you can use the numerical method of Gaussian elimination to reduce the matrix to a row-echelon form and, in so doing, find the determinant of the matrix.

Chapter 4

Vector Spaces

I n Chapter 1, you learned about n-vectors and their various operations. A similar development was given for matrices in Chapters 2 and 3. Perhaps you recognized a number of similarities between n-vectors and matrices, such as the ability to add and subtract them and the fact that these operations satisfy certain properties, such as commutativity. Because of these and other similarities, it is possible to unify n-vectors and matrices in a single framework, as illustrated in this chapter. The advantage is that unification (and generalization) allows you to study not only n-vectors and matrices, but also *many other mathematical objects that have similar properties*. The unification performed in this chapter requires several new mathematical techniques described in what follows. This chapter also provides new results useful in solving problems involving linear equations, such as the following one that was presented in Section 2.1.3

The Linear Programming Problem of Fiber Optics

Fiber Optics produces a low-density and a high-density fiber optic cable. The company has just received a one-time order for 2000 feet of the low-density cable and 4500 feet of the high-density cable for the following week. Each foot of the low-density cable costs $4 to produce and requires half a minute of machine time. Each foot of the high-density cable costs $6 to produce and requires three-fourths of a minute on the same machine. Because only 40 hours of machine time are available next week, Fiber Optics cannot meet these demands. However, the company can make up the difference by buying some amount of these cables from a Japanese company at a cost of $7 per foot for the low-density cable and $8 per foot for the high-density cable. You have been asked to find a production/purchase plan for the company that minimizes the cost of meeting the demand while not exceeding the available machine time.

4.1 Unifying *n*-Vectors and Matrices into a Vector Space

The first step in unifying *n*-vectors and matrices is to identify as many similarities and differences as possible. A special technique is then used to overcome the differences so that only common features remain. A unified framework is then developed. These steps are now illustrated.

4.1.1 Identifying Similarities and Differences Among *n*-Vectors and Matrices

In working with *n*-vectors and matrices, you may have noticed some of the following similarities and differences.

Similarities Among *n*-Vectors and Matrices

1. You can multiply both an *n*-vector and a matrix by a real number.
2. You can perform the binary operations of addition, subtraction, and multiplication on both *n*-vectors and matrices. These operations also satisfy certain properties, such as associativity.
3. There are certain vectors and matrices that are special with respect to these operations. For example, the zero vector and both the zero and identity matrices have special properties.

Differences Among *n*-Vectors and Matrices

1. There are certain operations that apply only to *n*-vectors and others that apply only to matrices. For example, the operation of computing length applies to *n*-vectors but not to matrices (although it is possible to create such a concept for matrices). Likewise, the operation of taking the transpose applies to a matrix but not to an *n*-vector (unless you think of an *n*-vector as a matrix).
2. The operation of multiplying two *n*-vectors differs in several significant ways from multiplying two matrices. For example, the result of multiplying two *n*-vectors is a real number whereas the result of multiplying two matrices of appropriate dimensions is a matrix.
3. The zero vector is the only vector that has special properties with regard to vector operations. For matrices, both the zero matrix and the identity matrix are special.
4. The specific way in which the operations are performed differs for *n*-vectors and for matrices. This is because *n*-vectors and matrices are themselves different mathematical objects.

   The objective now is to find ways to overcome the foregoing differences so that unification of *n*-vectors and matrices is possible.

Overcoming Some of the Differences

The first of the foregoing differences is that there are certain operations that apply only to *n*-vectors and others that apply only to matrices. One solution to this problem is to consider only those operations that apply to both n-vectors and matrices, namely, multiplication by a real number and the binary operations of addition, subtraction, and multiplication.

The second identified difference is that the operation of multiplying two n-vectors is significantly different from the corresponding operation on matrices. Once again, the solution is not to include multiplication. Thus, the only operations considered in the unified framework are multiplication by a real number and the binary operations of addition and subtraction.

Eliminating the operation of multiplication has also resolved the third difference. This is because, with regard to addition and subtraction, only the zero vector and the zero matrix are special. The identity matrix, which is special when multiplying matrices, need no longer be considered.

To overcome the final difference—due to the fact that n-vectors and matrices are inherently different objects—requires a new mathematical technique described next.

4.1.2 Using Abstraction to Create an Axiomatic System

The key to unifying different types of items, such as n-vectors and matrices, is to use the following mathematical technique.

Abstraction is the process of taking the focus farther and farther away from specific items by working with general "objects." Thus, you become more abstract—hence the term "abstract mathematics."

To illustrate the idea of abstraction in a non-mathematical setting, consider apples and oranges. You can unify these two items into the single comprehensive class of fruits (fruits include apples and oranges as special cases). You can then apply generalization by considering, instead of fruits, the more general class of foods (foods include fruits as a special case). With abstraction, you broaden the class even further by considering objects rather than specific items like foods or fruits. By thinking of objects, you can now include in the same group such diverse items as foods, computers, houses, and much more.

Turning to n-vectors and matrices, you can use abstraction to include these two diverse items in a single group by thinking of objects rather than n-vectors and matrices. That is, you can create a set of objects, say, V. The elements of V can all be n-vectors, matrices, or other items. Thus, the set V contains both n-vectors and matrices as special cases.

Abstraction allows you to unify n-vectors and matrices, but one of the disadvantages of doing so is that you lose the properties of the specific items that give rise to the abstraction. For example, you know how to add two n-vectors **u** and **v**, however, you cannot "add" two objects **u** and **v** from an arbitrary set V, so a syntax error results when you write

$$\mathbf{u} + \mathbf{v}$$

You will now see how to overcome this problem.

Creating an Abstract Concept of Addition

Having used abstraction to create a set V of objects that includes n-vectors and matrices as special cases, you now need a method for performing operations on those objects. For example, as with n-vectors and matrices, you would like to be able to "add" elements $\mathbf{u}$ and $\mathbf{v}$ of V. To do so, consider using a general binary operation on V, denoted by $\oplus$, that combines two elements $\mathbf{u}$ and $\mathbf{v}$ of V to create a new element of V, namely,

$$\mathbf{u} \oplus \mathbf{v} \tag{4.1}$$

The binary operation in (4.1) is called a **closed** binary operation on the set V, meaning that for any two elements $\mathbf{u}$ and $\mathbf{v}$ in V, $\mathbf{u} \oplus \mathbf{v} \in V$.

Observe that the details of how $\oplus$ in (4.1) is used to combine $\mathbf{u}$ and $\mathbf{v}$ are not specified. However, when working with a specific set V of objects, such as n-vectors, you need to specify precisely how the operation $\oplus$ is used to combine two elements.

Creating an Abstract System and an Axiomatic System

Abstraction has led to a unification of addition of n-vectors and matrices consisting of a set V together with a closed binary operation, $\oplus$, on V.

> The pair $(V, \oplus)$ is an example of an **abstract system**, meaning a set together with one or more ways to perform operations on the elements of the set.

The advantage of abstraction is that you can unify different mathematical items together in the framework of a single set. As you have just seen, a disadvantage of abstraction is that you lose properties of the specific items that give rise to the abstraction. As another example of this disadvantage, recall that $\mathbf{0}$ is an n-vector that is special with regard to adding n-vectors because,

$$\text{for all } n\text{-vectors } \mathbf{u}, \mathbf{u} + \mathbf{0} = \mathbf{0} + \mathbf{u} = \mathbf{u} \tag{4.2}$$

Likewise, the zero matrix, 0, is special with regard to adding matrices because,

$$\text{for all matrices } A, A + 0 = 0 + A = A \tag{4.3}$$

When you use abstraction to unify n-vectors and matrices in the set V of objects, you lose the existence of this special "zero" item. You can overcome this deficiency by *hypothesizing* the existence of a special element in V that has the same desirable properties as the corresponding element in the special cases. For example, you can hypothesize the existence of a special element $\mathbf{0} \in V$. For this element to have the same desirable properties as the zero vector in (4.2) and as the zero matrix in (4.3), $\mathbf{0}$ should satisfy the following property with respect to the operation $\oplus$ on V:

For all elements $\mathbf{v} \in V$, $\mathbf{v} \oplus \mathbf{0} = \mathbf{0} \oplus \mathbf{v} = \mathbf{v}$

In summary, although abstraction results in losing properties that apply to the special cases, one way to overcome this deficiency is by creating **axioms**, which are properties that are assumed to hold in the corresponding abstract system. For instance, in this example, you create the following axiom for the abstract system $(V, \oplus)$ to ensure the existence of an element of V that has the same properties as the zero vector and the zero matrix:

There is an element $\mathbf{0} \in V$ such that for all $\mathbf{v} \in V$, $\mathbf{v} \oplus \mathbf{0} = \mathbf{0} \oplus \mathbf{v} = \mathbf{v}$ (4.4)

You are free to choose the specific axioms to include with the abstract system. Which ones you choose depend on what properties of the special cases you want to study, the kind of results you eventually want to obtain about the abstract system, and more. In any event, when finished, the result is the following system.

> An **axiomatic system** is an abstract system together with a list of axioms that are assumed to hold true.

The remainder of this section is devoted to the development of additional operations and axioms to create an axiomatic system that has the same operations and properties as the special cases of *n*-vectors and matrices.

4.1.3 Identifying Additional Operations on *V*

Besides addition, it is also possible to subtract two *n*-vectors and two matrices and also to multiply an *n*-vector and a matrix by a real number. Finding a way to state these two operations in the context of the set *V* is accomplished next.

Subtracting Two Vectors

One approach to including the operation of subtracting *n*-vectors and subtracting matrices in the context of the set *V* is to create a corresponding binary operation on *V*. However, mathematicians have developed an alternative approach. The idea is to think of subtracting the *n*-vector **v** from the *n*-vector **u** as the operation of *adding* $-\mathbf{v}$ to **u**, that is,

$$\mathbf{u} - \mathbf{v} = \mathbf{u} + (-\mathbf{v}) \tag{4.5}$$

Turning to the abstract system, when working with elements $\mathbf{u}, \mathbf{v} \in V$, you can "subtract" **v** from **u** by "adding" $-\mathbf{v}$ to **u**. In terms of the operation $\oplus$ on V, subtraction in V is accomplished as follows:

$$\mathbf{u} - \mathbf{v} = \mathbf{u} \oplus (-\mathbf{v})$$

Unfortunately, the foregoing expression contains a syntax error because the element $-\mathbf{v}$ is undefined in V. To overcome this problem, look at (4.5) and, instead of thinking of $-\mathbf{v}$ as the operation of multiplying the *n*-vector **v** by -1, think of $-\mathbf{v}$ as an *n*-vector that, when added to **v**, results in the zero vector. For the abstract system, you can now hypothesize the existence of the element $-\mathbf{v}$ by creating the following axiom using the vector $\mathbf{0} \in V$ from (4.4):

> For each $\mathbf{v} \in V$, there is an element $-\mathbf{v} \in V$ such that
> $$\mathbf{v} \oplus (-\mathbf{v}) = (-\mathbf{v}) \oplus \mathbf{v} = \mathbf{0} \tag{4.6}$$

Thus far, the axiomatic system consists of the abstract system $(V, \oplus)$ together with the two axioms in (4.4) and (4.6).

Multiplying a Vector by a Real Number

The operation of multiplying an *n*-vector by a real number t is different from the binary operations of adding and subtracting *n*-vectors in that the former operation combines a real number with an *n*-vector and the latter operations combine two *n*-vectors. A new operation in V is therefore needed that corresponds to multiplying an *n*-vector (or a matrix) by a real number. To that end, consider an operation, denoted by $\odot$, that combines a real number t

with an element $\mathbf{v}$ in V to create an element $t \odot \mathbf{v}$ in V.

The result so far is the abstract system $(V, \oplus, \odot)$, which, in fact, is the unified framework for studying n-vectors and matrices. In this setting, an element of V is called a **vector** and a real number is referred to as a **scalar**. All that remains is to identify axioms—in addition to the ones in (4.4) and (4.6)—to ensure that the operations $\oplus$ and $\odot$ have the same properties as the corresponding operations on n-vectors and matrices. These additional axioms are given next.

4.1.4 The Axiomatic System of a Vector Space

To complete the axiomatic system, the following properties of addition of n-vectors and of matrices must be stated as axioms in terms of the abstract system $(V, \oplus, \odot)$:

(a) (*Associativity*) For all n-vectors $\mathbf{u}$, $\mathbf{v}$, and $\mathbf{w}$, $(\mathbf{u} + \mathbf{v}) + \mathbf{w} = \mathbf{u} + (\mathbf{v} + \mathbf{w})$.

(b) (*Commutativity*) For all n-vectors $\mathbf{u}$ and $\mathbf{v}$, $\mathbf{u} + \mathbf{v} = \mathbf{v} + \mathbf{u}$.

Likewise, the following properties of multiplying an n-vector or a matrix by a real number must be stated as axioms in the context of the abstract system:

(c) For all real numbers s and for all n-vectors $\mathbf{u}$ and $\mathbf{v}$, $s(\mathbf{u} + \mathbf{v}) = (s\mathbf{u}) + (s\mathbf{v})$.

(d) For all real numbers s and t and for all n-vectors $\mathbf{v}$, $(s + t)\mathbf{v} = (s\mathbf{v}) + (t\mathbf{v})$.

(e) For all real numbers s and t and for all n-vectors $\mathbf{v}$, $(st)\mathbf{v} = s(t\mathbf{v})$.

(f) For all n-vectors $\mathbf{v}$, $1\mathbf{v} = \mathbf{v}$.

Stating the foregoing list in terms of the abstract system $(V, \oplus, \odot)$ and including the axioms in (4.4) and (4.6), you obtain the axiomatic system in the following definition.

DEFINITION 4.1

A **vector space** over the real numbers is an abstract system $(V, \oplus, \odot)$, in which the elements of V are called *vectors* and the real numbers are called *scalars*, that satisfies all of the axioms in the following list:

Axioms for the Operations $\oplus$ and $\odot$

(1) For all $\mathbf{u}, \mathbf{v} \in V$, $\mathbf{u} \oplus \mathbf{v} \in V$, that is, $\oplus$ is a closed binary operation on V.

(2) For all scalars t and for all $\mathbf{v} \in V$, $t \odot \mathbf{v} \in V$, that is, $\odot$ combines a real number with an element in V to produce an element in V.

Axioms for Vectors

(3) There is an element $\mathbf{0} \in V$ such that for all $\mathbf{v} \in V$, $\mathbf{v} \oplus \mathbf{0} = \mathbf{0} \oplus \mathbf{v} = \mathbf{v}$.

(4) For all $\mathbf{v} \in V$, there is an element $-\mathbf{v} \in V$ such that $\mathbf{v} \oplus (-\mathbf{v}) = (-\mathbf{v}) \oplus \mathbf{v} = \mathbf{0}$.

(5) For all $\mathbf{u}, \mathbf{v}, \mathbf{w} \in V$, $(\mathbf{u} \oplus \mathbf{v}) \oplus \mathbf{w} = \mathbf{u} \oplus (\mathbf{v} \oplus \mathbf{w})$.

(6) For all $\mathbf{u}, \mathbf{v} \in V$, $\mathbf{u} \oplus \mathbf{v} = \mathbf{v} \oplus \mathbf{u}$.

Axioms for Scalars and Vectors

(7) For all scalars s and for all $\mathbf{u}, \mathbf{v} \in V$, $s \odot (\mathbf{u} \oplus \mathbf{v}) = (s \odot \mathbf{u}) \oplus (s \odot \mathbf{v})$.

(8) For all scalars s and t and for all $\mathbf{v} \in V$, $(s + t) \odot \mathbf{v} = (s \odot \mathbf{v}) \oplus (t \odot \mathbf{v})$.

(9) For all scalars s and t and for all $\mathbf{v} \in V$, $(st) \odot \mathbf{v} = s \odot (t \odot \mathbf{v})$.

(10) For all $\mathbf{v} \in V$, $1 \odot \mathbf{v} = \mathbf{v}$.

One of the advantages of Definition 4.1 is that, by working with the single unified framework of a vector space, you can study not only the special cases of n-vectors and matrices, but also *all other systems that have the properties of a vector space*. Several examples of such systems are given next as well as examples that are not vector spaces.

4.1.5 Examples and Nonexamples of Vector Spaces

In the following examples of vector spaces, (1) the set V is specified, (2) the operation $\oplus$ that combines two elements in V is described, (3) the operation $\odot$ that combines a real number and an element of V is described, and (4) the 10 axioms of Definition 4.1 are verified for the abstract system $(V, \oplus, \odot)$.

Example A: The Vector Space of n-Vectors

The set $V = R^n$ of n-vectors is a vector space when $\oplus$ is the operation of adding n-vectors and $\odot$ is the multiplication of a real number and an n-vector. For the n-vectors $\mathbf{u} = (u_1, \ldots, u_n)$ and $\mathbf{v} = (v_1, \ldots, v_n)$, the operation

$$\mathbf{u} \oplus \mathbf{v} = (u_1 + v_1, \ldots, u_n + v_n)$$

is closed on V and for the real number t, the operation

$$t \odot \mathbf{u} = (tu_1, \ldots, tu_n)$$

produces the n-vector $t \odot \mathbf{u} = t\mathbf{u}$. The remaining axioms in Definition 4.1 are established in Theorem 1.1 in Section 1.2.3, so $(V, \oplus, \odot)$ is a vector space.

Example B: The Vector Space of (m x n) Matrices

The set $V = R^{m \times n}$ consisting of all $(m \times n)$ matrices is a vector space when $\oplus$ is matrix addition and $\odot$ is the multiplication of a real number and a matrix. For $(m \times n)$ matrices A and B, the operation

$$(A \oplus B)_{ij} = A_{ij} + B_{ij}, \quad \text{for all } i = 1, \ldots, m \text{ and for all } j = 1, \ldots, n$$

is closed on V and for the real number t, the operation

$$(t \odot A)_{ij} = t A_{ij}, \quad \text{for all } i = 1, \ldots, m \text{ and for all } j = 1, \ldots, n$$

combines a real number t with an $(m \times n)$ matrix A to produce the $(m \times n)$ matrix tA. The remaining axioms in Definition 4.1 are established in Theorem 2.1 and Theorem 2.2 in Section 2.4.5, so $(V, \oplus, \odot)$ is a vector space.

Example C: The Vector Space of Real-Valued Functions

Another example of a vector space arises when V is the set of all real-valued functions of one variable, that is,

$$V = \{f : R \longrightarrow R\} \quad \text{(where } R \text{ is the set of real numbers)}$$

To create a vector space, two operations $\oplus$ and $\odot$ are needed. In this case, $\oplus$ combines two functions f and g in V to create the function $f \oplus g : R \to R$ defined as follows:

$$(f \oplus g)(x) = f(x) + g(x), \quad \text{for each } x \in R$$

The operation $\odot$ combines a real number t and a function $f \in V$ to create the function $t \odot f : R \to R$ defined as follows:

$$(t \odot f)(x) = tf(x), \quad \text{for each } x \in R$$

The remaining axioms in Definition 4.1 are now verified.

Axiom 3: The existence of a zero vector is needed—that is, a zero function, in this case. The function $z : R \to R$ defined by

$$z(x) = 0, \quad \text{for each } x \in R$$

is such a function because for any function $f \in V$, $f \oplus z = f$ since, for each $x \in R$,

$$(f \oplus z)(x) = f(x) + z(x) = f(x) + 0 = f(x)$$

Similarly, $z \oplus f = f$.

Axiom 4: Let $f \in V$. It is necessary to create a function that is the negative of f. In this case, define the function $-f$ as follows:

$$(-f)(x) = -f(x), \quad \text{for each } x \in R$$

You can see that $f \oplus (-f) = z$ because, for each $x \in R$,

$$[f \oplus (-f)](x) = f(x) + [(-f)(x)] = f(x) - f(x) = 0 = z(x)$$

Likewise, $(-f) \oplus f = z$.

Axiom 5: For $f, g, h \in V$, it follows that for each $x \in R$,

$$
\begin{aligned}
[(f \oplus g) \oplus h](x) &= (f \oplus g)(x) + h(x) &&\text{(definition of } \oplus) \\
&= [f(x) + g(x)] + h(x) &&\text{(definition of } \oplus) \\
&= f(x) + [g(x) + h(x)] &&\text{(associativity of } +) \\
&= f(x) + (g \oplus h)(x) &&\text{(definition of } \oplus) \\
&= [f \oplus (g \oplus h)](x) &&\text{(definition of } \oplus)
\end{aligned}
$$

Axiom 6: For $f, g \in V$, it follows that for each $x \in R$,

$$(f \oplus g)(x) = f(x) + g(x) = g(x) + f(x) = (g \oplus f)(x)$$

(Give reasons for each of the foregoing equalities.)

For the remaining axioms, let $s, t, x \in R$ and $f, g \in V$. Provide reasons for each step.

Axiom 7: You have that

$$[t \odot (f \oplus g)](x) = t[(f \oplus g)(x)]$$
$$= t[f(x) + g(x)]$$
$$= tf(x) + tg(x)$$
$$= (t \odot f)(x) + (t \odot g)(x)$$
$$= [(t \odot f) \oplus (t \odot g)](x)$$

Axiom 8: You have that

$$[(s + t) \odot f](x) = (s + t)f(x)$$
$$= sf(x) + tf(x)$$
$$= (s \odot f)(x) + (t \odot f)(x)$$
$$= [(s \odot f) \oplus (t \odot f)](x)$$

Axiom 9: You have that

$$[(st) \odot f](x) = (st)f(x) = s[tf(x)] = s[(t \odot f)(x)] = [s \odot (t \odot f)](x)$$

Axiom 10: You have that

$$(1 \odot f)(x) = 1f(x) = f(x)$$

Because all axioms in Definition 4.1 hold, the set V of real-valued functions of one variable is a vector space under these definitions of $\oplus$ and $\odot$.

Example D: A Vector Space Consisting of a Single Element

As you are asked to show in Exercise 5, the set V consisting of the single element denoted by $\mathbf{v}$ is a vector space under the operations $\oplus$ and $\odot$ defined as follows, for a real number t:

$$\mathbf{v} \oplus \mathbf{v} = \mathbf{v} \quad \text{and} \quad t \odot \mathbf{v} = \mathbf{v}$$

Example E: A Plane Through the Origin in 3-Space

If a, b, and c are given real numbers, then the plane through the origin in 3-space defined by the equation

$$ax + by + cz = 0$$

is a vector space. That is, the set $V = \{(x, y, z) : ax + by + cz = 0\}$ is a vector space when the operations $\oplus$ and $\odot$ are defined as the usual operations on vectors in 3-space. Specifically, for two vectors (x_1, y_1, z_1) and (x_2, y_2, z_2) in V and the real number t, define

$$(x_1, y_1, z_1) \oplus (x_2, y_2, z_2) = (x_1 + x_2, y_1 + y_2, z_1 + z_2)$$

and

$$t \odot (x_1, y_1, z_1) = (tx_1, ty_1, tz_1)$$

In Exercise 7, you are asked to verify that this abstract system $(V, \oplus, \odot)$ satisfies the axioms of a vector space.

Example F: A Digital Audio Signal

A continuous audio signal is represented mathematically by a function x whose value at time t is the amplitude of the signal at time t, as shown in Figure 4.1. A digital audio signal is obtained by sampling a continuous audio signal 44,100 times per second (the amount of time between successive samplings is one time unit). For each $k = 0, 1, 2, \ldots$, the signal sampled at time unit k is $x(k)$, hereafter denoted x_k. Thus, a digital audio signal is represented mathematically as the following infinite ordered list of real numbers:

$$\mathbf{X} = (x_0, x_1, x_2, \ldots)$$

As you are asked to show in Exercise 9, the set V of all such signals is a vector space when $\oplus$ and $\odot$ are defined for two signals $\mathbf{X} = (x_0, x_1, x_2, \ldots)$ and $\mathbf{Y} = (y_0, y_1, y_2, \ldots)$ and a real number t, as follows:

$$\mathbf{X} \oplus \mathbf{Y} = (x_0 + y_0, x_1 + y_1, x_2 + y_2, \ldots) \quad \text{and} \quad t\mathbf{X} = (tx_0, tx_1, tx_2, \ldots)$$

Your understanding of the properties of a vector space is further enhanced by considering the following examples of sets V with operations $\oplus$ and $\odot$ for which $(V, \oplus, \odot)$ is not a vector space. In each case, one or more of the axioms in Definition 4.1 fails to hold.

Example G: The Non-Vector Space of Nonnegative *n*-Vectors

The set V of n-vectors $\mathbf{x} = (x_1, \ldots, x_n)$ such that each component x_i of $\mathbf{x}$ is nonnegative is not a vector space under the usual operations of addition of n-vectors and multiplication of an n-vector by a real number. The reason is that Axiom 4 fails to hold. That is, for an n-vector $\mathbf{x} \in V$, there is no n-vector $-\mathbf{x} \in V$. For example, the negative of the vector $(2, 3)$ is $(-2, -3)$, but $(-2, -3)$ is not in V because the components of $(-2, -3)$ are negative.

Example H: The Non-Vector Space of a Line Not Through the Origin in 2-Space

Given a slope m and an intercept $b \neq 0$, the line $y = mx + b$ shown in Figure 4.2 is not a vector space. Specifically, the set

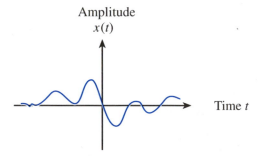

Figure 4.1 *A Continuous Audio Signal.*

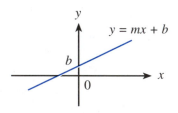

Figure 4.2 *A Line Not Going Through the Origin in 2-Space.*

$$V = \{(x, y) : y = mx + b\}$$

is not a vector space under the usual definition of adding vectors in R^2 and multiplying a vector in R^2 by a real number. The reason is that Axiom 3 fails to hold. That is, V has no zero vector because the foregoing line does not contain the origin in R^2, as seen in Figure 4.2.

In this section, you have seen how abstraction is used to unify n-vectors and matrices in a single axiomatic system of a vector space. One of the advantages of studying a vector space is that any properties satisfied by a vector space apply to all of the special cases. That is, by establishing that a vector space satsifies a particular property, you simultaneously know that all special cases of a vector space satisfy that same property. Some basic properties of a vector space are derived in Section 4.2.

Exercises for Section 4.1

1. Explain the technique of abstraction. What is the reason for using this technique?

2. What is the difference between an abstract system and an axiomatic system?

3. Explain how to apply abstraction to the components of an n-vector. Identify three different special cases of the elements of such a vector.

4. Suppose that x and y are real numbers.
   (a) Explain what syntax error arises in applying abstraction to the expression $x \cdot y$.
   (b) Explain how a binary operator overcomes the syntax error in part (a).

5. Show that if V consists of a single vector $\mathbf{v}$, then V is a vector space under the operations $\oplus$ and $\odot$ defined as follows for a real number t:

$$\mathbf{v} \oplus \mathbf{v} = \mathbf{v} \quad \text{and} \quad t \odot \mathbf{v} = \mathbf{v}$$

6. For a given integer $n \geq 0$, consider the set, F^n, of all functions p of the form

$$p(x) = a_0 + a_1 x + a_2 x^2 + \cdots + a_n x^n$$

   for some real numbers $a_0, a_1, \ldots, a_n$. Show that F^n is a vector space when the operations of $\oplus$ and $\odot$ are defined as follows for a real number t and the functions

$$p(x) = a_0 + a_1 x + a_2 x^2 + \cdots + a_n x^n$$
$$q(x) = b_0 + b_1 x + b_2 x^2 + \cdots + b_n x^n$$

$$(p \oplus q)(x) = (a_0 + b_0) + (a_1 + b_1)x + (a_2 + b_2)x^2 + \cdots + (a_n + b_n)x^n$$
$$(t \odot p)(x) = ta_0 + ta_1 x + ta_2 x^2 + \cdots + ta_n x^n$$

7. Show that the plane $ax + by + cz = 0$ (where a, b, and c are given real numbers) is a vector space. That is, show that the set $V = \{(x, y, z) : ax + by + cz = 0\}$ is a vector space when the operations $\oplus$ and $\odot$ are defined as the usual operations on vectors in 3-space. Specifically, for two vectors (x_1, y_1, z_1) and (x_2, y_2, z_2) in V and the real number t,

$$(x_1, y_1, z_1) \oplus (x_2, y_2, z_2) = (x_1 + x_2, y_1 + y_2, z_1 + z_2)$$

and

$$t \odot (x_1, y_1, z_1) = (tx_1, ty_1, tz_1)$$

8. Show that the line through the origin in 3-space in the given direction (a, b, c) is a vector space. That is, show that $\{(ta, tb, tc) : t$ is a real number$\}$ is a vector space using the usual operations on vectors in R^3.

9. Show that the set V of all digital audio signals $\mathbf{X} = (x_0, x_1, x_2, \ldots)$ is a vector space when the operations $\oplus$ and $\odot$ are defined as follows for two signals $\mathbf{X} = (x_0, x_1, x_2, \ldots)$ and $\mathbf{Y} = (y_0, y_1, y_2, \ldots)$ and a real number t:

$$\mathbf{X} \oplus \mathbf{Y} = (x_0 + y_0, x_1 + y_1, x_2 + y_2, \ldots) \quad \text{and} \quad t\mathbf{X} = (tx_0, tx_1, tx_2, \ldots)$$

10. Show that the set of complex numbers is a vector space when the following operations are defined for the complex numbers $a + bi$ and $c + di$, and for a real number t:

$$(a + bi) \oplus (c + di) = (a + c) + (b + d)i \quad \text{and} \quad t \odot (a + bi) = ta + tbi$$

11. Explain why the set V of all invertible (2×2) matrices is not a vector space under the usual operations of matrix addition and multiplying a matrix by a real number.

12. Suppose that A is an $(n \times n)$ matrix and $\mathbf{b} \in R^{n \times 1}$ is a column vector with $\mathbf{b} \neq \mathbf{0}$. Explain why the set $V = \{$column vectors $\mathbf{x} \in R^{n \times 1} : A\mathbf{x} = \mathbf{b}\}$ is not a vector space.

13. Create an abstract system consisting of a set and an operation on the elements of the set that serves as a unification of the following mathematical items: the absolute value of a number and the norm of an n-vector. Is the operation closed on the set?

14. Create an abstract system consisting of a set and an operation on the elements of the set that serves as a unification of the following mathematical items: computing the reciprocal of a nonzero real number and finding the inverse of an invertible $(n \times n)$ matrix. Is the operation closed on the set?

15. Consider an abstract system consisting of a set S and a closed binary operation $\odot$ on S that is designed to unify multiplication of real numbers and multiplication of $(n \times n)$ matrices in a single framework.
    (a) What special property does the number 1 satisfy with regard to multiplying real numbers?
    (b) What special property does the $(n \times n)$ identity matrix I satisfy with regard to multiplying $(n \times n)$ matrices?
    (c) On the basis of the similarities in parts (a) and (b), write an axiom for the abstract system $(S, \odot)$ to include an element in S that has the same desirable properties as the number 1 and the identity matrix I.

16. Recall the operation of the dot product of two n-vectors.
    (a) What n-vector is special with regard to this operation? Explain.
    (b) Create an appropriate abstract system for studying the dot product.
    (c) Create an axiom so that your abstract system in part (b) contains an element having the same properties as the one identified in part (a).

4.2 Basic Properties and Subspaces of Vector Spaces

In this section, some basic properties satisfied by a vector space are developed. These properties apply to each special case of a vector space.

4.2.1 Basic Properties of Vector Spaces

Not every property of a vector space is listed in the axioms in Definition 4.1 in Section 4.1.4. This is because it is possible to derive some properties from the axioms in Definition 4.1, as illustrated in the following theorem.

THEOREM 4.1

Suppose that $(V, \oplus, \odot)$ is a vector space in which $\mathbf{0}$ is the zero vector. If $\mathbf{u} \in V$ and t is a real number, then

(a) The vector $\mathbf{0}$ is the only zero vector.

(b) The vector $-\mathbf{u}$ is the only negative of the vector $\mathbf{u}$.

(c) $0 \odot \mathbf{u} = \mathbf{0}$ (the scalar zero times the vector $\mathbf{u}$ is the vector $\mathbf{0}$).

(d) $t \odot \mathbf{0} = \mathbf{0}$ (the scalar t times the vector $\mathbf{0}$ is the vector $\mathbf{0}$).

(e) $(-1) \odot \mathbf{u} = -\mathbf{u}$ (the scalar -1 times $\mathbf{u}$ is the negative of $\mathbf{u}$).

(f) If $t \odot \mathbf{u} = \mathbf{0}$, then either $t = 0$ or $\mathbf{u} = \mathbf{0}$.

Proof.

In each part, you will need to prove that two vectors are equal. This is accomplished in various ways using the axioms in Definition 4.1.

(a) To show that $\mathbf{0}$ is the only zero vector, according to the uniqueness method, suppose that $\mathbf{z}$ is also a vector such that

$$\text{for all } \mathbf{v} \in V, \mathbf{z} \oplus \mathbf{v} = \mathbf{v} \oplus \mathbf{z} = \mathbf{v} \tag{4.7}$$

You must show that $\mathbf{z} = \mathbf{0}$. Because (4.7) holds *for all* $\mathbf{v} \in V$, in particular, by specialization, (4.7) holds for $\mathbf{v} = \mathbf{0}$. By replacing $\mathbf{v}$ in (4.7) with $\mathbf{0}$, you obtain

$$\mathbf{z} \oplus \mathbf{0} = \mathbf{0} \oplus \mathbf{z} = \mathbf{0} \tag{4.8}$$

From Axiom 3, you know that $\mathbf{z} \oplus \mathbf{0} = \mathbf{z}$ so, from (4.8), it follows that $\mathbf{z} = \mathbf{0}$. This means that $\mathbf{0}$ is the only zero vector.

(b) To show that $-\mathbf{u}$ is the only negative of the vector $\mathbf{u}$, according to the uniqueness method, suppose that $\mathbf{v}$ also satisfies $\mathbf{u} \oplus \mathbf{v} = \mathbf{v} \oplus \mathbf{u} = \mathbf{0}$. You must show that $\mathbf{v} = -\mathbf{u}$. But now,

$$\mathbf{u} \oplus \mathbf{v} = \mathbf{0}$$

$$(-\mathbf{u}) \oplus (\mathbf{u} \oplus \mathbf{v}) = (-\mathbf{u}) \oplus \mathbf{0} \quad \text{(add } -\mathbf{u} \text{ to the left of both sides)}$$

$$[(-\mathbf{u}) \oplus \mathbf{u}] \oplus \mathbf{v} = (-\mathbf{u}) \oplus \mathbf{0} \quad \text{(by Axiom 5)}$$

$$0 \oplus v = (-u) \oplus 0 \quad \text{(by Axiom 4)}$$
$$v = -u \qquad \text{(by Axiom 3)}$$

(c) To show that $0 \odot u$ is the zero vector, consider $-(0 \odot u)$, the negative of $0 \odot u$, which exists by Axiom 4. Then,

$$
\begin{aligned}
0 \odot u &= (0 \odot u) \oplus 0 && \text{(Axiom 3)} \\
&= (0 \odot u) \oplus \{(0 \odot u) \oplus [-(0 \odot u)]\} && \text{(Axiom 4)} \\
&= [(0 \odot u) \oplus (0 \odot u)] \oplus [-(0 \odot u)] && \text{(Axiom 5)} \\
&= [(0 + 0) \odot u] \oplus [-(0 \odot u)] && \text{(Axiom 8)} \\
&= (0 \odot u) \oplus [-(0 \odot u)] && (0 + 0 = 0) \\
&= 0 && \text{(Axiom 4)}
\end{aligned}
$$

(d) The result follows by noting that

$$
\begin{aligned}
t \odot 0 &= t \odot (0 \odot 0) && \text{[part (c)]} \\
&= [t(0)] \odot 0 && \text{(Axiom 9)} \\
&= 0 \odot 0 && (t(0) = 0) \\
&= 0 && \text{[part (c)]}
\end{aligned}
$$

(e) To show that $(-1) \odot u = -u$, from part (b), it suffices to show that $u \oplus [(-1) \odot u] = 0$. But

$$
\begin{aligned}
u \oplus [(-1) \odot u] &= (1 \odot u) \oplus [(-1) \odot u] && \text{(Axiom 10)} \\
&= [1 + (-1)] \odot u && \text{(Axiom 8)} \\
&= 0 \odot u && (1 + (-1) = 0) \\
&= 0 && \text{[part (c)]}
\end{aligned}
$$

(f) To prove this part, observe that the key words *either/or* appear in the conclusion in the form

$$\text{either } t = 0 \text{ or } u = 0$$

It is necessary to show that at least one of these two statements is true. Because you do not know which one to show is true, suppose you were to make the additional assumption that $t \neq 0$. In this case, it must be that $u = 0$.

In other words, to use the **either/or method** to prove that

p implies (q or r)

is true (where p, q, and r are statements), assume that p is true and q is not true. You must use this information to reach the conclusion that r is true (see Appendix A.11).

For the current problem, with the either/or method, assume that

$$t \odot \mathbf{u} = \mathbf{0} \quad \text{and} \quad t \neq 0 \tag{4.9}$$

You must show that $\mathbf{u} = \mathbf{0}$. However, because $t \neq 0$, you can multiply the first equality in (4.9) through by the real number t^{-1} to obtain

$$t^{-1} \odot (t \odot \mathbf{u}) = t^{-1} \odot \mathbf{0} \quad \text{(multiply (4.9) through by } t^{-1}\text{)}$$
$$(t^{-1}t) \odot \mathbf{u} = \mathbf{0} \qquad \text{[Axiom 9 and part (d)]}$$
$$1 \odot \mathbf{u} = \mathbf{0} \qquad (t^{-1}t = 1)$$
$$\mathbf{u} = \mathbf{0} \qquad \text{(Axiom 10)}$$

This completes the proof.

One advantage of studying the unifying system of a vector space (and axiomatic systems in general) is that any result obtained applies to each special case with nothing more than an appropriate substitution of symbols. For example, Theorem 4.1(a) applies to the vector space of real-valued functions of one variable described in Example C of Section 4.1.5. Specifically, the zero function, z, is the only function with the property that, when added to any function f results in the function f. The use of Theorem 4.1 is illustrated now together with another important example of a vector space.

4.2.2 Subspaces

Example E in Section 4.1.5, a plane through the origin in 3-space, is a vector space that is contained in the larger vector space R^3. Equivalently stated, starting with the vector space $(V, \oplus, \odot)$ in which $V = R^3$, a subset of V—namely, a plane through the origin—produces a vector space using the same operations $\oplus$ and $\odot$ as in V. Such subsets of a vector space are often useful in problem solving and are formalized in the following definition.

DEFINITION 4.2

Given a vector space $(V, \oplus, \odot)$, a subset U of V is a **subspace** of V if and only if $(U, \oplus, \odot)$ is itself a vector space, meaning that U together with the operations $\oplus$ and $\odot$ from V satisfy the axioms of a vector space.

To verify that a subset U of a vector space V is a subspace of V, you must verify all 10 axioms in Definition 4.1. The following theorem reduces the work substantially by showing that you need verify only the first two axioms to prove that U is a subspace of V.

THEOREM 4.2

Suppose that $(V, \oplus, \odot)$ is a vector space and that U is a nonempty subset of V. Then $(U, \oplus, \odot)$ is a subspace of V if and only if

(a) For all $\mathbf{u}, \mathbf{v} \in U$, $\mathbf{u} \oplus \mathbf{v} \in U$.

(b) For all real numbers t and for all $\mathbf{u} \in U$, $t \odot \mathbf{u} \in U$.

Proof.

Suppose first that U is a subspace of V. By Definition 4.2, this means that $(U, \oplus, \odot)$ is a vector space and thus satisfies the first two axioms of Definition 4.1, which are precisely conditions (a) and (b).

For the converse, suppose that conditions (a) and (b) hold. It remains to verify that axioms (3) through (10) hold for $(U, \oplus, \odot)$.

For Axiom 3, it is necessary to construct a zero vector in U. That zero vector is the zero vector of V, say, $\mathbf{0}$. You can see that $\mathbf{0} \in U$ by applying specialization to (b) with $t = 0$ and using Theorem 4.1(c).

For Axiom 4, let $\mathbf{u} \in U$ for which it is necessary to construct a vector $-\mathbf{u} \in U$ such that $\mathbf{u} \oplus (-\mathbf{u}) = (-\mathbf{u}) \oplus \mathbf{u} = \mathbf{0}$. This vector $-\mathbf{u}$ is the same as the negative of $\mathbf{u}$ that exists in V because V is a vector space and $\mathbf{u} \in V$. You must be sure, however, that $-\mathbf{u} \in U$. This fact follows by specializing condition (b) to $t = -1$ and using Theorem 4.1(e).

All of the remaining axioms hold for vectors in U because those properties hold for all vectors in V, and U is a subset of V. ■

The advantage of Theorem 4.2 is that, given a vector space $(V, \oplus, \odot)$ and a nonempty subset U of V, you can prove that U is a subspace of V by verifying that $\oplus$ is closed on U and that $\odot$ combines a real number with an element of U to produce an element of U, as shown in the subsequent examples of subspaces.

Example A: The Subspace of a Plane Through the Origin

In Example E in Section 4.1.5, a plane through the origin in 3-space is presented as an example of a vector space, which requires verifying that all 10 axioms of Definition 4.1 hold for

$$U = \{(x, y, z) : ax + by + cz = 0\}$$

under the operations of addition of vectors in 3-space and multiplication of a vector in 3-space by a real number.

Alternatively, and more simply, you can verify that U is a subspace of the vector space $(V, \oplus, \odot)$ in which $V = R^3$ by checking that conditions (a) and (b) in Theorem 4.2 hold. As seen in Figure 4.3, when you add two vectors in a plane through the origin, you get a vector in the same plane. Likewise, when you multiply a vector in the plane by a real number, you again get a vector in that same plane.

You can translate the visual images in Figure 4.3 to symbolic form as follows. For $\mathbf{u} = (u_1, u_2, u_3)$ and $\mathbf{v} = (v_1, v_2, v_3)$ in U, the vector $\mathbf{u} \oplus \mathbf{v} = (u_1 + v_1, u_2 + v_2, u_3 + v_3)$ is in U because

$$a(u_1 + v_1) + b(u_2 + v_2) + c(u_3 + v_3) =$$
$$(au_1 + bu_2 + cu_3) + (av_1 + bv_2 + cv_3) = 0 + 0 = 0$$

Likewise, for a real number t, $t\mathbf{u} = t(u_1, u_2, u_3) = (tu_1, tu_2, tu_3)$ is in U because

$$a(tu_1) + b(tu_2) + c(tu_3) = t(au_1 + bu_2 + cu_3) = t(0) = 0$$

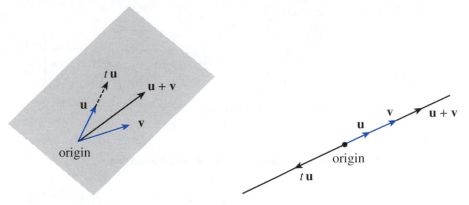

Figure 4.3 *Operations on Vectors in a Plane.*

Figure 4.4 *Operations on Vectors in a Line.*

Example B: A Line Through the Origin in 3-Space

Another example of a subspace of R^3 is a line through the origin. From Figure 4.4, you can see that adding two vectors that lie on a line through the origin results in a vector on that same line through the origin. Likewise, multiplying a vector on that line by a real number t also results in a vector on that same line. You are asked in Exercise 3 to translate these visual images to symbolic form, thus verifying that conditions (a) and (b) of Theorem 4.2 hold.

Example C: The Zero Subspace and the Vector Space Itself

A vector space $(V, \oplus, \odot)$ always has at least two subspaces. One is the **zero subspace** that consists of one single vector, namely, the zero vector $\mathbf{0}$ from V. That is,

$$U = \{\mathbf{0}\}$$

is a subspace of V. Also, the set V itself is a subspace of V. You are asked in Exercise 4 to verify that conditions (a) and (b) of Theorem 4.2 hold for both $U = \{\mathbf{0}\}$ and $U = V$.

Example D: The Subspace of Upper Triangular (3 x 3) Matrices

From Example B in Section 4.1.5, you know that the set V of all (3×3) matrices is a vector space under the operations of matrix addition and multiplying a matrix by a real number. The set

$$U = \{A \in V : A \text{ is upper triangular}\}$$

is a subspace of V. To see that this is so by Theorem 4.2, let A and B be two matrices in U, so A and B have the following form:

$$A = \begin{bmatrix} A_{11} & A_{12} & A_{13} \\ 0 & A_{22} & A_{23} \\ 0 & 0 & A_{33} \end{bmatrix} \quad \text{and} \quad B = \begin{bmatrix} B_{11} & B_{12} & B_{13} \\ 0 & B_{22} & B_{23} \\ 0 & 0 & B_{33} \end{bmatrix}$$

Then, for any real number t,

$$A + B = \begin{bmatrix} A_{11} + B_{11} & A_{12} + B_{12} & A_{13} + B_{13} \\ 0 & A_{22} + B_{22} & A_{23} + B_{23} \\ 0 & 0 & A_{33} + B_{33} \end{bmatrix} \text{ and } tA = \begin{bmatrix} tA_{11} & tA_{12} & tA_{13} \\ 0 & tA_{22} & tA_{23} \\ 0 & 0 & tA_{33} \end{bmatrix}$$

You can see that $A + B$ and tA are upper triangular, so $A + B \in U$ and $tA \in U$. Thus, by Theorem 4.2, U is a subspace of V.

Example E: The Subspace of Polynomial Functions

A polynomial is a function of the form

$$p(x) = a_0 + a_1 x + a_2 x^2 + \cdots + a_n x^n$$

for some real numbers $a_0, \ldots, a_n$. The *degree* of p is the highest power of x whose coefficient is not 0. If $p(x) = a_0 \neq 0$, then the degree of p is 0. The set P^n of all polynomials of degree $\leq n$, together with the zero polynomial, each of whose coefficients is 0, is a subspace of the vector space V of all real-valued functions (see Example C in Section 4.1.5).

To see that this is so by Theorem 4.2, let $p(x)$ and $q(x)$ be two polynomials in P^n, say,

$$p(x) = a_0 + a_1 x + a_2 x^2 + \cdots + a_n x^n$$
$$q(x) = b_0 + b_1 x + b_2 x^2 + \cdots + b_n x^n$$

Then $p + q \in P^n$ because

$$p(x) + q(x) = (a_0 + a_1 x + \cdots + a_n x^n) + (b_0 + b_1 x + \cdots + b_n x^n)$$
$$= (a_0 + b_0) + (a_1 + b_1)x + \cdots + (a_n + b_n)x^n$$

So $p(x) + q(x)$ is either the zero polynomial or else a polynomial of degree $\leq n$ and hence is in P^n.

Likewise, for a real number t, $tp(x) \in P^n$ because

$$tp(x) = t(a_0 + a_1 x + a_2 x^2 + \cdots + a_n x^n)$$
$$= (ta_0) + (ta_1)x + (ta_2)x^2 + \cdots + (ta_n)x^n$$

So, $tp(x)$ is either the zero polynomial or else a polynomial of degree $\leq n$ and hence is in P^n.

Example F: The Subspace of Solutions to a Homogeneous System of Linear Equations

For an $(m \times n)$ matrix A, recall that the problem of solving the homogeneous system of linear equations requires finding a column vector $\mathbf{x} \in R^{n \times 1}$ that satisfies

$$A\mathbf{x} = \mathbf{0}$$

The set of solutions to this system is a subspace of $R^{n \times 1}$. That is,

$$U = \{\mathbf{x} \in R^{n \times 1} : A\mathbf{x} = \mathbf{0}\}$$

is a subspace of $R^{n \times 1}$. To see that this is so by Theorem 4.2, let $\mathbf{x}, \mathbf{y} \in U$, so $A\mathbf{x} = \mathbf{0}$ and $A\mathbf{y} = \mathbf{0}$. Then $\mathbf{x} + \mathbf{y} \in U$ because

$$A(\mathbf{x} + \mathbf{y}) = A\mathbf{x} + A\mathbf{y} = \mathbf{0} + \mathbf{0} = \mathbf{0}$$

Likewise, for a real number t, $t\mathbf{x} \in U$ because

$$A(t\mathbf{x}) = t(A\mathbf{x}) = t\mathbf{0} = \mathbf{0}$$

Thus, by Theorem 4.2, U is a subspace of $R^{n \times 1}$.

In this section, you have seen a number of examples of subspaces of vector spaces. In the next section, results that are useful in solving problems involving vector spaces are derived.

Exercises for Section 4.2

1. Apply each part of Theorem 4.1 in Section 4.2.1 to the special case of $(m \times n)$ matrices. That is, write each part of the theorem specifically in the context of $(m \times n)$ matrices using appropriate notation.

2. Apply each part of Theorem 4.1 in Section 4.2.1 to the special case of all real-valued functions of a single variable. That is, write each part of the theorem specifically in the context of real-valued functions of a single variable using appropriate notation.

3. Suppose that a, b, and c are given real numbers. Use Theorem 4.2 in Section 4.2.2 to show that the line through the origin in R^3 described by $\{t(a, b, c) : t \text{ is a real number}\}$ is a subspace of R^3.

4. Suppose that $(V, \oplus, \odot)$ is a vector space whose zero vector is $\mathbf{0}$. Use Theorem 4.2 in Section 4.2.2 to show that each of the following subsets U is a subspace of V.
   (a) $U = \{\mathbf{0}\}$.   (b) $U = V$.

5. Use Theorem 4.2 in Section 4.2.2 to show that the set of all differentiable real-valued functions is a subspace of the vector space of all real-valued functions.

6. Prove that if U and W are subspaces of a vector space $(V, \oplus, \odot)$, then $U \cap W$ is a subspace of V.

7. Draw the set $U = \{(-1, 0) + t(2, 1) : t \text{ is a nonnegative real number}\}$. Then show that U is not a subspace of R^2 by providing specific numerical examples of vectors in U and scalars to illustrate that neither of the two properties in Theorem 4.2 in Section 4.2.2 holds.

8. Let $\mathbf{v}_0$ and $\mathbf{v}$ be given nonzero vectors in R^n that are not parallel. Show that $U = \{\mathbf{v}_0 + t\mathbf{v} : t \text{ is a real number}\}$ is not a subspace of R^n by determining which of the two properties in Theorem 4.2 in Section 4.2.2 fails to hold.

9. Draw the set $U = \{(x, y) : x^2 + y^2 \leq 4\}$, which is the disk of radius 2 in R^2. Then show that U is not a subspace of R^2 by providing specific numerical examples of vectors in U and scalars to illustrate that neither of the two properties in Theorem 4.2 in Section 4.2.2 holds.

10. Show that the set $\{(x_1, \ldots, x_n) : x_1^2 + \cdots + x_n^2 \leq r^2\}$, which is the closed ball of radius r in R^n, is not a subspace of R^n.

11. Use the axioms of a vector space in Definition 4.1 in Section 4.1.1 and the properties in Theorem 4.1 in Section 4.2.1 to prove that for all scalars t and for all vectors $\mathbf{v}$ in a vector space $(V, \oplus, \odot)$, $(-t) \odot \mathbf{v} = t \odot (-\mathbf{v}) = -(t \odot \mathbf{v})$.

12. Use the axioms of a vector space in Definition 4.1 in Section 4.1.1 and the properties in Theorem 4.1 in Section 4.2.1 to prove that for all vectors $\mathbf{v}$ and $\mathbf{w}$ in a vector space $(V, \oplus, \odot)$, there is a unique vector $\mathbf{x} \in V$ such that $\mathbf{x} \oplus \mathbf{v} = \mathbf{w}$.

13. Which of the following sets are subspaces of the vector space V of all real-valued functions? Explain.
    (a) $\{f \in V : f(0) = 0\}$.
    (b) $\{f \in V : f(0) = 3\}$.
    (c) $\{f \in V : f(1) = 0\}$.
    (d) $\{f \in V : f(x) = \text{a constant}\}$.

14. Prove or disprove that each of the following is a subspace of the vector space of all $(n \times n)$ matrices.
    (a) $\{D \in R^{n \times n} : D \text{ is a diagonal matrix}\}$.
    (b) $\{A \in R^{n \times n} : A \text{ is not invertible}\}$.
    (c) $\{A \in R^{n \times n} : \text{for all } i, j, A_{ij} = A_{ji}\}$.

4.3 Span, Linear Independence, and Basis

One advantage of the unifying nature of a vector space is that you can study many systems simultaneously while working with only one axiomatic system. A disadvantage, however, is that an axiomatic system loses many of the properties of the individual special cases that are useful in problem solving. Some of those properties—such as the existence of a zero vector—are included in an abstract system as appropriate axioms. As you will learn in this section, other properties are restored through the development of definitions and theorems pertaining to the axiomatic system.

4.3.1 The Components of a Vector

One of the nice features of the vector space R^n is that, for $n = 2$ and 3, you can visualize a vector as a point whose coordinates are those of the components of the vector. For example, the vector $\mathbf{v} = (3, 2)$ is a point in the plane whose x-coordinate is 3 and whose y-coordinate is 2. The ability to visualize vectors in 2-space and in 3-space is a valuable problem-solving tool, as illustrated in Section 1.4. When working with a vector $\mathbf{v}$ in an arbitrary vector space, it is not clear what the components of $\mathbf{v}$ are. One objective in this section is to show that vectors in certain vector spaces have components that you can use in solving problems.

Generalizing the Components of an *n*-Vector

To understand the nature of these components, first consider the n-vector

$$\mathbf{v} = (v_1, \ldots, v_n) \tag{4.10}$$

When you use abstraction to think of $\mathbf{v}$ as an object belonging to a vector space $(V, \oplus, \odot)$, (4.10) is no longer valid because a general vector $\mathbf{v} \in V$ does not have components. The issue, then, is how to generalize (4.10) to an abstract vector $\mathbf{v}$.

One way to do so is to write the components in (4.10) as an expression involving the addition of particular n-vectors and the multiplication of these n-vectors by real numbers.

Once such an expression is found, generalization to a vector space $(V, \oplus, \odot)$ is possible by using the operations $\oplus$ and $\odot$.

To illustrate this process, observe that you can express the vector $\mathbf{v} = (3, 2)$ in 2-space as follows:

$$\mathbf{v} = (3, 2) = 3(1, 0) + 2(0, 1) \tag{4.11}$$

The advantage of the expression in (4.11) is that generalization to an abstract vector space $(V, \oplus, \odot)$ is possible. To see how, let $\mathbf{v}_1 = (1, 0)$ and $\mathbf{v}_2 = (0, 1)$. Also, let $t_1 = 3$ and $t_2 = 2$. Then you can write (4.11) as

$$\mathbf{v} = t_1 \mathbf{v}_1 + t_2 \mathbf{v}_2 \tag{4.12}$$

The generalization of (4.12) to an abstract vector space $(V, \oplus, \odot)$ is accomplished by (1) thinking of $\mathbf{v}_1$ and $\mathbf{v}_2$ as two given vectors in V, (2) replacing in (4.12) the operation of $+$ that applies only to n-vectors with the operation $\oplus$ that applies to vectors in V, and (3) replacing in (4.12) the operation of multiplying a real number by an n-vector with the operation $\odot$ that applies in V. On so doing, the generalization of (4.12) to $(V, \oplus, \odot)$ is the following:

$$\mathbf{v} = (t_1 \odot \mathbf{v}_1) \oplus (t_2 \odot \mathbf{v}_2) \tag{4.13}$$

The real numbers t_1 and t_2 in (4.13) then become the components of the vector $\mathbf{v} \in V$ *relative to the two special vectors $\mathbf{v}_1$ and $\mathbf{v}_2$.*

In general, to find the components of a vector $\mathbf{v}$ in a vector space $(V, \oplus, \odot)$, it is necessary to do the following:

1. Identify a collection of special vectors, say, $\mathbf{v}_1, \ldots, \mathbf{v}_k$.
2. Find real numbers $t_1, \ldots, t_k$ so that

$$\mathbf{v} = (t_1 \odot \mathbf{v}_1) \oplus (t_2 \odot \mathbf{v}_2) \oplus \cdots \oplus (t_k \odot \mathbf{v}_k) \tag{4.14}$$

If successful, then $(t_1, \ldots, t_k)$ are the **components of v relative to $\mathbf{v}_1, \ldots, \mathbf{v}_k$.**

There are several questions that arise in trying to find the components of a vector $\mathbf{v}$ in a vector space:

1. What properties should the collection $\mathbf{v}_1, \ldots, \mathbf{v}_k$ have and how do you find these vectors?
2. Given the vector $\mathbf{v}$ and the special vectors $\mathbf{v}_1, \ldots, \mathbf{v}_k$, what problems arise in trying to find scalars $t_1, \ldots, t_k$ so that (4.14) holds?

Both of these issues are addressed in what follows.

Linear Combinations of Vectors

The answer to the first question is motivated by exploring the second question. To that end, suppose that you have a collection of k special vectors, say, $\mathbf{v}_1, \ldots, \mathbf{v}_k$. For a given vector $\mathbf{v}$, you may or may not be able to find real numbers $t_1, \ldots, t_k$ so that (4.14) holds. The favorable case in which you *can* find these real numbers gives rise to the following definition.

DEFINITION 4.3

A vector **v** in a vector space $(V, \oplus, \odot)$ is a **linear combination** of the vectors $\mathbf{v}_1, \ldots, \mathbf{v}_k$ if and only if there are real numbers $t_1, \ldots, t_k$ such that

$$\mathbf{v} = (t_1 \odot \mathbf{v}_1) \oplus \cdots \oplus (t_k \odot \mathbf{v}_k) \tag{4.15}$$

EXAMPLE 4.1 Linear Combinations of *n*-Vectors
The vector $\mathbf{v} = (3, 2)$ is a linear combination of the vectors $(1, 0)$ and $(0, 1)$ because

$$(3, 2) = 3(1, 0) + 2(0, 1)$$

In this case, the components of **v** relative to the special vectors $(1, 0)$ and $(0, 1)$ are 3 and 2.
The vector $(3, 2)$ is also a linear combination of the vectors $(1, 1)$ and $(-1, -2)$ because

$$(3, 2) = 4(1, 1) + 1(-1, -2)$$

Here, the components of **v** relative to the special vectors $(1, 1)$ and $(-1, -2)$ are 4 and 1.

Example 4.1 illustrates that the components of a vector depend on the choice of the special vectors. In general, to determine the components of an *n*-vector **v** relative to the special *n*-vectors $\mathbf{v}_1, \ldots, \mathbf{v}_k$, you can solve an associated system of linear equations, as shown in the following example.

EXAMPLE 4.2 Solving Linear Equations to Determine the Components of an *n*-Vector
To find the components of the vector $\mathbf{v} = (2, -3, 4)$ relative to the vectors $\mathbf{v}_1 = (0, 1, -1)$ and $\mathbf{v}_2 = (2, 0, 1)$, you must find scalars t_1 and t_2 so that

$$\begin{aligned}
(2, -3, 4) &= t_1(0, 1, -1) + t_2(2, 0, 1) \\
&= (0, t_1, -t_1) + (2t_2, 0, t_2) \\
&= (2t_2, t_1, -t_1 + t_2)
\end{aligned}$$

Equating corresponding components, you need to find values for t_1 and t_2 so that

$$\begin{aligned}
2 &= 2t_2 \\
-3 &= t_1 \\
4 &= -t_1 + t_2
\end{aligned}$$

Solving this system of linear equations results in $t_1 = -3$ and $t_2 = 1$. As a result, **v** is a linear combination of $\mathbf{v}_1$ and $\mathbf{v}_2$ because

$$\mathbf{v} = -3\mathbf{v}_1 + 1\mathbf{v}_2$$

Thus, the components of **v** relative to $\mathbf{v}_1$ and $\mathbf{v}_2$ are -3 and 1.

A system of linear equations may have either no solution, exactly one solution (as in the system in Example 4.2), or infinitely many solutions. With regard to finding the components of an n-vector, if the associated system of linear equations has no solution, the n-vector does not have components relative to $\mathbf{v}_1, \ldots, \mathbf{v}_k$, as shown in the following example.

EXAMPLE 4.3 An *n*-Vector that Does Not Have Components Relative to Given *n*-Vectors

To determine the components of the vector $\mathbf{v} = (2, -3, 0)$ relative to the given vectors $\mathbf{v}_1 = (0, 1, -1)$ and $\mathbf{v}_2 = (2, 0, 1)$, you must find real numbers t_1 and t_2 so that

$$(2, -3, 0) = t_1(0, 1, -1) + t_2(2, 0, 1)$$
$$= (0, t_1, -t_1) + (2t_2, 0, t_2)$$
$$= (2t_2, t_1, -t_1 + t_2)$$

Equating corresponding components, you need to find values for t_1 and t_2 so that

$$2 = 2t_2$$
$$-3 = t_1$$
$$0 = -t_1 + t_2$$

You can verify that this system is inconsistent. Therefore, $\mathbf{v}$ is not a linear combination of $\mathbf{v}_1$ and $\mathbf{v}_2$ and thus $\mathbf{v}$ does not have components relative to the vectors $\mathbf{v}_1$ and $\mathbf{v}_2$.

Utilizing the approach in Example 4.3 to find the components of an n-vector relative to $\mathbf{v}_1, \ldots, \mathbf{v}_k$, you may find that the associated system of linear equations has an infinite number of solutions. In this case, there are an infinite number of choices for the components of the n-vector, as shown in the following example.

EXAMPLE 4.4 Multiple Components of an *n*-Vector

To find the components of the vector $\mathbf{v} = (3, 2)$ relative to the vectors $\mathbf{v}_1 = (1, 0)$, $\mathbf{v}_2 = (0, 1)$, and $\mathbf{v}_3 = (-1, -1)$, you must find values for t_1, t_2, and t_3 so that $\mathbf{v} = t_1\mathbf{v}_1 + t_2\mathbf{v}_2 + t_3\mathbf{v}_3$, which gives rise to the following system of linear equations:

$$3 = t_1 \quad - t_3$$
$$2 = \quad t_2 - t_3$$

You can verify that this system has an infinite number of solutions. One solution is $t_1 = 3$, $t_2 = 2$, $t_3 = 0$. Hence, one set of components of $\mathbf{v}$ relative to the vectors $\mathbf{v}_1$, $\mathbf{v}_2$, and $\mathbf{v}_3$ is $(3, 2, 0)$. Another solution to the foregoing system is $t_1 = 0, t_2 = -1$ and $t_3 = -3$. Correspondingly, another set of components of $\mathbf{v}$ relative to $\mathbf{v}_1$, $\mathbf{v}_2$, and $\mathbf{v}_3$ is $(0, -1, -3)$.

From the foregoing examples, you can see that the choice of the special vectors $\mathbf{v}_1, \ldots, \mathbf{v}_k$ is critical for the determination of the components of a vector $\mathbf{v}$ for the following reasons:

1. The vector $\mathbf{v}$ may not have components relative to $\mathbf{v}_1, \ldots, \mathbf{v}_k$ because $\mathbf{v}$ is not a linear combination of these vectors. This situation arises for n-vectors when the associated system of linear equations has no solution.

2. Even if $\mathbf{v}$ is a linear combination of the vectors $\mathbf{v}_1, \ldots, \mathbf{v}_k$, there can be different sets of values for $t_1, \ldots, t_k$ that satisfy

$$\mathbf{v} = (t_1 \odot \mathbf{v}_1) \oplus \cdots \oplus (t_k \odot \mathbf{v}_k)$$

In this case, $\mathbf{v}$ has different sets of components relative to the vectors $\mathbf{v}_1, \ldots, \mathbf{v}_k$. This situation arises for n-vectors when the corresponding system of linear equations has an infinite number of solutions.

On the basis of this discussion, an ideal choice for the vectors $\mathbf{v}_1, \ldots, \mathbf{v}_k$ is one that has neither of the foregoing problems. The development of such a set of vectors is accomplished in the next two sections.

4.3.2 Spanning Vectors

As you have seen in Section 4.3.1, to ensure that each vector in V has components relative to $\mathbf{v}_1, \ldots, \mathbf{v}_k$, every vector $\mathbf{v}$ should be a linear combination of these special vectors. Such a set of special vectors $\mathbf{v}_1, \ldots, \mathbf{v}_k$ is formalized in the following definition.

DEFINITION 4.4

A collection $\mathbf{v}_1, \ldots, \mathbf{v}_k$ of vectors **spans** a vector space $(V, \oplus, \odot)$ if and only if each vector $\mathbf{v} \in V$ is a linear combination of $\mathbf{v}_1, \ldots, \mathbf{v}_k$.

EXAMPLE 4.5 Vectors that Span R^3
The vectors $\mathbf{i} = (1, 0, 0)$, $\mathbf{j} = (0, 1, 0)$, and $\mathbf{k} = (0, 0, 1)$ span R^3 because any vector $\mathbf{v} = (v_1, v_2, v_3)$ is a linear combination of $\mathbf{i}$, $\mathbf{j}$, and $\mathbf{k}$ since

$$\mathbf{v} = (v_1, v_2, v_3) = v_1 \mathbf{i} + v_2 \mathbf{j} + v_3 \mathbf{k}$$

EXAMPLE 4.6 Vectors that Do Not Span R^3
The vectors $\mathbf{v}_1 = (1, 0, -1)$, $\mathbf{v}_2 = (2, 0, 3)$, and $\mathbf{v}_3 = (-1, 0, 4)$ do not span R^3. If they did, any vector $\mathbf{v} = (v_1, v_2, v_3)$ would be a linear combination of $\mathbf{v}_1$, $\mathbf{v}_2$, and $\mathbf{v}_3$. That is, you would be able to find values for t_1, t_2, and t_3 so that

$$\mathbf{v} = t_1 \mathbf{v}_1 + t_2 \mathbf{v}_2 + t_3 \mathbf{v}_3$$

or equivalently,

$$(v_1, v_2, v_3) = t_1(1, 0, -1) + t_2(2, 0, 3) + t_3(-1, 0, 4)$$

In other words, you would be able to solve the following system:

$$v_1 = t_1 + 2t_2 - t_3$$
$$v_2 = 0t_1 + 0t_2 + 0t_3$$
$$v_3 = -t_1 + 3t_2 + 4t_3$$

for all values of v_1, v_2, and v_3. You cannot do so, however, when $v_2 \neq 0$.

The next two examples illustrate spanning vectors in vector spaces other than n-space.

EXAMPLE 4.7 A Collection of Spanning Matrices
The matrices

$$M^{11} = \begin{bmatrix} 1 & 0 \\ 0 & 0 \end{bmatrix}, \quad M^{12} = \begin{bmatrix} 0 & 1 \\ 0 & 0 \end{bmatrix}, \quad M^{21} = \begin{bmatrix} 0 & 0 \\ 1 & 0 \end{bmatrix}, \quad M^{22} = \begin{bmatrix} 0 & 0 \\ 0 & 1 \end{bmatrix}$$

span the vector space $R^{2 \times 2}$ of all (2×2) matrices. This is because any (2×2) matrix

$$M = \begin{bmatrix} a & b \\ c & d \end{bmatrix}$$

is a linear combination of M^{11}, M^{12}, M^{21}, and M^{22} since

$$M = \begin{bmatrix} a & b \\ c & d \end{bmatrix} = a \begin{bmatrix} 1 & 0 \\ 0 & 0 \end{bmatrix} + b \begin{bmatrix} 0 & 1 \\ 0 & 0 \end{bmatrix} + c \begin{bmatrix} 0 & 0 \\ 1 & 0 \end{bmatrix} + d \begin{bmatrix} 0 & 0 \\ 0 & 1 \end{bmatrix}$$

EXAMPLE 4.8 A Collection of Spanning Polynomials
The polynomials

$$p_0(x) = 1, \quad p_1(x) = x, \quad p_2(x) = x^2, \quad \ldots, \quad p_n(x) = x^n$$

span the vector space P^n of all polynomials of degree n or less. This is the case because any polynomial $p \in P^n$,

$$p(x) = a_0 + a_1 x + a_2 x^2 + \cdots + a_n x^n$$

is a linear combination of $p_0(x), \ldots, p_n(x)$ since you can write $p(x)$ as:

$$p(x) = a_0 p_0(x) + a_1 p_1(x) + \cdots + a_n p_n(x)$$

When the vectors $\mathbf{v}_1, \ldots, \mathbf{v}_k$ do not span V, some vectors $\mathbf{v}$ in V are a linear combination of $\mathbf{v}_1, \ldots, \mathbf{v}_k$ and others are not. The set of all vectors that are expressible as a linear combination of $\mathbf{v}_1, \ldots, \mathbf{v}_k$ gives rise to the following definition.

DEFINITION 4.5

Given a set of vectors $S = \{\mathbf{v}_1, \ldots, \mathbf{v}_k\}$ in a vector space $(V, \oplus, \odot)$, the **space spanned by** S is denoted by span(S) or span$\{\mathbf{v}_1, \ldots, \mathbf{v}_k\}$ and is the following set:

$$\text{span}(S) = \{\mathbf{u} \in V : \text{there are real numbers } t_1, \ldots, t_k \text{ with } \mathbf{u} = (t_1 \odot \mathbf{v}_1) \oplus \cdots \oplus (t_k \odot \mathbf{v}_k)\}$$
$$= \{(t_1 \odot \mathbf{v}_1) \oplus \cdots \oplus (t_k \odot \mathbf{v}_k) : t_1, \ldots, t_k \text{ are real numbers}\}$$

The next theorem shows that the space spanned by a set of vectors is a subspace of V.

THEOREM 4.3

If $\mathbf{v}_1, \ldots, \mathbf{v}_k$ are vectors in a vector space $(V, \oplus, \odot)$, then $U = \text{span}\{\mathbf{v}_1, \ldots, \mathbf{v}_k\}$ is a subspace of V.

Proof.
According to Theorem 4.2 in Section 4.2.2, you need show only that if $\mathbf{u}, \mathbf{v} \in U$ and t is a real number, then (1) $\mathbf{u} \oplus \mathbf{v} \in U$ and (2) $t \odot \mathbf{u} \in U$. To establish (1), you must show that an element (namely, $\mathbf{u} \oplus \mathbf{v}$) is in span$\{\mathbf{v}_1, \ldots, \mathbf{v}_k\}$. To do so using Definition 4.5, the quantifier *there is* means you must construct real numbers $t_1, \ldots, t_k$ such that

$$\mathbf{u} \oplus \mathbf{v} = (t_1 \odot \mathbf{v}_1) \oplus \cdots \oplus (t_k \odot \mathbf{v}_k) \tag{4.16}$$

To construct these real numbers, use the fact that $\mathbf{u}, \mathbf{v} \in U$, from which it follows that there are real numbers $r_1, \ldots, r_k$ and $s_1, \ldots, s_k$ such that

$$\mathbf{u} = (r_1 \odot \mathbf{v}_1) \oplus \cdots \oplus (r_k \odot \mathbf{v}_k) \tag{4.17}$$
$$\mathbf{v} = (s_1 \odot \mathbf{v}_1) \oplus \cdots \oplus (s_k \odot \mathbf{v}_k) \tag{4.18}$$

The desired values of $t_1, \ldots, t_k$ in (4.16) are $t_1 = r_1 + s_1, \ldots, t_k = r_k + s_k$. These values are correct because, from (4.17) and (4.18),

$$\mathbf{u} \oplus \mathbf{v} = (r_1 + s_1) \odot \mathbf{v}_1 \oplus \cdots \oplus (r_k + s_k) \odot \mathbf{v}_k$$
$$= (t_1 \odot \mathbf{v}_1) \oplus \cdots \oplus (t_k \odot \mathbf{v}_k)$$

To see that $t \odot \mathbf{u} \in U$, you must construct real numbers $t_1, \ldots, t_k$ so that

$$t \odot \mathbf{u} = (t_1 \odot \mathbf{v}_1) \oplus \cdots \oplus (t_k \odot \mathbf{v}_k) \tag{4.19}$$

Because $\mathbf{u} \in U$, there are real numbers $r_1, \ldots, r_k$ such that (4.17) holds. Construct $t_1 = tr_1, \ldots, t_k = tr_k$. Multiplying (4.17) through by t you can see that (4.19) holds. ∎

EXAMPLE 4.9 The Space Spanned by a Single Vector
The space spanned by a single vector $\mathbf{v}$ is the set of all multiples of $\mathbf{v}$. For example, the space spanned by the vector $\mathbf{v} = (2, 1)$ is

$$\text{span}\{\mathbf{v}\} = \{\mathbf{w} \in R^2 : \text{there is a number } t \text{ such that } \mathbf{w} = t\mathbf{v} = t(2, 1)\}$$
$$= \{t(2, 1) : t \text{ is a real number}\}$$

In this case, span{**v**} is a straight line through the origin in the plane in the direction **v** = (2, 1), as shown in Figure 4.5.

EXAMPLE 4.10 The Space Spanned by Two Vectors

The space spanned by the vectors $\mathbf{v}_1$ and $\mathbf{v}_2$ is the set of all vectors of the form $(t_1 \odot \mathbf{v}_1) \oplus (t_2 \odot \mathbf{v}_2)$. When $\mathbf{v}_1$ and $\mathbf{v}_2$ are vectors in 3-space that do not lie on the same line, span{$\mathbf{v}_1, \mathbf{v}_2$} is the plane containing $\mathbf{v}_1$ and $\mathbf{v}_2$, as shown in Figure 4.6.

The reason for trying to find vectors $\mathbf{v}_1, \ldots, \mathbf{v}_k$ that span a vector space V is so that each vector $\mathbf{v} \in V$ will be a linear combination of $\mathbf{v}_1, \ldots, \mathbf{v}_k$. In this case, there are real numbers $t_1, \ldots, t_k$ such that

$$\mathbf{v} = (t_1 \odot \mathbf{v}_1) \oplus \cdots \oplus (t_k \odot \mathbf{v}_k) \tag{4.20}$$

However, as mentioned previously, there can be many different sets of values for $t_1, \ldots, t_k$ that satisfy (4.20). A property of the vectors $\mathbf{v}_1, \ldots, \mathbf{v}_k$ is now developed so that there is a unique solution to (4.20).

4.3.3 Linearly Dependent and Independent Vectors

For n-vectors, the property needed to ensure a unique solution to (4.20) is that of vectors "opening up properly." For example, the vectors **u** and **v** in Figure 4.7(a) and in Figure 4.7(b) open up properly. In contrast, the vectors **u** and **v** in Figure 4.8(a) and in Figure 4.8(b) do not open up properly. This is because the vectors in Figure 4.8(a) lie on top of each other and the vectors in Figure 4.8(b) point in exactly opposite directions. The objective now is to create a mathematical definition of what it means for vectors to "open up properly" by translating the associated visual image to symbolic form.

One approach is to develop a definition for the property satisfied by the vectors in Figure 4.7. Alternatively, you can first develop a definition for the property of "not opening up properly," satisfied by the vectors in Figure 4.8, and then write the negation of that definition to capture the property of the vectors in Figure 4.7.

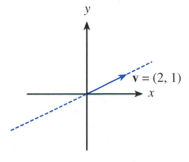

Figure 4.5 *The Subspace of R^2 Spanned by (2, 1).*

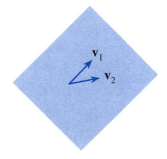

Figure 4.6 *The Subspace of R^3 Spanned by Two Vectors.*

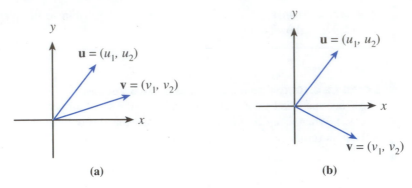

Figure 4.7 *Examples of Vectors that Open Up Properly.*

Linearly Dependent Vectors

Following the latter approach, try to identify similarities between the pair of vectors in Figure 4.8(a) and the pair in Figure 4.8(b). For example, in Figure 4.8(a), **u** points in the same direction as **v** but has a different length. Recall, from Section 1.2.2, that you can change the length of **v** (without affecting the direction) by multiplying **v** by a nonnegative real number, say, $a \geq 0$. Thus, by an appropriate choice of the real number $a \geq 0$, you can make the vectors **u** and a**v** equal. You can translate this observation to the following symbolic form using the quantifier *there is*:

There is a real number $a \geq 0$ such that $\mathbf{u} = a\mathbf{v}$.   (4.21)

Looking now at Figure 4.8(b), you will notice that **u** points in the opposite direction to **v**. In this case, by multiplying **v** by an appropriate choice of the nonpositive real number $a \leq 0$, you can again make **u** and a**v** equal. This observation translates to the following symbolic form:

There is a real number $a \leq 0$ such that $\mathbf{u} = a\mathbf{v}$.   (4.22)

Can you create a single statement that includes both (4.21) and (4.22) as special cases? One such approach is to allow a to be any real number—positive, negative, or 0. Thus, a

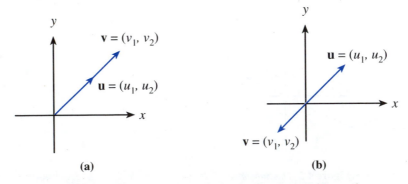

Figure 4.8 *Examples of Vectors that Do Not Open Up Properly.*

unification of (4.21) and (4.22) is the following:

There is a real number a such that $\mathbf{u} = a\mathbf{v}$. (4.23)

Generalizing the Definition to 3-Space. The next step is to generalize (4.23) when the vectors lie in 3-space. Indeed, (4.23) is still valid when $\mathbf{u}$ and $\mathbf{v}$ are vectors in 3-space that lie on top of, or opposite to, each other. However, what if you are working with a third vector, $\mathbf{w}$, in 3-space, such as the one in Figure 4.9? You might write the following statement for this property:

There are real numbers a and b such that $\mathbf{u} = a\mathbf{v} + b\mathbf{w}$ (4.24)

Is the property in (4.24) correct for all groups of three vectors in 3-space that do not open up properly? Only by extensive trials with other special cases of $\mathbf{u}$, $\mathbf{v}$, and $\mathbf{w}$ will you discover that (4.24) is not necessarily correct in all cases. For example, the vectors in Figure 4.10 do not satisfy (4.24) because $a\mathbf{v} + b\mathbf{w}$ always points in the same direction as $\mathbf{v}$ and $\mathbf{w}$. Rather, the vectors in Figure 4.10 satisfy the following property:

There are real numbers a and b such that $\mathbf{v} = a\mathbf{u} + b\mathbf{w}$ (4.25)

The vectors in Figure 4.10 also satisfy the property that

there are real numbers a and b such that $\mathbf{w} = a\mathbf{u} + b\mathbf{v}$ (4.26)

In fact, at least one of (4.24), (4.25), or (4.26) is always true for vectors $\mathbf{u}$, $\mathbf{v}$, and $\mathbf{w}$ in 3-space that do not open up properly. The challenge is to write a single statement that covers all three of the special cases in (4.24), (4.25), and (4.26). Such a unification requires cleverness by moving all vectors to the same side of the equality sign and then realizing that there are real numbers associated with each vector, that is,

There are real numbers a, b, and c such that $a\mathbf{u} + b\mathbf{v} + c\mathbf{w} = \mathbf{0}$ (4.27)

However, the real numbers in the special cases are not just *any* real numbers —they have some special properties: in (4.24), $a = 1$; in (4.25), $b = 1$; and in (4.26), $c = 1$. One way to capture this fact is to require that at least one of the real numbers in (4.27) be 1. An alternative, but equivalent, way to say this (as you are asked to verify in Exercise 25) is:

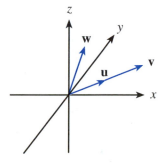

Figure 4.9 *Three Vectors in R^3 that Do Not Open Up Properly.*

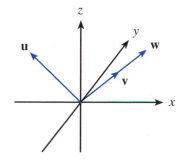

Figure 4.10 *Three Other Vectors in R^3 that Do Not Open Up Properly.*

There are real numbers a, b, c, not all 0, with $a\mathbf{u} + b\mathbf{v} + c\mathbf{w} = \mathbf{0}$ (4.28)

If $a \neq 0$, for example, then statement (4.28) is equivalent to saying that $\mathbf{u}$ is a linear combination of $\mathbf{v}$ and $\mathbf{w}$.

Generalizing the Definition to *n*-Space. The next step is to generalize (4.28) to the case where the vectors lie in n-space. So suppose you have a group of k vectors in n-space that do not open up properly. Subscript and superscript notation is helpful. Let each of $\mathbf{v}_1, \ldots, \mathbf{v}_k$ be an n-vector. Introducing the term "linearly dependent" for "not opening up properly" and generalizing (4.28) in the natural way, you obtain the following:

DEFINITION 4.6

The n-vectors $\mathbf{v}_1, \ldots, \mathbf{v}_k$ are **linearly dependent** if and only if there are real numbers $t_1, \ldots t_k$, not all zero, such that

$$t_1\mathbf{v}_1 + \cdots + t_k\mathbf{v}_k = \mathbf{0}$$

EXAMPLE 4.11 Linearly Dependent Vectors in *n*-Space

The vectors $\mathbf{v}_1 = (1, 0, -2)$ and $\mathbf{v}_2 = (2, 0, -4)$ are linearly dependent because for $t_1 = 2$ and $t_2 = -1$, it follows that

$$t_1\mathbf{v}_1 + t_2\mathbf{v}_2 = 2(1, 0, -2) + (-1)(2, 0, -4) = (0, 0, 0)$$

Using Matrix-Vector Notation. Matrix-vector notation provides an alternative way to write the definition of linearly dependent n-vectors. To see how, think of the n-vectors $\mathbf{v}_1, \ldots, \mathbf{v}_k$ as the columns of an $(n \times k)$ matrix V, so, for each column j of V, $V_{*j} = \mathbf{v}_j$. Also, think of $t_1, \ldots, t_k$ as the components of a column vector $\mathbf{t} \in R^{k \times 1}$. Then, as you were asked to show in Exercise 13(c) in Section 2.4,

$$
V\mathbf{t} \;=\;
\begin{bmatrix}
V_{11}t_1 & + & V_{12}t_2 & + & \cdots & + & V_{1k}t_k \\
V_{21}t_1 & + & V_{22}t_2 & + & \cdots & + & V_{2k}t_k \\
& & & \vdots & & & \\
V_{n1}t_1 & + & V_{n2}t_2 & + & \cdots & + & V_{nk}t_k
\end{bmatrix}
$$

$$
=\;
\begin{bmatrix}
V_{11} \\ V_{21} \\ \vdots \\ V_{n1}
\end{bmatrix} t_1 +
\begin{bmatrix}
V_{12} \\ V_{22} \\ \vdots \\ V_{n2}
\end{bmatrix} t_2 + \cdots +
\begin{bmatrix}
V_{1k} \\ V_{2k} \\ \vdots \\ V_{nk}
\end{bmatrix} t_k
$$

$$
=\; V_{*1}t_1 + V_{*2}t_2 + \cdots + V_{*k}t_k
$$

By using the fact that $V_{*j} = \mathbf{v}_j$, you can now see that

$$V\mathbf{t} = \mathbf{v}_1 t_1 + \mathbf{v}_2 t_2 + \cdots + \mathbf{v}_k t_k \tag{4.29}$$

The expression in (4.29) says that a matrix V times a column vector $\mathbf{t}$ is equal to the sum of each column of V times the corresponding component of $\mathbf{t}$. This result is used frequently from here on, such as in rewriting the conditions in Definition 4.6 to obtain the following definition of linearly dependent vectors in matrix-vector notation.

Definition of Linearly Dependent Vectors in Matrix-Vector Notation

The n-vectors $\mathbf{v}_1, \ldots, \mathbf{v}_k$ are *linearly dependent* if and only if there is a nonzero solution to the homogeneous system $V\mathbf{t} = \mathbf{0}$, that is, if and only if there is a column vector $\mathbf{t} \in R^{k \times 1}$ with $\mathbf{t} \neq \mathbf{0}$ such that

$$V\mathbf{t} = \mathbf{0}$$

where V is the $(n \times k)$ matrix in which $V_{*j} = \mathbf{v}_j$ (see Theorem 3.4(c) in Section 3.2.3).

Linearly Independent Vectors

Definition 4.6, that of linearly dependent vectors, applies to vectors in dimensions that you cannot visualize, which is one of the advantages of this symbolic definition. Another advantage is that you can state the negation of the definition. In particular, the negation of Definition 4.6 captures the concept of a group of *n*-vectors "opening up properly." By applying the rules for negating statements containing quantifiers to Definition 4.6, you obtain the following definition.

DEFINITION 4.7

The *n*-vectors $\mathbf{v}_1, \ldots, \mathbf{v}_k$ are **linearly independent** if and only if for all real numbers $t_1, \ldots, t_k$ with

$$t_1 \mathbf{v}_1 + \cdots + t_k \mathbf{v}_k = \mathbf{0}$$

it follows that

$$t_1 = \cdots = t_k = 0$$

EXAMPLE 4.12 Linear Independence of the Vectors i, j, and k

The vectors $\mathbf{i} = (1, 0, 0)$, $\mathbf{j} = (0, 1, 0)$, and $\mathbf{k} = (0, 0, 1)$ are linearly independent because the only solution to

$$t_1 \mathbf{i} + t_2 \mathbf{j} + t_3 \mathbf{k} = \mathbf{0}$$

is $t_1 = 0$, $t_2 = 0$, and $t_3 = 0$. This is the case because if $t_1 \mathbf{i} + t_2 \mathbf{j} + t_3 \mathbf{k} = \mathbf{0}$, then

$$t_1 \mathbf{i} + t_2 \mathbf{j} + t_3 \mathbf{k} = t_1(1, 0, 0) + t_2(0, 1, 0) + t_3(0, 0, 1) = (t_1, t_2, t_3) = (0, 0, 0)$$

Equating corresponding components of the vectors in the last two expressions yields $t_1 = 0$, $t_2 = 0$, and $t_3 = 0$, so $\mathbf{i}$, $\mathbf{j}$, and $\mathbf{k}$ are linearly independent.

EXAMPLE 4.13 Linearly Independent Vectors in *n*-Space

The vectors $\mathbf{v}_1 = (1, 0, -1)$ and $\mathbf{v}_2 = (-2, 3, 0)$ are linearly independent because the only solution to

$$t_1\mathbf{v}_1 + t_2\mathbf{v}_2 = \mathbf{0}$$

is $t_1 = 0$ and $t_2 = 0$. This is the case because if $t_1\mathbf{v}_1 + t_2\mathbf{v}_2 = \mathbf{0}$, then

$$
\begin{aligned}
t_1\mathbf{v}_1 + t_2\mathbf{v}_2 &= t_1(1, 0, -1) + t_2(-2, 3, 0) \\
&= (t_1, 0, -t_1) + (-2t_2, 3t_2, 0) \\
&= (t_1 - 2t_2, 3t_2, -t_1) \\
&= (0, 0, 0)
\end{aligned}
$$

Equating corresponding components of the vectors in the last two expressions yields $t_1 = 0$ and $t_2 = 0$, so $\mathbf{v}_1$ and $\mathbf{v}_2$ are linearly independent.

Matrix-vector notation provides an alternative way to write the definition of linearly independent vectors. To see how, think of the n-vectors $\mathbf{v}_1, \ldots, \mathbf{v}_k$ as the columns of an $(n \times k)$ matrix V and $t_1, \ldots, t_k$ as the components of a column vector $\mathbf{t} \in R^{k \times 1}$ of variables. Writing the conditions in Definition 4.7 using matrix-vector notation results in the following definition of linearly independent vectors.

Definition of Linearly Independent Vectors in Matrix-Vector Notation

The n-vectors $\mathbf{v}_1, \ldots, \mathbf{v}_k$ are *linearly independent* if and only if the only solution to the homogeneous system

$$V\mathbf{t} = \mathbf{0}$$

is

$$\mathbf{t} = \mathbf{0}$$

where V is the $(n \times k)$ matrix in which column j of V is the column vector $\mathbf{v}_j$ and $\mathbf{t} \in R^{k \times 1}$ is a column vector of variables.

Generalization of Linear Independence and Linear Dependence to Vector Spaces

You can generalize Definition 4.6 and Definition 4.7 to a vector space $(V, \oplus, \odot)$ using the operations $\oplus$ and $\odot$, as follows.

DEFINITION 4.8

The vectors $\mathbf{v}_1, \ldots, \mathbf{v}_k$ in a vector space $(V, \oplus, \odot)$ are **linearly dependent** if and only if there are real numbers $t_1, \ldots t_k$, not all zero, such that

$$(t_1 \odot \mathbf{v}_1) \oplus \cdots \oplus (t_k \odot \mathbf{v}_k) = \mathbf{0}$$

DEFINITION 4.9

The vectors $\mathbf{v}_1, \ldots, \mathbf{v}_k$ in a vector space $(V, \oplus, \odot)$ are **linearly independent** if and only if for all real numbers $t_1, \ldots, t_k$ with

$$(t_1 \odot \mathbf{v}_1) \oplus \cdots \oplus (t_k \odot \mathbf{v}_k) = \mathbf{0}$$

it follows that

$$t_1 = \cdots = t_k = 0$$

EXAMPLE 4.14 Linearly Independent Matrices

The (2×2) matrices M^{11}, M^{12}, M^{21}, and M^{22} in Example 4.7 are linearly independent because, if the real numbers a, b, c, and d satisfy

$$aM^{11} + bM^{12} + cM^{21} + dM^{22} = 0 \tag{4.30}$$

then, by performing the multiplications and additions, (4.30) becomes

$$\begin{bmatrix} a & b \\ c & d \end{bmatrix} = \begin{bmatrix} 0 & 0 \\ 0 & 0 \end{bmatrix}$$

Equating corresponding components, it then follows that $a = 0$, $b = 0$, $c = 0$, and $d = 0$, so the matrices M^{11}, M^{12}, M^{21}, and M^{22} are linearly independent.

EXAMPLE 4.15 Linearly Independent Polynomials

The polynomials $p_0(x) = 1, p_1(x) = x, \ldots, p_n(x) = x^n$ in Example 4.8 are linearly independent because, if the real numbers $a_0, a_1, \ldots, a_n$ satisfy

$$a_0 p_0(x) + a_1 p_1(x) + \cdots + a_n p_n(x) = z(x) \quad \text{(the zero polynomial)}$$

then

$$a_0 + a_1 x + \cdots + a_n x^n = 0 + 0x + \cdots + 0x^n \tag{4.31}$$

Equating coefficients of like powers on each side of (4.31), it follows that

$$a_0 = 0, \quad a_1 = 0, \quad \ldots, \quad a_n = 0$$

A relationship between the concepts of linear dependence and linear combination is given in the following theorem.

THEOREM 4.4

Suppose that $k \geq 2$ and that $\mathbf{v}_1, \ldots, \mathbf{v}_k$ are vectors in a vector space $(V, \oplus, \odot)$.

(a) The vectors $\mathbf{v}_1, \ldots, \mathbf{v}_k$ are linearly dependent if and only if at least one of these vectors is a linear combination of the other vectors.

(b) The vectors $\mathbf{v}_1, \ldots, \mathbf{v}_k$ are linearly independent if and only if none of these vectors is a linear combination of the remaining vectors.

Proof.

For (a), suppose first that $\mathbf{v}_1, \ldots, \mathbf{v}_k$ are linearly dependent. According to Definition 4.8, there are real numbers $t_1, \ldots, t_k$, not all zero, such that

$$(t_1 \odot \mathbf{v}_1) \oplus (t_2 \odot \mathbf{v}_2) \oplus \cdots \oplus (t_k \odot \mathbf{v}_k) = \mathbf{0} \tag{4.32}$$

Now either $t_1 \neq 0$ or $t_2 \neq 0$, ..., or $t_k \neq 0$. Because you do not know which of these numbers is not 0, you must proceed by cases, as follows.

Case 1: Assume that $t_1 \neq 0$. In this case, you can multiply (4.32) through by $1/t_1$ and solve for $\mathbf{v}_1$ in terms of the remaining vectors to obtain

$$\mathbf{v}_1 = \left(-\frac{t_2}{t_1}\right) \odot \mathbf{v}_2 \oplus \cdots \oplus \left(-\frac{t_k}{t_1}\right) \odot \mathbf{v}_k \tag{4.33}$$

Now, (4.33) indicates that $\mathbf{v}_1$ is a linear combination of the remaining vectors, as desired.

A similar proof is needed for each of the cases $t_2 \neq 0, \ldots, t_k \neq 0$. Specifically, in the case where you assume that $t_j \neq 0$, you can show that $\mathbf{v}_j$ is a linear combination of the remaining vectors.

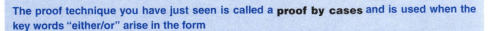

The proof technique you have just seen is called a **proof by cases** and is used when the key words "either/or" arise in the form

if *p* or *q*, then *r*,

where *p*, *q*, and *r* are statements. You proceed by cases: (Case 1) Assume that *p* is true and prove that *r* is true and then (Case 2) Assume that *q* is true and prove that *r* is true. It often happens that the proof in Case 2 is similar in spirit to that in Case 1 (this is the situation in the current proof). Typically, only Case 1 is presented in detail. In written proofs, a mathematician might say "Proceeding by cases, assume, without loss of generality, that Case 1 holds, ...," which means that only one of the cases is presented and the details of the remaining case are left to the reader. A further discussion of this proof technique is given in Appendix A.11.

Returning to the current proof, to complete part (a), assume that one of $\mathbf{v}_1, \ldots, \mathbf{v}_k$ is a linear combination of the remaining vectors. It must be shown that $\mathbf{v}_1, \ldots, \mathbf{v}_k$ are linearly dependent by constructing real numbers $t_1, \ldots, t_k$, not all 0, such that

$$(t_1 \odot \mathbf{v}_1) \oplus (t_2 \odot \mathbf{v}_2) \oplus \cdots \oplus (t_k \odot \mathbf{v}_k) = \mathbf{0} \tag{4.34}$$

To construct these real numbers, use the assumption that one of the vectors $\mathbf{v}_1, \ldots, \mathbf{v}_k$ is a linear combination of the remaining vectors. Because you do not know which vector is a linear combination of the remaining ones, you can again proceed by cases. As just discussed, assume, without loss of generality, that $\mathbf{v}_1$ is a linear combination of $\mathbf{v}_2, \ldots, \mathbf{v}_k$. By Definition 4.3, there are real numbers $s_2, \ldots, s_k$ such that

$$\mathbf{v}_1 = (s_2 \odot \mathbf{v}_2) \oplus \cdots \oplus (s_k \odot \mathbf{v}_k) \tag{4.35}$$

Rewriting (4.35), you have that

$$1 \odot \mathbf{v}_1 \oplus (-s_2) \odot \mathbf{v}_2 \oplus \cdots \oplus (-s_k) \odot \mathbf{v}_k = \mathbf{0} \tag{4.36}$$

Comparing (4.36) and (4.34), you can see that the desired real numbers satisfying (4.34) are

$$t_1 = 1, \quad t_2 = -s_2, \quad \ldots, \quad t_k = -s_k$$

which are not all zero because $t_1 = 1$. Thus, $\mathbf{v}_1, \ldots, \mathbf{v}_k$ are linearly dependent and this completes the proof of part (a).

The proof of part (b) follows by writing the negation of the hypothesis and the conclusion in part (a), thus completing the proof. ■

One of the advantages of the vectors $\mathbf{v}_1, \ldots, \mathbf{v}_k$ being linearly independent is that if a vector has components relative to $\mathbf{v}_1, \ldots, \mathbf{v}_k$, then those components are unique, as shown in the following theorem.

THEOREM 4.5

If $\mathbf{v}_1, \ldots, \mathbf{v}_k$ are linearly independent vectors in a vector space $(V, \oplus, \odot)$ and $\mathbf{v} \in V$ is a linear combination of $\mathbf{v}_1, \ldots, \mathbf{v}_k$, then there are unique real numbers $t_1, \ldots, t_k$ such that

$$\mathbf{v} = (t_1 \odot \mathbf{v}_1) \oplus \cdots \oplus (t_k \odot \mathbf{v}_k)$$

Proof.

Because $\mathbf{v}$ is a linear combination of $\mathbf{v}_1, \ldots, \mathbf{v}_k$, there are real numbers $t_1, \ldots, t_k$ such that

$$\mathbf{v} = (t_1 \odot \mathbf{v}_1) \oplus \cdots \oplus (t_k \odot \mathbf{v}_k) \tag{4.37}$$

Utilizing the uniqueness method, you can assume that $s_1, \ldots, s_k$ also satisfy

$$\mathbf{v} = (s_1 \odot \mathbf{v}_1) \oplus \cdots \oplus (s_k \odot \mathbf{v}_k) \tag{4.38}$$

It must be shown that $s_1 = t_1, \ldots, s_k = t_k$. So, subtracting (4.37) from (4.38) yields

$$\mathbf{0} = (s_1 - t_1) \odot \mathbf{v}_1 \oplus \cdots \oplus (s_k - t_k) \odot \mathbf{v}_k \tag{4.39}$$

You can now use the fact that $\mathbf{v}_1, \ldots, \mathbf{v}_k$ are linearly independent to conclude, from Definition 4.9 by specialization, that all of the real numbers in (4.39) are 0, that is,

$$s_1 - t_1 = 0, \ldots, s_k - t_k = 0$$

It then follows that $s_1 = t_1, \ldots, s_k = t_k$. This establishes the uniqueness and completes the proof. ■

4.3.4 A Basis for a Vector Space

Recall, from the end of Section 4.3.1, that the objective is to develop conditions to overcome the following two obstacles to finding components of a vector $\mathbf{v}$ in a vector space $(V, \oplus, \odot)$ relative to the special vectors $\mathbf{v}_1, \ldots, \mathbf{v}_k$:

1. The vector $\mathbf{v}$ may not have components relative to $\mathbf{v}_1, \ldots, \mathbf{v}_k$ because $\mathbf{v}$ is not a linear combination of these vectors.

2. Even if $\mathbf{v}$ is a linear combination of the vectors $\mathbf{v}_1, \ldots, \mathbf{v}_k$, there can be different sets of values for $t_1, \ldots, t_k$ that satisfy

$$\mathbf{v} = (t_1 \odot \mathbf{v}_1) \oplus \cdots \oplus (t_k \odot \mathbf{v}_k)$$

Such conditions have now been found. In particular, if the special vectors $\mathbf{v}_1, \ldots, \mathbf{v}_k$ span V (see Definition 4.4 in Section 4.3.2), then every vector in V is a linear combination of $\mathbf{v}_1, \ldots, \mathbf{v}_k$, thus resolving (1). Also, if $\mathbf{v}_1, \ldots, \mathbf{v}_k$ are linearly independent (see Definition 4.9 in Section 4.3.3), then each vector is a unique linear combination of $\mathbf{v}_1, \ldots, \mathbf{v}_k$, thus resolving (2). A collection of special vectors satisfying these two desirable properties is formalized in the following definition.

DEFINITION 4.10

The vectors $\mathbf{v}_1, \ldots, \mathbf{v}_k$ are a **basis** for a vector space $(V, \oplus, \odot)$ if and only if the following two conditions hold:

(1) The vectors $\mathbf{v}_1, \ldots, \mathbf{v}_k$ span V.

(2) The vectors $\mathbf{v}_1, \ldots, \mathbf{v}_k$ are linearly independent.

The fact that a basis $\mathbf{v}_1, \ldots, \mathbf{v}_k$ provides a unique set of components for a vector $\mathbf{v}$ relative to $\mathbf{v}_1, \ldots, \mathbf{v}_k$ is summarized in the following theorem whose proof is based on Definition 4.4, Theorem 4.5, and Definition 4.10.

THEOREM 4.6

If the vectors $\mathbf{v}_1, \ldots, \mathbf{v}_k$ in a vector space $(V, \oplus, \odot)$ are a basis for V, then, for any vector $\mathbf{v} \in V$, there are unique real numbers $t_1, \ldots, t_k$ such that

$$\mathbf{v} = (t_1 \odot \mathbf{v}_1) \oplus \cdots \oplus (t_k \odot \mathbf{v}_k)$$

In this case, $t_1, \ldots, t_k$ are the components of $\mathbf{v}$ relative to $\mathbf{v}_1, \ldots, \mathbf{v}_k$.

Proof.

The quantifier *for any* in the conclusion suggests using the choose method to choose a vector $\mathbf{v} \in V$. It must then be shown that there are unique real numbers $t_1, \ldots, t_k$ such that

$$\mathbf{v} = (t_1 \odot \mathbf{v}_1) \oplus \cdots \oplus (t_k \odot \mathbf{v}_k) \tag{4.40}$$

Condition (1) in Definition 4.10 ensures that there are real numbers $t_1, \ldots, t_k$ such that (4.40) is true. Condition (2) in Definition 4.10 ensures, by Theorem 4.5, that there is only one set of real numbers that satisfies (4.40). The proof is now complete. ■

The next four examples illustrate bases for different vector spaces, V. Each basis consists of a collection of linearly independent vectors that span V.

EXAMPLE 4.16 A Basis for R^3

The vectors $\mathbf{i} = (1, 0, 0)$, $\mathbf{j} = (0, 1, 0)$, and $\mathbf{k} = (0, 0, 1)$ form a basis for R^3 because these vectors span R^3 (see Example 4.5) and are linearly independent (see Example 4.12). Consequently, any vector $\mathbf{v} \in R^3$ has unique components relative to the basis $\mathbf{i}$, $\mathbf{j}$, and $\mathbf{k}$. For instance, the unique components of the vector $\mathbf{v} = (3, -1, 2)$ relative to $\mathbf{i}$, $\mathbf{j}$, and $\mathbf{k}$ are $(3, -1, 2)$ because

$$\mathbf{v} = (3, -1, 2) = 3(1, 0, 0) - 1(0, 1, 0) + 2(0, 0, 1) = 3\mathbf{i} - 1\mathbf{j} + 2\mathbf{k}$$

EXAMPLE 4.17 A Basis for R^n

As you are asked to show in Exercise 32, the following standard unit vectors are a basis for R^n (called the **standard basis for R^n**):

$$\mathbf{e}_1 = (1, 0, 0, \ldots, 0)$$
$$\mathbf{e}_2 = (0, 1, 0, \ldots, 0)$$
$$\vdots \quad \vdots \qquad \vdots$$
$$\mathbf{e}_n = (0, 0, 0, \ldots, 1)$$

EXAMPLE 4.18 A Basis for the Vector Space of (2 x 2) Matrices

The matrices

$$M^{11} = \begin{bmatrix} 1 & 0 \\ 0 & 0 \end{bmatrix}, \quad M^{12} = \begin{bmatrix} 0 & 1 \\ 0 & 0 \end{bmatrix}, \quad M^{21} = \begin{bmatrix} 0 & 0 \\ 1 & 0 \end{bmatrix}, \quad M^{22} = \begin{bmatrix} 0 & 0 \\ 0 & 1 \end{bmatrix}$$

are a basis for the vector space $R^{2 \times 2}$ of all (2×2) matrices under matrix addition and multiplication of a matrix by a real number. This is because these four matrices span $R^{2 \times 2}$ (see Example 4.7) and are linearly independent (see Example 4.14). Consequently, any (2×2) matrix M has unique components relative to the basis M^{11}, M^{12}, M^{21}, and M^{22}. For instance, the unique components of the matrix

$$M = \begin{bmatrix} 2 & -1 \\ -3 & 5 \end{bmatrix}$$

relative to M^{11}, M^{12}, M^{21}, and M^{22} are $(2, -1, -3, 5)$ because

$$M = \begin{bmatrix} 2 & -1 \\ -3 & 5 \end{bmatrix} = 2\begin{bmatrix} 1 & 0 \\ 0 & 0 \end{bmatrix} - 1\begin{bmatrix} 0 & 1 \\ 0 & 0 \end{bmatrix} - 3\begin{bmatrix} 0 & 0 \\ 1 & 0 \end{bmatrix} + 5\begin{bmatrix} 0 & 0 \\ 0 & 1 \end{bmatrix}$$

EXAMPLE 4.19 A Basis for the Vector Space of Polynomials
The polynomials

$$p_0(x) = 1, \quad p_1(x) = x, \quad p_2(x) = x^2, \quad \ldots, \quad p_n(x) = x^n$$

are a basis for the vector space P^n of all polynomials of degree n or less, under addition of polynomials and multiplying a polynomial by a real number. This is because these $n + 1$ polynomials span P^n (see Example 4.8) and are linearly independent (see Example 4.15). Consequently, any polynomial of degree $\leq n$ has unique components relative to the basis $p_0(x), p_1(x), \ldots, p_n(x)$. For instance, the unique components of the polynomial

$$p(x) = a_0 + a_1 x + a_2 x^2 + \cdots + a_n x^n$$

relative to $p_0(x), p_1(x), \ldots, p_n(x)$ are $(a_0, a_1, \ldots, a_n)$ because

$$p(x) = a_0 p_0(x) + a_1 p_1(x) + a_2 p_2(x) + \cdots + a_n p_n(x)$$

One use of the components $(a_0, a_1, a_2, \ldots, a_n)$ of the polynomial $p(x) = a_0 + a_1 x + a_2 x^2 + \cdots + a_n x^n$ described in Example 4.19 is to represent $p(x)$ as an n-vector that can be stored in the memory of a computer. This approach for working with polynomials on computers was presented in Section 1.1.3.

You now know that a basis for a vector space $(V, \oplus, \odot)$, which is a collection of linearly independent vectors that span V, provides unique components for each vector $\mathbf{v} \in V$. In the next section, you will learn how to find a basis for certain vector spaces.

Exercises for Section 4.3

1. Show that the vector $(4, 2)$ is a linear combination of the vectors $(2, 0)$ and $(0, -2)$.

2. Is the vector $(-1, 1)$ a linear combination of the vectors $(1, 1)$ and $(-2, -2)$? Why or why not? Explain.

3. Show that the polynomial $-1 - 2x + x^2$ is a linear combination of the polynomials $1 + x, x + x^2$, and x^2.

4. Is the polynomial $x^2 - 2x + 1$ a linear combination of the polynomials $1 + x + x^2$ and $2x$? Why or why not? Explain.

5. Write a system of linear equations to determine if the vector $(1, -2, 0, 1)$ is a linear combination of the vectors $(1, 0, 3, -1), (0, -1, 0, -2), (0, 0, -1, 2)$. You need not solve the system.

6. Write a system of linear equations to determine if an n-vector $\mathbf{u} = (u_1, \ldots, u_n)$ is a linear combination of the given n-vectors $\mathbf{v}_1, \ldots, \mathbf{v}_k$.

7. Show that the vectors $(2, 0)$ and $(0, -2)$ span R^2.

8. Do the vectors $(1, 1)$ and $(-2, -2)$ span R^2? Why or why not? Explain.

9. Show that the following matrices span the set of all (2×2) matrices:

$$A = \begin{bmatrix} 1 & 0 \\ 0 & 0 \end{bmatrix}, \quad B = \begin{bmatrix} 0 & -1 \\ 0 & 0 \end{bmatrix}, \quad C = \begin{bmatrix} 0 & 0 \\ -2 & 0 \end{bmatrix}, \quad D = \begin{bmatrix} 0 & 1 \\ 1 & 1 \end{bmatrix}$$

10. Show that $1 + x, x + x^2$, and x^2 span P^2, the set of all polynomials of degree 2 or less.

11. Prove that if the vectors $\mathbf{v}_1, \ldots, \mathbf{v}_{k+1}$ belong to a vector space $(V, \oplus, \odot)$, then span $\{\mathbf{v}_1, \ldots, \mathbf{v}_k\} \subseteq$ span $\{\mathbf{v}_1, \ldots, \mathbf{v}_{k+1}\}$.

12. Prove that if $\mathbf{v}_1, \ldots, \mathbf{v}_k$ span a vector space $(V, \oplus, \odot)$ and $\mathbf{w}$ is a vector that is not in span$\{\mathbf{v}_1, \ldots, \mathbf{v}_k\}$, then $\mathbf{v}_1, \ldots, \mathbf{v}_k, \mathbf{w}$ span V.

13. Show that the vectors $(-1, 1)$ and $(2, 1)$ are linearly independent.

14. Show that the vectors $(1, 0, -1), (0, 2, 0)$, and $(1, -1, 1)$ are linearly independent.

15. Show that the matrices in Exercise 9 are linearly independent.

16. Show that the polynomials $1 + x, x + x^2$, and x^2 are linearly independent.

17. (a) Use Definition 4.7 in Section 4.3.3 to prove that if $\mathbf{u}$ and $\mathbf{v}$ are nonzero n-vectors for which $\mathbf{u} \cdot \mathbf{v} = 0$, then $\mathbf{u}$ and $\mathbf{v}$ are linearly independent.
    (b) Prove that if $\mathbf{u}$ and $\mathbf{v}$ are nonzero vectors in 3-space for which $\mathbf{u} \times \mathbf{v} \neq \mathbf{0}$, then $\mathbf{u}$ and $\mathbf{u} \times \mathbf{v}$ are linearly independent.

18. Use Definition 4.7 in Section 4.3.3 to prove that if the n-vectors $\mathbf{v}_1, \ldots, \mathbf{v}_k$ are linearly independent, then any nonempty subset of these contains linearly independent vectors.

19. Suppose that $\mathbf{v}_1, \ldots, \mathbf{v}_k$ are nonzero vectors in a vector space $(V, \oplus, \odot)$ such that no vector is a linear combination of its predecessors. Prove that $\mathbf{v}_1, \ldots, \mathbf{v}_k$ are linearly independent vectors.

20. Suppose the columns of an $(m \times n)$ matrix A are linearly independent. Prove that if $\mathbf{v}_1, \ldots, \mathbf{v}_k$ are linearly independent n-vectors, then $A\mathbf{v}_1, \ldots, A\mathbf{v}_k$ are linearly independent m-vectors.

21. Show that the vectors $(1, 1)$ and $(-2, -2)$ are linearly dependent.

22. Show that the vectors $(1, 0, -1), (2, 1, -3)$, and $(0, 1, -1)$ are linearly dependent.

23. Show that the following matrices are linearly dependent by expressing C as a linear combination of A and B:

$$A = \begin{bmatrix} 1 & -2 \\ -1 & 0 \end{bmatrix}, \quad B = \begin{bmatrix} -3 & 0 \\ 2 & -4 \end{bmatrix}, \quad C = \begin{bmatrix} 6 & -6 \\ -5 & 4 \end{bmatrix}$$

24. Show that the polynomials $1 + x + x^2, -2 + 3x^2$, and $-2x - 5x^2$ are linearly dependent by expressing $-2x - 5x^2$ as a linear combination of the other two polynomials.

25. Prove that the following two statements are equivalent for the n-vectors $\mathbf{u}, \mathbf{v}$, and $\mathbf{w}$:
    (i) There are real numbers a, b, and c, at least one of which is 1, such that $a\mathbf{u} + b\mathbf{v} + c\mathbf{w} = \mathbf{0}$.
    (ii) There are real numbers d, e, and f, not all 0, such that $d\mathbf{u} + e\mathbf{v} + f\mathbf{w} = \mathbf{0}$.

26. Prove that if $\mathbf{v}_1, \ldots, \mathbf{v}_k$ are vectors in a vector space $(V, \oplus, \odot)$, then $\mathbf{v}_1, \ldots, \mathbf{v}_k, \mathbf{0}$ are linearly dependent.

27. Suppose that A is an $(m \times n)$ matrix. Prove that if $\mathbf{v}_1, \ldots, \mathbf{v}_k$ are linearly dependent n-vectors, then $A\mathbf{v}_1, \ldots, A\mathbf{v}_k$ are linearly dependent m-vectors.

28. (a) Create an alternate definition of two vectors in 2-space being linearly dependent by using the angle between the two vectors. Be sure to verify your definition.

(b) Explain what difficulties arise in trying to generalize your definition in part (a) to a collection of k vectors in n-space.

29. Prove that $(2, 0)$ and $(2, 1)$ form a basis for R^2.

30. Prove that $(1, 0, 1)$, $(-1, 1, 0)$, and $(0, 0, 1)$ form a basis for R^3.

31. Prove that $(1, 2)$ and $(-2, 4)$ form a basis for R^2.

32. Prove that the standard unit n-vectors $\mathbf{e}_1, \ldots, \mathbf{e}_n$ form a basis for R^n.

33. Suppose that $\mathbf{v}_1, \mathbf{v}_2, \mathbf{v}_3$ form a basis for a vector space $(V, \oplus, \odot)$. Prove that the vectors $\mathbf{v}_1, \mathbf{v}_1 \oplus \mathbf{v}_2, \mathbf{v}_1 \oplus \mathbf{v}_2 \oplus \mathbf{v}_3$ form a basis for V.

34. Let $\mathbf{v}_1, \ldots, \mathbf{v}_k$ be linearly independent vectors in a vector space $(V, \oplus, \odot)$ and let $U = \text{span}\{\mathbf{v}_1, \ldots, \mathbf{v}_k\}$. Prove that if $\mathbf{w}$ is a vector not in U, then $\mathbf{v}_1, \ldots, \mathbf{v}_k, \mathbf{w}$ form a basis for $\text{span}\{\mathbf{v}_1, \ldots, \mathbf{v}_k, \mathbf{w}\}$.

35. Prove that if $\mathbf{v}_1, \ldots, \mathbf{v}_k$ form a basis for a vector space $(V, \oplus, \odot)$ and $\mathbf{w}$ is a vector that is not in $\text{span}\{\mathbf{v}_1, \ldots, \mathbf{v}_{k-1}\}$, then $\mathbf{v}_1, \ldots, \mathbf{v}_{k-1}, \mathbf{w}$ form a basis for V.

36. Find the components of the given vectors relative to the basis $\mathbf{v}_1 = (-1, 1)$, $\mathbf{v}_2 = (1, 1)$.
    (a) $\mathbf{v} = (1, -1)$.   (b) $\mathbf{v} = (1, 0)$.

37. Find the components of each of the given matrices relative to the following basis for the set of all (2×2) matrices:

$$A = \begin{bmatrix} 1 & 0 \\ 0 & 0 \end{bmatrix}, \quad B = \begin{bmatrix} 0 & -1 \\ 0 & 0 \end{bmatrix}, \quad C = \begin{bmatrix} 0 & 0 \\ -2 & 0 \end{bmatrix}, \quad D = \begin{bmatrix} 0 & 1 \\ 1 & 1 \end{bmatrix}$$

(a) $\begin{bmatrix} 1 & 0 \\ 0 & 1 \end{bmatrix}$ (b) $\begin{bmatrix} -1 & 1 \\ 2 & 0 \end{bmatrix}$

4.4 The Dimension of a Vector Space

A line is a one-dimensional object, a plane is two dimensional, and the world is three dimensional. These notions of dimension, as they pertain to a vector space, are now made precise using the concept of a basis.

4.4.1 Using a Basis to Define Dimension

Some vector spaces—such as R^n—have a basis consisting of a finite number of vectors. Other vector spaces—such as the vector space of all real-valued functions of a single variable—do not. This difference gives rise to the following definition.

DEFINITION 4.11

A vector space $(V, \oplus, \odot)$ for which there is a positive integer n and vectors $\mathbf{v}_1, \ldots, \mathbf{v}_n$ in V that form a basis for V is a **finite-dimensional vector space**. By convention, $V = \{\mathbf{0}\}$ is finite dimensional, even though this vector space has no basis. In all other cases, V is **infinite dimensional**, meaning that V has no basis consisting of a finite number of vectors.

From Examples 4.17, 4.18, and 4.19, you can see that R^n, $R^{2\times2}$, and P^n are finite-dimensional vector spaces. In contrast, the vector space of all polynomials of all degrees, the vector space of all real-valued functions of a single variable, and the vector space of all digital audio signals (see Example F in Section 4.1.5) are examples of infinite-dimensional vector spaces.

To determine the dimension of a finite-dimensional vector space, consider the special cases of R^2 and R^3. One way to explain the fact that R^2 is two dimensional is to note that there are two vectors in the basis $(1, 0)$ and $(0, 1)$. Likewise, R^3 is three dimensional because the basis $(1, 0, 0)$, $(0, 1, 0)$, and $(0, 0, 1)$ contains three vectors. Applying generalization, a reasonable way to define the dimension of a finite-dimensional vector space is by the number of vectors in a basis. But what if a vector space has two bases, each with a different number of vectors? Such a situation cannot happen, the proof of which is based on the following property of a basis.

THEOREM 4.7

If $\mathbf{v}_1, \ldots, \mathbf{v}_n$ form a basis for a vector space $(V, \oplus, \odot)$ and $\mathbf{w}_1, \ldots, \mathbf{w}_k$ are vectors in V with $k > n$, then the vectors $\mathbf{w}_1, \ldots, \mathbf{w}_k$ are linearly dependent.

Proof.

To show that the vectors $\mathbf{w}_1, \ldots, \mathbf{w}_k$ are linearly dependent, by Definition 4.8, it is necessary to construct real numbers $t_1, \ldots, t_k$, not all 0, such that

$$(t_1 \odot \mathbf{w}_1) \oplus \cdots \oplus (t_k \odot \mathbf{w}_k) = \mathbf{0} \tag{4.41}$$

These real numbers are constructed using the assumption that $\mathbf{v}_1, \ldots, \mathbf{v}_n$ is a basis, which, as a result, means that each vector $\mathbf{w}_i$ is a linear combination of $\mathbf{v}_1, \ldots, \mathbf{v}_n$. That is, there are real numbers s_{ij} such that

$$\mathbf{w}_1 = (s_{11} \odot \mathbf{v}_1) \oplus (s_{12} \odot \mathbf{v}_2) \oplus \cdots \oplus (s_{1n} \odot \mathbf{v}_n)$$
$$\mathbf{w}_2 = (s_{21} \odot \mathbf{v}_1) \oplus (s_{22} \odot \mathbf{v}_2) \oplus \cdots \oplus (s_{2n} \odot \mathbf{v}_n)$$
$$\vdots \qquad \vdots \qquad \qquad \vdots \qquad \qquad \vdots$$
$$\mathbf{w}_k = (s_{k1} \odot \mathbf{v}_1) \oplus (s_{k2} \odot \mathbf{v}_2) \oplus \cdots \oplus (s_{kn} \odot \mathbf{v}_n)$$

Multiplying each of the foregoing equations i by the corresponding real number t_i (as yet unknown), adding up the equations, and comparing the result with (4.41), you want $t_1, \ldots, t_k$ to satisfy

$$s_{11}t_1 + s_{21}t_2 + \cdots + s_{k1}t_k = 0$$
$$s_{12}t_1 + s_{22}t_2 + \cdots + s_{k2}t_k = 0$$
$$\vdots$$
$$s_{1n}t_1 + s_{2n}t_2 + \cdots + s_{kn}t_k = 0$$

The foregoing is a consistent system of n equations in the k unknowns $t_1, \ldots, t_k$ in which $k > n$. By Theorem 2.6 in Section 2.4.6, this system has an infinite number of solutions and so has a nontrivial solution, thus completing the proof. ■

A consequence of Theorem 4.7 is that every basis of a finite-dimensional vector space has the same number of vectors, as is now shown.

THEOREM 4.8

Any two bases in a finite-dimensional vector space $(V, \oplus, \odot)$ have the same number of vectors.

Proof.

Let $\mathbf{v}_1, \ldots, \mathbf{v}_n$ and $\mathbf{w}_1, \ldots, \mathbf{w}_k$ be two bases for a finite-dimensional vector space V. It is necessary to show that $n = k$, which is accomplished by showing that both $k \leq n$ and $n \leq k$. From Theorem 4.7, you can conclude that $k \leq n$ because $\mathbf{v}_1, \ldots, \mathbf{v}_n$ form a basis and $\mathbf{w}_1, \ldots, \mathbf{w}_k$ are linearly independent vectors. Likewise, $n \leq k$ because $\mathbf{w}_1, \ldots, \mathbf{w}_k$ form a basis and $\mathbf{v}_1, \ldots, \mathbf{v}_n$ are linearly independent vectors. Thus $n = k$ and the proof is complete. ■

As a result of Theorem 4.8, the number of vectors in a basis provides the dimension of a vector space, as formalized in the following definition.

DEFINITION 4.12

The **dimension** of a finite-dimensional vector space V, denoted by dim V, is the number of vectors in any basis of V. By agreement, the dimension of the zero vector space is 0.

The dimension of the vector space R^n is n because the n standard unit vectors in Example 4.17 form a basis. Analogously, the dimension of the vector space P^n of all polynomials of degree n or less is $n + 1$ because the basis in Example 4.19 consists of $n + 1$ polynomials.

To show that a collection of vectors is a basis for a vector space V, by Definition 4.10, it is necessary to show that the vectors are linearly independent and that they span V. In the spirit of using mathematics to reduce the amount of work needed to solve a problem, if you know that the dimension of a vector space is n, then, to verify that a collection of n vectors is a basis, you need show only that these vectors are linearly independent or that the vectors span V, not both. This fact is summarized in the following theorem, whose proof is left to Exercise 5 and Exercise 6.

THEOREM 4.9

Let $(V, \oplus, \odot)$ be an n-dimensional vector space.

(a) If $\mathbf{v}_1, \ldots, \mathbf{v}_n$ are linearly independent vectors in V, then $\mathbf{v}_1, \ldots, \mathbf{v}_n$ form a basis for V.

(b) If $\mathbf{v}_1, \ldots, \mathbf{v}_n$ are vectors in V that span V, then $\mathbf{v}_1, \ldots, \mathbf{v}_n$ form a basis for V.

One task that still remains is to develop a systematic procedure for finding a basis in a finite-dimensional vector space. Such a procedure is developed now for certain subspaces of R^n.

4.4.2 Row Space, Column Space, and Null Space

Suppose you are working with all linear combinations of the vectors

$$\mathbf{v}_1 = (-1, 0, \quad 2, -3, \quad 1), \quad \mathbf{v}_2 = (0, \quad 1, 0, 1, 1)$$
$$\mathbf{v}_3 = (\quad 2, 0, -4, \quad 6, -2), \quad \mathbf{v}_4 = (1, -4, 2, 7, 7)$$

that is, with span(S), where $S = \{\mathbf{v}_1, \mathbf{v}_2, \mathbf{v}_3, \mathbf{v}_4\}$. It is usually necessary to find a basis for span(S) so you can express a general vector $\mathbf{v} \in$ span(S) as a linear combination of these basis vectors.

Each vector in span(S) belongs to R^5, so you might think that the five standard unit vectors in R^5, which are linearly independent and span span(S), constitute a basis for span(S). This is not the case, however, because only the vector $(0, 0, 0, 0, 1)$ belongs to span(S), as you are asked to show in Exercise 9. All of the vectors in a basis of a vector space V must belong to V, so, how can you find a basis for span(S)?

The Row Space of a Matrix

One way to do so for this example is to construct a matrix consisting of four rows, one for each of the four vectors, that is:

$$A = \begin{bmatrix} \mathbf{v}_1 \\ \mathbf{v}_2 \\ \mathbf{v}_3 \\ \mathbf{v}_4 \end{bmatrix} = \begin{bmatrix} -1 & 0 & 2 & -3 & 1 \\ 0 & 1 & 0 & 1 & 1 \\ 2 & 0 & -4 & 6 & -2 \\ 1 & -4 & 2 & 7 & 7 \end{bmatrix}$$

Observe that span(S) is the set of all linear combinations of the rows of A. In general, for an ($m \times n$) matrix A, the **row space** of A is the subspace of R^n spanned by the m rows of A, that is, the set of all linear combinations of the rows of A.

One approach to finding a basis for the row space of a matrix A consists of using elementary row operations to reduce A to a row-echelon form, R. The nonzero rows of R then constitute a basis for the row space of A. This approach is illustrated in the following example and then proved to be correct in Theorem 4.10 and Theorem 4.11.

EXAMPLE 4.20 Using Elementary Row Operations to Find a Basis for the Row Space of a Matrix

To find a basis for the row space of the matrix

$$A = \begin{bmatrix} -1 & 0 & 2 & -3 & 1 \\ 0 & 1 & 0 & 1 & 1 \\ 2 & 0 & -4 & 6 & -2 \\ 1 & -4 & 2 & 7 & 7 \end{bmatrix}$$

first apply elementary row operations to obtain the following matrix in row-echelon form:

$$R = \begin{bmatrix} 1 & 0 & -2 & 3 & -1 \\ 0 & 1 & 0 & 1 & 1 \\ 0 & 0 & 1 & 2 & 3 \\ 0 & 0 & 0 & 0 & 0 \end{bmatrix}$$

The nonzero rows of R, namely,

$$R_{1*} = [\, 1 \;\; 0 \;\; -2 \;\;\; 3 \;\; -1 \,], \quad R_{2*} = [\, 0 \;\; 1 \;\; 0 \;\; 1 \;\; 1 \,], \quad R_{3*} = [\, 0 \;\; 0 \;\; 1 \;\; 2 \;\; 3 \,]$$

constitute a basis for the row space of both R and A.

The approach in Example 4.20 works because (1) row operations do not affect the row space of A, which means that a basis for the row space of R is also a basis for the row space of A and (2) the nonzero rows of R form a basis for the row space of R and hence for the row space of A. These two facts are formalized in the next two theorems.

THEOREM 4.10

If B is an $(m \times n)$ matrix obtained from an $(m \times n)$ matrix A by performing a sequence of elementary row operations, then the row space of B is the same as the row space of A.

Proof.

Suppose first that B is obtained from A by performing a single elementary row operation. The key question associated with showing that the row space of A is the same as the row space of B is, "How can I show that two sets are equal?" To that end, it is shown both that the row space of A is a subset of the row space of B, and vice versa.

Using the definition of subset, you must now show that every element of the row space of A is a member of the row space of B. Recognize the key word *every* and use the choose method to choose an element $\mathbf{x}$ in the row space of A. It must be shown that $\mathbf{x}$ is in the row space of B. Because $\mathbf{x}$ is in the row space of A, $\mathbf{x}$ is a linear combination of the rows of A, that is, there are real numbers $s_1, \ldots, s_m$ such that

$$\mathbf{x} = s_1 A_{1*} + \cdots + s_i A_{i*} + \cdots + s_j A_{j*} + \cdots + s_m A_{m*} \tag{4.42}$$

To show that $\mathbf{x}$ is in the row space of B, it must be shown that $\mathbf{x}$ is a linear combination of the rows of B, that is, that there are real numbers $t_1, \ldots, t_m$ such that

$$\mathbf{x} = t_1 B_{1*} + \cdots + t_i B_{i*} + \cdots + t_j B_{j*} + \cdots + t_m B_{m*} \tag{4.43}$$

The way in which the values for $t_1, \ldots, t_m$ are constructed depends on the specific elementary row operation used to create B. For example, if B is obtained by multiplying row i of A by the nonzero constant c, then $B_{i*} = cA_{i*}$ and hence $A_{i*} = (1/c)B_{i*}$. You can then write (4.42) as follows:

$$\mathbf{x} = s_1 B_{1*} + \cdots + (s_i/c) B_{i*} + \cdots + s_j B_{j*} + \cdots + s_m B_{m*} \tag{4.44}$$

Compare (4.44) with (4.43). You can see that the desired values of the real numbers $t_1, \ldots, t_m$

are $t_1 = s_1, \ldots, t_i = s_i/c, \ldots, t_j = s_j, \ldots, t_m = s_m$.

As another example, if B is obtained from A by interchanging row i of A with row j of A, then $B_{i*} = A_{j*}$ and $B_{j*} = A_{i*}$, so you can write (4.42) as follows:

$$\begin{aligned}
\mathbf{x} &= s_1 B_{1*} + \cdots + s_i B_{j*} + \cdots + s_j B_{i*} + \cdots + s_m B_{m*} \\
&= s_1 B_{1*} + \cdots + s_j B_{i*} + \cdots + s_i B_{j*} + \cdots + s_m B_{m*}
\end{aligned} \tag{4.45}$$

Comparing (4.45) with (4.43), you can see that the values of the real numbers $t_1, \ldots, t_m$ are $t_1 = s_1, \ldots, t_i = s_j, \ldots, t_j = s_i, \ldots, t_m = s_m$.

Finally, if B is obtained from A by adding c times row i of A to row j of A, then $B_{j*} = A_{j*} + cA_{i*}$ and hence $A_{j*} = B_{j*} - cA_{i*} = B_{j*} - cB_{i*}$. You can then write (4.42) as follows:

$$\begin{aligned}
\mathbf{x} &= s_1 B_{1*} + \cdots + s_i B_{i*} + \cdots + s_j (B_{j*} - cB_{i*}) + \cdots + s_m B_{m*} \\
&= s_1 B_{1*} + \cdots + (s_i - cs_j) B_{i*} + \cdots + s_j B_{j*} + \cdots + s_m B_{m*}
\end{aligned} \tag{4.46}$$

Comparing (4.46) with (4.43), you can see that the values of the real numbers $t_1, \ldots, t_m$ are $t_1 = s_1, \ldots, t_i = s_i - cs_j, \ldots, t_j = s_j, \ldots, t_m = s_m$.

A similar proof is needed to establish that the row space of B is a subset of the row space of A. After doing so, you can conclude that the row space of A is the same as the row space of B. The proof is completed by repeating this argument for any finite number of row operations performed on A. ∎

As a result of Theorem 4.10, the row space of a row-echelon form, R, of a matrix A is the same as the row space of A. The advantage of working with R is that the nonzero rows are linearly independent and thus form a basis for the row space of R and hence, by Theorem 4.10, of A. This fact is summarized in the following theorem, whose proof is left to Exercise 10.

THEOREM 4.11

Let R be a row-echelon form of a matrix A. The nonzero rows of R form a basis for the row space of both R and A.

The Column and Null Space of a Matrix

Observe that not all of the vectors in the basis in Example 4.20 are rows of A. If you want to find a basis for the row space of A that consists exclusively of rows of A, then you need to work with the **column space** of A, which is the subspace of R^m spanned by the n columns of A.

A row-echelon form R is again used to find a basis for the column space of a matrix A. This is because it is easy to identify a group of columns of R that form a basis for the column space of R, as indicated in the next theorem, whose proof is left to Exercise 19.

THEOREM 4.12

If R is a matrix in row-echelon form, then the columns of R that contain the leading values in the rows of R form a basis for the column space of R.

EXAMPLE 4.21 Finding a Basis for the Column Space of a Matrix in Row-Echelon Form

The matrix

$$R = \begin{bmatrix} 1 & -1 & 2 & 0 & 3 \\ 0 & 0 & 1 & 1 & -1 \\ 0 & 0 & 0 & 0 & 1 \\ 0 & 0 & 0 & 0 & 0 \end{bmatrix}$$

is in row-echelon form, with leading values in rows 1, 2, and 3. According to Theorem 4.12, columns 1, 3, and 5 that contain the leading values form a basis for the column space of R. That is, the following vectors form a basis for the column space of R:

$$R_{*1} = \begin{bmatrix} 1 \\ 0 \\ 0 \\ 0 \end{bmatrix}, \quad R_{*3} = \begin{bmatrix} 2 \\ 1 \\ 0 \\ 0 \end{bmatrix}, \quad R_{*5} = \begin{bmatrix} 3 \\ -1 \\ 1 \\ 0 \end{bmatrix}$$

From Theorem 4.12 and Example 4.21, you can see that once a matrix A is reduced to a row-echelon form, R, it is easy to identify a basis for the column space of R. Unfortunately, these vectors need not form a basis for the column space of A but, as you will now see, the column space of A is related to the column space of R. The key to this relationship lies in the following definition:

DEFINITION 4.13

The **null space** of an $(m \times n)$ matrix A is the set of all solutions to the homogeneous system $A\mathbf{x} = \mathbf{0}$.

The null space of A indicates whether the columns of A are linearly independent or not because the columns of A are linearly independent if and only if the only solution to

$$A_{*1}x_1 + \cdots + A_{*n}x_n = \mathbf{0}$$

is $x_1 = 0, \ldots, x_n = 0$. Equivalently stated in matrix-vector notation, the columns of A are linearly independent if and only if the only solution to

$$A\mathbf{x} = \mathbf{0}$$

is $\mathbf{x} = \mathbf{0}$. In terms of the null space of A, the columns of A are linearly independent if and only if the null space of A consists of only the zero vector.

Performing an elementary row operation on a matrix A does not affect the null space of A. To see why, suppose that B is a matrix obtained by performing an elementary row operation on A, so

$$B = EA$$

where E is an (invertible) elementary matrix. Then $\mathbf{x}$ is a solution to

$$A\mathbf{x} = \mathbf{0} \quad \text{if and only if} \quad EA\mathbf{x} = E\mathbf{0} \quad \text{if and only if} \quad B\mathbf{x} = \mathbf{0}$$

In other words, the columns of A are linearly independent if and only if the columns of B are linearly independent. An extension of these facts is given in the next theorem, whose proof is omitted.

THEOREM 4.13

Suppose that B is a matrix obtained from a matrix A by elementary row operations.

(a) A set of columns of B is linearly independent if and only if the corresponding set of columns of A is linearly independent.

(b) A set of columns of B is a basis for the column space of B if and only if the corresponding columns of A form a basis for the column space of A.

In view of Theorem 4.13, one way to find a basis for the column space of a matrix A is to find a row-echelon form, R, and then to identify (by Theorem 4.12) a set of columns of R that forms a basis for the column space of R. Then, by Theorem 4.13, the corresponding columns of A form a basis for the column space of A. This process is illustrated in the following example.

EXAMPLE 4.22 Finding a Basis for the Column Space of a Matrix
To find a basis for the column space of

$$A = \begin{bmatrix} 1 & -1 & 2 & 0 & 3 \\ 2 & -2 & 5 & 1 & 5 \\ 1 & -1 & 3 & 1 & 4 \\ 3 & -3 & 6 & 0 & 9 \end{bmatrix}$$

first reduce the matrix to a row-echelon form. You can verify that the matrix R given in Example 4.21 is such a matrix. As indicated in Example 4.21, columns 1, 3, and 5 of R form a basis for the column space of R. By Theorem 4.13, the corresponding columns of A form a basis for the column space of A. Thus, the following vectors form a basis for the column space of A:

$$A_{*1} = \begin{bmatrix} 1 \\ 2 \\ 1 \\ 3 \end{bmatrix}, \quad A_{*3} = \begin{bmatrix} 2 \\ 5 \\ 3 \\ 6 \end{bmatrix}, \quad A_{*5} = \begin{bmatrix} 3 \\ 5 \\ 4 \\ 9 \end{bmatrix}$$

A final example brings together your knowledge of finding bases for the row and column space of a matrix.

EXAMPLE 4.23 Finding a Basis for the Row Space of a Matrix that Consists of Rows of the Matrix

To find a basis for the row space of the matrix

$$A = \begin{bmatrix} -1 & 0 & 2 & -3 & 1 \\ 0 & 1 & 0 & 1 & 1 \\ 2 & 0 & -4 & 6 & -2 \\ 1 & -4 & 2 & 7 & 7 \end{bmatrix}$$

that consists of rows of A, first transpose the matrix to obtain

$$A^t = \begin{bmatrix} -1 & 0 & 2 & 1 \\ 0 & 1 & 0 & -4 \\ 2 & 0 & -4 & 2 \\ -3 & 1 & 6 & 7 \\ 1 & 1 & -2 & 7 \end{bmatrix}$$

The row space of A is the same as the column space of A^t, so a basis for the column space of A^t is a basis for the row space of A. To find a basis for the column space of A^t, use the approach in Example 4.21 of finding a row-echelon form of A^t. Doing so results in the following matrix:

$$R = \begin{bmatrix} 1 & 0 & -2 & -1 \\ 0 & 1 & 0 & -4 \\ 0 & 0 & 0 & 1 \\ 0 & 0 & 0 & 0 \\ 0 & 0 & 0 & 0 \end{bmatrix}$$

By Theorem 4.12, columns 1, 2, and 4 of R form a basis for the column space of R, so by Theorem 4.13, columns 1, 2, and 4 of A^t form a basis for the column space of A^t. Consequently, rows 1, 2, and 4 of A form a basis for the row space of A. That is, the following rows of A form a basis for the row space of A:

$$A_{1*} = [\; -1 \quad 0 \quad 2 \quad -3 \quad 1\;]$$
$$A_{2*} = [\; 0 \quad 1 \quad 0 \quad 1 \quad 1\;]$$
$$A_{4*} = [\; 1 \quad -4 \quad 2 \quad 7 \quad 7\;]$$

These three rows of A are linearly independent and span the row space of A and thus constitute the desired basis.

4.4.3 Rank and Nullity: A Relation Between a Homogeneous and Nonhomogeneous System of Linear Equations

From Example 4.20 and Example 4.23, you can see that bases for the row and column space of the matrix

$$A = \begin{bmatrix} -1 & 0 & 2 & -3 & 1 \\ 0 & 1 & 0 & 1 & 1 \\ 2 & 0 & -4 & 6 & -2 \\ 1 & -4 & 2 & 7 & 7 \end{bmatrix}$$

each contain three vectors and so both the row and column space are three dimensional. The next theorem establishes that this is the case for any matrix.

THEOREM 4.14

The dimensions of the row and column spaces of a matrix A are the same.

Proof.

Let R be the reduced row-echelon form of A. From Theorem 4.11, you know that

dimension of row space of A = dimension of row space of R

Likewise, from Theorem 4.13,

dimension of column space of A = dimension of column space of R

The proof is completed by showing that the dimension of the row space of R is equal to the dimension of the column space of R. The dimension of the row space of R is the number of rows of R that contain leading 1's (see Theorem 4.11). But each nonzero row of R contains a leading 1 and so, by Theorem 4.12, each column containing a leading 1 is used in forming a basis for the column space of A. Thus, the dimension of the row space of R is the same as the dimension of the column space of R, and likewise for A, thus completing the proof. ∎

The common value of the dimension of the row and column space of a matrix A is called the **rank** of A and is denoted rank(A). This number is related to the dimension of the null space of A, called the **nullity** of A and is denoted nullity(A). The computation of rank(A) and nullity(A) are illustrated in the next two examples.

EXAMPLE 4.24 Finding the Rank of a Matrix

To find the rank of the matrix

$$A = \begin{bmatrix} 1 & 0 & -2 & 1 & -1 \\ -4 & -1 & 6 & -1 & 8 \\ 0 & 1 & 2 & -3 & -4 \\ -4 & 0 & 8 & -4 & 4 \end{bmatrix}$$

first obtain the reduced row-echelon form:

$$R = \begin{bmatrix} 1 & 0 & -2 & 1 & -1 \\ 0 & 1 & 2 & -3 & -4 \\ 0 & 0 & 0 & 0 & 0 \\ 0 & 0 & 0 & 0 & 0 \end{bmatrix}$$

There are two nonzero rows of R (or equivalently, two leading 1's), so the dimension of the row space of R, and hence A, is 2, so rank$(A) = 2$.

EXAMPLE 4.25 Finding the Nullity of a Matrix

To find the nullity of the matrix A in Example 4.24, you must find the dimension of the null space, that is, the dimension of the solution space to $A\mathbf{x} = \mathbf{0}$. To solve this system, reduce the augmented matrix $[A \mid \mathbf{0}]$ to reduced row-echelon form, which results in the matrix R in Example 4.24, with one additional column of zeros. The corresponding system of equations is:

$$
\begin{aligned}
x_1 + \quad - 2x_3 + \; x_4 - \; x_5 &= 0 \\
x_2 + 2x_3 - 3x_4 - 4x_5 &= 0
\end{aligned}
$$

Solving for the leading variables x_1 and x_2 yields

$$
\begin{aligned}
x_1 &= \quad 2x_3 - \; x_4 + \; x_5 \\
x_2 &= -2x_3 + 3x_4 + 4x_5
\end{aligned}
$$

By assigning arbitrary values of a, b, and c to the free variables x_3, x_4, and x_5, respectively, the general solution to $A\mathbf{x} = \mathbf{0}$ is

$$\begin{bmatrix} x_1 \\ x_2 \\ x_3 \\ x_4 \\ x_5 \end{bmatrix} = a \begin{bmatrix} 2 \\ -2 \\ 1 \\ 0 \\ 0 \end{bmatrix} + b \begin{bmatrix} -1 \\ 3 \\ 0 \\ 1 \\ 0 \end{bmatrix} + c \begin{bmatrix} 1 \\ 4 \\ 0 \\ 0 \\ 1 \end{bmatrix} \tag{4.47}$$

The three vectors on the right side of (4.47) form a basis for the null space of A, so nullity$(A) = 3$.

From Example 4.24 and Example 4.25, the matrix A in Example 4.24 that has 5 columns satisfies

$$\text{rank}(A) + \text{nullity}(A) = 5 = \text{number of columns of } A$$

This relationship is true for any matrix, as stated in the next theorem.

THEOREM 4.15

For any matrix A with n columns, $\text{rank}(A) + \text{nullity}(A) = n$.

Proof.

In solving the homogeneous system $A\mathbf{x} = \mathbf{0}$, let R be the reduced row-echelon form of A. There are n columns and each of the n variables is either a leading variable or a free variable, so,

$$\begin{pmatrix} \text{number of} \\ \text{leading variables} \end{pmatrix} + \begin{pmatrix} \text{number of} \\ \text{free variables} \end{pmatrix} = n \qquad (4.48)$$

But the number of leading variables is equal to the number of leading 1's in R, which is equal to the dimension of the row space of R, which is equal to the dimension of the row space of A, which is equal to the rank of A. Also, as you are asked to show in Exercise 20, the number of free variables is equal to nullity(A). Thus, (4.48) becomes

$$\text{rank}(A) + \text{nullity}(A) = n$$

and so the proof is complete. ■

4.4.4   More Results on Systems of Linear Equations

Several useful results pertaining to the solution of a system of linear equations are presented in this section.

Results for an (m × n) System of Linear Equations

As you know, a system of m linear equations in n unknowns may or may not have a solution. The next theorem relates consistency of a system of linear equations to concepts you have learned in this chapter.

THEOREM 4.16

For a given $(m \times n)$ matrix A and a column vector $\mathbf{b} \in R^{m \times 1}$, the following are equivalent:

(a) The system $A\mathbf{x} = \mathbf{b}$ is consistent.

(b) The column vector $\mathbf{b}$ is in the column space of A.

(c) $\text{rank}([A \mid \mathbf{b}]) = \text{rank}(A)$.

Proof.

To see that (a) implies (b), assume that (a) is true, so, there is a column vector $\mathbf{x} \in R^{n \times 1}$ such that $A\mathbf{x} = \mathbf{b}$. Expanding this expression using the components of $\mathbf{x}$ shows that $\mathbf{b}$ is in the column space of A because

$$A\mathbf{x} = A_{*1}x_1 + A_{*2}x_2 + \cdots + A_{*n}x_n = \mathbf{b}$$

To see that (b) implies (c), assume that (b) is true, that is, that $\mathbf{b}$ is in the column space of A. Then there are real numbers $x_1, \ldots, x_n$ such that

$$A_{*1}x_1 + \cdots + A_{*n}x_n = \mathbf{b}$$

In other words, the system $A\mathbf{x} = \mathbf{b}$ is consistent. Now let $E_1, \ldots, E_k$ be elementary matrices that reduce A to R, the reduced row-echelon form of A. That is,

$$R = E_k \cdots E_1 A$$

so

$$E_k \cdots E_1 [A \mid \mathbf{b}] = [R \mid E_k \cdots E_1 \mathbf{b}]$$

The rank of A is the number of nonzero rows in R and the rank of $[A \mid \mathbf{b}]$ is the number of nonzero rows in $[R \mid E_k \cdots E_1 \mathbf{b}]$. The only way the latter matrix can have a different number of nonzero rows than R is if $[R \mid E_k \cdots E_1 \mathbf{b}]$ contains a row consisting of all zeros except for a nonzero value in the final column. But if this happens, then the system $A\mathbf{x} = \mathbf{b}$ is inconsistent, which is impossible because it has already been shown that this system is consistent.

Finally, to see that (c) implies (a), as in the proof that (b) implies (c), if $\text{rank}([A \mid \mathbf{b}]) = \text{rank}(A)$, then the reduced row-echelon form of $[A \mid \mathbf{b}]$ cannot have a row of all zeros except for a nonzero value in the last column. Consequently, the system $A\mathbf{x} = \mathbf{b}$ is consistent. Thus the proof is complete. ■

In Chapter 2, you learned how to use Gauss-Jordan elimination to find the general solution to the system

$$A\mathbf{x} = \mathbf{b}$$

when this system is consistent. Specifically, after obtaining the reduced row-echelon form of the augmented matrix, you solve for the leading variables in terms of the free variables. Then, any set of values for the free variables, when substituted in the final equations, provides values for the leading variables and thus a solution to the original system of linear equations.

An alternative way to describe the general solution to a system of linear equations is given in the following theorem.

THEOREM 4.17

Suppose that $\mathbf{x}_0$ is a solution to the system

$$A\mathbf{x} = \mathbf{b}$$

and that $\mathbf{v}_1, \ldots, \mathbf{v}_k$ form a basis for the null space of A. A column vector $\mathbf{x} \in R^{n \times 1}$ satisfies $A\mathbf{x} = \mathbf{b}$ if and only if there are real numbers $t_1, \ldots, t_k$ such that

$$\mathbf{x} = \mathbf{x}_0 + t_1 \mathbf{v}_1 + \cdots t_k \mathbf{v}_k$$

Proof.
Suppose first that $\mathbf{x}$ satisfies $A\mathbf{x} = \mathbf{b}$. Then you can see that the vector $\mathbf{x} - \mathbf{x}_0$ is in the null space of A because

$$A(\mathbf{x} - \mathbf{x}_0) = A\mathbf{x} - A\mathbf{x}_0 = \mathbf{b} - \mathbf{b} = \mathbf{0}$$

Because $\mathbf{v}_1, \ldots, \mathbf{v}_k$ form a basis for the null space of A, $\mathbf{x} - \mathbf{x}_0$ is a linear combination of these basis vectors. Thus, there are real numbers $t_1, \ldots, t_k$ such that

$$\mathbf{x} - \mathbf{x}_0 = t_1 \mathbf{v}_1 + \cdots + t_k \mathbf{v}_k$$

or equivalently,

$$\mathbf{x} = \mathbf{x}_0 + t_1 \mathbf{v}_1 + \cdots + t_k \mathbf{v}_k$$

For the converse, suppose that there are real numbers $t_1, \ldots, t_k$ such that

$$\mathbf{x} = \mathbf{x}_0 + t_1 \mathbf{v}_1 + \cdots + t_k \mathbf{v}_k \tag{4.49}$$

To see that $\mathbf{x}$ satisfies $A\mathbf{x} = \mathbf{b}$, multiply (4.49) through by A and use the fact that for each i, $A\mathbf{v}_i = \mathbf{0}$ (because $\mathbf{v}_i$ is in the null space of A) to obtain

$$\begin{aligned}
A\mathbf{x} &= A(\mathbf{x}_0 + t_1 \mathbf{v}_1 + \cdots + t_k \mathbf{v}_k) \\
&= A\mathbf{x}_0 + t_1 A\mathbf{v}_1 + \cdots + t_k A\mathbf{v}_k \\
&= \mathbf{b} + t_1 \mathbf{0} + \cdots + t_k \mathbf{0} \\
&= \mathbf{b}
\end{aligned}$$

The proof is now complete. ■

The solution $\mathbf{x}_0$ in Theorem 4.17 is called a **particular solution** of $A\mathbf{x} = \mathbf{b}$. The expression

$$t_1 \mathbf{v}_1 + \cdots + t_k \mathbf{v}_k$$

is called the **general solution of** $A\mathbf{x} = \mathbf{0}$. The expression

$$\mathbf{x} = \mathbf{x}_0 + t_1 \mathbf{v}_1 + \cdots + t_k \mathbf{v}_k$$

is called the **general solution of** $A\mathbf{x} = \mathbf{b}$. These concepts are illustrated in the following example.

EXAMPLE 4.26 An Expression for the General Solution to a System of Linear Equations

To find the general solution to the system $A\mathbf{x} = \mathbf{b}$ in which

$$A = \begin{bmatrix} 1 & 0 & -2 & 1 & -1 \\ -4 & -1 & 6 & -1 & 8 \\ 0 & 1 & 2 & -3 & -4 \\ -4 & 0 & 8 & -4 & 4 \end{bmatrix} \quad \text{and} \quad \mathbf{b} = \begin{bmatrix} -1 \\ 2 \\ 2 \\ 4 \end{bmatrix}$$

according to Theorem 4.17, you need one particular solution together with a basis for the null space of A. Applying Gauss-Jordan elimination and setting all of the free variables to zero yields the particular solution $\mathbf{x}_0 = (-1, 2, 0, 0, 0)$. Also, the following basis for the null space of A is given in Example 4.25:

$$\mathbf{v}_1 = \begin{bmatrix} 2 \\ -2 \\ 1 \\ 0 \\ 0 \end{bmatrix}, \quad \mathbf{v}_2 = \begin{bmatrix} -1 \\ 3 \\ 0 \\ 1 \\ 0 \end{bmatrix}, \quad \mathbf{v}_3 = \begin{bmatrix} 1 \\ 4 \\ 0 \\ 0 \\ 1 \end{bmatrix}$$

Thus, a general solution to $A\mathbf{x} = \mathbf{b}$ is

$$\mathbf{x} = \begin{bmatrix} -1 \\ 2 \\ 0 \\ 0 \\ 0 \end{bmatrix} + t_1 \begin{bmatrix} 2 \\ -2 \\ 1 \\ 0 \\ 0 \end{bmatrix} + t_2 \begin{bmatrix} -1 \\ 3 \\ 0 \\ 1 \\ 0 \end{bmatrix} + t_3 \begin{bmatrix} 1 \\ 4 \\ 0 \\ 0 \\ 1 \end{bmatrix}$$

In this section, you have seen how to find a basis for certain subspaces of R^n, using the column, row, and null space of an associated matrix. Some of these ideas are used in the next section to solve the problem presented at the beginning of the chapter.

Exercises for Section 4.4

1. Explain what you would have to do to prove that a line through the origin in R^n is one dimensional. (Do not prove this.)

2. Explain what you would have to do to prove that a plane through the origin in R^n is two dimensional. (Do not prove this.)

3. Find a basis for the row space of each of the following matrices. What is the dimension of that row space?

   (a) $\begin{bmatrix} 1 & 0 & -2 \\ 0 & 0 & 1 \end{bmatrix}$

   (b) $\begin{bmatrix} 1 & 0 & -1 \\ 0 & 1 & 2 \\ 0 & 0 & 0 \end{bmatrix}$

4. Use the reduced row-echelon matrices in the previous exercise and Theorem 4.11 in Section 4.4.2 to find a basis for the row space of the following matrices.

   (a) $\begin{bmatrix} 5 & 0 & -10 \\ 1 & 0 & 3 \end{bmatrix}$

   (b) $\begin{bmatrix} 3 & 0 & -3 \\ -5 & 1 & 7 \\ -6 & 0 & 6 \end{bmatrix}$

5. Complete part (a) of Theorem 4.9 in Section 4.4.1 by proving that if $\mathbf{v}_1, \ldots, \mathbf{v}_n$ are linearly independent vectors in a vector space $(V, \oplus, \odot)$ whose dimension is n, then these vectors form a basis for V.

6. Complete part (b) of Theorem 4.9 in Section 4.4.1 by proving that if $\mathbf{v}_1, \ldots, \mathbf{v}_n$ span a vector space $(V, \oplus, \odot)$ whose dimension is n, then these vectors form a basis for V.

7. Prove that if $\mathbf{v}_1, \ldots, \mathbf{v}_k$ span a vector space $(V, \oplus, \odot)$, then dim $V \leq k$.

8. Prove that if U is a subspace of an n-dimensional vector space $(V, \oplus, \odot)$ and that dim $U = n$, then $U = V$.

9. Let S consist of the following four vectors:

$$\mathbf{v}_1 = (-1, 0, \quad 2, -3, 1), \quad \mathbf{v}_2 = (0, \quad 1, 0, 1, 1)$$
$$\mathbf{v}_3 = (\quad 2, 0, -4, \quad 6, 2), \quad \mathbf{v}_4 = (1, -4, 2, 7, 7)$$

Use MATLAB, a graphing calculator, or other technology to verify that of the five standard unit vectors in R^5, only $(0, 0, 0, 0, 1)$ belongs to span(S).

10. Complete Theorem 4.11 in Section 4.4.2 by proving that the nonzero rows of a row-echelon form R of a matrix A form a basis for the row space of R and thus of A.

11. Find a basis for the column space of each of the matrices in Exercise 3. What is the dimension of that column space?

12. Use the results of the previous exercise and Theorem 4.13 in Section 4.4.2 to find a basis for the column space of the matrices in Exercise 4.

13. Find a basis for the row space of the following matrix A that consists of some of the rows of A:

$$A = \begin{bmatrix} -2 & 4 & 6 \\ 4 & -8 & -12 \\ 0 & 3 & 0 \end{bmatrix}$$

14. Find a basis for the row space of the following matrix A that consists of some of the rows of A:

$$A = \begin{bmatrix} 0 & 1 & 0 & -2 \\ 0 & -2 & 0 & 4 \\ 0 & 3 & 0 & -6 \\ 0 & -4 & 1 & 9 \end{bmatrix}$$

15. What are the rank and nullity of the matrix in Exercise 13?

16. What are the rank and nullity of the matrix in Exercise 14?

17. Find the general solution to the system of linear equations $A\mathbf{x} = \mathbf{b}$, where A is the matrix in Exercise 13 and the three elements of $\mathbf{b}$ are -4, 8, and -3. Use the following approach to do so:
    (a) Use Gauss-Jordan elimination to find a particular solution $\mathbf{x}_0$ by setting all the free variables to 0.
    (b) Find a basis for the null space of A by applying Gauss-Jordan elimination to $[A \mid \mathbf{0}]$.
    (c) Combine parts (a) and (b) to write the general solution to the system $A\mathbf{x} = \mathbf{b}$ as $\mathbf{x}_0 + t_1\mathbf{v}_1$.

18. Find the general solution to the system of linear equations $A\mathbf{x} = \mathbf{b}$, where A is the matrix in Exercise 14 and the four elements of $\mathbf{b}$ are $1, -2, 3$, and -2. Use the following approach to do so:
    (a) Use Gauss-Jordan elimination to find a particular solution $\mathbf{x}_0$ by setting all the free variables to 0.
    (b) Find a basis $\mathbf{v}_1, \mathbf{v}_2$ for the null space of A by applying Gauss-Jordan elimination to $[A \mid \mathbf{0}]$.
    (c) Combine parts (a) and (b) to write the general solution to the system $A\mathbf{x} = \mathbf{b}$ as $\mathbf{x}_0 + t_1\mathbf{v}_1 + t_2\mathbf{v}_2$.

19. Complete Theorem 4.12 in Section 4.4.2 by proving that if R is a matrix in row-echelon form, then the columns of R that contain the leading values in the rows of R form a basis for the column space of R.

20. Complete Theorem 4.15 in Section 4.4.3 by proving that nullity(A) is equal to the number of free variables in the reduced row-echelon form R of A.

21. In statistical analysis of certain problems involving a matrix A, it is often necessary that A has *full rank*, that is, that the rank of A be as large as possible. Prove that an $(m \times n)$ matrix in which $m > n$ has full rank if and only if the columns of A are linearly independent.

22. Prove that for an $(m \times n)$ matrix A:
    (a) The dimension of the row space of A plus the dimension of the null space of A equals n (the number of columns of A).
    (b) The dimension of the column space of A plus the dimension of the null space of A^t equals m (the number of rows of A).

4.5 Problem Solving with Vector Spaces

The material in this chapter is brought together by solving the linear programming problem of Fiber Optics introduced at the beginning of the chapter. Doing so involves formulating and solving a linear programming problem.

4.5.1 The Problem of Fiber Optics

Fiber Optics produces a low-density and a high-density fiber-optic cable. The company has just received a one-time order for 2000 feet of the low-density cable and 4500 feet of the high-density cable for the following week. Each foot of the low-density cable costs $4 to produce and requires half a minute of machine time. Each foot of the high-density cable costs $6 to produce and requires three-fourths of a minute on the same machine. Because only 40 hours of machine time are available next week, Fiber Optics cannot meet these demands. However, the company can make up the difference by buying some amount of these cables from a Japanese company at a cost of $7 per foot for the low-density cable and $8 per foot for the high-density cable. You have been asked to find a production/purchase plan for the company that minimizes the cost of meeting the demand while not exceeding the available machine time.

4.5.2 Problem Formulation

In Section 2.1.3, a mathematical model of this problem was developed by identifying the following four variables:

$$LP = \text{the number of feet of low-density cable to produce}$$
$$HP = \text{the number of feet of high-density cable to produce}$$
$$LJ = \text{the number of feet of low-density cable to purchase}$$
$$HJ = \text{the number of feet of high-density cable to purchase}$$

The objective of the problem is to find values for these variables that minimize the total production and purchase costs while simultaneously satisfying all of the demand and resource constraints. Writing mathematical expressions in terms of the variables and other problem data gives rise to the following linear programming problem:

minimize	$4LP +$	$6HP + 7LJ + 8HJ$		(objective function)
subject to	LP	$+ LJ$	$= 2000$	(low-density demand)
	HP	$+ HJ = 4500$		(high-density demand)
	$0.5LP + 0.75HP$	$= 2400$		(machine time)
	LP , HP , LJ , $HJ \geq$	0		(nonnegativity)

Temporarily ignoring the objective function and the nonnegativity constraints, you have the following system of 3 linear equations in 4 unknowns:

$$\begin{bmatrix} 1 \\ 0 \\ 0.5 \end{bmatrix} LP + \begin{bmatrix} 0 \\ 1 \\ 0.75 \end{bmatrix} HP + \begin{bmatrix} 1 \\ 0 \\ 0 \end{bmatrix} LJ + \begin{bmatrix} 0 \\ 1 \\ 0 \end{bmatrix} HJ = \begin{bmatrix} 2000 \\ 4500 \\ 2400 \end{bmatrix} \qquad (4.50)$$

Equivalently stated in matrix-vector notation, the system is:

$$\underbrace{\begin{bmatrix} 1 & 0 & 1 & 0 \\ 0 & 1 & 0 & 1 \\ 0.5 & 0.75 & 0 & 0 \end{bmatrix}}_{A} \underbrace{\begin{bmatrix} LP \\ HP \\ LJ \\ HJ \end{bmatrix}}_{\mathbf{x}} = \underbrace{\begin{bmatrix} 2000 \\ 4500 \\ 2400 \end{bmatrix}}_{\mathbf{b}} \qquad (4.51)$$

In the context of the terminology of this chapter, finding values for the variables that satisfy (4.50) and (4.51) is equivalent to expressing $\mathbf{b}$ as a linear combination of the columns of A. To satisfy the nonnegativity requirements, you want the multipliers in this linear combination—that is, the values of the variables—to be nonnegative. Further, among all such nonnegative values of the variables, you want ones that produce the best value of the objective function. One systematic approach for attempting to find such values is presented next.

4.5.3 Problem Solution

To find a nonnegative combination of the columns of A that yield $\mathbf{b}$, you might first solve the system of equations by applying Gauss-Jordan elimination to the augmented matrix $[A \mid \mathbf{b}]$ associated with (4.51), obtaining the following reduced row-echelon form:

$$R = \begin{bmatrix} 1 & 0 & 0 & -1.5 & -1950 \\ 0 & 1 & 0 & 1.0 & 4500 \\ 0 & 0 & 1 & 1.5 & 3950 \end{bmatrix} \qquad (4.52)$$

You can see from (4.52) that the system of equations is consistent. In fact, there are an infinite number of solutions because the last variable, HJ, is free. In the context of a linear programming problem, a free variable is referred to as a *nonbasic variable* and the leading variables are called *basic variables*.

One solution to the system is obtained by setting the value of the nonbasic variable HJ to 0 and using (4.52) to obtain:

$$LP = -1950, \quad HP = 4500, \quad LJ = 3950, \quad HJ = 0$$

Unfortunately, the value of LP in the foregoing solution is less than 0 and therefore does not satisfy the nonnegativity requirement.

Nevertheless, the result in (4.52) provides some valuable information. For example, because columns 1, 2, and 3 of R contain the leading 1's, columns 1, 2, and 3 of A form a basis for the column space of A (see Theorem 4.13 in Section 4.4.2). This basis allows you to obtain values for the basic variables by expressing $\mathbf{b}$ as a unique linear combination of the vectors in this basis.

This particular basis for the column space of A does not produce values for the variables that satisfy the nonnegativity requirements, but perhaps a different basis for the column space of A does. Mathematicians have proved that, for a system such as the one in this example, if there is a solution to the linear programming problem, then such a solution can be found by identifying one particular basis consisting of 3 columns of A. The following numerical method attempts to find this special basis for a linear programming problem having m equality constraints and n variables.

Numerical-method Solution for Linear Programming Problems

Step 1. List all choices of m columns from A, each of which might form a basis for the column space of A. In the current example, this means listing all choices of 3 columns of A.

Step 2. For each choice in (1), set the values of the $n - m$ nonbasic variables not associated with the chosen columns to 0. Find the values of the remaining m basic variables by solving a system of m linear equations in m unknowns. In this problem, each choice consists of $m = 3$ of the $n = 4$ columns of A, so there is $n - m = 1$ nonbasic variable. The values of the 3 remaining basic variables are found by solving a system of 3 linear equations in 3 unknowns.

Step 3. Among all choices for bases in (2) that result in nonnegative values of all m basic variables, identify one in which the objective function value is the best.

To illustrate this numerical method for the problem of Fiber Optics, consider a possible basis consisting of columns 1, 2, and 4 of A in (4.51). For this choice, variable 3 (LJ) is nonbasic and its value is set to 0. The values of the three remaining basic variables LP, HP, and HJ are found by solving the following system of linear equations obtained from (4.50):

$$
\begin{bmatrix} 1 \\ 0 \\ 0.5 \end{bmatrix} LP + \begin{bmatrix} 0 \\ 1 \\ 0.75 \end{bmatrix} HP + \begin{bmatrix} 0 \\ 1 \\ 0 \end{bmatrix} HJ = \begin{bmatrix} 2000 \\ 4500 \\ 2400 \end{bmatrix}
$$

Solving this system yields $LP = 2000$, $HP = 1866.67$, and $HJ = 2633.33$. Note that these values satisfy the nonnegativity requirements and provide the following value of the objective function:

$$
4LP + 6HP + 7LJ + 8HP = 4(2000) + 6(1866.67) + 7(0) + 8(2633.33) = 40267
$$

Performing similar computations for each choice of 3 columns from A yields the results in the following table, in which the variable whose value is 0 is nonbasic:

Columns in the Basis	Values of the Variables				Value of the Objective Function
	LP	HP	LJ	HJ	
1, 2, 3	−1950	4500	3950	0	46850
1, 2, 4	2000	1866.67	0	2633.33	40267
1, 3, 4	4800	0	−2800	4500	35600
2, 3, 4	0	3200	2000	1300	43600

From this table, you can see that the choices consisting of columns 1, 2, and 3 and columns 1, 3, and 4 produce values for the variables that do not satisfy the nonnegativity requirements and are therefore discarded. The choice consisting of columns 1, 2, and 4 yields nonnegative values of the variables with an associated objective function value of 40267. The choice consisting of columns 2, 3, and 4 yields nonnegative values for the variables with an associated objective function value of 43600.

These computations lead to the conclusion that the solution to this linear programming problem is the one associated with the basis consisting of columns 1, 2, and 4, in which

$$
LP = 2000, \quad HP = 1866.67, \quad LJ = 0, \quad HJ = 2633.33
$$

This is because the values of these variables are nonnegative and yield the smallest value for the objective function, namely, 40267. In other words, the optimal production/purchase plan for Fiber Optics is to produce 2000 feet of the low-density cable and 1866.67 feet of the high-density cable and purchase none of the low-density cable and 2633.33 feet of the high-density cable from the Japanese company. This plan incurs a total cost of $40,267.

4.5.4 Some Comments On Linear Programming

The numerical method presented in Section 4.5.3 for solving a linear programming problem having m equality constraints and n variables requires listing all choices of m columns from the ($m \times n$) matrix A associated with the equality constraints. Doing so for the problem of Fiber Optics is computationally feasible because there are only four such choices. In general, however, there are the following number of choices for m of the n columns of A:

$$\binom{n}{m} = \frac{n!}{n!(n-m)!} \quad \text{(where } k! = k(k-1)\cdots 1) \tag{4.53}$$

(The expression $k!$ is pronounced "k factorial.") As n and m increase, the approach of listing all the choices in (4.53) becomes computationally infeasible.

In 1951, George Dantzig developed a systematic method, called the *simplex algorithm*, for attempting to find the right basis for the column space of A without having to list all possible choices. The details are outside the scope of this book but the idea is to move from one basis of A to another, each time finding nonnegative values for the variables that result in a better value of the objective function, as stated in the following steps.

Idea of the Simplex Algorithm for Linear Programming

Step 1. Attempt to find an initial basis consisting of m columns of A that produce nonnegative values for all of the associated basic variables. If successful, go to Step 2.

Step 2. Test the current basis for optimality. That is, perform a relatively simple algebraic computation to determine if the values of the variables associated with the current basis matrix, together with a value of 0 for each of the nonbasic variables, provide the best value of the objective function. If so, stop, otherwise, go to Step 3.

Step 3. Use the fact that the current basis does not produce the best values for the objective function to create a new basis of m columns of A that yields nonnegative values for the associated basic variables and a better objective function value than the current values of the variables. (The new basis is obtained by replacing one column of the current basis with a column of A not in the current basis.) Return to Step 2.

The simplex algorithm has been used successfully since its inception to solve large real-world linear programming problems.

Exercises for Section 4.5

PROJECT 4.1: The Linear Programming Problem of Fiber Optics

For the problem of Fiber Optics described in Section 4.5.1, suppose that the cost of purchasing low-density cable from the Japanese company is $5 per foot and the cost for purchasing high-density cable is $9 per foot, thus giving rise to the following linear programming problem:

$$
\begin{array}{llll}
\text{minimize} & 4LP + \quad 6HP + 5LJ + 9HJ & & \text{(objective function)} \\
\text{subject to} & LP \qquad\qquad\quad + \;\; LJ \qquad\quad\; = 2000 & \text{(low-density demand)} \\
& \qquad\qquad HP \qquad\quad + \;\; HJ = 4500 & \text{(high-density demand)} \\
& 0.5LP + 0.75HP \qquad\qquad\;\; = 2400 & \text{(machine time)} \\
& \quad LP\;, \qquad HP\;, \;\; LJ\;, \;\; HJ \geq \quad 0 & \text{(nonnegativity)}
\end{array}
$$

Use a graphing calculator, MATLAB, or other software package, as indicated by your instructor, to perform the computations in Exercises 1, 2, and 3. Whenever possible, use matrix and vector operations available on your system. Write the answer and the sequence of operations you performed to obtain the answer.

1. Store the data for this linear programming problem as follows. Create a 4-vector, call it **c**, containing the coefficients of the four variables in the objective function. Create a column vector **b** containing the constant values on the right side of the equality constraints. Create the (3×4) matrix A containing the coefficients of the variables in the equality constraints.

2. Find one set of values for the variables as follows:

   (a) Create a (3×3) matrix B consisting of columns 1, 2, and 3 of A.

   (b) Because column 4 of A is not included in the matrix B, the value of the nonbasic variable HJ is 0. Find the values of the remaining basic variables LP, HP, and LJ associated with the columns chosen in part (a) by solving the system $B\mathbf{x} = \mathbf{b}$.

   (c) Compute the value of the objective function using the values of the variables obtained in part (b).

3. Repeat the previous exercise for each choice of three columns of A. Enter your results in a table like the following one:

Columns in the Basis	Values of the Variables LP	HP	LJ	HJ	Value of the Objective Function
1, 2, 3				0	
1, 2, 4			0		
1, 3, 4		0			
2, 3, 4	0				

What is the solution to this modified linear programming problem of Fiber Optics?

PROJECT 4.2: The Problem of Clear Chemicals

Clear Chemicals needs to prepare a combined total of 5000 liters of two of their products, H_2SO_4 and H-CL, which are made from concentrated acids, as follows. Each liter of concentrated sulfuric acid cost $12 and is diluted with 20 liters of distilled water to produce 21 liters of H_2SO_4 that then sells for $1 per liter. Each liter of concentrated hydrochloric acid cost $18 and is diluted with 30 liters of distilled water to produce 31 liters of H-CL that then sells for $1 per liter. Given that distilled water costs $0.15 per liter and that 200 liters of concentrated sulfuric acid and 150 liters of concentrated hydrochloric acid are available, the following linear programming problem was formulated to maximize the profit of the company, in which

SA = the number of liters of concentrated sulfuric acid to use

USA = the number of liters of unused concentrated sulfuric acid

HA = the number of liters of concentrated hydrochloric acid to use

UHA = the number of liters of unused concentrated hydrochloric acid

DW = the number of liters of distilled water to use

$$
\begin{array}{llll}
\text{max} & 9SA & + 13HA & - 0.15DW & \text{(objective function)} \\
\text{s.t.} & -20SA & - 30HA & + \quad DW = & 0 \text{ (distilled water)} \\
& 21SA & + 31HA & = 5000 \text{ (demand)} \\
& SA + USA & & = \quad 200 \text{ (sulfuric acid)} \\
& & HA + UHA & = \quad 150 \text{ (hydrochloric acid)} \\
& SA \ , \ \ USA \ , & HA \ , \ UHA \ , & DW \geq \quad 0
\end{array}
$$

Use a graphing calculator, MATLAB, or other software package, as indicated by your instructor, to perform the computations in Exercise 4 through Exercise 8. Whenever possible, use matrix and vector operations available on your system. Write the answer and the sequence of operations you performed to obtain the answer.

4. Store the data for this linear programming problem as follows. Create a 5-vector, call it $\mathbf{c}$, containing the coefficients of the five variables in the objective function. Create a column vector $\mathbf{b}$ containing the constant values on the right side of the equality constraints. Create the (4×5) matrix A containing the coefficients of the variables in the equality constraints.

5. Find one set of values for the variables as follows:

    (a) Create a (4×4) matrix B consisting of columns 1, 2, 3, and 4 of A.

    (b) Because column 5 of A is not included in the matrix B, the value of the nonbasic variable DW is 0. Find the values of the remaining basic variables SA, USA, HA, and UHA associated with the columns chosen in part (a) by solving the system $B\mathbf{x} = \mathbf{b}$.

    (c) Compute the value of the objective function using the values of the variables obtained in part (b).

6. Repeat the previous exercise for each choice of four columns of A. Enter your results in a table like the following one:

Columns in the Basis	SA	USA	HA	UHA	DW	Value of the Objective Function
1, 2, 3, 4					0	
1, 2, 3, 5				0		
1, 2, 4, 5			0			
1, 3, 4, 5		0				
2, 3, 4, 5	0					

What is the solution to the problem of Clear Chemicals? How much profit will the company realize?

7. The coefficient of 9 for SA in the objective function of the linear program indicates that each liter of concentrated sulfuric acid used results in a net profit of $9 for the company. Use a trial-and-error approach to determine by how much this coefficient would have

to decrease, to the nearest cent, before the solution you found in the previous exercise is no longer the best one.

8. Return to the original problem in which the coefficient of SA in the objective function is 9. Management wants to achieve a net profit of $1500. Use trial and error to determine how much total combined H_2SO_4 and H-CL the company must produce and sell to reach this profit level. How much of the concentrated acids and distilled water are needed?

Chapter Summary

In this chapter, you have seen how the axiomatic system of a vector space unifies the special cases of n-vectors and matrices in a single framework. The advantage of doing so is that you can study these systems, and all similar ones, simultaneously while only working with the one system, a vector space. To accomplish this unification requires the following mathematical techniques and concepts.

General Mathematical Concepts

1. Abstraction   The process of thinking in terms of general objects rather than specific items. Doing so allows you to combine different mathematical items as elements of a single set.

2. Abstract System   A set together with one or more ways to perform operations on the elements of the set. These operations are generalizations of the ones in the special cases you want to study.

3. **Axiom**   A property that is assumed to hold in an abstract system. Axioms are used to restore properties of the special cases under consideration that are lost in the process of abstraction.

4. Axiomatic System   An abstract system together with a list of axioms that are assumed to hold. Any results obtained for an axiomatic system apply to all of the special cases by a simple substitution of the symbols from the special cases.

5. Creating Mathematical Definitions   The process of translating a visual image of a desirable property satisfied by certain objects to symbolic form. Doing so involves the following steps:

Step 1.   Identify a common property shared by all desirable objects you are considering.

Step 2.   Choose a representative name for the common desirable property.

Step 3.   Translate the visual image of this property to a symbolic form that then becomes the definition.

Step 4.   Verify that the resulting definition works in two ways: (1) all desirable objects should satisfy the stated property and (2) any object satisfying the stated property should be a desirable object.

Vector Spaces and Subspaces

A vector space $(V, \oplus, \odot)$ is an axiomatic system that consists of a set V of objects together with two operations $\oplus$ and $\odot$. The closed binary operation $\oplus$ combines two elements of V

to produce an element of V. The operation $\odot$ combines a real number with an element in V to produce an element in V. To be a vector space, the abstract system $(V, \oplus, \odot)$ must satisfy the 10 axioms in Definition 4.1 in Section 4.1.4. The axiomatic system of a vector space includes not only n-vectors and $(m \times n)$ matrices but also many other systems, such as the set of all real-valued functions of one variable. A vector space satisfies many properties in addition to the 10 axioms in the definition, such as the fact that there is only one zero vector. These properties, once proved from the axioms, apply to all special cases of vector spaces.

A subspace of a vector space $(V, \oplus, \odot)$ is a subset U of V that is itself a vector space under the operations $\oplus$ and $\odot$ from V. A subset U is a subspace of $(V, \oplus, \odot)$ if and only if $\oplus$ is closed on U and $\odot$ combines a real number with an element of U and produces an element of U.

An n-vector $\mathbf{v}$ has components $(v_1, \ldots, v_n)$. A general vector $\mathbf{v}$ in a vector space $(V, \oplus, \odot)$ may or may not have components. To determine the components of a particular vector $\mathbf{v}$, it is first necessary to find a basis for V consisting of n linearly independent vectors that span V, say, $\mathbf{v}_1, \ldots, \mathbf{v}_n$. If successful, then any vector $\mathbf{v} \in V$ is a unique linear combination of the basis vectors. That is, for any vector $\mathbf{v} \in V$, there are unique real numbers $t_1, \ldots, t_n$ such that

$$\mathbf{v} = (t_1 \odot \mathbf{v}_1) \oplus (t_2 \odot \mathbf{v}_2) \oplus \cdots \oplus (t_n \odot \mathbf{v}_n)$$

In this case, the components of $\mathbf{v}$ relative to $\mathbf{v}_1, \ldots, \mathbf{v}_n$ are $(t_1, \ldots, t_n)$. The existence of these real numbers is ensured by the fact that the vectors in a basis span V. The uniqueness of the components is ensured by the fact that the vectors in a basis are linearly independent.

A basis not only provides the components of a vector but also the dimension of a vector space. Specifically, the dimension of a vector space having a basis $\mathbf{v}_1, \ldots, \mathbf{v}_n$ is n, the number of vectors in a basis.

The row space of an $(m \times n)$ matrix A is the set of all linear combinations of the rows of A and the column space of A is set of all linear combinations of the columns of A. To find a basis for these vector spaces, first obtain a row-echelon form R of A. The nonzero rows of R form a basis for the row space of A. The columns of A corresponding to the columns of R having the leading 1's provide a basis for the column space of A.

Chapter 5

Linear Transformations

You know that the operation of multiplying an $(m \times n)$ matrix A by a column vector $\mathbf{x} \in R^{n \times 1}$ results in the m-vector $A\mathbf{x}$ (see Section 2.4.4). In this chapter, you will learn to think of this operation as one particular way to *transform* the n-vector $\mathbf{x}$ to a new m-vector. By learning to think in this manner, you can extend many of the results you already know to a more general framework, thereby allowing you to solve even more problems, such as the following one.

The Filtering Problem of Audio Technologies

At Audio Technologies, music is recorded digitally by sampling a continuous signal 44,100 times per second. The amount of time between successive samplings is one time unit. For each $k = 0, 1, 2, \ldots$, the signal sampled at time unit k is a real number, say, x_k. Taking all of these measurements collectively, the digital signal is represented mathematically as the following infinite ordered list of real numbers:

$$\mathbf{X} = (x_0, x_1, x_2, \ldots)$$

Such a signal $\mathbf{X}$ is *filtered* by combining terms from $\mathbf{X}$ in a special way to create a new signal $\mathbf{Y}$. For example, one particular filter uses the values of x_k to create a new signal $\mathbf{Y}$ in which

$$y_k = x_{k+3} + x_{k+2} - 10x_{k+1} + 8x_k, \quad \text{for } k = 0, 1, 2, \ldots \tag{5.1}$$

The audio engineers know that when some signals $\mathbf{X}$ are fed into the filter in (5.1), the result is the zero signal, that is, the signal $\mathbf{Y} = (y_0, y_1, y_2, \ldots,) = (0, 0, 0, \ldots)$. As a member of the technical staff, you have been asked to find a representation of all signals for which the filter in (5.1) results in the zero signal so that design engineers can determine if a particular signal can be eliminated by using this filter.

5.1 A Review of Functions

You have seen many examples of functions in your previous studies, such as the function $F(x) = 2x + 1$. Through the technique of abstraction presented in Chapter 4, you can think of a function as a method for associating to each object x, an object $F(x)$ (pronounced "F of x"), as you will now see.

5.1.1 A Sequential Generalization of Functions

You know that a real-valued function of one variable, F, is a rule that associates to each real number, x, a unique real number, $F(x)$. Two visual images of F are shown in Figure 5.1. Figure 5.1(a) is the graph of F, which you recognize from previous experience. The image in Figure 5.1(b) is that of F as a *black box* into which you put a real number x and, after performing some computations, obtain the real number $F(x)$. The element that goes into the box is called the **input value** and that which comes out is called the **output value** of the function. The operations that go on inside the box constitute the rule associated with the function.

Sequential generalization is used now to develop a more general concept of a function that is useful in problem solving and, in particular, for showing how the matrix multiplication $A\mathbf{x}$ is a type of function that associates to each column vector $\mathbf{x} \in R^{n \times 1}$, the m-vector $A\mathbf{x}$.

Functions from R^n to R^1

One generalization arises when considering a function whose input is an n-vector, $\mathbf{x} = (x_1, \ldots, x_n)$, rather than just a single number. For example, consider the objective function of the linear programming problem of Fiber Optics described in Section 2.1.3, in which you want to find values for the variables LP, HP, LJ, and HJ so as to minimize the total cost represented by the expression

$$4LP + 6HP + 7LJ + 8HJ \tag{5.2}$$

You can view the computation in (5.2) as a function that associates to each 4-vector (LP, HP, LJ, HJ), the real number $4LP + 6HP + 7LJ + 8HJ$.

In general, a function F that associates to each n-vector $\mathbf{x}$, a unique real number $F(\mathbf{x})$

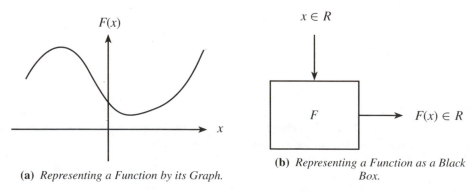

(a) *Representing a Function by its Graph.*

(b) *Representing a Function as a Black Box.*

Figure 5.1 *Visualizations of a Real-valued Function of One Variable.*

is written $F : R^n \rightarrow R$ or $F : R^n \rightarrow R^1$ and is read as "F maps R^n to R^1." The value of F at the point $\mathbf{x} = (x_1, \ldots, x_n)$ is written $F(x_1, \ldots, x_n)$ instead of $F((x_1, \ldots, x_n))$. The graph of such a function, when $n = 2$, is shown in Figure 5.2(a). The black-box image associated with a function $F : R^n \rightarrow R$ is given in Figure 5.2(b). Another example of this type of function follows.

EXAMPLE 5.1 A Function from R^n to R

An example of a function $L : R^n \rightarrow R$ is the one that associates to each n-vector $\mathbf{x} = (x_1, \ldots, x_n)$, the length of $\mathbf{x}$, that is,

$$L(\mathbf{x}) = \|\mathbf{x}\| = \sqrt{x_1^2 + \cdots + x_n^2}$$

Functions from R^n to R^m

A further generalization of a function arises when the output of the function F is an m-vector. For example, the input to the function

$$F(x, y) = (\sqrt{|x + y|}, 2^x, x^2 + 2xy - y^2) \tag{5.3}$$

is a 2-vector (x, y) and the output is the 3-vector $(\sqrt{|x + y|}, 2^x, x^2 + 2xy - y^2)$. In general, a function $F : R^n \rightarrow R^m$ means that the input to F is an n-vector $\mathbf{x}$ and the unique output is the m-vector $F(\mathbf{x}) = (F_1(\mathbf{x}), \ldots, F_m(\mathbf{x}))$. You can think of each of the m outputs, F_i, as a **coordinate function** that associates to each n-vector $\mathbf{x}$, the real number $F_i(\mathbf{x})$. For example, the three coordinate functions of F in (5.3) are

$$F_1(x, y) = \sqrt{|x + y|}, \quad F_2(x, y) = 2^x, \quad F_3(x, y) = x^2 + 2xy - y^2$$

A function $F : R^n \rightarrow R^1$ is a special case of a function $F : R^n \rightarrow R^m$ obtained by substituting $m = 1$. The following is another example of a function $F : R^n \rightarrow R^m$ that involves a matrix multiplication.

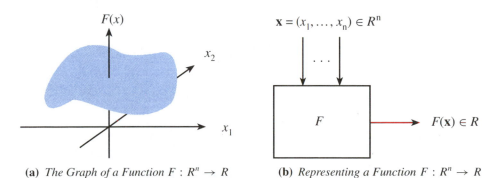

(a) *The Graph of a Function $F : R^n \rightarrow R$ for $n = 2$.*

(b) *Representing a Function $F : R^n \rightarrow R$ as a Black Box.*

Figure 5.2 *Visualizations of a Function $F : R^n \rightarrow R^1$.*

EXAMPLE 5.2 A Function from R^n to R^m

Given an $(m \times n)$ matrix A, the function $F : R^n \to R^m$ defined by

$$F(\mathbf{x}) = A\mathbf{x}$$

associates with each n-vector $\mathbf{x}$, the m-vector $A\mathbf{x}$ obtained by thinking of $\mathbf{x}$ as a column vector. For example, for

$$A = \begin{bmatrix} 1 & -1 & -2 \\ -2 & 0 & 3 \end{bmatrix} \quad \text{and} \quad \mathbf{x} = \begin{bmatrix} x_1 \\ x_2 \\ x_3 \end{bmatrix}$$

$F(\mathbf{x}) = A\mathbf{x}$ is the function from R^3 to R^2 in which

$$F(\mathbf{x}) = A\mathbf{x} = \begin{bmatrix} 1 & -1 & -2 \\ -2 & 0 & 3 \end{bmatrix} \begin{bmatrix} x_1 \\ x_2 \\ x_3 \end{bmatrix} = \begin{bmatrix} x_1 - x_2 - 2x_3 \\ -2x_1 + 3x_3 \end{bmatrix}$$

The two coordinate functions of F are

$$F_1(x_1, x_2, x_3) = x_1 - x_2 - 2x_3 \quad \text{and} \quad F_2(x_1, x_2, x_3) = -2x_1 + 3x_3$$

One visual image of $F : R^n \to R^m$ is the black box in Figure 5.3 in which the input to F is an n-vector $\mathbf{x}$ and the output is the m-vector $F(\mathbf{x}) = (F_1(\mathbf{x}), \ldots, F_m(\mathbf{x}))$.

The graph of a function $F : R^n \to R^m$ is more challenging to visualize. One way to do so is to visualize the graphs of the m coordinate functions of F. For $n = 2$, each of the m coordinate functions is a function from R^2 to R^1 whose graph is like the one in Figure 5.2(a).

When $n = m = 2$, you can create a visual image of $F : R^2 \to R^2$ in a single picture because, in this case, F associates to each point (x, y) in the plane, a point $(F_1(x, y), F_2(x, y))$ in the plane. To do so, represent the function input and output as two points in the plane and use an arrow pointing from (x, y) to $(F_1(x, y), F_2(x, y))$, as shown in Figure 5.4.

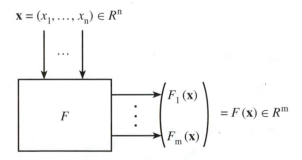

Figure 5.3 *Visualizing a Function $F : R^n \to R^m$ as a Black Box.*

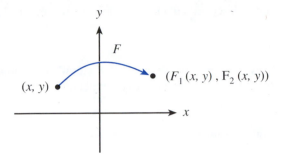

Figure 5.4 *Visualizing the Graph of a Function $F : R^2 \to R^2$.*

A Function from a Set to a Set

Recall, from Section 4.1.2, that abstraction is the process of thinking in terms of general objects rather than specific items. You can apply this technique to the input and output of a function to create a concept even more general than a function from R^n to R^m. To see how, think of the input to a function F as an object, x, belonging to a set V, rather than as a real number or as an n-vector. Likewise, think of the output of the function as an object, $F(x)$, belonging to a set W. Abstraction thus leads to the following definition of a function from a set V to a set W, a visual image of which is given in Figure 5.5.

DEFINITION 5.1

Given two sets V and W, a **function** (also called a **transformation** or a **mapping**) from V to W, written $F : V \to W$, is a rule that associates to each element $x \in V$, a unique element $F(x) \in W$. The set V is the **domain** of F. For each $x \in V$, $F(x)$ is the **image** of x under F.

A function from R^n to R^m is a special case of Definition 5.1 in which $V = R^n$ and $W = R^m$. In many cases, however, the sets V and W are not identified explicitly. Rather,

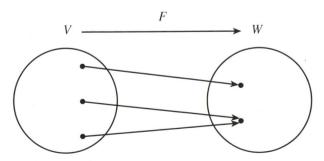

Figure 5.5 *Visualizing a Function from a Set V to a set W.*

the function is described only by the rule used to compute $F(x)$, such as the function

$$F(x) = \frac{1}{x} \tag{5.4}$$

In such cases, be sure to identify the sets V and W. The domain V is the set of all allowable input values for F. For example, for the function in (5.4),

$$V = \{x \in R^1 : x \neq 0\} \quad \text{and} \quad W = R^1$$

As another example, consider the function

$$F(x) = \sqrt{x} \tag{5.5}$$

The domain of the function F in (5.5) is the set of all allowable input values for F. For instance, you can restrict the input to F to be a nonnegative real number x and so the result of evaluating F at x is a real number. In this case,

$$V = \{x \in R^1 : x \geq 0\} \quad \text{and} \quad W = R^1$$

In contrast, you can allow any real number x as an input to F, in which case, $F(x)$ might be a complex number. For example, if $x = -2$, then $F(x) = \sqrt{-2}$. In this case,

$$V = R^1 \quad \text{and} \quad W = \{\text{complex numbers}\}$$

Several other examples of functions from a set to a set follow.

EXAMPLE 5.3 Representing a Signal as a Function
Recall that a digital signal is an infinite ordered list of real numbers obtained by sampling a continuous signal at the time intervals $0, 1, 2, \ldots$, that is,

$$\mathbf{X} = (x_0, x_1, x_2, \ldots) \tag{5.6}$$

Another way to think of the signal in (5.6) is as a function $x : \{0, 1, 2, \ldots\} \to R$. The input to the function x is a nonnegative integer time k. The output of the function x is the value of the continuous signal at time k. Thus, as seen in Figure 5.6, the value of the signal x at time 0 is $x(0)$, which is written as x_0 in (5.6). The value of the signal at time 1 is $x(1)$, which is written as x_1 in (5.6). In general, the value of the signal at the nonnegative integer time k is $x(k)$, written as x_k in (5.6).

EXAMPLE 5.4 Representing a Filter as a Function
Let S be the set of all signals, as described in Example 5.3. A **filter** F is a function $F : S \to S$ that associates to each signal $\mathbf{X} \in S$, a signal $\mathbf{Y} = F(\mathbf{X}) \in S$ by using n real numbers $a_1, a_2, \ldots, a_n$, called **filter coefficients**, as follows. From a given a signal, $\mathbf{X}$, create a new signal, $\mathbf{Y}$, in which y_k is defined as follows:

$$y_k = x_{k+n} + a_1 x_{k+n-1} + \cdots + a_n x_k, \quad \text{for } k = 0, 1, 2, \ldots$$

To illustrate a filter, recall the one in the problem of Audio Technologies at the beginning of this section, whose coefficients are $a_1 = 1$, $a_2 = -10$, and $a_3 = 8$. This filter creates,

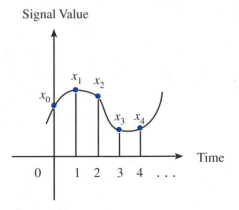

Figure 5.6 *Visualizing a Signal as a Function.*

from an input signal $\mathbf{X} = (x_0, x_1, x_2, \ldots)$, the output signal $\mathbf{Y} = (y_0, y_1, y_2, \ldots)$ in which

$$y_k = x_{k+3} + x_{k+2} - 10x_{k+1} + 8x_k$$

Now that you know some basic information about functions from a set to a set, you will learn several important operations you can perform on such functions.

5.1.2 Operations on Functions

Whenever you encounter a new mathematical concept (such as a function), you know to develop an associated image (such as a black box). The reason for creating such a concept is to help you solve problems. To do so with functions, you need to know how to compare two functions to see if they are the same and how to perform operations on functions to create new functions. Both of these topics are addressed now.

Comparing Two Functions for Equality

You often want to know if two functions, described by different rules, are really the same, the meaning of which is made precise in the following definition.

> **DEFINITION 5.2**
>
> Two functions $F : U \to V$ and $G : W \to X$ are **equal**, written $F = G$, if and only if the domains are equal (that is, $U = W$) and, for each $x \in U$, $F(x) = G(x)$.

You can also use Definition 5.2 to define the meaning of the statement "F is not equal to G." Assuming that the domains of F and G are both U, by using the rules for negating the quantifier *for all*, the statement that $F \neq G$ means that there is an element $x \in U$ such that $F(x) \neq G(x)$.

Function Composition

Suppose that $F : U \to V$ and $G : X \to W$. One common operation is to apply these two functions sequentially, as shown in Figure 5.7. That is, first apply F to a point x and then

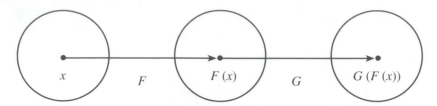

Figure 5.7 *The Composition of Two Functions.*

apply G to the resulting point, $F(x)$. The objective now is to translate this image and verbal description to symbolic form by creating an appropriate definition.

Observe in Figure 5.7 that if $x \in U$ and $F(x)$ is not in the domain of G, then a syntax error results when you write

$$G(F(x)) \tag{5.7}$$

This is because you cannot evaluate G at the point $F(x)$. Recalling that $F : U \to V$ and $G : X \to W$, one way to ensure that $F(x)$ is in the domain of G is to require that $V = X$, as is assumed from here on. As a result, (5.7) is valid and gives rise to the binary operation on two functions described in the following definition.

DEFINITION 5.3

Suppose that $F : U \to V$ and $G : V \to W$. The **composition** of F and G, denoted by $G \circ F$, is the function $G \circ F : U \to W$ defined by

$$(G \circ F)(x) = G(F(x))$$

EXAMPLE 5.5 The Composition of Two Functions

If $F : R \to R$ is defined by $F(x) = 2x + 3$ and $G : R \to R$ is defined by $G(x) = x^2$, then $G \circ F : R \to R$ is defined by

$$(G \circ F)(x) = G(F(x)) = G(2x + 3) = (2x + 3)^2 = 4x^2 + 12x + 9$$

Thus, for instance, for $x = -5$, you have that

$$F(-5) = 2(-5) + 3 = -7$$

and

$$G(-7) = (-7)^2 = 49$$

so

$$(G \circ F)(-5) = G(F(-5)) = G(-7) = 49 = 4(-5)^2 + 12(-5) + 9$$

EXAMPLE 5.6 The Composition of Two Functions

Suppose that $F : \{1, 2, 3\} \to \{1, 2\}$ is defined by

$$F(1) = 2, \quad F(2) = 1, \quad F(3) = 1$$

and $G : \{1, 2\} \to \{4, 5\}$ is defined by

$$G(1) = 5, \quad G(2) = 4$$

Then the function $G \circ F : \{1, 2, 3\} \to \{4, 5\}$ is defined by

$$(G \circ F)(1) = G(F(1)) = G(2) = 4$$
$$(G \circ F)(2) = G(F(2)) = G(1) = 5$$
$$(G \circ F)(3) = G(F(3)) = G(1) = 5$$

Suppose that $F : V \to W$ and $x, y \in V$. To solve some problems, it is necessary to add $F(x)$ and $F(y)$. However, a syntax error arises unless the operation of addition is defined in W so that you can compute $F(x) + F(y)$. Such an error does not occur when W is a vector space because the operation of addition is defined for W in this case. The rest of this chapter deals with a function $F : V \to W$, where V and W are vector spaces and F satisfies the properties described in Section 5.2.

Exercises for Section 5.1

1. Identify the domain of each of the following functions in such a way that you can evaluate the function at the given input value. What type of value (integer, vector, matrix, and so on) is output by the function? Evaluate the function at the given input value.
   (a) $F(x)$ = the reciprocal of the real number x. Given input: 1/4.
   (b) $F(\mathbf{x})$ = the subscript of the component of the n-vector $\mathbf{x}$ having the largest absolute value. Given input: $(1, 0, -2, -1)$.
   (c) Given a set A containing a finite number of elements, define $|A|$ as the number of elements of A. Given input: $A = \{0\}$.

2. Repeat the previous exercise for the following functions:
   (a) For a real number x, define $F(x) = \sqrt{1 - x^2}$. Given input: $x = 2$.
   (b) The function that computes the determinant of a matrix. Given input: I_n (the $(n \times n)$ identity matrix).
   (c) Let $F : R \to R$ be a given function. Define the function G by $G(F) =$ the function whose value at a real number x is $F(x) + 3$. Given input: $F(x) = x^2 + 2x - 3$.

3. Draw a visual image of the function F that associates to each point (x, y) in the plane, the point $F(x, y)$ directly opposite from (x, y) on the other side of the x-axis. Find a closed-form expression for F, that is, express $F(x, y)$ in terms of x and y.

4. Draw a visual image of the function F that associates to each point (x, y) in the plane, the point $F(x, y)$ directly opposite from (x, y) on the other side of the y-axis. Find a closed-form expression for F, that is, express $F(x, y)$ in terms of x and y.

5. Identify the coordinate functions of each of the following functions.
    (a) $F(x, y, z) = (zx + y, yz + x)$.
    (b) $F(x) = (1, x, x^2)$.
    (c) $F(x_1, \ldots, x_n) = (x_1, \ldots, x_n)$.

6. Identify the coordinate functions of each of the following functions.
    (a) $F(\theta) = (\sin(\theta), \cos(\theta))$.
    (b) $F(x_1, \ldots, x_n) = \max\{|x_i| : i = 1, \ldots, n\}$.
    (c) $F(x_1, \ldots, x_n) = (x_2, \ldots, x_n, x_1)$.

7. For the given functions F and G, find the value of $G \circ F$ at the given number x. If possible, find $F \circ G$ at x or explain why you cannot do so.
    (a) $F(1) = -1$, $F(2) = 0$, $F(3) = 1$; $G(-2) = -3$, $G(-1) = -2$, $G(0) = -1$, $G(1) = 0$, $G(2) = 1$. Given number: $x = 1$.
    (b) $F(x) = 1 - x^2$; $G(x) = \sqrt{x}$. Given number: $x = 1$.

8. Suppose that X, Y, and Z are sets and that $F : X \to Y$ and $G : Y \to Z$.
    (a) Can you evaluate $G \circ F$ at a point $x \in X$? If not, what conditions on X, Y, and Z would allow you to do so?
    (b) Can you evaluate $F \circ G$ at a point $x \in X$? If not, what conditions on X, Y, and Z would allow you to do so?

9. Write a closed-form expression for $G \circ F$ when F is the function defined in Exercise 3 and G is the function in Exercise 4.

10. Write a closed-form expression for $G \circ F$ when F and G are the functions defined in Exercise 7(b).

5.2 Linear Transformations and Their Applications

As you learned in Section 5.1, a transformation T from R^n to R^m describes a relationship between an n-vector $\mathbf{x}$ and an m-vector $\mathbf{y} = T(\mathbf{x})$ in which the components of $\mathbf{x}$ are transformed by T to those of $\mathbf{y}$. For example, if x represents the number of hours a truck is driven at an average speed of 55 miles per hour and y represents the total number of miles traveled, then the transformation

$$y = 55x \tag{5.8}$$

transforms a driving time of $x = 8$ hours to a distance of $y = 55(8) = 440$ miles driven.

The transformation described in (5.8) is the equation of a line through the origin and is therefore a *linear* transformation, as is the transformation $T : R^2 \to R^1$ defined by

$$T(x_1, x_2) = 3x_1 - 2x_2 \tag{5.9}$$

In contrast, the following transformation $T : R^2 \to R^1$ is not linear:

$$T(x_1, x_2) = \sqrt{x_1^2 + x_2^2} \tag{5.10}$$

Linear transformations have many desirable properties that are useful in problem solving, as you will discover throughout the rest of this chapter. When the transformations involved are not linear, more complex approaches are needed, and they do not always result in the solution to the problem. An alternative is to approximate the nonlinear transformation using a linear transformation. Doing so enables you to obtain an approximate solution to your original problem. The quality of this approximate solution depends on how accurately the linear transformation approximates the nonlinear one.

The objective now is to create a definition of a linear transformation that includes the ones in (5.8) and (5.9)—and all other transformations with similar properties—while excluding nonlinear transformations, such as the one in (5.10).

5.2.1   The Definition of a Linear Transformation

As you learned in Chapter 4, to create a definition, you must identify a common property shared by all objects being defined. For a linear transformation $T : R^n \rightarrow R^1$, you might describe the common property by the form of T. For example,

$T : R^n \rightarrow R^1$ is *linear* if and only if there are real numbers $a_1, \ldots, a_n$ such that for all $\mathbf{x} \in R^n$, $T(\mathbf{x}) = a_1 x_1 + \cdots + a_n x_n$.

If you generalize this definition by replacing $T : R^n \rightarrow R^1$ with $T : R^n \rightarrow R^m$, a syntax error arises because, in the latter case, $T(\mathbf{x})$ is an m-vector and, in the definition, $a_1 x_1 + \cdots + a_n x_n$ is a real number. One way to avoid this error is to generalize this definition to $T : R^n \rightarrow R^m$ by requiring that each coordinate function $T_i : R^n \rightarrow R^1$ have the foregoing form.

However, additional syntax errors arise when you then attempt to generalize this definition to a transformation $T : V \rightarrow W$, where V and W are sets. For example, you cannot perform the operations on the right side of the equality. As you are asked to show in Exercise 13, there are difficulties generalizing this definition even when V and W are vector spaces. Another approach is needed.

To understand the new approach, suppose that V and W are vector spaces and that $T : V \rightarrow W$. Rather than use the form of T (which may not be available in closed-form), the approach mathematicians have developed is to identify *properties* that T must satisfy to be linear. These properties are described in terms of the operations in V and W.

To simplify the subsequent notation, the operation of multiplying a real number a by a vector $\mathbf{v} \in V$ is written

$$a\mathbf{v} \quad \text{(where } a \text{ is a real number and } \mathbf{v} \in V) \tag{5.11}$$

Likewise, multiplying a by a vector $\mathbf{w} \in W$ is also written as

$$a\mathbf{w} \quad \text{(where } a \text{ is a real number } a \text{ and } \mathbf{w} \in W) \tag{5.12}$$

It is important to keep in mind that, although (5.11) and (5.12) are the same in form, they represent different operations. The operation in (5.11) combines a real number with a vector in V. The operation in (5.12) combines a real number with a vector in W.

The operation of adding vectors $\mathbf{u}, \mathbf{v} \in V$ is denoted

$$\mathbf{u} + \mathbf{v} \quad \text{(where } \mathbf{u}, \mathbf{v} \in V) \tag{5.13}$$

Likewise, the operation of adding vectors $\mathbf{x}, \mathbf{y} \in W$ is denoted

$$\mathbf{x} + \mathbf{y} \quad \text{(where } \mathbf{x}, \mathbf{y} \in W) \tag{5.14}$$

Once again, although the operations denoted by "+" in both (5.13) and (5.14) have the same form, they represent two different operations—one for adding vectors in V and one for adding vectors in W. Be sure to identify in your mind how the symbol "+" is used.

With these notational understandings, the properties that a transformation T must satisfy to be linear are stated in the following definition.

DEFINITION 5.4

Suppose that V and W are vector spaces and that $T : V \rightarrow W$. Then T is a **linear transformation** if and only if

(1) For all vectors $\mathbf{u}, \mathbf{v} \in V$, $T(\mathbf{u} + \mathbf{v}) = T(\mathbf{u}) + T(\mathbf{v})$.

(2) For all scalars a and for all vectors $\mathbf{u} \in V$, $T(a\mathbf{u}) = aT(\mathbf{u})$.

The first occurrence of the symbol "+" in (1) of Definition 5.4 refers to adding the two vectors $\mathbf{u}$ and $\mathbf{v}$ in V and the second occurrence of "+" refers to the operation of adding the two vectors $T(\mathbf{u})$ and $T(\mathbf{v})$ in W. Likewise, the operation $a\mathbf{u}$ on the left side of the equality in (2) of Definition 5.4 refers to multiplication in V and the operation $aT(\mathbf{u})$ on the right side refers to multiplication in W.

5.2.2 Examples and Applications of Linear Transformations

Now you will see examples of linear transformations and how they arise in solving problems.

Example A: A Linear Transformation from R^n to R^1

Recall the objective function of the linear programming problem of Fiber Optics described in Section 2.1.3, in which you want to find values for the variables LP, HP, LJ, and HJ so as to minimize the total cost, represented by the expression

$$4LP + 6HP + 7LJ + 8HJ \tag{5.15}$$

You can view the computation in (5.15) as a function that associates to each 4-vector (LP, HP, LJ, HJ), the real number $4LP + 6HP + 7LJ + 8HJ$. Using subscript notation, you can write (5.15) as the following transformation $T : R^4 \rightarrow R^1$:

$$T(x_1, x_2, x_3, x_4) = 4x_1 + 6x_2 + 7x_3 + 8x_4 \tag{5.16}$$

The transformation in (5.16) satisfies the two properties in Definition 5.4 and is therefore linear. This is because, for $\mathbf{u} = (u_1, u_2, u_3, u_4)$ and $\mathbf{v} = (v_1, v_2, v_3, v_4)$, you have from (5.16) that

$$
\begin{aligned}
T(\mathbf{u} + \mathbf{v}) &= T(u_1 + v_1, u_2 + v_2, u_3 + v_3, u_4 + v_4) \\
&= 4(u_1 + v_1) + 6(u_2 + v_2) + 7(u_3 + v_3) + 8(u_4 + v_4) \\
&= (4u_1 + 6u_2 + 7u_3 + 8u_4) + (4v_1 + 6v_2 + 7v_3 + 8v_4) \\
&= T(\mathbf{u}) + T(\mathbf{v})
\end{aligned}
$$

and also, for any scalar a,

$$T(a\mathbf{u}) = T(au_1, au_2, au_3, au_4)$$
$$= 4(au_1) + 6(au_2) + 7(au_3) + 8(au_4)$$
$$= a(4u_1 + 6u_2 + 7u_3 + 8u_4)$$
$$= aT(\mathbf{u})$$

A generalization of the linear transformation in (5.16) arises when considering an n-vector $\mathbf{x} = (x_1, \ldots, x_n)$ of variables and a given n-vector $\mathbf{c} = (c_1, \ldots, c_n)$. Then $T : R^n \rightarrow R^1$ defined by

$$T(\mathbf{x}) = \mathbf{c} \cdot \mathbf{x} = c_1 x_1 + \cdots + c_n x_n$$

is a linear transformation, as you are asked to show in Exercise 9. An application of this transformation is the objective function in a linear programming problem with n variables, denoted by the n-vector $\mathbf{x} = (x_1, \ldots, x_n)$, in which the components of $\mathbf{c} = (c_1, \ldots, c_n)$ are the known coefficients of the variables in the objective function. The objective of the linear programming problem is to find values for the components of $\mathbf{x}$ that minimize or maximize the linear function $\mathbf{c} \cdot \mathbf{x}$.

Example B: Matrix Transformations

Given an $(m \times n)$ matrix A and a column vector $\mathbf{x}$ of variables, the transformation $T : R^n \rightarrow R^m$ defined by

$$T(\mathbf{x}) = A\mathbf{x} \tag{5.17}$$

is called a **matrix transformation** and is linear because, for $\mathbf{u}, \mathbf{v} \in R^n$,

$$T(\mathbf{u} + \mathbf{v}) = A(\mathbf{u} + \mathbf{v}) = (A\mathbf{u}) + (A\mathbf{v}) = T(\mathbf{u}) + T(\mathbf{v})$$

and for any scalar a,

$$T(a\mathbf{u}) = A(a\mathbf{u}) = a(A\mathbf{u}) = aT(\mathbf{u})$$

When working with a matrix transformation $T : R^n \rightarrow R^m$ defined by $T(\mathbf{x}) = A\mathbf{x}$, the input to T is a column vector $\mathbf{x} \in R^{n \times 1}$ and the output is the column vector $A\mathbf{x} \in R^{m \times 1}$.

A Matrix Transformation for Population Migration.

All of the applications in Chapters 2 and 3 involve matrix transformations. For yet another example, suppose that in a particular year, x people are living in a city and y people are living in the suburbs. Given that 5% of those living in the city move to the suburbs and that 3% of those living in the suburbs move to the city, you want to determine how many people are living in the city and how many in the suburbs after one year, assuming that the combined population in the city and suburbs remains constant.

This problem involves a transformation $T : R^2 \rightarrow R^2$. Specifically, the migration data—consisting of the given percentages—transform the number of people, x, living in the city and the number of people, y, living in the suburbs at the beginning of the year to $T(x, y)$ after one year. The first component of $T(x, y)$ is the number of people living in the city after one year and the second component of $T(x, y)$ is the number of people living in the suburbs after one year.

You can find a closed-form expression for T by using the migration data to create two linear equations in terms of x and y, as follows:

$$\begin{pmatrix} \text{Number in city} \\ \text{after one year} \end{pmatrix} = \begin{pmatrix} \text{Number staying} \\ \text{in the city} \end{pmatrix} + \begin{pmatrix} \text{Number moving} \\ \text{to the city} \end{pmatrix}$$

$$= 0.95x + 0.03y$$

$$\begin{pmatrix} \text{Number in suburbs} \\ \text{after one year} \end{pmatrix} = \begin{pmatrix} \text{Number moving} \\ \text{to the suburbs} \end{pmatrix} + \begin{pmatrix} \text{Number staying} \\ \text{in the suburbs} \end{pmatrix}$$

$$= 0.05x + 0.97y$$

You can express these two linear equations in matrix-vector notation by creating the following (2×2) matrix A and column vector $\mathbf{v}$:

$$A = \begin{bmatrix} 0.95 & 0.03 \\ 0.05 & 0.97 \end{bmatrix} \quad \text{and} \quad \mathbf{v} = \begin{bmatrix} x \\ y \end{bmatrix}$$

The size of the populations living in both the city and the suburbs after one year are now determined by the following matrix transformation $T : R^2 \to R^2$:

$$T(\mathbf{v}) = A\mathbf{v} = \begin{bmatrix} 0.95 & 0.03 \\ 0.05 & 0.97 \end{bmatrix} \begin{bmatrix} x \\ y \end{bmatrix} = \begin{bmatrix} 0.95x + 0.03y \\ 0.05x + 0.97y \end{bmatrix}$$

Matrix Transformations in Linear Programming. Another example of the matrix transformation in (5.17) arises in a linear programming problem with n variables, denoted by the column vector $\mathbf{x} \in R^{n \times 1}$. If A represents the coefficients of the variables on the left side of the m equality constraints and $\mathbf{b} \in R^{m \times 1}$ is the column vector of known values on the right side, then these m constraints are described in matrix-vector notation by

$$A\mathbf{x} = \mathbf{b} \tag{5.18}$$

The left side of (5.18) is an example of a matrix transformation $T : R^n \to R^m$. In fact, letting $\mathbf{c}$ denote the known coefficients of the variables in an objective function that is to be minimized, you can write a linear programming problem in matrix-vector notation, as follows:

minimize $\mathbf{c} \cdot \mathbf{x}$

subject to $A\mathbf{x} = \mathbf{b}$

$\mathbf{x} \geq \mathbf{0}$ (all variables nonnegative)

Example C: Reflections and Rotations in the Plane

Another important example of a transformation arises when you need to find the components of the point obtained by performing a specific operation on a given point (x, y) in the plane. Several such examples are presented now.

Reflections About a Line. One common operation is a **reflection about a line** L, in which a point is transformed to its mirror image on the opposite side of L. For example, in Figure 5.8(a), the reflection of the point (x, y) about the x-axis results in the point whose components are $(x, -y)$, thus, $T(x, y) = (x, -y)$. You can verify that another way to describe T is as the matrix transformation $T(\mathbf{x}) = A\mathbf{x}$, in which

$$A = \begin{bmatrix} 1 & 0 \\ 0 & -1 \end{bmatrix} \qquad \text{(reflection about the } x\text{-axis)}$$

The reflection of the point (x, y) about the y-axis is shown in Figure 5.8(b) and results in the point whose components are $(-x, y)$. You can verify that the corresponding matrix transformation is $T(\mathbf{x}) = A\mathbf{x}$, in which

$$A = \begin{bmatrix} -1 & 0 \\ 0 & 1 \end{bmatrix} \qquad \text{(reflection about the } y\text{-axis)}$$

The reflection of the point (x, y) about the line $y = x$ is shown in Figure 5.8(c) and results in the point whose components are (y, x). The corresponding matrix transformation is $T(\mathbf{x}) = A\mathbf{x}$, in which

$$A = \begin{bmatrix} 0 & 1 \\ 1 & 0 \end{bmatrix} \qquad \text{(reflection about the line } y = x)$$

Rotations in the Plane. Another common operation in the plane is to rotate a point (x, y) counterclockwise through an angle of θ radians, as shown in Figure 5.9. The components (x', y') of the rotated point are obtained by applying the matrix transformation $T(\mathbf{v}) = A\mathbf{v}$,

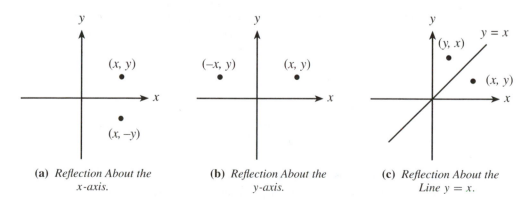

(a) *Reflection About the x-axis.*   **(b)** *Reflection About the y-axis.*   **(c)** *Reflection About the Line y = x.*

Figure 5.8 *Reflections About a Line in the Plane.*

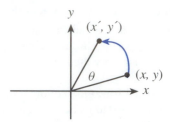

Figure 5.9 *Rotating a Point Counterclockwise Through an Angle θ.*

where

$$\mathbf{v} = \begin{bmatrix} x \\ y \end{bmatrix} \quad \text{and} \quad A = \begin{bmatrix} \cos(\theta) & -\sin(\theta) \\ \sin(\theta) & \cos(\theta) \end{bmatrix}$$

That is, given the original point whose components are (x, y), the components (x', y') of the point obtained by rotating (x, y) counterclockwise through the angle θ are

$$\begin{bmatrix} x' \\ y' \end{bmatrix} = A\mathbf{v} = \begin{bmatrix} \cos(\theta) & -\sin(\theta) \\ \sin(\theta) & \cos(\theta) \end{bmatrix} \begin{bmatrix} x \\ y \end{bmatrix} = \begin{bmatrix} x\cos(\theta) - y\sin(\theta) \\ x\sin(\theta) + y\cos(\theta) \end{bmatrix}$$

The justification that this matrix transformation does in fact provide the desired rotation of (x, y) through the angle θ is given in Section 5.3.2.

Example D: Shears

Another linear transformation in the plane that arises in computer graphics, physics, and crystallography is a **shear**. Such a transformation has the effect of converting a square to a parallelogram in which one side of the square remains fixed. For example, the shear $T : R^2 \rightarrow R^2$ defined by

$$T(\mathbf{x}) = A\mathbf{x} = \begin{bmatrix} 1 & 3 \\ 0 & 1 \end{bmatrix} \begin{bmatrix} x \\ y \end{bmatrix} = \begin{bmatrix} x + 3y \\ y \end{bmatrix}$$

transforms the 1×1 square in Figure 5.10 to the parallelogram. Observe that the side of the square on the x-axis remains unchanged by this shear. Another example is the shear $T : R^2 \rightarrow R^2$ defined by

$$T(\mathbf{x}) = A\mathbf{x} = \begin{bmatrix} 1 & 0 \\ -2 & 1 \end{bmatrix} \begin{bmatrix} x \\ y \end{bmatrix} = \begin{bmatrix} x \\ -2x + y \end{bmatrix}$$

that transforms the 2×2 square in Figure 5.11 to the parallelogram. Observe that the side of the square on the y-axis remains unchanged by this shear.

Example E: Dilations and Contractions

Another useful linear transformation is one that multiplies each vector by a fixed real number k. Such a transformation is used, for example, to change the units of the components of an

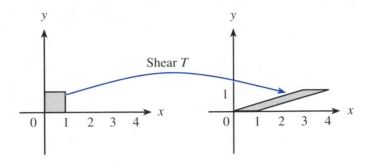

Figure 5.10 *A Shear that Does Not Change Points on the x-axis.*

n-vector **x**. To illustrate, suppose the components of **x** are the costs of n items in dollars and the current exchange rate in Japan is $k = 106$ yen per dollar, then the components of the n-vector $k\mathbf{x}$ are the costs of the n items in yen. You are asked to show in Exercise 7 that the transformation $T : R^n \to R^n$ defined by $T(\mathbf{x}) = k\mathbf{x}$ is linear.

One special case of this transformation is when $k = 0$. In this case, the result of applying T to any n-vector **x** is the zero vector, that is, $T(\mathbf{x}) = 0\mathbf{x} = \mathbf{0}$. This transformation is called the **zero transformation**.

Another special case is when $k = 1$. In this case, the result of applying T to any n-vector **x** is that same vector **x**, that is, $T(\mathbf{x}) = 1\mathbf{x} = \mathbf{x}$. This transformation is called the **identity transformation**. For a general vector space V, the identity transformation $i : V \to V$ is defined by $i(\mathbf{x}) = \mathbf{x}$, for each $\mathbf{x} \in V$.

When $0 < k < 1$, the transformation $T(\mathbf{x}) = k\mathbf{x}$ is called a **contraction** because the result of applying T to **x** is the n-vector that points in the same direction as **x** but is shorter in length, as shown by the visual representation in Figure 5.12(a) for $n = 2$. When $k > 1$, this transformation is called a **dilation** and serves to expand a vector **x**, as shown in Figure 5.12(b) for $n = 2$.

Example F: Projections Onto the Coordinate Axes

Another example of a linear transformation from R^n to R^1 is a **projection onto the coordinate axes**. Specifically, for the n-vector $\mathbf{x} = (x_1, \ldots, x_n)$ and an integer i satisfying

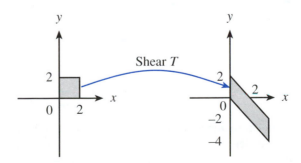

Figure 5.11 *A Shear that Does Not Change Points on the y-axis.*

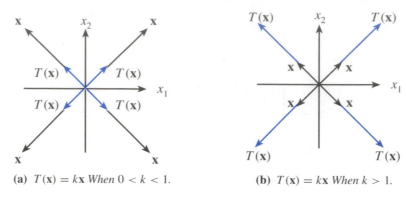

(a) $T(\mathbf{x}) = k\mathbf{x}$ *When* $0 < k < 1$.　　**(b)** $T(\mathbf{x}) = k\mathbf{x}$ *When* $k > 1$.

Figure 5.12 *Contractions and Dilations in* R^2.

$1 \le i \le n$, the projection of $\mathbf{x}$ onto the coordinate axis corresponding to x_i is the transformation $T_i : R^n \to R^1$ whose value is component i of $\mathbf{x}$, that is, $T_i(\mathbf{x}) = x_i$. You are asked to prove that these transformations are linear and to create visual representations in Exercise 5 and Exercise 6.

All of the linear transformations so far have been from R^n to R^m. The domains of the transformations in the next two examples are general vector spaces.

Example G: The Components of a Vector in a Vector Space

Recall, from Theorem 4.6 in Section 4.3.4, that each vector $\mathbf{v}$ in an n-dimensional vector space V has n components relative to a given basis $B = \{\mathbf{v}_1, \ldots, \mathbf{v}_n\}$. Those components are the unique real numbers $t_1, \ldots, t_n$ such that

$$\mathbf{v} = t_1\mathbf{v}_1 + \cdots + t_n\mathbf{v}_n$$

That is, for a given basis B, associated with each vector $\mathbf{v} \in V$ are the components of $\mathbf{v}$ relative to B. You can think of this association as a transformation $T : V \to R^n$ defined for each $\mathbf{v} \in V$ by

$$T(\mathbf{v}) = (t_1, \ldots, t_n) \quad \text{(the components of } \mathbf{v} \text{ relative to } B) \tag{5.19}$$

EXAMPLE 5.7 The Components of an n-Vector Relative to a Basis B

Consider the vector $\mathbf{v} = (5, -1)$ in the plane and the basis B consisting of the vectors $(1, 1)$ and $(-1, 1)$. The components of $\mathbf{v}$ relative to B are $(2, -3)$ because

$$\mathbf{v} = (5, -1) = 2(1, 1) - 3(-1, 1)$$

so $T(\mathbf{v}) = (2, -3)$.

The transformation in (5.19) is linear. To convince yourself that this is so, use Definition 5.4 and consider vectors $\mathbf{u}, \mathbf{v} \in V$ whose components relative to $B = \{\mathbf{v}_1, \ldots, \mathbf{v}_n\}$ are, respectively, $s_1, \ldots, s_n$ and $t_1, \ldots, t_n$. You must show that $T(\mathbf{u} + \mathbf{v}) = T(\mathbf{u}) + T(\mathbf{v})$, that is, that the components of $\mathbf{u} + \mathbf{v}$ relative to B are equal to the sum of the components of $\mathbf{u}$

relative to B and the components of $\mathbf{v}$ relative to B. But this is true because $\mathbf{u}$ has the form

$$\mathbf{u} = s_1 \mathbf{v}_1 + \cdots + s_n \mathbf{v}_n \tag{5.20}$$

and $\mathbf{v}$ has the form

$$\mathbf{v} = t_1 \mathbf{v}_1 + \cdots + t_n \mathbf{v}_n \tag{5.21}$$

Hence, the addition of (5.20) and (5.21) yields

$$\mathbf{u} + \mathbf{v} = (s_1 + t_1)\mathbf{v}_1 + \cdots + (s_n + t_n)\mathbf{v}_n \tag{5.22}$$

The result in (5.22) indicates that the components of $\mathbf{u} + \mathbf{v}$ relative to the basis B are $(s_1 + t_1, \ldots, s_n + t_n)$, which is the sum of the components of $\mathbf{u}$ relative to B and the components of $\mathbf{v}$ relative to B.

To complete the proof that the transformation in (5.19) is linear, let $\mathbf{u} \in V$ have components $(s_1, \ldots, s_n)$ relative to B and let a be a scalar. You must show that $T(a\mathbf{u}) = aT(\mathbf{u})$, that is, that the components of $a\mathbf{u}$ relative to B are equal to a times the components of $\mathbf{u}$ relative to B. This is true because $\mathbf{u}$ has the form in (5.20). Multiplying (5.20) through by the scalar a, you have

$$a\mathbf{u} = a(s_1 \mathbf{v}_1 + \cdots + s_n \mathbf{v}_n) = (as_1)\mathbf{v}_1 + \cdots + (as_n)\mathbf{v}_n \tag{5.23}$$

The result in (5.23) indicates that the components of $a\mathbf{u}$ relative to B are equal to a times the components of $\mathbf{u}$ relative to B. You now know that the transformation T in (5.19) is linear.

Notation for the Components of a Vector Relative to a Basis

It is often convenient to denote the components of a vector $\mathbf{v}$ relative to a basis B by the following column vector:

$$[\mathbf{v}]_B \quad \text{(components of } \mathbf{v} \text{ relative to } B) \tag{5.24}$$

Example H: Linear Filters

Recall, from Section 5.1.1, that a signal is an infinite ordered list of the form

$$\mathbf{X} = (x_0, x_1, x_2, \ldots)$$

Let S denote the set of all such signals. A filter whose coefficients are the real numbers $a_1, \ldots, a_n$ is a transformation $F : S \to S$ that asssociates to each signal $\mathbf{X} \in S$, a new signal $\mathbf{Y} = F(\mathbf{X}) \in S$, in which the term y_k of $\mathbf{Y}$ is defined by

$$y_k = x_{k+n} + a_1 x_{k+n-1} + \cdots + a_n x_k \tag{5.25}$$

To show that this transformation F is linear, you must first understand the operations in the vector space S. Specifically, for a real number a and two signals $\mathbf{U} = (u_0, u_1, u_2, \ldots)$ and $\mathbf{V} = (v_0, v_1, v_2, \ldots)$ in S, define the following two operations:

$$\mathbf{U} + \mathbf{V} = (u_0 + v_0, u_1 + v_1, \ldots)$$
$$a\mathbf{U} = (au_0, au_1, \ldots)$$

The transformation F representing the filter with coefficients $a_1, \ldots, a_n$ is linear because, from (5.25), for two signals $\mathbf{U}$ and $\mathbf{V}$, the signal $F(\mathbf{U} + \mathbf{V})$ is the signal $\mathbf{Y}$ in which

$$y_k = (u_{k+n} + v_{k+n}) + a_1(u_{k+n-1} + v_{k+n-1}) + \cdots + a_n(u_k + v_k)$$
$$= (u_{k+n} + a_1 u_{k+n-1} + \cdots + a_n u_k) + (v_{k+n} + a_1 v_{k+n-1} + \cdots + a_n v_k)$$

which is the sum of the corresponding terms of the signals $F(\mathbf{U})$ and $F(\mathbf{V})$. Likewise, for a real number a, the signal $F(a\mathbf{U})$ is the signal $\mathbf{Y}$ in which

$$y_k = (au_{k+n}) + (aa_1 u_{k+n-1}) + \cdots + (aa_n u_k)$$
$$= a(u_{k+n} + a_1 u_{k+n-1} + \cdots + a_n u_k)$$

which is a times the corresponding term of the signal $F(\mathbf{U})$.

In this section, you have learned the properties of a linear transformation and seen numerous applications of such transformations. In Section 5.3, additional properties of linear transformations are developed. It is these properties that enable you to solve problems involving linear transformations. Many of these properties fail to hold when the transformation is not linear.

Exercises for Section 5.2

1. Identify a matrix transformation and the associated matrix A in each of the following systems of linear equations.
   (a) The system for the Leontief economic model in Section 2.1.1.
   (b) The system in (2.4) for the network-flow problem in Section 2.1.2.

2. Identify a matrix transformation and the associated matrix A in each of the following system of equations.
   (a) The system in (3.1) for the Leontief model in Section 3.1.1.
   (b) The system for the electrical network problem in Section 3.1.2.

3. Use the form of the following transformations to determine which are linear.
   (a) $T(x, y) = -x + 2y$.   (b) $T(r, \theta) = r \cos(\theta)$.

4. Use the form of the following transformations to determine which are linear.
   (a) The transformation that associates to the length, width, and height of a box, the surface area of the box.
   (b) For a given n-vector $\mathbf{x} = (x_1, \ldots, x_n)$, define $T(\mathbf{x}) = \sum_{i=1}^{n} x_i$.

5. Consider the transformation $T_1 : R^2 \to R^1$ that projects a vector $\mathbf{x} = (x_1, x_2)$ onto the x_1-axis (see Example F in Section 5.2.2).
   (a) Use Definition 5.4 to show that T_1 is linear.   (b) Draw a visual representation of T_1.

6. Repeat the previous exercise for the transformation $T_2 : R^2 \to R^1$ that projects a vector $\mathbf{x} = (x_1, x_2)$ onto the x_2-axis (see Example F in Section 5.2.2).

7. Let k be a given real number. Use Definition 5.4 to prove that the transformation $T : R^n \to R^n$ defined by $T(\mathbf{x}) = k\mathbf{x}$ is linear.

8. Explain why the transformation that rotates a point (x, y) in the plane clockwise through a fixed angle θ is linear.

9. Let $\mathbf{c} = (c_1, \ldots, c_n)$ be a given n-vector. Use Definition 5.4 to prove that the transformation $T : R^n \to R$ defined by $T(\mathbf{x}) = \mathbf{c} \cdot \mathbf{x}$ is linear.

10. Suppose that $F, G : R^n \to R^n$ are linear transformations. Use Definition 5.4 to prove that the transformation $H : R^n \to R^n$ defined by $H = G \circ F$ is linear.

11. Use Definition 5.4 to prove that the transformation T that associates to each polynomial $p(x) = a_0 + a_1 x + a_2 x^2 + \cdots + a_n x^n$, the polynomial $q(x) = a_1 + 2a_2 x + \cdots + na_n x^{n-1}$ is linear.

12. Is the transformation that associates the determinant to each $(n \times n)$ matrix A linear? If so, use Definition 5.4 to prove it. If not, explain why not.

13. Suppose you were to define a transformation $T : R^n \to R^1$ to be linear if there are real numbers $a_1, \ldots, a_n$ such that for each n-vector $\mathbf{x} = (x_1, \ldots, x_n)$, $T(\mathbf{x}) = a_1 x_1 + \cdots + a_n x_n$. Explain what difficulties and syntax errors arise in trying to generalize this definition to a transformation $T : V \to W$, where V and W are vector spaces.

14. Let $a_1, \ldots, a_n$ be the coefficients of a filter F, $\mathbf{0}$ be the zero signal, and define

$$S = \{\text{signals } \mathbf{X} : F(\mathbf{X}) = \mathbf{0}\}$$

Show that the transformation $T : S \to R^n$ defined by $T(\mathbf{X}) = (x_0, \ldots, x_{n-1})$ is linear.

15. Suppose that $B = \{\mathbf{v}_1, \ldots, \mathbf{v}_n\}$ is a basis for an n-dimensional vector space V, B' is a basis for an m-dimensional vector space W, and that $T : V \to W$ is a linear transformation. For any vector $\mathbf{v} \in V$, let $[\mathbf{v}]_B$ be the components of $\mathbf{v}$ relative to B and let $[T(\mathbf{v})]_{B'}$ be the components of $T(\mathbf{v})$ relative to B'. Prove that the transformation $U : R^n \to R^m$ defined by $U(t_1, \ldots, t_n) = [T(\mathbf{v})]_{B'}$ is linear, where $\mathbf{v} = t_1 \mathbf{v}_1 + \cdots + t_n \mathbf{v}_n$.

16. Explain why the transformation $T(x, y) = 5 - x + 2y$ is not linear.

5.3 The Matrix of a Linear Transformation

One important linear transformation you have seen is the matrix transformation in Example B in Section 5.2.2, in which, given an $(m \times n)$ matrix A, $T : R^n \to R^m$ is defined by $T(\mathbf{x}) = A\mathbf{x}$, for each column vector $\mathbf{x} \in R^{n \times 1}$. You have also seen other linear transformations from R^n to R^n, such as $T(\mathbf{x}) = k\mathbf{x}$, in which k is a real number. The transformation $T(\mathbf{x}) = k\mathbf{x}$ is a matrix transformation $T(\mathbf{x}) = A\mathbf{x}$ in which $A = kI_n$, where I_n is the $(n \times n)$ identity matrix. One of the major results in this section is that *every* linear transformation T from R^n to R^m, regardless of form, is a matrix transformation. You will also learn how to find the matrix of such a transformation. To do so requires the knowledge of various properties satisfied by all linear transformations, some of which are described next.

5.3.1 Basic Properties of Linear Transformations

From Definition 5.4, a linear transformation T from a vector space V to a vector space W satisfies the following two properties:

1. For all vectors $\mathbf{u}, \mathbf{v} \in V$, $T(\mathbf{u} + \mathbf{v}) = T(\mathbf{u}) + T(\mathbf{v})$.

2. For all scalars a and for all vectors $\mathbf{u} \in V$, $T(a\mathbf{u}) = aT(\mathbf{u})$.

Linear transformations satisfy many other properties useful for problem solving. These additional properties are derived from the two properties in the definition, as shown in the proof of the following theorem.

THEOREM 5.1

A linear transformation $T : V \to W$, in which $\mathbf{0}$ denotes the zero vector of appropriate dimension in both V and W, satisfies the following properties:

(a) $T(\mathbf{0}) = \mathbf{0}$ (T applied to the zero vector of V results in the zero vector of W).

(b) For any vector $\mathbf{v} \in V$, $T(-\mathbf{v}) = -T(\mathbf{v})$ (T applied to the negative of a vector $\mathbf{v}$ in V is the negative (in W) of T applied to $\mathbf{v}$).

(c) For any integer $n \geq 1$ and for any vectors $\mathbf{v}_1, \ldots, \mathbf{v}_n$ and scalars $s_1, \ldots, s_n$, $T(s_1\mathbf{v}_1 + \cdots + s_n\mathbf{v}_n) = s_1 T(\mathbf{v}_1) + \cdots + s_n T(\mathbf{v}_n)$ (T applied to a linear combination of vectors in V is that same linear combination (in W) of T applied to those vectors).

(d) For any vectors $\mathbf{u}, \mathbf{v} \in V$, $T(\mathbf{u} - \mathbf{v}) = T(\mathbf{u}) - T(\mathbf{v})$.

Proof.

Each part of this theorem is proved by applying specialization to properties (1) and (2) from the definition of a linear transformation, as stated prior to this theorem. In reading these proofs, keep in mind whether the operations and vectors are in V or in W.

To see that (a) is true, specialize property (2) to the particular scalar $a = 0$ and to any vector $\mathbf{u} \in V$ to obtain $T(\mathbf{0}) = T(0\mathbf{u}) = 0T(\mathbf{u})$. Now the scalar 0 times the vector $T(\mathbf{u})$ in W is the zero vector in W, so

$$T(\mathbf{0}) = T(0\mathbf{u}) = 0T(\mathbf{u}) = \mathbf{0}$$

To see that (b) is true, choose a vector $\mathbf{v} \in V$. Using the fact that in any vector space, $-\mathbf{v} = (-1)\mathbf{v}$ and specializing property (2) of a linear transformation to the particular case $a = -1$, you have that

$$T(-\mathbf{v}) = T((-1)\mathbf{v}) = (-1)T(\mathbf{v}) = -T(\mathbf{v})$$

Because of the form of the statement in part (c), the following special proof technique is used.

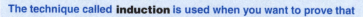

The technique called **induction** is used when you want to prove that

for all integers $n \geq 1$, some statement $S(n)$ is true

To use induction, begin by showing that the first statement, $S(1)$, is true. After establishing that $S(1)$ is true, you can now use this fact to prove that $S(2)$ is true, after which, you want to prove that $S(3)$ is true, and so on. That is, assuming you have already proved that $S(n)$ is true, you want to use that fact to prove that the next statement, $S(n + 1)$, is true. Thus, you can assume that $S(n)$ is true and use this assumption to prove that $S(n + 1)$ is true. The act of using the assumption that $S(n)$ is true is called using the **induction hypothesis**.

To illustrate how this technique is used in the current proof, observe that part (c) is in a form suitable for induction. In this case, the statement $S(n)$ that must be proved true for each integer $n \geq 1$ is

$S(n)$: for any vectors $\mathbf{v}_1, \ldots, \mathbf{v}_n$ and scalars $s_1, \ldots, s_n$,
$$T(s_1\mathbf{v}_1 + \cdots + s_n\mathbf{v}_n) = s_1 T(\mathbf{v}_1) + \cdots + s_n T(\mathbf{v}_n)$$

According to the induction method, begin by showing that $S(1)$ is true. In this case, $S(1)$ is the statement that for any vector $\mathbf{v}_1 \in V$, and for any scalar s_1, $T(s_1\mathbf{v}_1) = s_1 T(\mathbf{v}_1)$. This follows immediately by specializing property (2) to the particular scalar $a = s_1$ and the particular vector $\mathbf{u} = \mathbf{v}_1$.

In the next step of induction, you can assume that

$S(n)$: for any vectors $\mathbf{v}_1, \ldots, \mathbf{v}_n$ and scalars $s_1, \ldots, s_n$,
$$T(s_1\mathbf{v}_1 + \cdots + s_n\mathbf{v}_n) = s_1 T(\mathbf{v}_1) + \cdots + s_n T(\mathbf{v}_n)$$

You must use this assumption to prove that

$S(n + 1)$: for any vectors $\mathbf{v}_1, \ldots, \mathbf{v}_{n+1}$ and scalars $s_1, \ldots, s_{n+1}$,
$$T(s_1\mathbf{v}_1 + \cdots + s_{n+1}\mathbf{v}_{n+1}) = s_1 T(\mathbf{v}_1) + \cdots + s_{n+1}T(\mathbf{v}_{n+1})$$

So, choose vectors $\mathbf{v}_1, \ldots, \mathbf{v}_{n+1}$ and scalars $s_1, \ldots, s_{n+1}$. The key now is to write $T(s_1\mathbf{v}_1 + \cdots + s_{n+1}\mathbf{v}_{n+1})$ in terms of $T(s_1\mathbf{v}_1 + \cdots + s_n\mathbf{v}_n)$ so that you can use the induction hypothesis, $S(n)$. Specifically,

$$
\begin{aligned}
T(s_1\mathbf{v}_1 &+ \cdots + s_{n+1}\mathbf{v}_{n+1}) \\
&= T((s_1\mathbf{v}_1 + \cdots + s_n\mathbf{v}_n) + s_{n+1}\mathbf{v}_{n+1}) \\
&= T(s_1\mathbf{v}_1 + \cdots + s_n\mathbf{v}_n) + T(s_{n+1}\mathbf{v}_{n+1}) && \text{(property (1) of } T) \\
&= T(s_1\mathbf{v}_1 + \cdots + s_n\mathbf{v}_n) + s_{n+1}T(\mathbf{v}_{n+1}) && \text{(property (2) of } T) \\
&= s_1 T(\mathbf{v}_1) + \cdots + s_n T(\mathbf{v}_n) + s_{n+1}T(\mathbf{v}_{n+1}) && \text{(induction hypothesis)}
\end{aligned}
$$

This completes the proof of part (c).

To see that part (d) is true, choose vectors $\mathbf{u}, \mathbf{v} \in V$. The desired conclusion that $T(\mathbf{u} - \mathbf{v}) = T(\mathbf{u}) - T(\mathbf{v})$ follows by specializing part (c) to the case when $n = 2$, $\mathbf{v}_1 = \mathbf{u}$, $\mathbf{v}_2 = \mathbf{v}$, $s_1 = 1$, and $s_2 = -1$.

The proof of the theorem is now complete. ■

5.3.2 Finding the Matrix of a Linear Transformation from R^n to R^m

As mentioned at the beginning of this section, every linear transformation $T : R^n \to R^m$ is a matrix transformation that has the form $T(\mathbf{x}) = A\mathbf{x}$, for some $(m \times n)$ matrix A. The objective now is to explain why this statement is true and to show how to find such a matrix.

To illustrate, consider a linear transformation $T : R^2 \to R^3$ whose values, when applied to the two standard unit vectors

$$
\mathbf{e}_1 = \begin{bmatrix} 1 \\ 0 \end{bmatrix} \quad \text{and} \quad \mathbf{e}_2 = \begin{bmatrix} 0 \\ 1 \end{bmatrix}
$$

in R^2, are:

$$T(\mathbf{e}_1) = T\left(\begin{bmatrix} 1 \\ 0 \end{bmatrix}\right) = \begin{bmatrix} -2 \\ 5 \\ 4 \end{bmatrix}, \quad T(\mathbf{e}_2) = T\left(\begin{bmatrix} 0 \\ 1 \end{bmatrix}\right) = \begin{bmatrix} 3 \\ 0 \\ -6 \end{bmatrix} \tag{5.26}$$

The foregoing linear transformation T is a matrix transformation. That is, there is a (3×2) matrix A such that for any 2-vector $\mathbf{x}$, $T(\mathbf{x}) = A\mathbf{x}$. The key to finding A is to write $\mathbf{x}$ as a linear combination of the standard unit vectors $\mathbf{e}_1$ and $\mathbf{e}_2$ and use property (c) of Theorem 5.1, as follows:

$$T(\mathbf{x}) = T(x_1\mathbf{e}_1 + x_2\mathbf{e}_2) = x_1 T(\mathbf{e}_1) + x_2 T(\mathbf{e}_2) \tag{5.27}$$

You can rewrite (5.27) using (5.26) and, in so doing, identify the (3×2) matrix A associated with this linear transformation T, as follows:

$$T(\mathbf{x}) = x_1 \begin{bmatrix} -2 \\ 5 \\ 4 \end{bmatrix} + x_2 \begin{bmatrix} 3 \\ 0 \\ -6 \end{bmatrix} = \underbrace{\begin{bmatrix} -2 & 3 \\ 5 & 0 \\ 4 & -6 \end{bmatrix}}_{A} \begin{bmatrix} x_1 \\ x_2 \end{bmatrix} = A\mathbf{x} \tag{5.28}$$

For the linear transformation $T : R^2 \to R^3$ in this example, the columns of the associated (3×2) matrix A are the images of the standard unit vectors in R^2, that is, $A = [T(\mathbf{e}_1) \quad T(\mathbf{e}_2)]$. This result generalizes to any linear transformation $T : R^n \to R^m$, as stated in the following theorem, whose proof is left to Exercise 5.

> **THEOREM 5.2**
>
> For any linear transformation $T : R^n \to R^m$, there is a unique $(m \times n)$ matrix A such that for all column vectors $\mathbf{x} \in R^{n \times 1}$, $T(\mathbf{x}) = A\mathbf{x}$. In particular, letting $\mathbf{e}_1, \ldots, \mathbf{e}_n$ be the standard unit vectors in R^n,
>
> $$A = [\ T(\mathbf{e}_1) \quad \cdots \quad T(\mathbf{e}_n)\ ] \tag{5.29}$$

The matrix A in (5.29) is called the **standard matrix for the linear transformation** T. Finding this standard matrix is illustrated again in the following examples.

EXAMPLE 5.8 The Standard Matrix for $T(\mathbf{x}) = k\mathbf{x}$

For a given real number k, recall the linear transformation $T : R^n \to R^n$ in Example E in Section 5.2.2, defined by $T(\mathbf{x}) = k\mathbf{x}$, where k is a real number. According to Theorem 5.2, the columns of the $(n \times n)$ standard matrix A associated with T are the results of applying T to the standard unit vectors in R^n. In this case, because $T(\mathbf{x}) = k\mathbf{x}$, you have that for each standard unit vector $\mathbf{e}_i$,

$$T(\mathbf{e}_i) = k\mathbf{e}_i, \quad \text{for } i = 1, \ldots, n$$

Consequently, the standard matrix for T is

$$A = [\ T(\mathbf{e}_1) \quad \cdots \quad T(\mathbf{e}_n)\ ] = [\ k\mathbf{e}_1 \quad \cdots \quad k\mathbf{e}_n\ ] = k[\ \mathbf{e}_1 \quad \cdots \quad \mathbf{e}_n\ ] = kI_n$$

EXAMPLE 5.9 The Standard Matrix of a Rotation in the Plane

Recall the linear transformation $T : R^2 \to R^2$ in Example C in Section 5.2.2 that rotates a given point (x, y) in the plane through an angle θ in the counterclockwise direction. According to Theorem 5.2, the columns of the (2×2) standard matrix A associated with T are the results of applying T to the standard unit vectors in R^2. In this case, from Figure 5.13, you can see that

$$T\left(\begin{bmatrix} 1 \\ 0 \end{bmatrix}\right) = \begin{bmatrix} \cos(\theta) \\ \sin(\theta) \end{bmatrix} \quad \text{and} \quad T\left(\begin{bmatrix} 0 \\ 1 \end{bmatrix}\right) = \begin{bmatrix} -\sin(\theta) \\ \cos(\theta) \end{bmatrix}$$

So, the standard matrix for the rotation transformation $T : R^2 \to R^2$ is

$$A = [T(\mathbf{e}_1) \quad T(\mathbf{e}_2)] = \begin{bmatrix} \cos(\theta) & -\sin(\theta) \\ \sin(\theta) & \cos(\theta) \end{bmatrix}$$

You now know some of the basic properties of a linear transformation $T : R^n \to R^m$. You also know that the columns of the standard matrix associated with T are obtained by applying T to the the standard unit vectors in R^n. This approach is now generalized to a linear transformation from a vector space V to a vector space W.

5.3.3 The Matrix of a Linear Transformation from a Vector Space *V* to a Vector Space *W*

The standard unit vectors do not exist in a general n-dimensional vector space V, but you can generalize the concept of a standard matrix to $T : V \to W$ by replacing the standard unit vectors with a basis for V. To see how, choose a basis B consisting of n vectors in V, say, $B = \{\mathbf{v}_1, \ldots, \mathbf{v}_n\}$. Any vector $\mathbf{v} \in V$ is a unique linear combination of these basis vectors, so, there are unique real numbers $s_1, \ldots, s_n$ such that

$$\mathbf{v} = s_1 \mathbf{v}_1 + \cdots + s_n \mathbf{v}_n \tag{5.30}$$

That is, $s_1, \ldots, s_n$ are the components of $\mathbf{v}$ relative to the basis B, which, as you will recall from Example G in Section 5.2.2, are denoted by the column vector $[\mathbf{v}]_B$.

Likewise, the vector $T(\mathbf{v})$ is a vector in W. Choosing a basis B' consisting of m vectors

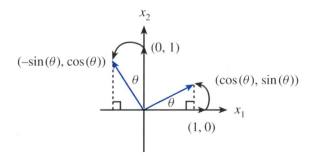

Figure 5.13 *Rotating the Standard Unit Vectors in R^2 Counterclockwise Through an Angle θ.*

in W, say, $B' = \{\mathbf{w}_1, \ldots, \mathbf{w}_m\}$, any vector $\mathbf{w} \in W$ is a linear combination of these basis vectors. In particular, $T(\mathbf{v}) \in W$, so there are unique real numbers $t_1, \ldots, t_m$ that satisfy the following property:

$$T(\mathbf{v}) = t_1 \mathbf{w}_1 + \cdots + t_m \mathbf{w}_m \tag{5.31}$$

In other words, $t_1, \ldots, t_m$ are the components of $T(\mathbf{v})$ relative to the basis B' of W, that is, $[T(\mathbf{v})]_{B'}$.

What this means in terms of components is that, for the transformation $T : V \to W$, the n components of $\mathbf{v}$ relative to B, namely, $[\mathbf{v}]_B$, are transformed to the m components of $T(\mathbf{v})$ relative to B', namely, $[T(\mathbf{v})]_{B'}$, as shown in Figure 5.14. You can think of this transformation of components as a function $U : R^n \to R^m$. As a result of Exercise 15 in Section 5.2, U is a linear transformation. Thus, by Theorem 5.2, there is an $(m \times n)$ standard matrix A for U. Letting $\mathbf{e}_1, \ldots, \mathbf{e}_n$ be the standard unit vectors in R^n, the columns of A are $U(\mathbf{e}_1), \ldots, U(\mathbf{e}_n)$. What remains is to find $U(\mathbf{e}_1), \ldots, U(\mathbf{e}_n)$.

To find $U(\mathbf{e}_1)$, note from Figure 5.14 that $\mathbf{e}_1$ corresponds to the vector $\mathbf{v} \in V$ whose components relative to the basis $B = \{\mathbf{v}_1, \ldots, \mathbf{v}_n\}$ are $(1, 0, \ldots, 0)$, so,

$$\mathbf{v} = 1\mathbf{v}_1 + 0\mathbf{v}_2 + \cdots + 0\mathbf{v}_n = \mathbf{v}_1 \tag{5.32}$$

From (5.32), $U(\mathbf{e}_1)$ are the components of $T(\mathbf{v}_1)$ relative to the basis B' in W, that is,

$$U(\mathbf{e}_1) = [T(\mathbf{v}_1)]_{B'}$$

Likewise, for each $i = 1, \ldots, n$, $U(\mathbf{e}_i)$ are the components of $T(\mathbf{v}_i)$ relative to the basis B' in W, that is,

$$U(\mathbf{e}_i) = [T(\mathbf{v}_i)]_{B'}$$

In summary, constructing the matrix A of the linear transformation $U : R^n \to R^m$ in Figure 5.14 requires finding, for each basis vector $\mathbf{v}_i$ in B, the components of $T(\mathbf{v}_i)$ relative to the basis B' in W, for then,

$$A = \begin{bmatrix} [T(\mathbf{v}_1)]_{B'} & [T(\mathbf{v}_2)]_{B'} & \cdots & [T(\mathbf{v}_n)]_{B'} \end{bmatrix} \tag{5.33}$$

The matrix in (5.33) is called the **matrix for T relative to the bases B and B'**. This matrix A is used to find $T(\mathbf{v})$, for any vector $\mathbf{v} \in V$, as follows:

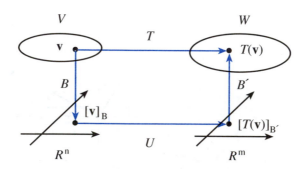

Figure 5.14 *The Component Representation of a Linear Transformation from V to W.*

1. Find the components of $\mathbf{v}$ relative to the basis B, that is, find $[\mathbf{v}]_B$.

2. Use A to find the components of $T(\mathbf{v})$ relative to the basis B' of W by computing $[T(\mathbf{v})]_{B'} = A[\mathbf{v}]_B$. Suppose these components are the real numbers $t_1, \ldots, t_m$.

3. Then $T(\mathbf{v}) = t_1 \mathbf{w}_1 + \cdots + t_m \mathbf{w}_m$.

An example of finding and using the matrix of a linear transformation T from V to W relative to B and B' follows.

EXAMPLE 5.10 Finding and Using the Matrix of a Linear Transformation $T : V \to W$ Relative to B and B'

Suppose that $B = \{\mathbf{v}_1, \mathbf{v}_2, \mathbf{v}_3\}$ is a basis for a vector space V and $B' = \{\mathbf{w}_1, \mathbf{w}_2\}$ is a basis for a vector space W. Let $T : V \to W$ be a linear transformation for which

$$T(\mathbf{v}_1) = 2\mathbf{w}_1 - 3\mathbf{w}_2, \quad T(\mathbf{v}_2) = -\mathbf{w}_1 + 4\mathbf{w}_2, \quad T(\mathbf{v}_3) = 5\mathbf{w}_1 - 2\mathbf{w}_2$$

Then,

$$[T(\mathbf{v}_1)]_{B'} = \begin{bmatrix} 2 \\ -3 \end{bmatrix}, \quad [T(\mathbf{v}_2)]_{B'} = \begin{bmatrix} -1 \\ 4 \end{bmatrix}, \quad [T(\mathbf{v}_3)]_{B'} = \begin{bmatrix} 5 \\ -2 \end{bmatrix}$$

So the matrix A for T relative to the bases B and B' is

$$A = \begin{bmatrix} [T(\mathbf{v}_1)]_{B'} & [T(\mathbf{v}_2)]_{B'} & [T(\mathbf{v}_3)]_{B'} \end{bmatrix} = \begin{bmatrix} 2 & -1 & 5 \\ -3 & 4 & -2 \end{bmatrix}$$

To use A to compute $T(\mathbf{v})$ for $\mathbf{v} = -2\mathbf{v}_1 - 3\mathbf{v}_2 + \mathbf{v}_3$, for example, you have

$$[\mathbf{v}]_B = \begin{bmatrix} -2 \\ -3 \\ 1 \end{bmatrix}, \quad \text{so} \quad [T(\mathbf{v})]_{B'} = A[\mathbf{v}]_B = \begin{bmatrix} 2 & -1 & 5 \\ -3 & 4 & -2 \end{bmatrix} \begin{bmatrix} -2 \\ -3 \\ 1 \end{bmatrix} = \begin{bmatrix} 4 \\ -8 \end{bmatrix}$$

Then,

$$T(\mathbf{v}) = 4\mathbf{w}_1 - 8\mathbf{w}_2$$

In this section, you have learned that every linear transformation from R^n to R^m is a matrix transformation of the form $T(\mathbf{x}) = A\mathbf{x}$. The columns of A are the result of T applied to the n standard unit vectors in R^n. To generalize this idea to a linear transformation T from an n-dimensional vector space V to an m-dimensional vector space W, you choose a basis B for V and another basis B' for W. The columns of the matrix for T relative to B and B' are the components of T applied to the vectors in B, expressed relative to the basis B'. Several special cases of a linear transformation from V to W arise when $W = V$. The application of the results in this section to such transformations is described in Section 5.4.

Exercises for Section 5.3

1. Find the standard matrix associated with the linear transformation $T(x, y) = (x, -y)$ by finding the value of T applied to the standard unit vectors.

2. Find the standard matrix associated with the linear transformation $T(x_1, x_2, x_3, x_4) = 4x_1 + 6x_2 + 7x_3 + 8x_4$ by finding the value of T applied to the standard unit vectors.

3. Find the standard matrix associated with the linear transformation T in which

$$T(-1, 0) = (3, 0 - 1) \quad \text{and} \quad T(0, 2) = (0, -2, 4)$$

Do so by finding the value of T applied to the standard unit vectors. (Hint: Express the standard unit vectors as a linear combination of $(-1, 0)$ and $(0, 2)$.)

4. Find the standard matrix associated with the linear transformation T in which

$$T(-1, 1) = (3, 0, -1) \quad \text{and} \quad T(2, -1) = (0, -2, 4)$$

Do so by finding the value of T applied to the standard unit vectors.

5. Prove Theorem 5.2 that for any linear transformation $T : R^n \rightarrow R^m$, there is a unique $(m \times n)$ matrix A such that

for all column vectors $\mathbf{x} \in R^{n \times 1}$, $T(\mathbf{x}) = A\mathbf{x}$

In particular, letting $\mathbf{e}_1, \ldots, \mathbf{e}_n$ be the standard unit vectors in R^n,

$$A = [\ T(\mathbf{e}_1) \quad \cdots \quad T(\mathbf{e}_n)\ ]$$

(Hint: To prove the uniqueness, assume that there is another $(m \times n)$ matrix B such that for all column vectors $\mathbf{x} \in R^{n \times 1}$, $T(\mathbf{x}) = B\mathbf{x}$. Then show that $A = B$ by applying specialization.)

6. Show that the result of computing $T(\mathbf{v})$ in Example 5.10 by using the matrix of T relative to the bases B and B' is the same as computing $T(\mathbf{v})$ by using the linearity property of T.

7. Find the matrix of $T(x, y, z) = (2x + y + z, x - 2y)$ relative to the bases $B = \{(1, 0, -1), (0, 2, 0), (1, -2, 0)\}$ for R^3 and $B' = \{(1, 1), (-1, 0)\}$ for R^2.

8. Let $B = \{(1, 0), (0, 1)\}$ be a basis for R^2 and let B' be the following basis for $R^{2 \times 2}$, the set of all (2×2) matrices:

$$E = \begin{bmatrix} 1 & 0 \\ 0 & 0 \end{bmatrix}, \quad F = \begin{bmatrix} 0 & 0 \\ 1 & 0 \end{bmatrix}, \quad G = \begin{bmatrix} 0 & 1 \\ 0 & 0 \end{bmatrix}, \quad H = \begin{bmatrix} 0 & 0 \\ 0 & 1 \end{bmatrix}$$

Find the matrix of the following linear transformation $T : R^2 \rightarrow R^{2 \times 2}$ relative to the given bases B and B':

$$T(a, b) = \begin{bmatrix} a & 0 \\ 0 & b \end{bmatrix}$$

5.4 Linear Transformations from *V* to *V*

In this section, two special cases of a linear transformation $T : V \rightarrow W$ in which $W = V$ are presented. The first uses a basis, B, for V and a different basis, B', for W. The second example uses the same basis B for both V and W.

5.4.1 Linear Transformations from R^n to R^n

Matrix-vector notation is useful for working with a linear transformation $T : R^n \to R^n$. For example, from Theorem 5.2 in Section 5.3.2, you know that the columns of the $(n \times n)$ standard matrix A associated with T are T applied to the standard unit vectors $\mathbf{e}_1, \ldots, \mathbf{e}_n$ and that for any column vector $\mathbf{x} \in R^{n \times 1}$, $T(\mathbf{x}) = A\mathbf{x}$. In this section, the use of matrix-vector notation simplifies the process of working with T when using a basis B for R^n other than the standard unit vectors.

So suppose that B is a basis for R^n consisting of the vectors $\mathbf{v}_1, \ldots, \mathbf{v}_n$. Throughout this section, it is convenient to think of B as an $(n \times n)$ matrix in which column i of B is $\mathbf{v}_i$. Such a matrix is called a **basis matrix**. For example, the basis matrix corresponding to the basis $\{(3, -1), (-2, 1)\}$ for R^2 is:

$$B = \begin{bmatrix} 3 & -2 \\ -1 & 1 \end{bmatrix} \tag{5.34}$$

An $(n \times n)$ basis matrix B is invertible because the vectors in a basis are linearly independent. For example, the basis matrix in (5.34) is invertible and

$$B^{-1} = \begin{bmatrix} 3 & -2 \\ -1 & 1 \end{bmatrix}^{-1} = \begin{bmatrix} 1 & 2 \\ 1 & 3 \end{bmatrix} \tag{5.35}$$

A basis matrix B provides a convenient way for finding $[\mathbf{x}]_B$, the components of an n-vector $\mathbf{x}$ relative to the basis B. Specifically, the components of $[\mathbf{x}]_B$ are the multipliers for the columns of B that express $\mathbf{x}$ as a linear combination of the columns of B. In matrix-vector notation,

$$\mathbf{x} = B[\mathbf{x}]_B \tag{5.36}$$

You can find $[\mathbf{x}]_B$ for a given n-vector $\mathbf{x}$ by multiplying both sides of (5.36) by B^{-1}, that is,

$$[\mathbf{x}]_B = B^{-1}\mathbf{x} \tag{5.37}$$

EXAMPLE 5.11 Using a Basis Matrix to Find the Components of an n-Vector Relative to a Given Basis

For the basis B in (5.34), the components of the vector $(4, -2)$ relative to B are computed using (5.37) and (5.35), as follows:

$$\begin{bmatrix} 4 \\ -2 \end{bmatrix}_B = B^{-1} \begin{bmatrix} 4 \\ -2 \end{bmatrix} = \begin{bmatrix} 1 & 2 \\ 1 & 3 \end{bmatrix} \begin{bmatrix} 4 \\ -2 \end{bmatrix} = \begin{bmatrix} 0 \\ -2 \end{bmatrix}$$

As another example, the components of the vector $(-5, 3)$ relative to B are computed using (5.37) and (5.35), as follows:

$$\begin{bmatrix} -5 \\ 3 \end{bmatrix}_B = B^{-1} \begin{bmatrix} -5 \\ 3 \end{bmatrix} = \begin{bmatrix} 1 & 2 \\ 1 & 3 \end{bmatrix} \begin{bmatrix} -5 \\ 3 \end{bmatrix} = \begin{bmatrix} 1 \\ 4 \end{bmatrix}$$

5.4.2 Change of Bases in R^n

Suppose you are solving a problem involving vectors in R^n whose components are expressed relative to a given basis B, but that it is easier to solve the problem when the components of those vectors are expressed relative to a basis B'. (Such a problem is presented in Chapter 6.) The need to do so gives rise to the following **change-of-basis problem in R^n**.

> **PROBLEM 5.1 Change-of-Basis Problem in R^n**
> Find the components $[\mathbf{v}]_{B'}$ of an n-vector $\mathbf{v}$ relative to a given basis matrix B' using the known components $[\mathbf{v}]_B$ of $\mathbf{v}$ relative to a given basis matrix B.

In this problem, the given data are the basis matrices B and B' together with the components $[\mathbf{v}]_B$ relative to the basis B. The desired output are the components of $\mathbf{v}$ relative to B', namely, $[\mathbf{v}]_{B'}$.

As a numerical example, suppose that the original basis matrix B and the new basis matrix B' for R^2 are

$$B = \begin{bmatrix} 4 & -5 \\ -2 & 3 \end{bmatrix} \quad \text{and} \quad B' = \begin{bmatrix} 3 & -2 \\ -1 & 1 \end{bmatrix} \tag{5.38}$$

Given the components of a vector $\mathbf{v}$ relative to B, how do you find the components of $\mathbf{v}$ relative to B'?

Solving the Problem with Matrix-Vector Notation

Using the matrix-vector notation in Section 5.4.1, the given components of the vector $\mathbf{v}$ relative to B, namely, $[\mathbf{v}]_B$, satisfy

$$\mathbf{v} = B[\mathbf{v}]_B \tag{5.39}$$

Likewise, the unknown components of $\mathbf{v}$ relative to B', namely, $[\mathbf{v}]_{B'}$, must satisfy

$$\mathbf{v} = B'[\mathbf{v}]_{B'} \tag{5.40}$$

Equating the right sides of the equalities in (5.39) and (5.40) yields

$$B[\mathbf{v}]_B = B'[\mathbf{v}]_{B'} \tag{5.41}$$

You can solve (5.41) for the desired vector $[\mathbf{v}]_{B'}$ by multiplying both sides of (5.41) on the left by $(B')^{-1}$, which exists because B' is a basis matrix, to obtain

$$[\mathbf{v}]_{B'} = ((B')^{-1}B)[\mathbf{v}]_B \tag{5.42}$$

For the foregoing numerical example, $(B')^{-1}$ is given in (5.35) and so

$$(B')^{-1}B = \begin{bmatrix} 1 & 2 \\ 1 & 3 \end{bmatrix} \begin{bmatrix} 4 & -5 \\ -2 & 3 \end{bmatrix} = \begin{bmatrix} 0 & 1 \\ -2 & 4 \end{bmatrix} \tag{5.43}$$

Thus, if the known components of $\mathbf{v}$ relative to the basis B are

$$[\mathbf{v}]_B = \begin{bmatrix} x \\ y \end{bmatrix}$$

then, from (5.42) and (5.43), you have that

$$[\mathbf{v}]_{B'} = ((B')^{-1}B)[\mathbf{v}]_B = \begin{bmatrix} 0 & 1 \\ -2 & 4 \end{bmatrix}\begin{bmatrix} x \\ y \end{bmatrix} = \begin{bmatrix} y \\ -2x + 4y \end{bmatrix}$$

In summary, the solution to the change-of-basis problem in R^n is the following:

SOLUTION to Problem 5.1

Given the components $[\mathbf{v}]_B$ of an n-vector $\mathbf{v}$ relative to an $(n \times n)$ basis matrix B, the components of $\mathbf{v}$ relative to an $(n \times n)$ basis matrix B' are:

$$[\mathbf{v}]_{B'} = ((B')^{-1}B)[\mathbf{v}]_B$$

Solving the Problem with Linear Transformations

In the context of the material in Section 5.3, the matrix $(B')^{-1}B$ in the solution to Problem 5.1 is the standard matrix A of the identity transformation $i : R^n \to R^n$, defined by $i(\mathbf{v}) = \mathbf{v}$, relative to the bases B and B'. To see why this is so, let the columns of the basis matrix B be $\mathbf{v}_1, \ldots, \mathbf{v}_n$. Recall, from Section 5.3.2, that the columns of A are the components of the vectors $i(\mathbf{v}_1) = \mathbf{v}_1, \ldots, i(\mathbf{v}_n) = \mathbf{v}_n$ expressed relative to the components of the given basis vectors in B'. In matrix-vector notation, you have

$$\begin{aligned}
A &= [\ [i(\mathbf{v}_1)]_{B'} \cdots [i(\mathbf{v}_n)]_{B'}\ ] && \text{(definition of } A\text{)} \\
&= [\ [\mathbf{v}_1]_{B'} \cdots [\mathbf{v}_n]_{B'}\ ] && \text{(definition of } i\text{)} \\
&= [\ (B')^{-1}\mathbf{v}_1 \cdots (B')^{-1}\mathbf{v}_n\ ] && \text{(from (5.37))} \\
&= (B')^{-1}[\mathbf{v}_1 \cdots \mathbf{v}_n] && \text{(matrix algebra)} \\
&= (B')^{-1}B && \text{(definition of } B\text{)}
\end{aligned}$$

To illustrate with the foregoing numerical example, the columns of A are the basis vectors $(4, -2)$ and $(-5, 3)$ of B expressed relative to the vectors $(3, -1)$ and $(-2, 1)$ in the basis B'. From the results in Example 5.11 in Section 5.4.1, you therefore have that

$$A = \begin{bmatrix} \begin{bmatrix} 4 \\ -2 \end{bmatrix}_{B'} & \begin{bmatrix} -5 \\ 3 \end{bmatrix}_{B'} \end{bmatrix} = \begin{bmatrix} 0 & 1 \\ -2 & 4 \end{bmatrix}$$

This matrix A is the same as that computed by $(B')^{-1}B$ in (5.43).

5.4.3 The Matrix of a Linear Transformation Relative to a Single Basis

For a linear transformation $T : R^n \to R^n$, you know that there is an $(n \times n)$ matrix A such that for each column vector $\mathbf{v} \in R^{n \times 1}$, $T(\mathbf{v}) = A\mathbf{v}$. The columns of A are obtained by applying T to the standard unit vectors in R^n.

For a linear transformation $T : V \to W$, where V is an n-dimensional vector space and W is an m-dimensional vector space, you can use the approach in Section 5.3.3 to evaluate T at a vector $\mathbf{v} \in V$. That approach is to choose a basis B for V and a basis B' for W and find the matrix for T relative to B and B' by computing the components of T applied to each vector in B, relative to the basis B'.

When $V = W$, you can choose a single basis $B = \{\mathbf{v}_1, \ldots, \mathbf{v}_n\}$ for both V and W. In this case, the matrix for T relative to B and B' is called the **matrix for T relative to the basis B** and is denoted by $[T]_B$. Column i of this matrix is found by computing the components of $T(\mathbf{v}_i)$ relative to B. You can then use $[T]_B$ to compute $T(\mathbf{v})$ for any vector $\mathbf{v} \in V$, as follows (see Figure 5.14 in Section 5.3.3):

> **Using The Matrix T Relative to B to Compute $T(v)$**
>
> **Step 1.** Find $[\mathbf{v}]_B$, the components of $\mathbf{v}$ relative to B.
>
> **Step 2.** Compute the components of $T(\mathbf{v})$ relative to B, as follows: $[T(\mathbf{v})]_B = [T]_B[\mathbf{v}]_B$.
>
> **Step 3.** Compute $T(\mathbf{v})$ as a linear combination of the basis vectors in B using the scalars $[T(\mathbf{v})]_B$ found in Step 2.

Finding and Using $[T]_B$ for $T : V \to V$

To illustrate the process of finding and using the matrix $[T]_B$, consider the vector space P^2 consisting of all polynomials of degree 2 or less and the linear transformation $T : P^2 \to P^2$ defined by

$$T(a_0 + a_1 x + a_2 x^2) = a_1 + 2a_2 x$$

Suppose you are using the basis B consisting of the polynomials $\{1, x, x^2\}$ for P^2. To find the columns of the (3×3) matrix $[T]_B$, you must find the components of $T(1)$, $T(x)$, and $T(x^2)$ relative to B. So,

$$T(1) = T(1 + 0x + 0x^2) = 0 + 2(0)x = 0 \quad \text{(the zero polynomial)}$$
$$T(x) = T(0 + 1x + 0x^2) = 1 + 2(0)x = 1 \quad \text{(the polynomial whose value is 1)}$$
$$T(x^2) = T(0 + 0x + 1x^2) = 0 + 2(1)x = 2x$$

Thus,

$$[T(1)]_B = \begin{bmatrix} 0 \\ 0 \\ 0 \end{bmatrix}, \quad [T(x)]_B = \begin{bmatrix} 1 \\ 0 \\ 0 \end{bmatrix}, \quad \text{and} \quad [T(x^2)]_B = \begin{bmatrix} 0 \\ 2 \\ 0 \end{bmatrix}$$

So, the matrix $[T]_B$ is

$$[T]_B = \begin{bmatrix} 0 & 1 & 0 \\ 0 & 0 & 2 \\ 0 & 0 & 0 \end{bmatrix}$$

For any polynomial of degree 2, say, $p(x) = a_0 + a_1 x + a_2 x^2$, you use $[T]_B$ to compute $T(p)$, as follows:

Step 1. Identify $[p]_B$, namely,

$$[p]_B = \begin{bmatrix} a_0 \\ a_1 \\ a_2 \end{bmatrix}$$

Step 2. Use $[T]_B$ to find the components of the polynomial $T(p)$ relative to the basis B, that is, $[T(p)]_B$, by computing

$$[T(p)]_B = [T]_B[p]_B = \begin{bmatrix} 0 & 1 & 0 \\ 0 & 0 & 2 \\ 0 & 0 & 0 \end{bmatrix} \begin{bmatrix} a_0 \\ a_1 \\ a_2 \end{bmatrix} = \begin{bmatrix} a_1 \\ 2a_2 \\ 0 \end{bmatrix}$$

Step 3. Compute $T(p)$ as a linear combination of the basis vectors $1, x, x^2$ using $[T(p)]_B$ from Step 2, as follows:

$$T(p) = (a_1)1 + (2a_2)x + (0)x^2 = a_1 + 2a_2x$$

Finding and Using $[T]_B$ for $T : R^n \to R^n$

When $V = R^n$, you can use the standard matrix A associated with T and the chosen basis matrix, B, whose columns, say, are the n-vectors $\mathbf{v}_1, \ldots, \mathbf{v}_n$, to find $[T]_B$. To see how, notice that for each basis vector $\mathbf{v}_i$, you have from (5.37) that

$$[T(\mathbf{v}_i)]_B = B^{-1}T(\mathbf{v}_i) = B^{-1}(A\mathbf{v}_i) \tag{5.44}$$

Thus,

$$\begin{aligned} [T]_B &= [\ [T(\mathbf{v}_1)]_B \cdots [T(\mathbf{v}_n)]_B\ ] && \text{(definition of } [T]_B) \\ &= [B^{-1}A\mathbf{v}_1 \cdots B^{-1}A\mathbf{v}_n] && \text{(from (5.44))} \\ &= B^{-1}A[\mathbf{v}_1 \cdots \mathbf{v}_n] && \text{(matrix algebra)} \\ &= B^{-1}AB && \text{(definition of } B) \end{aligned}$$

You can then use $[T]_B$ to compute $T(\mathbf{v})$ for any column vector $\mathbf{v} \in R^{n \times 1}$, as follows:

$$\begin{aligned} T(\mathbf{v}) &= B[T(\mathbf{v})]_B && \text{(the column vector } T(\mathbf{v}) \text{ is a linear} \\ & && \text{combination of the basis vectors in } B) \\ &= B[T]_B[\mathbf{v}]_B && \text{(using } [T]_B \text{ to find } [T(\mathbf{v})]_B) \\ &= (B[T]_B B^{-1})\mathbf{v} && \text{[from (5.37)]} \end{aligned}$$

An example of finding and using $[T]_B$ follows.

EXAMPLE 5.12 Finding and Using the Matrix for *T* Relative to *B*

Suppose that the standard matrix for $T : R^2 \to R^2$ is

$$A = \begin{bmatrix} 2 & -4 \\ -1 & 0 \end{bmatrix}$$

and that you choose to work with the following basis matrix for R^2:

$$B = \begin{bmatrix} 3 & -2 \\ -1 & 1 \end{bmatrix}$$

Then

$$[T]_B = B^{-1}AB = \begin{bmatrix} 1 & 2 \\ 1 & 3 \end{bmatrix}\begin{bmatrix} 2 & -4 \\ -1 & 0 \end{bmatrix}\begin{bmatrix} 3 & -2 \\ -1 & 1 \end{bmatrix} = \begin{bmatrix} 4 & -4 \\ 1 & -2 \end{bmatrix}$$

and for

$$\mathbf{v} = \begin{bmatrix} 2 \\ 1 \end{bmatrix}$$

it follows that

$$T(\mathbf{v}) = (B[T]_B B^{-1})\mathbf{v} = \begin{bmatrix} 3 & -2 \\ -1 & 1 \end{bmatrix}\begin{bmatrix} 4 & -4 \\ 1 & -2 \end{bmatrix}\begin{bmatrix} 1 & 2 \\ 1 & 3 \end{bmatrix}\begin{bmatrix} 2 \\ 1 \end{bmatrix} = \begin{bmatrix} 0 \\ -2 \end{bmatrix}$$

which is the same as using the standard matrix *A* because

$$T(\mathbf{v}) = A\mathbf{v} = \begin{bmatrix} 2 & -4 \\ -1 & 0 \end{bmatrix}\begin{bmatrix} 2 \\ 1 \end{bmatrix} = \begin{bmatrix} 0 \\ -2 \end{bmatrix}$$

Similarity

You now have two different methods for evaluating a linear transformation $T : R^n \to R^n$ at a column vector $\mathbf{v} \in R^{n \times 1}$, namely,

1. Use the standard matrix *A* to compute $T(\mathbf{v}) = A\mathbf{v}$.
2. Choose a basis matrix *B*, find the matrix $[T]_B$, and compute $T(\mathbf{v}) = (B[T]_B B^{-1})\mathbf{v}$.

You also know how to use *A* and *B* to find $[T]_B$, namely,

$$[T]_B = B^{-1}AB \tag{5.45}$$

Likewise, if you know $[T]_B$, then you can use (5.45) to find *A*. Specifically, multiplying both sides of (5.45) on the left by *B* and then on the right by B^{-1} results in

$$A = B[T]_B B^{-1} \tag{5.46}$$

From (5.45) and (5.46) you can see that A and $[T]_B$ are related to each other in a special way. That relationship is formalized in the following definition.

DEFINITION 5.5

An $(n \times n)$ matrix A is **similar** to an $(n \times n)$ matrix C if and only if there is an invertible $(n \times n)$ matrix B such that

$$A = BCB^{-1}$$

If A is similar to C, then, from Definition 5.5, there is an invertible $(n \times n)$ matrix B such that

$$A = BCB^{-1}$$

or, equivalently,

$$C = B^{-1}AB \qquad (5.47)$$

Letting $P = B^{-1}$, (5.47) becomes

$$C = PAP^{-1}$$

so C is similar to A. In other words, if A is similar to C, then C is similar to A. Thus, A and C are said to be *similar*. You have already seen in (5.45) and (5.46) that the $(n \times n)$ standard matrix A for a linear transformation $T : R^n \to R^n$ is similar to the $(n \times n)$ matrix $[T]_B$.

A natural response is to ask why you would want to use $[T]_B$ to compute $T(\mathbf{v})$ instead of the standard matrix A for T. One reason is that similarity is used to simplify certain computations performed with a matrix A. Such an example is given in Section 6.4.

In this section, you have learned how to work with a linear transformation $T : V \to V$ in problem solving, such as the change-of-basis problem in R^n. In Section 5.5, you will see how to solve another problem that often arises when working with linear transformations.

Exercises for Section 5.4

1. Find the components of each of the following vectors relative to the basis of R^2 consisting of $(1, 1)$ and $(-1, 0)$ by using the associated basis matrix.
   (a) $(1, 0)$. (b) $(0, 1)$. (c) $(2, -1)$.

2. Find the components of the vector $(1, -1)$ relative to each of the following basis matrices.

   (a) $\begin{bmatrix} 1/2 & 0 \\ 0 & 1/3 \end{bmatrix}$ (b) $\begin{bmatrix} -2 & -1 \\ 1 & 0 \end{bmatrix}$

3. Suppose that A is the standard matrix for a linear transformation $T : R^n \to R^n$ and that B is a given basis matrix for R^n. Show how to use A and B to compute $T(\mathbf{x})$ when you know $[\mathbf{x}]_B$.

4. Suppose that T, A, and B are as in the previous exercise. Use matrix-vector notation to show how to compute the components of $T(\mathbf{x})$ relative to the basis B, that is, $[T(\mathbf{x})]_B$.

5. Suppose that the components of a vector **v** relative to the the basis B in Exercise 1 are $(3, -2)$.
   (a) What are the components of **v** relative to the basis consisting of the two standard unit vectors in R^2?
   (b) What are the components of **v** relative to the basis B' in Exercise 2(b)?

6. Suppose that the components of a vector **v** relative to the the basis B in Exercise 2(b) are $(3, -2)$.
   (a) What are the components of **v** relative to the basis consisting of the two standard unit vectors in R^2?
   (b) What are the components of **v** relative to the basis B' in Exercise 1?

7. For the basis matrix B in Exercise 1 and the linear transformation $T : R^2 \rightarrow R^2$ defined by $T(x, y) = (3x, y)$:
   (a) Find the matrix $[T]_B$.
   (b) Use $[T]_B$ to find $T(-1, -2)$.
   (c) Verify your results by using the standard matrix for T to compute $T(-1, -2)$.

8. Repeat the previous exercise for the basis matrix B in Exercise 2(b) and the linear transformation $T : R^2 \rightarrow R^2$ defined by $T(x, y) = (y, x)$,

9. For the basis matrix B in Exercise 1, find the standard matrix of the linear transformation $T : R^2 \rightarrow R^2$ when

$$[T]_B = \begin{bmatrix} -2 & 0 \\ 1 & -3 \end{bmatrix}$$

10. For the basis matrix B in Exercise 2(b), find the standard matrix of the linear transformation $T : R^2 \rightarrow R^2$ when

$$[T]_B = \begin{bmatrix} -2 & 0 \\ 1 & -3 \end{bmatrix}$$

11. (a) Use the standard basis $B = \{M^{11}, M^{12}, M^{21}, M^{22}\}$ for $R^{2\times2}$ to find $[T]_B$ for the linear transformation $T : R^{2\times2} \rightarrow R^{2\times2}$, in which $T(M^{11}) = M^{12}$, $T(M^{12}) = M^{21}$, $T(M^{21}) = M^{22}$, and $T(M^{22}) = M^{11}$.
    (b) Show how to use $[T]_B$ to compute $T(M)$, where

$$M = \begin{bmatrix} 1 & 2 \\ 3 & 4 \end{bmatrix}$$

    (c) Verify your results by expressing M as a linear combination of the standard basis matrices and using the fact that T is linear.

12. For a linear transformation $T : R^n \rightarrow R^n$ and a chosen basis matrix B, create a visual image similar to Figure 5.14 in Section 5.3.3 to represent the computation of $T(\mathbf{x})$ by $B^{-1}AB[\mathbf{x}]_B$ and also by $A\mathbf{x}$.

13. Use the matrix B in Exercise 1 to determine if the following pairs of matrices are similar.

(a) $A = \begin{bmatrix} 4 & 0 \\ 5 & 2 \end{bmatrix}$, $C = \begin{bmatrix} 2 & -3 \\ 0 & 4 \end{bmatrix}$ (b) $A = \begin{bmatrix} -5 & 2 \\ -4 & 1 \end{bmatrix}$, $C = \begin{bmatrix} -3 & 4 \\ 0 & -1 \end{bmatrix}$

14. Use the matrix B in Exercise 2(b) to determine a matrix C that is similar to the matrix A in Exercise 13(a)?

15. Suppose that A is the standard matrix of a linear transformation $T : R^n \rightarrow R^n$ and that B and B' are basis matrices for R^n. Show how to use A, B, and B' to compute $[T(\mathbf{x})]_{B'}$ given $[x]_B$.

5.5 Solving Equations with Linear Transformations

Given an $(m \times n)$ matrix A and a column vector $\mathbf{b} \in R^{m \times 1}$, the problem of solving a system of linear equations is to find a column vector $\mathbf{x} \in R^{n \times 1}$ such that

$$A\mathbf{x} = \mathbf{b} \tag{5.48}$$

Thinking of the left side of (5.48) as the value of a linear transformation T applied to $\mathbf{x}$, a generalization is the following problem of solving equations with a linear transformation:

PROBLEM 5.2 Solving Equations with Linear Transformations
Suppose that V and W are vector spaces and $T : V \rightarrow W$ is a linear transformation. For a given vector $\mathbf{b} \in W$, find a vector $\mathbf{x} \in V$ such that $T(\mathbf{x}) = \mathbf{b}$.

Recalling that T transforms a vector $\mathbf{x} \in V$ to a vector $T(\mathbf{x}) \in W$, Problem 5.2 arises when you want to determine an input $\mathbf{x} \in V$ that T transforms to a given desired output $\mathbf{b} \in W$. The filtering problem of Audio Technologies, presented at the beginning of this chapter, is related to Problem 5.2, as explained in Section 5.6.

When solving a system of linear equations, difficulties arise if the system has no solution or infinitely many solutions. Correspondingly, there are two reasons why solving Problem 5.2 can be challenging (see Figure 5.15):

1. For the point $\mathbf{a} \in W$, there is no value of $\mathbf{x} \in V$ for which $T(\mathbf{x}) = \mathbf{a}$.

2. For the point $\mathbf{b} \in W$, there are at least two values of $\mathbf{x} \in V$ that satisfy $T(\mathbf{x}) = \mathbf{b}$, say, $\mathbf{x} = \mathbf{u}$ and $\mathbf{x} = \mathbf{v}$.

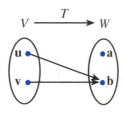

Figure 5.15 *Reasons Why Solving the Equation $T(\mathbf{x}) = \mathbf{b}$ for $\mathbf{x} \in V$ Can Be Challenging.*

The objective now is to develop conditions on T, V, and W so that for each value of $\mathbf{b} \in W$, there is one and only one value of $\mathbf{x} \in V$ that satisfies $T(\mathbf{x}) = \mathbf{b}$. Throughout this section, V and W are vector spaces.

5.5.1 Existence of a Solution

When you apply the transformation T shown in Figure 5.15 to all points $\mathbf{x} \in V$, you do not "cover" every point in W. In particular, you do not cover $\mathbf{a} \in W$. In contrast, a transformation that does cover all of W is illustrated in Figure 5.16. The objective is to translate the visual image in Figure 5.16, and the corresponding verbal description of "covering every point in W," to a symbolic definition.

The Property of Being Onto

The approach is based on the observation that, for the transformation in Figure 5.16, every point in the set W is covered by the transformation. Recognizing the word "every" should encourage you to use the quantifier *for all* to describe the desirable property in the following way:

For all vectors $\mathbf{y} \in W$, $\mathbf{y}$ is *covered* by T.

All that remains is to translate the concept "$\mathbf{y}$ is covered by T" to symbolic form. One way to do so is to use the quantifier *there is*, as follows:

There is a vector $\mathbf{x} \in V$ such that $T(\mathbf{x}) = \mathbf{y}$.

Combining these two statements results in the following definition.

DEFINITION 5.6

A linear transformation $T : V \rightarrow W$ is **onto** if and only if for all vectors $\mathbf{y} \in W$, there is a vector $\mathbf{x} \in V$ such that $T(\mathbf{x}) = \mathbf{y}$.

The property of being onto in Definition 5.6 ensures that there is at least one solution to Problem 5.2, as shown in the following example.

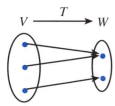

Figure 5.16 *A Transformation that "Covers Every Point in W."*

EXAMPLE 5.13 A Linear Transformation that Is Onto

The linear transformation $T : R^3 \to R^2$ whose standard matrix is

$$A = \begin{bmatrix} 1 & 2 & 4 \\ 0 & -1 & -2 \end{bmatrix}$$

is onto because T satisfies Definition 5.6. To see why, let $\mathbf{y} \in R^2$. For T to be onto, you want to find a column vector $\mathbf{x} \in R^3$ such that

$$T(\mathbf{x}) = A\mathbf{x} = \begin{bmatrix} 1 & 2 & 4 \\ 0 & -1 & -2 \end{bmatrix} \begin{bmatrix} x_1 \\ x_2 \\ x_3 \end{bmatrix} = \begin{bmatrix} y_1 \\ y_2 \end{bmatrix} = \mathbf{y}$$

That is, you want to find x_1, x_2, and x_3 such that

$$x_1 + 2x_2 + 4x_3 = y_1$$
$$- x_2 - 2x_3 = y_2$$

The foregoing system has an infinite number of solutions whose general solution—obtained by Gauss-Jordan elimination—is as follows:

$$x_1 = y_1 + 2y_2$$
$$x_2 = - y_2 - 2x_3$$

One solution, obtained by setting the free variable x_3 to 0, is

$$\mathbf{x} = \begin{bmatrix} x_1 \\ x_2 \\ x_3 \end{bmatrix} = \begin{bmatrix} y_1 + 2y_2 \\ -y_2 \\ 0 \end{bmatrix}$$

EXAMPLE 5.14 A Linear Transformation that Is Not Onto

The linear transformation $T : R^2 \to R^3$ whose standard matrix is

$$A = \begin{bmatrix} 1 & 0 \\ 0 & -2 \\ 0 & 0 \end{bmatrix}$$

is not onto because, for the column vector

$$\mathbf{y} = \begin{bmatrix} 1 \\ 2 \\ 3 \end{bmatrix}$$

no column vector $\mathbf{x} \in R^2$ satisfies $T(\mathbf{x}) = A\mathbf{x} = \mathbf{y}$. This is because no column vector $\mathbf{x} \in R^2$ satisfies

$$
A\mathbf{x} = \begin{bmatrix} 1 & 0 \\ 0 & -2 \\ 0 & 0 \end{bmatrix} \begin{bmatrix} x_1 \\ x_2 \end{bmatrix} = \begin{bmatrix} x_1 \\ -2x_2 \\ 0 \end{bmatrix} = \begin{bmatrix} 1 \\ 2 \\ 3 \end{bmatrix} = \mathbf{y}
$$

since $0 \neq 3$.

In view of Example 5.13 and Example 5.14, a natural question to ask is how you can determine if a linear transformation T is onto. When $V = R^n$ and $W = R^m$, the answer is related to a property of the columns of the associated standard matrix, as shown in the following theorem.

THEOREM 5.3

The linear transformation $T : R^n \rightarrow R^m$ is onto if and only if the columns of the standard matrix A associated with T span R^m.

Proof.
Recall, from Definition 4.4 in Section 4.3.2, that the columns of A span R^m if and only if every vector in R^m is a linear combination of the columns of A, that is, if and only if, for each column vector $\mathbf{y} \in R^{m \times 1}$, there is a column vector $\mathbf{x} \in R^{n \times 1}$ such that $A\mathbf{x} = T(\mathbf{x}) = \mathbf{y}$, that is, if and only if T is onto. ∎

Range of a Linear Transformation

Theorem 5.3 states that for a linear transformation $T : R^n \rightarrow R^m$, being onto is equivalent to the columns of the standard matrix A spanning R^m. As you will recall from Section 4.4.2, the space spanned by the columns of an $(m \times n)$ matrix A is the column space of A, that is,

$$
\text{column space of } A = \{A\mathbf{x} : \mathbf{x} \in R^{n \times 1}\}
$$

You can generalize the concept of the column space of A to a linear transformation by replacing $A\mathbf{x}$ with $T(\mathbf{x})$ and $R^{n \times 1}$ with a vector space V, thus giving rise to the following definition.

DEFINITION 5.7

The **range** of a linear transformation $T : V \rightarrow W$ is the set of all images of vectors $\mathbf{x} \in V$, that is,

$$
\text{range}(T) = \{T(\mathbf{x}) : \mathbf{x} \in V\}
$$

The range of a linear transformation $T : V \rightarrow W$ is a subspace of W, as you are asked to show in Exercise 17. An example of the range of a linear transformation follows.

EXAMPLE 5.15 The Range of a Linear Transformation

The range of the linear transformation $T : R^2 \rightarrow R^3$ defined by

$$T(x_1, x_2) = (x_1, -2x_2, 0)$$

whose standard matrix is given in Example 5.14, is

$$\text{range}(T) = \{T(\mathbf{x}) : \mathbf{x} \in R^2\}$$
$$= \{(x_1, -2x_2, 0) : x_1 \text{ and } x_2 \text{ are real numbers}\}$$

In Theorem 5.3, the property of a linear transformation T being onto is the same as the columns of the standard matrix A associated with T spanning R^m. In terms of Definition 5.7, T is onto if and only if the range of T is W. These observations—in relation to solving Problem 5.2—are summarized in the following theorem, whose proof is left to Exercise 18.

THEOREM 5.4

For a linear transformation $T : V \rightarrow W$, the following are equivalent:

(a) For any vector $\mathbf{b} \in W$, there is a vector $\mathbf{x} \in V$ such that $T(\mathbf{x}) = \mathbf{b}$.

(b) T is onto.

(c) $\text{range}(T) = W$.

The property of T being onto in Definition 5.6 ensures that there is at least one solution to Problem 5.2. The property developed next ensures that there is *only* one solution.

5.5.2 Uniqueness of a Solution

To ensure a unique solution to $T(\mathbf{x}) = \mathbf{b}$, it is necessary to rule out the situation in Figure 5.15 that both $T(\mathbf{u}) = \mathbf{b}$ and $T(\mathbf{v}) = \mathbf{b}$. That is, you want T to satisfy the property that "no two points in the domain of T result in the same value." The objective now is to translate this property—hereafter called *one-to-one*— to a symbolic definition.

The Property of Being One-to-One

You might begin by using the quantifier *there is* to write the following definition of a linear transformation being one-to-one:

The linear transformation $T : V \rightarrow W$ is *one-to-one* if and only if there do not exist vectors $\mathbf{u}, \mathbf{v} \in V$ such that $T(\mathbf{u}) = T(\mathbf{v})$.

However, recall that you must verify this definition to ensure its correctness. In so doing, you should discover a problem because no linear transformation satisfies this property. Specifically, for a linear transformation T, there always are vectors $\mathbf{u}, \mathbf{v} \in V$ for which $T(\mathbf{u}) = T(\mathbf{v})$, namely, $\mathbf{u} = \mathbf{v}$. To avoid this one case, the statement of the property should be changed to the following:

there do not exist $\mathbf{u}, \mathbf{v} \in V$ with $\mathbf{u} \neq \mathbf{v}$ such that $T(\mathbf{u}) = T(\mathbf{v})$

Equivalently stated,

for all $\mathbf{u}, \mathbf{v} \in V$ with $\mathbf{u} \neq \mathbf{v}$, $T(\mathbf{u}) \neq T(\mathbf{v})$

You can eliminate the word *not* by rewriting the foregoing statement in its equivalent **contrapositive** form. Specifically, as described in Section A.1, for given statements p and q, the contrapositive of the statement

if p is true, then q is true

is the statement

if q is not true, then p is not true

Using the contrapositive form results in the following definition of a one-to-one linear transformation.

DEFINITION 5.8

A linear transformation $T : V \rightarrow W$ is **one-to-one** if and only if for all vectors $\mathbf{u}, \mathbf{v} \in V$ with $T(\mathbf{u}) = T(\mathbf{v})$, it follows that $\mathbf{u} = \mathbf{v}$.

For instance, for a real number x, the linear transformation $T(x) = -2x$ is one-to-one because, if x and y are real numbers with $T(x) = T(y)$, then $-2x = -2y$ and so, $x = y$. In contrast, the linear transformation $T(x) = 0$ is not one-to-one because, for $x = 4$ and $y = -4$, you have $T(x) = T(y) = 0$ and yet $x \neq y$.

A visual image of a linear transformation that is one-to-one is shown in Figure 5.17, which is in contrast to the one in Figure 5.15. Other numerical examples follow.

EXAMPLE 5.16 A Linear Transformation that Is One-to-One

The linear transformation $T : R^2 \rightarrow R^3$ in Example 5.14 whose standard matrix is

$$A = \begin{bmatrix} 1 & 0 \\ 0 & -2 \\ 0 & 0 \end{bmatrix}$$

is one-to-one because, according to Definition 5.8, if $\mathbf{u}, \mathbf{v} \in R^2$ are vectors for which $T(\mathbf{u}) = T(\mathbf{v})$, then

$$\begin{bmatrix} 1 & 0 \\ 0 & -2 \\ 0 & 0 \end{bmatrix} \begin{bmatrix} u_1 \\ u_2 \end{bmatrix} = \begin{bmatrix} 1 & 0 \\ 0 & -2 \\ 0 & 0 \end{bmatrix} \begin{bmatrix} v_1 \\ v_2 \end{bmatrix}, \quad \text{so} \quad \begin{bmatrix} u_1 \\ -2u_2 \\ 0 \end{bmatrix} = \begin{bmatrix} v_1 \\ -2v_2 \\ 0 \end{bmatrix}$$

Equating corresponding components and simplifying yields $u_1 = v_1$ and $u_2 = v_2$, so $\mathbf{u} = \mathbf{v}$.

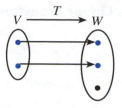

Figure 5.17 *A Transformation that Is One-to-One.*

EXAMPLE 5.17 A Linear Transformation that Is Not One-to-One

The linear transformation $T : R^3 \rightarrow R^2$ in Example 5.13 whose standard matrix is

$$A = \begin{bmatrix} 1 & 2 & 4 \\ 0 & -1 & -2 \end{bmatrix}$$

is not one-to-one because, for the column vectors

$$\mathbf{u} = \begin{bmatrix} 0 \\ -2 \\ 1 \end{bmatrix} \quad \text{and} \quad \mathbf{v} = \begin{bmatrix} 0 \\ -4 \\ 2 \end{bmatrix}$$

you can verify that $T(\mathbf{u}) = T(\mathbf{v}) = \mathbf{0}$ and yet $\mathbf{u} \neq \mathbf{v}$.

In view of Example 5.16 and Example 5.17, a natural question to ask is how to determine if a linear transformation T is one-to-one. When $V = R^n$ and $W = R^m$, the answer is to use the associated standard matrix, as shown in the following theorem.

THEOREM 5.5

The linear transformation $T : R^n \rightarrow R^m$ is one-to-one if and only if the columns of the standard matrix A associated with T are linearly independent.

Proof.

The transformation $T(\mathbf{x}) = A\mathbf{x}$ is one-to-one if and only if

for all vectors $\mathbf{u}, \mathbf{v} \in R^n$ with $A\mathbf{u} = A\mathbf{v}$, it follows that $\mathbf{u} = \mathbf{v}$

that is, if and only if

for all vectors $\mathbf{u}, \mathbf{v} \in R^n$ with $A(\mathbf{u} - \mathbf{v}) = \mathbf{0}$, it follows that $\mathbf{u} - \mathbf{v} = \mathbf{0}$

Letting $\mathbf{w} = \mathbf{u} - \mathbf{v}$, the foregoing statement is true if and only if

for all vectors $\mathbf{w} \in R^n$ with $A\mathbf{w} = \mathbf{0}$, it follows that $\mathbf{w} = \mathbf{0}$

that is, if and only if the columns of A are linearly independent. ■

Null Space of a Linear Transformation

From Theorem 5.5, when $T : R^n \rightarrow R^m$, the property of a linear transformation being one-to-one is equivalent to the columns of the standard matrix A being linearly independent. As you will recall from Section 4.4.2, the columns of a matrix A are linearly independent if and only if the null space of A consists of only the zero vector. Recalling that

$$\text{null space of } A = \{\mathbf{x} \in R^{n \times 1} : A\mathbf{x} = \mathbf{0}\}$$

you can generalize the concept of the null space to a linear transformation by replacing $A\mathbf{x}$ with $T(\mathbf{x})$ and $R^{n \times 1}$ with a vector space V, thus giving rise to the following definition.

DEFINITION 5.9

The **null space**, also called the **kernel**, of a linear transformation $T : V \rightarrow W$, denoted by $\text{null}(T)$, is the set of all vectors $\mathbf{x} \in V$ whose image under T is the zero vector in W, that is,

$\text{null}(T) = \{\mathbf{x} \in V : T(\mathbf{x}) = \mathbf{0}\}$

The null space of T is a subspace of V, as seen in Figure 5.18 and stated in the following theorem, whose proof is left to Exercise 19.

THEOREM 5.6

If $T : V \rightarrow W$ is a linear transformation, then $\text{null}(T)$ is a subspace of the vector space V.

From Theorem 5.5, you can conclude that a linear transformation T being one-to-one is the same as the columns of the standard matrix A associated with T being linearly independent. In terms of Definition 5.9, T is one-to-one if and only if the null space of T is the zero vector. These observations—in relation to solving Problem 5.2—are summarized in the following theorem, whose proof is left to Exercise 20.

THEOREM 5.7

For a linear transformation $T : V \rightarrow W$, the following are equivalent:

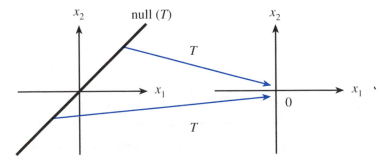

Figure 5.18 *The Null Space of a Linear Transformation $T : R^2 \rightarrow R^2$.*

(a) For each $\mathbf{b} \in W$, there is at most one vector $\mathbf{x} \in V$ such that $T(\mathbf{x}) = \mathbf{b}$.

(b) T is one-to-one.

(c) $\text{null}(T) = \{\mathbf{0}\}$.

The properties of being one-to-one and onto are used together in the following theorem.

THEOREM 5.8

Suppose that V is an n-dimensional vector space, W is an m-dimensional vector space, and that $T : V \to W$ is a linear transformation. If T is one-to-one and onto, then $\dim(V) = \dim(W)$.

Proof.

Recall, from Section 4.4, that the dimension of a vector space is the number of vectors in a basis. Now $\dim(V) = n$, so, let $\mathbf{v}_1, \ldots, \mathbf{v}_n$ be a basis for V. The conclusion follows by showing that $T(\mathbf{v}_1), \ldots, T(\mathbf{v}_n)$ is a basis for W. To that end, it must be shown that $T(\mathbf{v}_1), \ldots, T(\mathbf{v}_n)$ are linearly independent and span W. The fact that T is one-to-one ensures that these vectors are linearly independent and the fact that T is onto ensures that these vectors span W, as is now shown.

To see that $T(\mathbf{v}_1), \ldots, T(\mathbf{v}_n)$ are linearly independent, let $s_1, \ldots, s_n$ be scalars for which

$$s_1 T(\mathbf{v}_1) + \cdots + s_n T(\mathbf{v}_n) = \mathbf{0} \quad \text{(the zero vector of } W) \tag{5.49}$$

It must be shown that $s_1 = s_2 = \cdots = s_n = 0$. To that end, use the fact that T is linear to write (5.49) as

$$T(s_1 \mathbf{v}_1 + \cdots + s_n \mathbf{v}_n) = \mathbf{0} \tag{5.50}$$

Now (5.50) means that the vector $s_1 \mathbf{v}_1 + \cdots + s_n \mathbf{v}_n$ is in the null space of T. Because T is one-to-one, by Theorem 5.7(c), you can state that

$$s_1 \mathbf{v}_1 + \cdots + s_n \mathbf{v}_n = \mathbf{0} \quad \text{(the zero vector of } V) \tag{5.51}$$

Recalling that $\mathbf{v}_1, \ldots, \mathbf{v}_n$ form a basis for V, and hence are linearly independent, it now follows from (5.51) that $s_1 = s_2 = \cdots = s_n = 0$ and so $T(\mathbf{v}_1), \ldots, T(\mathbf{v}_n)$ are linearly independent.

To complete the proof, it is necessary to show that $T(\mathbf{v}_1), \ldots, T(\mathbf{v}_n)$ span W. To that end, let $\mathbf{w} \in W$, for which you must show that there are scalars $s_1, \ldots, s_n$ such that

$$s_1 T(\mathbf{v}_1) + \cdots + s_n T(\mathbf{v}_n) = \mathbf{w} \tag{5.52}$$

The fact that T is onto is used to construct these scalars. Specifically, because T is onto, there is a vector $\mathbf{v} \in V$ such that

$$T(\mathbf{v}) = \mathbf{w} \tag{5.53}$$

Now $\mathbf{v}_1, \ldots, \mathbf{v}_n$ form a basis for V, so you can express $\mathbf{v}$ as a unique linear combination of these basis vectors. Thus, there are scalars $s_1, \ldots, s_n$ such that

$$\mathbf{v} = s_1 \mathbf{v}_1 + \cdots + s_n \mathbf{v}_n \tag{5.54}$$

These scalars satisfy (5.52) because, from (5.53), (5.54), and the linearity of T,

$$\mathbf{w} = T(\mathbf{v}) = T(s_1\mathbf{v}_1 + \cdots + s_n\mathbf{v}_n) = s_1 T(\mathbf{v}_1) + \cdots + s_n T(\mathbf{v}_n)$$

Thus, $T(\mathbf{v}_1), \ldots, T(\mathbf{v}_n)$ span W.

It has thus been shown that $T(\mathbf{v}_1), \ldots, T(\mathbf{v}_n)$ form a basis for W consisting of n vectors, so $\dim(W) = n$, completing the proof. ■

The conclusion from Definition 5.6 and Definition 5.9 is that if a linear transformation $T : V \to W$ is onto and one-to-one, then you can solve Problem 5.2 because, for every vector $\mathbf{b} \in W$, there is a unique vector $\mathbf{x} \in V$ such that $T(\mathbf{x}) = \mathbf{b}$. The process of doing so for the special case of $V = W = R^n$ ends this section.

5.5.3 Solving Equations with $T : R^n \to R^n$

By the previous results, when a linear transformation $T : R^n \to R^n$ is one-to-one and onto, for each n-vector $\mathbf{b}$, there is a unique n-vector $\mathbf{x}$ such that

$$T(\mathbf{x}) = \mathbf{b} \tag{5.55}$$

The only question is how to find $\mathbf{x}$.

The answer, as usual, involves the $(n \times n)$ standard matrix A associated with T, for then, finding $\mathbf{x}$ in (5.55) is the same as finding a column vector $\mathbf{x} \in R^{n \times 1}$ that satisfies

$$A\mathbf{x} = \mathbf{b} \tag{5.56}$$

It is important to realize that the ability to find a unique solution to (5.56) depends not on the fact that A is $(n \times n)$, but rather, on the linear independence properties of the columns of A. For example, there is a unique solution to (5.56) when

$$A = \begin{bmatrix} 1 & 2 \\ -2 & -3 \end{bmatrix}$$

because the columns of A are linearly independent. In contrast, there is no unique solution to (5.56) when

$$A = \begin{bmatrix} 1 & 2 \\ -2 & -4 \end{bmatrix}$$

because the columns of A are not linearly independent.

In general, you know from Chapter 3 that when an $(n \times n)$ matrix A is invertible, the unique value of $\mathbf{x}$ that satisfies (5.56) is $\mathbf{x} = A^{-1}\mathbf{b}$. The next theorem establishes that the property of A being invertible is equivalent to the linear transformation T being one-to-one and onto.

THEOREM 5.9

The $(n \times n)$ standard matrix A associated with a linear transformation $T : R^n \to R^n$ is invertible if and only if T is one-to-one and onto.

Proof.

From Theorem 3.4 in Section 3.2.3, A is invertible if and only if for any column vector $\mathbf{b} \in R^{n \times 1}$, there is a unique column vector $\mathbf{x} \in R^{n \times 1}$ such that $A\mathbf{x} = \mathbf{b}$, that is, if and only if there is a unique n-vector $\mathbf{x}$ such that $T(\mathbf{x}) = \mathbf{b}$, that is, if and only if T is one-to-one and onto. This completes the proof. ■

You now know that if a linear transformation $T : V \to W$ is one-to-one and onto, then there is a unique solution to Problem 5.2, that is, for any vector $\mathbf{b} \in W$, there is a unique vector $\mathbf{x} \in V$ such that $T(\mathbf{x}) = \mathbf{b}$. Furthermore, when $T : R^n \to R^n$ is one-to-one and onto, the $(n \times n)$ standard matrix A associated with T is invertible and so the solution to $T(\mathbf{x}) = A\mathbf{x} = \mathbf{b}$ is $\mathbf{x} = A^{-1}\mathbf{b}$.

Exercises for Section 5.5

1. Determine if the transformation $T : R^1 \to R^1$ defined by $T(x) = 0$ is onto and justify your answer. What is the range of T?

2. Determine if the transformation $T : R^1 \to R^1$ defined by $T(x) = 2x$ is onto and justify your answer. What is the range of T?

3. Determine if the transformation $T : R^2 \to R^2$ defined by $T(x, y) = (3x + 3y, -x - y)$ is onto and justify your answer. What is the range of T?

4. Determine if the transformation $T : R^2 \to R^1$ defined by $T(x, y) = x + 2y$ is onto and justify your answer. What is the range of T?

5. Determine if the transformation $T : R^3 \to R^3$ defined by $T(x, y, z) = (2x - y + z, 3y - 4z, 2z)$ is onto and justify your answer. What is the range of T?

6. Determine if the transformation $T : R^3 \to R^2$ defined by $T(x, y, z) = (x + 2z, 2x - y)$ is onto and justify your answer. What is the range of T?

7. Determine if the transformation $T(x) = 0$ is one-to-one and justify your answer.

8. Determine if the transformation $T(x) = 2x$ is one-to-one and justify your answer.

9. Determine if the transformation $T(x, y) = (y, x)$ is one-to-one and justify your answer.

10. Determine if the transformation $T(x, y) = (3x + 3y, -x - y)$ is one-to-one and justify your answer.

11. Determine if the transformation $T(x, y, z) = (2x - y + z, 3y - 4z, 2z)$ is one-to-one and justify your answer.

12. Determine if the transformation $T(x, y) = (y, -x + 2y, 2x - 3y)$ is one-to-one and justify your answer.

13. Let P^2 be the set of all polynomials of degree 2 or less and define the linear transformation $T : P^2 \to P^1$ by $T(a_0 + a_1 x + a_2 x^2) = a_1 + 2a_2 x$.
    (a) Is T onto? If so, prove it. If not, explain why not.
    (b) Is T one-to-one? If so, prove it. If not, explain why not.
    (c) Find a polynomial p of degree 2 such that $T(p) = 3x + 4$. Is the polynomial p unique? Why or why not? Explain.

14. Let $T : R^2 \to R^2$ be a linear transformation that is one-to-one and onto.
    (a) Is the linear transformation $T(\mathbf{x}) - \mathbf{x}$ onto? If so, prove it. If not, explain why not.

(b) Is the linear transformation $T(\mathbf{x}) - \mathbf{x}$ one-to-one? If so, prove it. If not, explain why not.

(c) Find a vector $\mathbf{v} = (v_1, v_2)$ for which $T(\mathbf{v}) = \mathbf{v}$. Is $\mathbf{v}$ unique? Why or why not? Explain.

15. Let S be the vector space of all signals and T be a linear filter with coefficients $a_1, \ldots, a_n$. For a signal $\mathbf{X} \in \text{null}(T)$, define the linear transformation $F : \text{null(T)} \to R^n$ by $F(\mathbf{X}) = (x_0, x_1, \ldots, x_{n-1})$. Prove that F is one-to-one and onto.

16. Generalize the definitions of onto and one-to-one to a transformation T that might not be linear.

17. Prove that if $T : V \to W$ is a linear transformation, then range(T) is a subspace of W.

18. Prove Theorem 5.4 in Section 5.5.1.

19. Prove Theorem 5.6 in Section 5.5.2 by showing that the null space of a linear transformation $T : V \to W$ is a subspace of V.

20. Prove Theorem 5.7 in Section 5.5.2.

5.6 Problem Solving with Linear Transformations

Your knowledge of the properties of linear transformations is now brought together by solving the following problem of Audio Technologies presented at the beginning of this chapter.

5.6.1 The Problem of Audio Technologies

At Audio Technologies, music is recorded digitally by sampling a continuous signal 44,100 times per second (the amount of time between successive samplings is one time unit). For each $k = 0, 1, 2, \ldots$, the digital signal sampled at time unit k is a real number, say, x_k. Taking all of these measurements collectively, the digital signal is respresented mathematically as the following ordered list of real numbers:

$$\mathbf{X} = (x_0, x_1, x_2, \ldots)$$

Such a signal $\mathbf{X}$ is *filtered* by combining terms from $\mathbf{X}$ in a special way to create a new signal $\mathbf{Y}$. For example, one particular filter uses the values of x_k to create a new signal $\mathbf{Y}$ in which

$$y_k = x_{k+3} + x_{k+2} - 10x_{k+1} + 8x_k, \quad \text{for } k = 0, 1, 2, \ldots \tag{5.57}$$

The engineers know that when some signals $\mathbf{X}$ are fed into the filter in (5.57), the result is the zero signal, that is, the signal $\mathbf{Y} = (y_0, y_1, y_2, \ldots,) = (0, 0, 0, \ldots)$. As a member of the technical staff, you have been asked to find a mathematical representation of all signals for which the filter in (5.57) results in the zero signal so that design engineers can determine if a particular signal can be eliminated by using this filter.

5.6.2 Formulating the Problem

When formulating this problem mathematically, it is helpful to think of the process of filtering a signal $\mathbf{X}$ to create the new signal $\mathbf{Y}$ as a transformation. Specifically, let S be the vector space of all signals. Then you can define the transformation $T : S \rightarrow S$ to be the signal $T(\mathbf{X}) = \mathbf{Y} \in S$ obtained from the signal $\mathbf{X}$ by (5.57). In Example H in Section 5.2.2, it is shown that T is a linear transformation.

As stated in the problem description, your objective is to find a mathematical representation for those signals $\mathbf{X}$ for which the filter in (5.57) results in the zero signal. In terms of T, you want to find a mathematical representation for those signals $\mathbf{X} \in S$ for which $T(\mathbf{X}) = \mathbf{0}$, where $\mathbf{0}$ is the zero vector of S. That is, you want a mathematical representation for

$$\text{null}(T) = \{\mathbf{X} \in S : T(\mathbf{X}) = \mathbf{0}\}$$

5.6.3 Solving the Problem

Recall, from Theorem 5.6 in Section 5.5.2, that $\text{null}(T)$ is a subspace of S. As you learned in Chapter 4, if you can find a basis for a vector space, then any vector in that vector space is a linear combination of the basis vectors. In other words, you can represent $\text{null}(T)$ by finding a basis for $\text{null}(T)$. The question is: How many vectors are there in a basis for $\text{null}(T)$ and how can you find such vectors?

Finding the Number of Vectors in a Basis for Null(*T*)

First it is shown that $\text{null}(T)$ is a 3-dimensional vector space. The approach for doing so is provided in Theorem 5.8 in Section 5.5.2, which states that if U and W are finite-dimensional vector spaces and $F : U \rightarrow W$ is a linear transformation that is one-to-one and onto, then $\dim(U) = \dim(W)$. For the problem of Audio Technologies, let $U = \text{null}(T)$, $W = R^3$, and create the following transformation F from U to W. For any signal $\mathbf{X} \in U = \text{null}(T)$, define $F(\mathbf{X}) = (x_0, x_1, x_2)$. In Exercise 14 in Section 5.2, you showed that F is a linear transformation. As a result of Exercise 15 in Section 5.5, $F : \text{null}(T) \rightarrow R^3$ is one-to-one and onto. Hence, by Theorem 5.8 in Section 5.2.2, $\dim(\text{null}(T)) = \dim(R^3) = 3$.

Finding a Basis for Null(*T*)

To solve the problem of Audio Technologies it remains to find a basis for $\text{null}(T)$. Because you know that $\dim(\text{null}(T)) = 3$, by Theorem 4.9 in Section 4.4.1, it suffices to find three linearly independent vectors in $\text{null}(T)$. That is, you want to find three signals $\mathbf{U}, \mathbf{V}, \mathbf{W} \in \text{null}(T)$ that are linearly independent.

Temporarily forgetting the linear independence, for the signals $\mathbf{U} = (u_0, u_1, u_2, \ldots)$, $\mathbf{V} = (v_0, v_1, v_2, \ldots)$, and $\mathbf{W} = (w_0, w_1, w_2, \ldots)$ to be in the null space of T, each of these signals must satisfy (5.57) with $y_k = 0$. For example, $\mathbf{U} = (u_0, u_1, u_2, \ldots)$ must satisfy

$$u_{k+3} + u_{k+2} - 10u_{k+1} + 8u_k = 0, \quad \text{for } k = 0, 1, 2, \ldots \tag{5.58}$$

Likewise, $\mathbf{V}$ and $\mathbf{W}$ must satisfy the same property.

Solutions to (5.58) often have the form $u_k = r^k$, for some real number r. Substituting $u_k = r^k$ in (5.58) yields

$$r^{k+3} + r^{k+2} - 10r^{k+1} + 8r^k = 0 \tag{5.59}$$

Assuming that $r \neq 0$, you can divide both sides of (5.59) by r^k, obtaining

$$r^3 + r^2 - 10r + 8 = 0 \qquad (5.60)$$

Factoring (5.60), you get

$$(r - 1)(r - 2)(r + 4) = 0 \qquad (5.61)$$

The three roots in (5.61) are 1, 2, and -4. These roots are used to create the following three basis vectors for null(T):

$$
\begin{aligned}
\mathbf{U} &= (1^0, 1^1, 1^2, \ldots) = (1, 1, 1, \ldots) \\
\mathbf{V} &= (2^0, 2^1, 2^2, \ldots) = (1, 2, 4, \ldots) \\
\mathbf{W} &= ((-4)^0, (-4)^1, (-4)^2, \ldots) = (1, -4, 16, \ldots)
\end{aligned}
\qquad (5.62)
$$

It remains to show that these three vectors, which are in the null space of T, are linearly independent. With the definition of linearly independent vectors in mind, choose three scalars, s_1, s_2, and s_3, for which

$$s_1 \mathbf{U} + s_2 \mathbf{V} + s_3 \mathbf{W} = \mathbf{0} \quad \text{(the zero signal)}$$

that is,

$$s_1 u_k + s_2 v_k + s_3 w_k = 0, \quad \text{for } k = 0, 1, 2, \ldots \qquad (5.63)$$

You must show that $s_1 = s_2 = s_3 = 0$.

The goal is accomplished by using (5.62) and (5.63) to write a system of three linear equations in the three unknowns s_1, s_2, and s_3 and then showing that the only solution to this system is $s_1 = s_2 = s_3 = 0$. So, for $k = 0$, 1, and 2, you have from (5.62) and (5.63) that

$$
\begin{aligned}
(k = 0) \quad & s_1 u_0 + s_2 v_0 + s_3 w_0 = s_1 + s_2 + s_3 = 0 \\
(k = 1) \quad & s_1 u_1 + s_2 v_1 + s_3 w_1 = s_1 + 2s_2 - 4s_3 = 0 \\
(k = 2) \quad & s_1 u_2 + s_2 v_2 + s_3 w_2 = s_1 + 4s_2 + 16s_3 = 0
\end{aligned}
\qquad (5.64)
$$

Letting A be the (3×3) matrix associated with the left side of (5.64) and $\mathbf{s} \in R^{n \times 1}$ be the column vector whose three elements are s_1, s_2, and s_3, the system in (5.64) becomes

$$A\mathbf{s} = \mathbf{0} \quad \text{where} \quad A = \begin{bmatrix} 1 & 1 & 1 \\ 1 & 2 & -4 \\ 1 & 4 & 16 \end{bmatrix} \quad \text{and} \quad \mathbf{s} = \begin{bmatrix} s_1 \\ s_2 \\ s_3 \end{bmatrix} \qquad (5.65)$$

The last step is to show that A is invertible, for then, $\mathbf{s} = \mathbf{0}$ is the only solution to (5.64). You can see that A in (5.65) is invertible by using Gauss-Jordan elimination to find the following inverse of A:

$$A^{-1} = \frac{1}{30} \begin{bmatrix} 48 & -12 & -6 \\ -20 & 15 & 5 \\ 2 & -3 & 1 \end{bmatrix}$$

In summary, the solution to the problem of Audio Technologies is the following mathematical representation of those signals for which the filter in (5.57) results in the zero signal:

$$\text{null}(T) = \{s_1\mathbf{U} + s_2\mathbf{V} + s_3\mathbf{W}\}$$

where s_1, s_2, and s_3 are arbitrary scalars and

$$
\begin{aligned}
\mathbf{U} &= \left(1^k\right) &&= (1, 1, 1, \ldots) \\
\mathbf{V} &= \left(2^k\right) &&= (1, 2, 4, \ldots) \\
\mathbf{W} &= \left((-4)^k\right) &&= (1, -4, 16, \ldots)
\end{aligned}
$$

Exercises for Section 5.6

PROJECT 5.1: The Filtering Problem of Audio Technologies

Consider the problem of Audio Technologies in which you want to find a mathematical representation of all signals $\mathbf{X}$ for which the following filter for obtaining a signal $\mathbf{Y}$ from $\mathbf{X}$ results in $\mathbf{Y}$ being the zero signal:

$$y_k = x_{k+3} + x_{k+2} - 10x_{k+1} + 8x_k, \quad \text{for } k = 0, 1, 2, \ldots$$

Use a graphing calculator, MATLAB, or other software package, as indicated by your instructor, to perform the computations in Exercises 1 through 3. Whenever possible, use matrix and vector operations available on your system. Write the answer and the sequence of operations you performed to obtain the answer.

1. Find the roots of the polynomial $r^3 + r^2 - 10r + 8$ associated with this filter. Use these roots to write three signals in the null space of the filter.

2. After writing a system of three linear equations in three unknowns whose solution indicates whether the three signals in Exercise 1 are linearly independent or not [see (5.64)], use the roots of the polynomial in Exercise 1 to create a (3×3) matrix A that corresponds to the left side of this system.

3. Use the matrix A in Exercise 2 to show that the three signals in Exercise 1 are linearly independent by (a) row reducing A, (b) computing the determinant of A, and (c) finding the inverse of A.

PROJECT 5.2: Another Filtering Problem

As a member of the technical staff of Audio Technologies, you have been asked to find a mathematical representation of all signals $\mathbf{X}$ for which the filter whose coefficients are 3, -13, and -15 results in the signal $\mathbf{Y}$ in which $y_k = -48k + 8$ for each $k = 0, 1, , 2, \ldots$, that is, you want $\mathbf{X}$ to satisfy:

$$-48k + 8 = x_{k+3} + 3x_{k+2} - 13x_{k+1} - 15x_k, \quad \text{for } k = 0, 1, 2, \ldots$$

Use a graphing calculator, MATLAB, or other software package, as indicated by your instructor, to perform the computations in Exercises 4 through 6. Whenever possible, use matrix and vector operations available on your system. Write the answer and the sequence of operations you performed to obtain the answer.

4. Verify by hand that when the signal $\mathbf{X}$ in which $x_k = 2k$ for $k = 0, 1, 2, \ldots$ is fed into the foregoing filter, the result is the signal $\mathbf{Y}$ in which $y_k = -48k + 8$ for each $k = 0, 1, , 2, \ldots$.

5. Repeat Exercise 1, 2, and 3 in Project 5.1 to find three linearly independent vectors in the null space of this filter.

6. Solve the problem by representing a signal $\mathbf{X}$ that, when fed into the foregoing filter, results in the signal $\mathbf{Y}$ in which $y_k = -48k + 8$ for each $k = 0, 1, , 2, \ldots$ as the sum of the particular solution in Exercise 4 and a linear combination of the three linearly independent signals in Exercise 5.

Chapter Summary

Given two sets V and W, a function $F : V \to W$, also called a transformation or a mapping from V to W, is a rule that associates to each value of x in the domain V of F, a unique value $F(x) \in W$. You can visualize a function as a black box that takes an input value $x \in V$ and, after performing some computations, produces the unique output value $F(x) \in W$.

A linear transformation T from a vector space V to a vector space W is a transformation for which the following two properties hold for all real numbers a and vectors $\mathbf{u}, \mathbf{v} \in V$:

1. $T(\mathbf{u} + \mathbf{v}) = T(\mathbf{u}) + T(\mathbf{v})$.
2. $T(a\mathbf{u}) = aT(\mathbf{u})$.

Linear transformations arise in many applications, including population migration, linear programming, electrical networks, linear filters, and operations on points in the plane, such as reflections, rotations, and shears. All of these are examples of a matrix transformation defined by $T(\mathbf{x}) = A\mathbf{x}$, where A is an $(m \times n)$ matrix and $\mathbf{x} \in R^{n \times 1}$ is a column vector. In fact, every linear transformation $T : R^n \to R^m$ is a matrix transformation, regardless of the form of T, because you can always find a unique $(m \times n)$ matrix A such that for each column vector $\mathbf{x} \in R^{n \times 1}$, $T(\mathbf{x}) = A\mathbf{x}$. The columns of the standard matrix A are obtained by applying T to the standard unit vectors in R^n, that is,

$$A = [\, T(\mathbf{e}_1) \cdots T(\mathbf{e}_n)\,]$$

The concept of a standard matrix is extended to a linear transformation from an n-dimensional vector space V to an m-dimensional vector space W by first choosing a basis B for V and a basis B' for W. The matrix A for T relative to B and B' is the standard matrix of the linear transformation that associates to the n components of a vector $\mathbf{v} \in V$ relative to B, the m components of $T(\mathbf{v})$ relative to B'. That is, if $B = [\mathbf{v}_1, \ldots, \mathbf{v}_n]$, then

$$A = [\, [T(\mathbf{v}_1)]_{B'} \cdots [T(\mathbf{v}_n)]_{B'}\,]$$

One useful application of the foregoing concept is to solve the change-of-basis problem in R^n, in which you are given the components of an n-vector $\mathbf{x}$ relative to a basis matrix B and you want to find the components of $\mathbf{x}$ relative to another basis matrix B'. The solution is

$$[\mathbf{x}]_{B'} = ((B')^{-1}B)[\mathbf{x}]_B$$

Another example arises when considering a transformation $T : R^n \to R^n$ in which you use a single basis matrix B for R^n. In this case, you use B together with the standard matrix A for T to compute the matrix of T relative to B, as follows:

$$[T]_B = B^{-1}AB$$

Similarly, given $[T]_B$, you can find the standard matrix for T by computing

$$A = B[T]_B B^{-1}$$

The foregoing relationship between A and $[T]_B$ gives rise to the concept of two ($n \times n$) matrices A and C being similar, meaning that there is an invertible ($n \times n$) matrix B such that $A = BCB^{-1}$. One use of similarity is to improve efficiency when performing computations with A, as shown in Chapter 6.

In Section 5.5, solving a system of linear equations is generalized to solving a linear transformation problem. That is, for a linear transformation $T : V \to W$, given a desired output $\mathbf{b} \in W$, you want to find an input $\mathbf{x} \in V$ for which $T(\mathbf{x}) = \mathbf{b}$. You are always able to find a unique such vector $\mathbf{x} \in V$ for every $\mathbf{b} \in W$ when T is one-to-one and onto. For a linear transformation $T : R^n \to R^m$ whose standard matrix is A, the property of being one-to-one is equivalent to the condition that the columns of A are linearly independent. The property of T being onto is the same as the condition that the columns of A span R^m. In particular, when $m = n$, T is one-to-one and onto if and only if A is invertible and so, for a given column vector $\mathbf{b} \in R^{n \times 1}$, the unique solution to $T(\mathbf{x}) = \mathbf{b}$ is $\mathbf{x} = A^{-1}\mathbf{b}$.

Chapter 6

Eigenvalues and Eigenvectors

U sing linear algebra to solve most real-world problems involves substantial computational effort due to the large size of the matrices involved. The material in this chapter reduces significantly the effort required to solve some of these problems. One class of such problems arises in the study of **dynamical systems**, which are systems that evolve over time in a predictable way. An example of such a system is presented in the following problem.

The Predator-Prey Problem of the National Wildlife Foundation

The National Wildlife Foundation is trying to determine the importance of the rabbit population to that of the timber wolves, whose primary diet in Yellowstone National Park is the rabbit. Given the number of wolves at the beginning of year k, say, W_k, and the number of hundreds of rabbits at the beginning of that year, say, R_k, historical data indicate that the number of wolves at the beginning of the next year, W_{k+1}, and the number of hundreds of rabbits at the beginning of the next year, R_{k+1}, depend on W_k and R_k according to the following relationships:

$$W_{k+1} = 0.72W_k + 0.2R_k$$
$$R_{k+1} = -0.12W_k + 1.1R_k$$

As a consultant, you have been asked to determine what will happen to the wolf and rabbit populations over a long period of time, assuming that nothing happens to change the foregoing equations.

6.1 What Are Eigenvalues and Eigenvectors?

From Chapter 5, you know that a linear transformation $T : R^n \rightarrow R^n$ is a matrix transformation. That is, there is an $(n \times n)$ matrix A such that for each n-vector $\mathbf{x}$, $T(\mathbf{x}) = A\mathbf{x}$. (Throughout the rest of this chapter, A is an $(n \times n)$ matrix and $\mathbf{x}$ is a column vector.) Some matrix transformations are easier to evaluate computationally at a point $\mathbf{x}$ than others. For example,

$$T(\mathbf{x}) = \begin{bmatrix} 2 & 0 & 0 \\ 0 & 2 & 0 \\ 0 & 0 & 2 \end{bmatrix} \begin{bmatrix} x_1 \\ x_2 \\ x_3 \end{bmatrix} = \begin{bmatrix} 2x_1 \\ 2x_2 \\ 2x_3 \end{bmatrix} = 2\mathbf{x} \tag{6.1}$$

is easier to evaluate computationally than

$$T(\mathbf{x}) = \begin{bmatrix} 2 & 1 & -2 \\ 1 & -4 & -1 \\ -5 & 2 & 3 \end{bmatrix} \begin{bmatrix} x_1 \\ x_2 \\ x_3 \end{bmatrix} = \begin{bmatrix} 2x_1 + x_2 - 2x_3 \\ x_1 - 4x_2 - x_3 \\ -5x_1 + 2x_2 + 3x_3 \end{bmatrix}$$

The matrix transformation T in (6.1) is particularly simple because $T(\mathbf{x}) = 2\mathbf{x}$, for each vector $\mathbf{x}$. Although a general matrix transformation $T : R^n \rightarrow R^n$ might not have this simple form, in many cases, there are certain n-vectors $\mathbf{x}$ for which $T(\mathbf{x}) = \lambda\mathbf{x}$, for some scalar λ. As shown in Section 6.3, these special vectors are used in finding the solution to dynamical systems, such as the one in the predator-prey problem of the National Wildlife Foundation presented at the beginning of the chapter. These special types of vectors give rise to the following definitions.

DEFINITION 6.1

An **eigenvector** of an $(n \times n)$ matrix A is a nonzero n-vector $\mathbf{x}$ for which there exists a scalar λ such that $A\mathbf{x} = \lambda\mathbf{x}$. A scalar λ is an **eigenvalue** of A if and only if there is a nonzero n-vector $\mathbf{x}$ such that $A\mathbf{x} = \lambda\mathbf{x}$. In this case, $\mathbf{x}$ is an *eigenvector corresponding to* λ.

EXAMPLE 6.1 An Eigenvector and an Eigenvalue of a Matrix
The nonzero vector

$$\mathbf{x} = \begin{bmatrix} -2 \\ 3 \end{bmatrix}$$

is an eigenvector of

$$A = \begin{bmatrix} 0 & 2 \\ 3 & -1 \end{bmatrix}$$

because, for $\lambda = -3$, $A\mathbf{x} = \lambda\mathbf{x}$ since

$$A\mathbf{x} = \begin{bmatrix} 0 & 2 \\ 3 & -1 \end{bmatrix} \begin{bmatrix} -2 \\ 3 \end{bmatrix} = \begin{bmatrix} 6 \\ -9 \end{bmatrix} = -3 \begin{bmatrix} -2 \\ 3 \end{bmatrix} = -3\mathbf{x} = \lambda\mathbf{x}$$

The foregoing computation also shows that $\lambda = -3$ is an eigenvalue of A. In contrast, the vector

$$\mathbf{y} = \begin{bmatrix} 1 \\ -1 \end{bmatrix}$$

is not an eigenvector of A because there is no scalar λ for which $A\mathbf{y} = \lambda\mathbf{y}$ since

$$A\mathbf{y} = \begin{bmatrix} 0 & 2 \\ 3 & -1 \end{bmatrix} \begin{bmatrix} 1 \\ -1 \end{bmatrix} = \begin{bmatrix} -2 \\ 4 \end{bmatrix} \neq \lambda \begin{bmatrix} 1 \\ -1 \end{bmatrix} = \lambda\mathbf{y}$$

In view of Example 6.1, you might ask the following questions associated with an $(n \times n)$ matrix A:

1. How can you check if a given n-vector $\mathbf{x}$ is an eigenvector of A?
2. How can you check if a given scalar λ is an eigenvalue of A?
3. How can you find all eigenvectors of A?
4. How can you find all eigenvalues of A?

After learning the answers to these questions, you will see how eigenvectors and eigenvalues are used in problem solving. In the discussion that follows, sometimes the value of λ is known and sometimes the value is unknown. Likewise, sometimes the n-vector $\mathbf{x}$ is known and at other times $\mathbf{x}$ is an n-vector of variables. Be aware at all times whether the values of λ and $\mathbf{x}$ are known or unknown.

6.1.1 Checking Eigenvectors and Eigenvalues

The answers to the first two questions are now addressed.

Checking Eigenvectors

To determine if a given n-vector $\mathbf{x}$ is an eigenvector of a matrix A, you need to compute $A\mathbf{x}$ and see if you can find a scalar λ for which $A\mathbf{x} = \lambda\mathbf{x}$. This is done by equating the first component of $A\mathbf{x}$ to the first component of $\lambda\mathbf{x}$ and solving for λ. You must then determine whether $A\mathbf{x} = \lambda\mathbf{x}$ for this value of λ. If so, $\mathbf{x}$ is an eigenvector of A; otherwise, $\mathbf{x}$ is not an eigenvector of A. These steps are illustrated in the following example.

EXAMPLE 6.2 Checking if a Given Vector is an Eigenvector of a Matrix

To check if the vector

$$\mathbf{x} = \begin{bmatrix} -1 \\ 1 \\ 4 \end{bmatrix}$$

is an eigenvector of

$$A = \begin{bmatrix} 2 & 0 & 1 \\ 1 & -1 & 0 \\ 1 & 1 & -2 \end{bmatrix}$$

think of λ as a variable and compute

$$A\mathbf{x} = \begin{bmatrix} 2 & 0 & 1 \\ 1 & -1 & 0 \\ 1 & 1 & -2 \end{bmatrix} \begin{bmatrix} -1 \\ 1 \\ 4 \end{bmatrix} = \begin{bmatrix} 2 \\ -2 \\ -8 \end{bmatrix} \quad \text{and} \quad \lambda\mathbf{x} = \begin{bmatrix} -\lambda \\ \lambda \\ 4\lambda \end{bmatrix} \tag{6.2}$$

Now set the first element of $A\mathbf{x}$ in (6.2) equal to the first element of $\lambda\mathbf{x}$ to obtain $2 = -\lambda$. Solving for λ yields $\lambda = -2$. Substituting this value of λ in (6.2), you see that $A\mathbf{x} = \lambda\mathbf{x}$, so the given vector $\mathbf{x}$ is an eigenvector of A.

Checking Eigenvalues

To determine if a given scalar λ is an eigenvalue of a matrix A, using Definition 6.1, you need to see if there is a nonzero n-vector $\mathbf{x}$ such that

$$A\mathbf{x} = \lambda\mathbf{x}, \quad \text{that is,} \quad A\mathbf{x} - \lambda\mathbf{x} = \mathbf{0}$$

If you factor out the common vector $\mathbf{x}$ from the two terms in $A\mathbf{x} - \lambda\mathbf{x}$ and write $(A - \lambda)\mathbf{x} = \mathbf{0}$, a syntax error arises because you cannot subtract the real number λ from the matrix A. To avoid this syntax error, use the $(n \times n)$ identity matrix I to write $A\mathbf{x} - \lambda\mathbf{x} = \mathbf{0}$ as follows:

$$A\mathbf{x} - \lambda I\mathbf{x} = \mathbf{0}$$

You can now factor out $\mathbf{x}$ to state that you must see if there is a nonzero n-vector $\mathbf{x}$ such that

$$(A - \lambda I)\mathbf{x} = \mathbf{0} \tag{6.3}$$

In other words, thinking of $\mathbf{x}$ as an n-vector of variables, to check if a given scalar λ is an eigenvalue of A, you need to see if there is a nonzero solution to the system of linear equations in (6.3). You can do this by using Gauss-Jordan elimination, as shown in the following example.

EXAMPLE 6.3 Checking if a Given Scalar is an Eigenvalue of a Matrix

To check if $\lambda = 2$ is an eigenvalue of the matrix

$$A = \begin{bmatrix} 0 & 2 \\ 3 & -1 \end{bmatrix}$$

see if there is a nonzero solution to $(A - 2I)\mathbf{x} = \mathbf{0}$, that is, to the following system:

$$(A - \lambda I)\mathbf{x} = \left(\begin{bmatrix} 0 & 2 \\ 3 & -1 \end{bmatrix} - 2 \begin{bmatrix} 1 & 0 \\ 0 & 1 \end{bmatrix} \right) \begin{bmatrix} x_1 \\ x_2 \end{bmatrix}$$

$$= \begin{bmatrix} -2 & 2 \\ 3 & -3 \end{bmatrix} \begin{bmatrix} x_1 \\ x_2 \end{bmatrix} = \begin{bmatrix} 0 \\ 0 \end{bmatrix} \tag{6.4}$$

Reducing the augmented matrix associated with the system in (6.4) yields

$$\begin{bmatrix} 1 & -1 & | & 0 \\ 0 & 0 & | & 0 \end{bmatrix} \tag{6.5}$$

From (6.5), the general solution to (6.4) is $x_1 = x_2$, with x_2 free. The fact that x_2 is a free variable indicates that there is a nonzero solution to (6.4). One such solution is

$$\mathbf{x} = \begin{bmatrix} 1 \\ 1 \end{bmatrix}$$

Verify that $\mathbf{x}$ is an eigenvector of A corresponding to $\lambda = 2$ by checking that $A\mathbf{x} = 2\mathbf{x}$.

6.1.2 Finding All Eigenvectors and Eigenvalues

Now that you know how to check if a given n-vector is an eigenvector of a matrix A and if a given scalar λ is an eigenvalue of A, the next task is to find all eigenvectors and eigenvalues of A, as is now discussed.

Using the Characteristic Equation to Find the Eigenvalues

Recall that checking if a given scalar λ is an eigenvalue of A requires seeing if

there is an n-vector $\mathbf{x} \neq \mathbf{0}$ such that $(A - \lambda I)\mathbf{x} = \mathbf{0}$ \hfill (6.6)

From your knowledge of solving linear equations, the following are equivalent ways to do so:

1. Check if the columns of $A - \lambda I$ are linearly dependent.
2. Check if the matrix $A - \lambda I$ is not invertible.
3. Check if the determinant (see Section 3.3.1) of $A - \lambda I$ is equal to zero.

Thus, to find all eigenvalues of A, you want to find those values of λ for which any of the three foregoing statements is true. In particular, (3) provides an approach for finding all eigenvalues of a given matrix. To see how, consider the matrix A in Example 6.3 and think of λ as an unknown. Then, your goal is to find all values of λ for which

$$\det(A - \lambda I) = \det\left(\begin{bmatrix} 0 & 2 \\ 3 & -1 \end{bmatrix} - \lambda \begin{bmatrix} 1 & 0 \\ 0 & 1 \end{bmatrix}\right) = \det\begin{bmatrix} -\lambda & 2 \\ 3 & -1 - \lambda \end{bmatrix}$$

$$= (-\lambda)(-1 - \lambda) - 3(2) = \lambda^2 + \lambda - 6 = 0 \tag{6.7}$$

You can solve the quadratic equation in (6.7) and conclude that $\lambda = -3$ and $\lambda = 2$ are eigenvalues of A. (Use the technique in Example 6.3 to verify this.)

The foregoing approach generalizes to an $(n \times n)$ matrix A. That is, to find the eigenvalues of A, think of λ as an unknown quantity and find all values of λ for which

$$\det(A - \lambda I) = 0 \tag{6.8}$$

It can be shown that (6.8) is a single nonlinear equation in terms of λ and is called the **characteristic equation** of A.

In the foregoing numerical example, the characteristic equation is a quadratic equation because A is (2×2). When A is $(n \times n)$, it can be shown that the characteristic equation is a polynomial equation of degree n. The expression $\det(A - \lambda I)$ is called the **characteristic polynomial** of A. The use of the characteristic polynomial is illustrated in the next example.

EXAMPLE 6.4 Using the Characteristic Polynomial to Find the Eigenvalues of a Matrix

To find the eigenvalues of

$$A = \begin{bmatrix} 4 & 2 & -1 \\ 0 & -3 & 5 \\ 0 & 0 & 4 \end{bmatrix}$$

form the matrix $A - \lambda I$ by subtracting λ from each of the diagonal elements of A and then setting the determinant of the resulting matrix equal to 0, that is,

$$\det(A - \lambda I) = \det\begin{bmatrix} 4 - \lambda & 2 & -1 \\ 0 & -3 - \lambda & 5 \\ 0 & 0 & 4 - \lambda \end{bmatrix} = 0 \tag{6.9}$$

The matrix in (6.9) is upper triangular (see Definition 3.4 in Section 3.3.1), so, by Theorem 3.6, $\det(A - \lambda I)$ is the product of the diagonal elements, that is,

$$\det(A - \lambda I) = (4 - \lambda)(-3 - \lambda)(4 - \lambda) \tag{6.10}$$

The roots of the characteristic polynomial in (6.10), and hence the eigenvalues of A, are $\lambda = 4$ and $\lambda = -3$.

The root $\lambda = 4$ in Example 6.4 is said to have *multiplicity* 2 because the term $(4 - \lambda)$ appears twice in the characteristic polynomial in (6.10). In general, the **multiplicity** of an eigenvalue λ is the number of times that value is repeated as a root in the characteristic polynomial.

In general, there are n (not necessarily distinct) roots of a polynomial of degree n, some of which can be complex numbers. Because the characteristic polynomial of an $(n \times n)$ matrix A has degree n, there are n (not necessarily distinct) eigenvalues of A, some of which can be complex numbers. The study of complex eigenvalues is outside the scope of this book. So, by design, all eigenvalues of matrices from here on will be real numbers.

Unfortunately, there is no closed-form expression for finding the n roots of the characteristic polynomial when $n \geq 5$. So, the next best approach is to develop a numerical method. Several such methods are presented in Chapter 8. These methods, though not finite, are used to approximate the eigenvalues to any desired degree of accuracy. Observe that the reason it is possible to find the roots of the characteristic polynomial in Example 6.4 is because the upper-triangular property of A leads easily to the factors of the characteristic polynomial, as given in (6.10).

Finding All the Eigenvectors

Given an eigenvalue λ of A, you can find an eigenvector corresponding to λ by finding a nonzero solution to the following system of linear equations:

$$(A - \lambda I)\mathbf{x} = \mathbf{0} \tag{6.11}$$

The process of doing so is illustrated in Example 6.3. A key observation is that there are an infinite number of solutions to (6.11) because there is always a free variable in the reduced row-echelon solution. In fact, for a given value of λ, the set of all solutions to (6.11), including $\mathbf{x} = \mathbf{0}$, defines the null space (see Section 4.4.2) of the matrix $A - \lambda I$, which is called the **eigenspace** of A corresponding to the eigenvalue λ.

The eigenspace of A corresponding to any eigenvalue λ consists of the zero vector and all vectors that satisfy (6.11) for that particular value of λ. As just discussed, there are an infinite number of such eigenvectors, so you cannot list them all. An alternative is to find a basis for the eigenspace—as shown in the following example—for then, any vector in the eigenspace is a linear combination of the basis vectors.

EXAMPLE 6.5 Finding a Basis for the Eigenspace Corresponding to an Eigenvalue
To find a basis for the eigenspace of the matrix

$$A = \begin{bmatrix} 0 & 2 \\ 3 & -1 \end{bmatrix}$$

corresponding to the eigenvalue $\lambda = -3$, solve the system of linear equations $(A - \lambda I)\mathbf{x} = \mathbf{0}$. In so doing, you obtain the following reduced row-echelon form of the associated augmented matrix:

$$\begin{bmatrix} 1 & 2/3 & | & 0 \\ 0 & 0 & | & 0 \end{bmatrix}$$

Here, x_2 is a free variable and the general solution is

$$\begin{bmatrix} x_1 \\ x_2 \end{bmatrix} = x_2 \begin{bmatrix} -2/3 \\ 1 \end{bmatrix}, \qquad x_2 \text{ is free}$$

In this case, a basis for the eigenspace is the vector on the right side of the foregoing equality, or, equivalently, multiplying each component by 3, the following is a basis for this eigenspace:

$$\begin{bmatrix} -2 \\ 3 \end{bmatrix}$$

In summary, if you can find the n eigenvalues of A, some of which may be repeated and some of which may be complex numbers, you can find a basis for each of the associated eigenspaces. In the event that you do not know the eigenvalues of A, numerical methods are used to approximate these values, from which you can then approximate the eigenvectors. The process of finding all eigenvalues and eigenvectors of A is illustrated again in the following section, which also presents the geometry of eigenvectors and eigenspaces.

6.1.3 The Geometry of Eigenvectors and Eigenspaces

Visual representations of eigenvectors and eigenspaces are now developed. To that end, consider again the matrix

$$A = \begin{bmatrix} 0 & 2 \\ 3 & -1 \end{bmatrix} \tag{6.12}$$

The characteristic equation of A is

$$\det(A - \lambda I) = \det \begin{bmatrix} -\lambda & 2 \\ 3 & -1-\lambda \end{bmatrix} = \lambda^2 + \lambda - 6 = (\lambda - 2)(\lambda + 3) = 0$$

You can conclude therefore that A has two distinct eigenvalues, namely, $\lambda = 2$ and $\lambda = -3$.

A Visualization of an Eigenvector

From Example 6.3 in Section 6.1.1, you know that

$$\mathbf{v} = \begin{bmatrix} 1 \\ 1 \end{bmatrix}$$

is an eigenvector of A corresponding to $\lambda = 2$. What this means algebraically is that

$$A\mathbf{v} = \lambda\mathbf{v} = 2\mathbf{v} \tag{6.13}$$

Geometrically, (6.13) indicates that A multiplied by the eigenvector $\mathbf{v}$ results in the vector $2\mathbf{v}$, which is the vector that points in the same direction as $\mathbf{v}$ and is twice as long, as shown in Figure 6.1.

In general, A multiplied by an eigenvector $\mathbf{v}$ corresponding to an eigenvalue λ results in $\lambda\mathbf{v}$, which is a multiple of $\mathbf{v}$. As you learned in Section 1.2.2, the vector $\lambda\mathbf{v}$ points in the same direction as $\mathbf{v}$ if $\lambda > 0$ and in the opposite direction to $\mathbf{v}$ if $\lambda < 0$. The length of $\lambda\mathbf{v}$ is greater than that of $\mathbf{v}$ if $|\lambda| > 1$ and less than that of $\mathbf{v}$ if $|\lambda| < 1$. These various possibilities for a (2×2) matrix A are depicted in Figure 6.2.

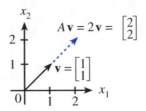

Figure 6.1 *A Geometric Visualization of an Eigenvector Corresponding to an Eigenvalue.*

A Visualization of an Eigenspace

You have just seen that for an eigenvector **v** corresponding to an eigenvalue λ of A, $A\mathbf{v}$ is related geometrically to **v** because $A\mathbf{v} = \lambda\mathbf{v}$. In contrast, for a general vector **u**, there may be no simple geometric relation between $A\mathbf{u}$ and **u**. These facts are illustrated for a (2×2) matrix A in Figure 6.3.

For a given eigenvalue λ of a matrix A, there are an infinite number of eigenvectors, namely, all nonzero vectors in the eigenspace corresponding to λ, that is,

$$\{\mathbf{x} \in R^n : \mathbf{x} \neq \mathbf{0} \text{ and } (A - \lambda I)\mathbf{x} = \mathbf{0}\} = \{\mathbf{x} \in R^n : \mathbf{x} \neq \mathbf{0} \text{ and } A\mathbf{x} = \lambda\mathbf{x}\}$$

A geometric visualization can be created because the eigenspace is the null space of the matrix $A - \lambda I$, which is a subspace of R^n. By finding a basis, you can visualize the eigenspace as the set of all linear combinations of the basis vectors, as is now illustrated for the matrix A in (6.12), whose two eigenvalues are $\lambda = 2$ and $\lambda = -3$.

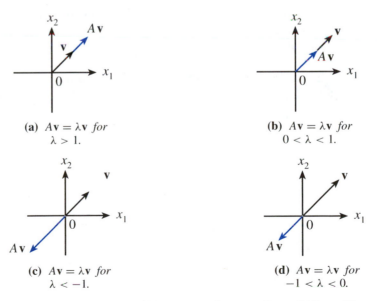

(a) $A\mathbf{v} = \lambda\mathbf{v}$ *for* $\lambda > 1$.

(b) $A\mathbf{v} = \lambda\mathbf{v}$ *for* $0 < \lambda < 1$.

(c) $A\mathbf{v} = \lambda\mathbf{v}$ *for* $\lambda < -1$.

(d) $A\mathbf{v} = \lambda\mathbf{v}$ *for* $-1 < \lambda < 0$.

Figure 6.2 *A Geometric Visualization of Eigenvectors Corresponding to Different Eigenvalues of A.*

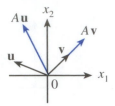

Figure 6.3 *The Geometry of an Eigenvector* **v** *and a Vector* **u** *that is Not an Eigenvector.*

The Eigenspace Corresponding to the Eigenvalue $\lambda = 2$. Following the technique in Example 6.5, you find a basis for the eigenspace of A corresponding to the eigenvalue $\lambda = 2$ by reducing the augmented matrix associated with the system $(A - \lambda I)\mathbf{x} = \mathbf{0}$. In this example, after applying Gauss-Jordan elimination to

$$\left[\begin{array}{cc|c} -2 & 2 & 0 \\ 3 & -3 & 0 \end{array} \right]$$

you obtain

$$\left[\begin{array}{cc|c} 1 & -1 & 0 \\ 0 & 0 & 0 \end{array} \right]$$

The general solution to $(A - \lambda I)\mathbf{x} = \mathbf{0}$ is $x_1 = x_2$, with x_2 free, that is,

$$\left[\begin{array}{c} x_1 \\ x_2 \end{array} \right] = x_2 \left[\begin{array}{c} 1 \\ 1 \end{array} \right], \quad x_2 \text{ is free}$$

The vector on the right side of the foregoing equality is therefore a basis for the eigenspace. The eigenspace itself is the set of all linear combinations of this basis vector. This one-dimensional eigenspace is represented by the dotted line in Figure 6.4.

The Eigenspace Corresponding to the Eigenvalue $\lambda = -3$. From Example 6.5, the following vector is a basis for the eigenspace of A corresponding to $\lambda = -3$:

$$\mathbf{v} = \left[\begin{array}{c} -2 \\ 3 \end{array} \right]$$

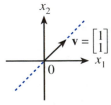

Figure 6.4 *The Eigenspace Corresponding to the Eigenvalue* $\lambda = 2$.

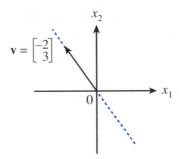

Figure 6.5 *The Eigenspace Corresponding to the Eigenvalue $\lambda = -3$.*

The resulting one-dimensional eigenspace is represented by the dotted line in Figure 6.5.

In this section, you have learned about eigenvalues, eigenvectors, and eigenspaces. In Section 6.2, you will see an important class of problems to which eigenvalues and eigenvectors are applied.

Exercises for Section 6.1

1. Is the following vector **x** an eigenvector of the given matrix A? If so, identify the associated eigenvalue. If not, explain why not.

$$\mathbf{x} = \begin{bmatrix} -1 \\ 3 \end{bmatrix} \quad \text{and} \quad A = \begin{bmatrix} 2 & 1 \\ 3 & 4 \end{bmatrix}$$

2. Is the following vector **x** an eigenvector of the given matrix A? If so, identify the associated eigenvalue. If not, explain why not.

$$\mathbf{x} = \begin{bmatrix} -1 \\ -1 \\ 1 \end{bmatrix} \quad \text{and} \quad A = \begin{bmatrix} 3 & -1 & 0 \\ 0 & 2 & 0 \\ -2 & 0 & 2 \end{bmatrix}$$

3. Verify that the following vector **x** is an eigenvector of the given matrix A by producing the associated eigenvalue.

$$\mathbf{x} = \begin{bmatrix} 1 \\ 1 \\ 1 \end{bmatrix} \quad \text{and} \quad A = \begin{bmatrix} 4 & 0 & -2 \\ 1 & 3 & -2 \\ 1 & 0 & 1 \end{bmatrix}$$

4. For what value of the real number a is the following vector **x** an eigenvector of the matrix in Exercise 2?

$$\mathbf{x} = \begin{bmatrix} -2 \\ -2 \\ a \end{bmatrix}$$

5. Determine whether $\lambda = 2$ is an eigenvalue of the following matrix:

$$A = \begin{bmatrix} 2 & 1 \\ 3 & 4 \end{bmatrix}$$

6. Determine whether $\lambda = 3$ is an eigenvalue of the following matrix:

$$A = \begin{bmatrix} 4 & 2 & -1 \\ 0 & -3 & 5 \\ 0 & 0 & 4 \end{bmatrix}$$

7. Verify that $\lambda = 0$ is an eigenvalue of the following matrix by producing an associated eigenvector:

$$A = \begin{bmatrix} 1 & -1 & 2 \\ 2 & -2 & 4 \\ 0 & 0 & 3 \end{bmatrix}$$

8. Explain why there is no value of the real number a for which $\lambda = 1$ is an eigenvalue of the following matrix:

$$A = \begin{bmatrix} 1 & -1 \\ -2 & a \end{bmatrix}$$

9. Perform the indicated instructions using the following matrix:

$$A = \begin{bmatrix} 1 & -1 \\ -2 & 2 \end{bmatrix}$$

(a) Write the characteristic equation of A.
(b) Use the characteristic equation to find the eigenvalues of A. Then find an eigenvector corresponding to each eigenvalue.
(c) Find a basis for the eigenspace associated with each of the eigenvalues of A. Draw a single graph that shows these eigenspaces.

10. Repeat the previous exercise for the following matrix:

$$A = \begin{bmatrix} 2 & 0 & 0 \\ -1 & -3 & 0 \\ 1 & 0 & 4 \end{bmatrix}$$

11. Write the characteristic equation of the following matrix:

$$A = \begin{bmatrix} 1 & 2 \\ -1 & 3 \end{bmatrix}$$

12. Write the characteristic equation of the following matrix:

$$A = \begin{bmatrix} a & 0 & 0 \\ -1 & b & 0 \\ 1 & 0 & c \end{bmatrix}$$

13. A student said that $\lambda = 0$ is always an eigenvalue of an $(n \times n)$ matrix A because, for $\mathbf{x} = \mathbf{0}$, $A\mathbf{x} = A\mathbf{0} = \mathbf{0} = 0\mathbf{x} = \lambda\mathbf{x}$. Do you agree with this statement? Why or why not? Explain.

14. Prove that 0 is an eigenvalue of A if and only if A is not invertible.

15. How are the eigenvalues of A related to those of A^t? Prove your result. (Hint: Determine how $\det(A - \lambda I)$ and $\det(A^t - \lambda I)$ are related.)

16. How are the eigenvalues of an invertible matrix A related to those of A^{-1}? Prove your result.

6.2 Applications to Dynamical Systems

Dynamical systems are systems that evolve over time in a predictable way. Such systems arise in the fields of ecology, biology, physics, and engineering. Models of these systems consist of mathematical representations for (1) the state of the system at time t and (2) changes in the system over time. The type of model depends on whether you are trying to monitor the state of the system continuously over time—such as the flight path of the space shuttle—or at fixed points in time—such as the state of the economy at the beginning of each month, for example. The former type of system is a **continuous dynamical system** and the latter is a **discrete dynamical system**. Discrete dynamical systems, hereafter called dynamical systems, are the focus of this chapter.

6.2.1 Components of a Dynamical System

An n-vector is generally used to represent the state of a dynamical system at a given point in time. To illustrate, recall the predator-prey problem presented at the beginning of this chapter, in which you are trying to understand the relationship of the rabbit population to that of the timber wolf. The state of this ecological system at the beginning of year k ($k = 0, 1, 2, \ldots$) is represented by the vector

$$\mathbf{x}_k = \begin{bmatrix} W_k \\ R_k \end{bmatrix} = \begin{bmatrix} \text{number of wolves at the start of year } k \\ \text{number of hundreds of rabbits at the start of year } k \end{bmatrix}$$

In many cases, $\mathbf{x}_0$ is known or can be determined by conducting a species count for a particular area. However, the values of $\mathbf{x}_1, \mathbf{x}_2, \ldots$, that represent the state of the ecological system at future points in time, are not known, although there is an assumption that $\mathbf{x}_{k+1}$ is related to $\mathbf{x}_k$. The specific relationship constitutes the next component of a model of a dynamical system, as now discussed.

This second component is a description of how the dynamical system changes over time. The approach used in this section is to assume that there is a linear relationship between the state of the system at time k, $\mathbf{x}_k$, and the state of the system at time $k + 1$,

$\mathbf{x}_{k+1}$. This linear relationship manifests itself in the form of a known $(n \times n)$ matrix A that is used to compute $\mathbf{x}_{k+1}$ from $\mathbf{x}_k$ by the following **linear difference equation**:

$$\mathbf{x}_{k+1} = A\mathbf{x}_k \qquad (6.14)$$

To illustrate, recall the following relationships between the number of wolves and hundreds of rabbits at the beginning of year $k + 1$ and year k:

$$\begin{aligned} W_{k+1} &= 0.72W_k + 0.2R_k \\ R_{k+1} &= -0.12W_k + 1.1R_k \end{aligned} \qquad (6.15)$$

Letting A be the (2×2) matrix of the coefficients of W_k and R_k on the right side of (6.15), the linear difference equation for this system is

$$\underbrace{\begin{bmatrix} W_{k+1} \\ R_{k+1} \end{bmatrix}}_{\mathbf{x}_{k+1}} = \underbrace{\begin{bmatrix} 0.72 & 0.2 \\ -0.12 & 1.1 \end{bmatrix}}_{A} \underbrace{\begin{bmatrix} W_k \\ R_k \end{bmatrix}}_{\mathbf{x}_k} \qquad (6.16)$$

To see how (6.16) is used, suppose there are $W_0 = 100$ wolves and $R_0 = 60$ hundred rabbits at the beginning of the study, in year 0. Then, from (6.16), you compute the number of wolves and hundreds of rabbits at the beginning of year 1, as follows:

$$\mathbf{x}_1 = \begin{bmatrix} W_1 \\ R_1 \end{bmatrix} = \begin{bmatrix} 0.72 & 0.2 \\ -0.12 & 1.1 \end{bmatrix} \begin{bmatrix} W_0 \\ R_0 \end{bmatrix} = \begin{bmatrix} 0.72 & 0.2 \\ -0.12 & 1.1 \end{bmatrix} \begin{bmatrix} 100 \\ 60 \end{bmatrix} = \begin{bmatrix} 84 \\ 54 \end{bmatrix}$$

The typical overall objective is to determine the *long-term behavior* of the dynamical system. That is, given the initial state of the system at time 0, $\mathbf{x}_0$, you want to see what happens to the state of the system, $\mathbf{x}_k$, as k becomes large. In mathematical terms, you are interested in the value of $\mathbf{x}_k$ as k *approaches infinity*, written, $k \rightarrow \infty$. It is in this area that eigenvectors and eigenvalues play a significant role, as shown in Section 6.3 and Section 6.4. Notationally, a sequence of vectors is written $\mathbf{X} = (\mathbf{x}_0, \mathbf{x}_1, \mathbf{x}_2, \ldots)$. In the remainder of this section, a variety of dynamical systems and their corresponding mathematical models are presented.

6.2.2 Markov Chains

A **Markov chain** is a dynamical system in which each element of the matrix A in the linear difference equation is nonnegative and each column of A sums to 1. Such a matrix is called a **stochastic** or **transition matrix**.

Two examples of Markov chains are now presented. In each case, a description is given of the vector, $\mathbf{x}_k$, representing the state of the system at time k and of the stochastic matrix, A, in the linear difference equation, describing how the system changes over time.

Using a Markov Chain to Model Population Movement

Recall the population model presented in Section 5.2.2 in which a_0 people are living in a city and b_0 people are living in the suburbs at the beginning of a particular year. If each year 5% of those living in the city move to the suburbs and 3% of those living in the suburbs move to the city, then the populations in the city and suburbs from year to year constitute a dynamical system. The vector that represents the state of the system at the beginning of year k is

$$\mathbf{x}_k = \begin{bmatrix} a_k \\ b_k \end{bmatrix} = \begin{bmatrix} \text{population living in the city at the start of year } k \\ \text{population living in the suburbs at the start of year } k \end{bmatrix}$$

You can identify the linear difference equation that describes how this system changes over time by writing the following two linear equations in terms of a_k and b_k with the aid of the given migration data:

$$\begin{pmatrix} \text{Number in city} \\ \text{after one year} \end{pmatrix} = \begin{pmatrix} \text{Number staying} \\ \text{in the city} \end{pmatrix} + \begin{pmatrix} \text{Number moving} \\ \text{to the city} \end{pmatrix}$$

$$= 0.95a_k + 0.03b_k$$

$$\begin{pmatrix} \text{Number in suburbs} \\ \text{after one year} \end{pmatrix} = \begin{pmatrix} \text{Number moving} \\ \text{to the suburbs} \end{pmatrix} + \begin{pmatrix} \text{Number staying} \\ \text{in the suburbs} \end{pmatrix}$$

$$= 0.05a_k + 0.97b_k$$

Letting A be the (2×2) matrix of the coefficients on the right side of the foregoing equations, you now have the following linear difference equation:

$$\mathbf{x}_{k+1} = \begin{bmatrix} a_{k+1} \\ b_{k+1} \end{bmatrix} = \begin{bmatrix} 0.95a_k + 0.03b_k \\ 0.05a_k + 0.97b_k \end{bmatrix} = \begin{bmatrix} 0.95 & 0.03 \\ 0.05 & 0.97 \end{bmatrix} \begin{bmatrix} a_k \\ b_k \end{bmatrix} = A\mathbf{x}_k$$

This model is a Markov chain because all elements in A are nonnegative and each column of A sums to 1.

Given the initial populations in the city and suburbs, $\mathbf{x}_0$, you want to know what happens to these populations, $\mathbf{x}_k$, as $k \to \infty$.

Using a Markov Chain to Model Brand Loyalty

The components of the vector $\mathbf{x}_k$ of a Markov chain often represent the probability of an event occurring or the fraction of a population having a particular property. In such a case, the vector $\mathbf{x}_k$ is called a **probability vector**, meaning that each component is a number between 0 and 1 and that the components sum to 1.

To illustrate this type of Markov chain, consider the problem of determining the fraction of households that use AT&T, MCI, and Sprint for their long distance service. Suppose that at the beginning of some month, the fraction of households using AT&T is a_0, of those using MCI is b_0, and of those using Sprint is c_0. In general, let a_k, b_k, and c_k be the fraction of households using AT&T, MCI, and Sprint, respectively, at the beginning of month k, for $k = 0, 1, 2, \ldots$.

Suppose further that historical data show that the fraction, a_{k+1}, of households using

AT&T at the beginning of month $k + 1$ is given by the following linear equation:

$$a_{k+1} = 0.8a_k + 0.2b_k + 0.2c_k$$

Likewise, suppose the following equations describe b_{k+1} and c_{k+1}:

$$b_{k+1} = 0.1a_k + 0.7b_k + 0.2c_k$$
$$c_{k+1} = 0.1a_k + 0.1b_k + 0.6c_k$$

Letting

$$\mathbf{x}_k = \begin{bmatrix} a_k \\ b_k \\ c_k \end{bmatrix}$$

and A be the matrix of coefficients on the right side of the three foregoing linear equations, the difference equation for this dynamical system is:

$$\mathbf{x}_{k+1} = \begin{bmatrix} a_{k+1} \\ b_{k+1} \\ c_{k+1} \end{bmatrix} = \begin{bmatrix} 0.8a_k + 0.2b_k + 0.2c_k \\ 0.1a_k + 0.7b_k + 0.2c_k \\ 0.1a_k + 0.1b_k + 0.6c_k \end{bmatrix} = \begin{bmatrix} 0.8 & 0.2 & 0.2 \\ 0.1 & 0.7 & 0.2 \\ 0.1 & 0.1 & 0.6 \end{bmatrix} \begin{bmatrix} a_k \\ b_k \\ c_k \end{bmatrix} = A\mathbf{x}_k$$

This dynamical system is a Markov chain because A is a stochastic matrix. Also, note that $\mathbf{x}_k$ is a probability vector whose three components represent the fractions of households using AT&T, MCI, and Sprint. The objective is to determine what happens to these fractions over a long period of time, assuming that nothing happens to change the foregoing linear difference equation that describes how the system evolves.

6.2.3 Representing the Filter of a Signal as a Dynamical System

As a final example of a dynamical system, recall, from Section 5.6, that a digital audio signal is represented as a sequence $\mathbf{Y} = (y_k)$ of real numbers at the points $k = 0, 1, 2, \ldots$ in time. In the filtering problem, you are given the numerical coefficients of a filter and a desired output signal $\mathbf{Z} = (z_k)$. The objective is to find an input signal $\mathbf{Y} = (y_k)$ that, when passed through the filter, results in the output signal $\mathbf{Z}$. For example, recall from Section 5.6, the problem of Audio Technologies in which you want to find an input signal $\mathbf{Y} = (y_k)$ that, when passed through the filter whose coefficients are $1, -10$, and 8, results in the zero signal, that is, $\mathbf{Y} = (y_k)$ should satisfy:

$$y_{k+3} + y_{k+2} - 10y_{k+1} + 8y_k = 0, \quad \text{for } k = 0, 1, 2, \ldots \tag{6.17}$$

You can view this filter as a dynamical system that evolves over time according to (6.17). Given y_k, y_{k+1}, and y_{k+2}, you can use (6.17) to compute $y_{k+3} = -8y_k + 10y_{k+1} - y_{k+2}$.

Thus, the state of this system at time k is represented by the following vector:

$$\mathbf{x}_k = \begin{bmatrix} y_k \\ y_{k+1} \\ y_{k+2} \end{bmatrix}$$

The associated linear difference equation that describes how $\mathbf{x}_k$ determines $\mathbf{x}_{k+1}$ arises from (6.17) because

$$\mathbf{x}_{k+1} = \begin{bmatrix} y_{k+1} \\ y_{k+2} \\ y_{k+3} \end{bmatrix} = \begin{bmatrix} 0y_k + y_{k+1} + 0y_{k+2} \\ 0y_k + 0y_{k+1} + y_{k+2} \\ -8y_k + 10y_{k+1} - y_{k+2} \end{bmatrix} \qquad (6.18)$$

Letting A be the matrix of coefficients of y_k, y_{k+1}, and y_{k+2} on the right side of (6.18), you can identify the following linear difference equation:

$$\mathbf{x}_{k+1} = \begin{bmatrix} y_{k+1} \\ y_{k+2} \\ y_{k+3} \end{bmatrix} = \begin{bmatrix} 0 & 1 & 0 \\ 0 & 0 & 1 \\ -8 & 10 & -1 \end{bmatrix} \begin{bmatrix} y_k \\ y_{k+1} \\ y_{k+2} \end{bmatrix} = A\mathbf{x}_k \qquad (6.19)$$

The objective of this dynamical system is different from the others presented in this section because, for the problem of Audio Technologies, you are not interested in the long-term behavior of $\mathbf{x}_k$. Rather, you are interested in finding a closed-form expression for a signal $\mathbf{Y} = (y_k)$ that satisfies the linear difference equation in (6.19) for each $k = 0, 1, 2, \ldots$.

In general, given n real numbers $a_1, \ldots, a_n$, finding a sequence $\mathbf{Y} = (y_k)$ that satisfies

$$y_{k+n} + a_1 y_{k+n-1} + \cdots + a_n y_k = 0, \qquad \text{for } k = 0, 1, 2, \ldots$$

is equivalent to finding a sequence of n-vectors $\mathbf{X} = (\mathbf{x}_0, \mathbf{x}_1, \mathbf{x}_2, \ldots)$ that satisfy $\mathbf{x}_{k+1} = A\mathbf{x}_k$, where

$$\mathbf{x}_k = \begin{bmatrix} y_k \\ y_{k+1} \\ \vdots \\ y_{k+n-1} \end{bmatrix} \quad \text{and} \quad A = \begin{bmatrix} 0 & 1 & 0 & \cdots & 0 \\ 0 & 0 & 1 & \cdots & 0 \\ \vdots & \vdots & \vdots & \ddots & \vdots \\ 0 & 0 & 0 & \cdots & 1 \\ -a_n & -a_{n-1} & -a_{n-2} & \cdots & -a_1 \end{bmatrix}$$

Now that you have seen various applications of dynamical systems and how to build appropriate models, you will learn how to solve such models.

Exercises for Section 6.2

1. In the following dynamical system, identify (a) the meaning of the components of the vector $\mathbf{x}_k$ that represents the state of the system at time k and the initial state of the system, (b) the matrix A in the linear difference equation $\mathbf{x}_{k+1} = A\mathbf{x}_k$ that describes how the system evolves from one period to the next, and (c) the overall objective of the

problem. Is the model a Markov chain?

U-Rent-A-Car rents cars that are returned either at the downtown office or at the airport. Historical data indicate that, on a weekly basis, about 60% of those cars rented at the downtown office are returned there and 95% of those rented at the airport are returned there. If there are 80 cars at the airport and 50 cars at the downtown office, how long will it take before there are fewer than 20 cars at the downtown office?

2. Repeat Exercise 1 for the following dynamical system. At any given hour, some of the 90 weaving machines at Weavers International are working and others are not. With probability 0.8, a machine that is working in one hour will be working in the next hour, and hence there is a probability of 0.2 that such a machine is not working in the next hour. Likewise, the repair crew estimates that there is a probability of 0.7 that a machine that is not working in a given hour will be repaired and working in the next hour. On average, how many machines are working in a given hour?

3. Repeat Exercise 1 for the following dynamical system. Advertising data indicate that at the current levels of expenditure, 90% of those who drink Coke in a given month will continue to do so, 5% will switch to Pepsi, and 5% to other brands. Likewise, 85% of the those who drink Pepsi continue to do so, 5% switch to Coke, and the remainder switch to other brands. Of those who drink other brands, 80% do not change, 9% switch to Coke and 11% switch to Pepsi. What share of the market is devoted to each group in the long run?

4. Repeat Exercise 1 for the following dynamical system. The annual production, in millions of dollars, of Germany, Japan, and the U. S. varies from year to year. Historical data indicate that the annual production of Germany in a given year is approximately the sum of 50% of Germany's annual production in the previous year, 20% of that of Japan, and 20% of that of the U. S. Likewise, the annual production of Japan in a given year is estimated as the sum of 40% of their own output in the previous year, 10% of the output of Germany in the previous year, and 10% of that of the U. S. Finally, the annual production of the U. S. in a given year is approximately 70% of its own production in the previous year plus 40% of Germany's and Japan's production in the previous year, combined. If the annual production of Germany is currently 75 billion dollars, that of Japan is 50 billion dollars, and that of the U.S. is 100 billion dollars, what will these figures be in the long run?

5. Repeat Exercise 1 for the following dynamical system. The American Steel Company has fixed demands of 8000 tons of steel for each of the next 24 months. At the beginning of each month, this demand is met from that month's production together with the available inventory, if necessary. Any excess amount produced is stored in inventory for use the following month. Management is considering a policy of producing, each month, 50% of the monthly demand plus 50% of the amount produced in the previous month. Assuming there are currently 4000 tons of steel in inventory and that last month, 10,000 tons of steel were produced, what is the inventory level at the end of the planning period?

6. Repeat Exercise 1 for the following dynamical system. One model of the population of the California condor is based on the number of baby, preadult, and adult female birds each year. Data collected by researchers indicate that from one year to the next, (1) about 30% of the adult females give birth each year, (2) 40% of the baby birds survive

to the preadult stage and the rest die, (3) 70% of the preadult females become adults and the rest die, and (4) 90% of the adult females survive. There are estimated to be 50 baby female condors, 75 preadult females, and 150 adult females. Will the condor population eventually die out and, if so, what must the survival rate from the baby to the preadult stage be for the condor population to grow over time?

6.3 Solving a Dynamical System

You now know that a dynamical system consists of (1) a given n-vector $\mathbf{x}_0$ that describes the initial state of the system at time 0, (2) an $(n \times n)$ matrix A for the linear difference equation

$$\mathbf{x}_{k+1} = A\mathbf{x}_k \qquad (6.20)$$

that describes how the system evolves from time k to time $k + 1$ and, for the purposes of this section, (3) an objective of determining what happens to the sequence $(\mathbf{x}_k)$ as $k \to \infty$. In this section, you will see how eigenvalues and eigenvectors play a role in providing both numerical-method and closed-form solutions.

6.3.1 The Direct Approach to Solving a Dynamical System

Perhaps the obvious numerical-method approach to solving a dynamical system is to use (6.20) to compute the vectors $\mathbf{x}_1, \mathbf{x}_2, \ldots$, in order. Specifically, use the initial n-vector $\mathbf{x}_0$ to compute $\mathbf{x}_1 = A\mathbf{x}_0$. Then use $\mathbf{x}_1$ to compute $\mathbf{x}_2 = A\mathbf{x}_1$, and so on, and "see what happens."

To illustrate this approach, recall the dynamical system in Section 6.2.2 for modeling the fraction of households that use AT&T, MCI, and Sprint at the beginning of month k, denoted by $\mathbf{x}_k$. Assume that the following initial fractions and associated linear difference equation hold:

$$\mathbf{x}_0 = \begin{bmatrix} 0.4 \\ 0.3 \\ 0.3 \end{bmatrix} \quad \text{and} \quad \mathbf{x}_{k+1} = A\mathbf{x}_k = \begin{bmatrix} 0.8 & 0.2 & 0.2 \\ 0.1 & 0.7 & 0.2 \\ 0.1 & 0.1 & 0.6 \end{bmatrix} \begin{bmatrix} a_k \\ b_k \\ c_k \end{bmatrix} \qquad (6.21)$$

The results of computing $A\mathbf{x}_k$, for $k = 1, 2, \ldots, 8$, rounded to three decimal places, are given in Table 6.1. On the basis of these results, you might guess that the three elements of $\mathbf{x}_k$ are getting closer to $0.5, 0.3,$ and 0.2, respectively, but you cannot be sure of this. That

	$\mathbf{x}_0$	$\mathbf{x}_1$	$\mathbf{x}_2$	$\mathbf{x}_3$	$\mathbf{x}_4$	$\mathbf{x}_5$	$\mathbf{x}_6$	$\mathbf{x}_7$	$\mathbf{x}_8$
a_k	0.400	0.440	0.464	0.478	0.487	0.492	0.495	0.497	0.498
b_k	0.300	0.310	0.311	0.309	0.307	0.305	0.304	0.303	0.302
c_k	0.300	0.250	0.225	0.213	0.207	0.204	0.202	0.201	0.201

Table 6.1 *Computing* $\mathbf{x}_1, \ldots, \mathbf{x}_8$ *for the Brand-Loyalty Dynamical System.*

is one of the drawbacks of this numerical method. Another disadvantage is the significant computational effort involved in computing $x_1, x_2, \ldots$. Imagine the effort needed to compute A^{500} for a (100×100) matrix.

In view of these drawbacks, a fruitful approach is to find conditions on the problem data, namely, the initial state x_0 and the matrix A, under which you can solve the problem more efficiently. This is one of the uses of mathematics. For the study of dynamical systems, one special case when it is possible to develop a more efficient numerical-method solution is a Markov chain, as discussed now.

Numerical-method Solutions for Markov Chains

The current model under consideration is a Markov chain because A is a stochastic matrix, since all elements are nonnegative and each column sums to 1. Also, the vectors representing the fractions of households using AT&T, MCI, and Sprint at the beginning of each month k are probability vectors. For a Markov chain, it is often possible to determine what happens to the sequence of vectors $X = (x_0, x_1, x_2, \ldots)$ *without having to compute the sequence explicitly*. This is because, under certain circumstances, the sequence of vectors is getting "closer and closer" to a final vector, say, x, as k approaches infinity.

The formal definition of "closer and closer" is beyond the scope of this book but, informally, a sequence of n-vectors $X = (x_0, x_1, x_2, \ldots)$ is said to **converge** to a given n-vector x as $k \to \infty$ if and only if the components of x_k can be made as close in value as desired to the corresponding components of x, provided that k is sufficiently large. This idea, together with the concept of a **regular stochastic matrix**—a stochastic matrix A such that all the elements of some power of A are strictly positive—is the basis for the determination of a solution to the long-term behavior of a Markov chain. The result is provided in following theorem, whose proof is given in most texts on Markov chains.

THEOREM 6.1

If A is an $(n \times n)$ regular stochastic matrix, then there is a unique probability vector x, called a **steady-state vector**, such that for all probability vectors x_0, the sequence of n-vectors $X = (x_0, x_1, x_2, \ldots)$ in which $x_{k+1} = Ax_k$ converges to x as $k \to \infty$. Furthemore, x is the unique probability vector satisfying

$$Ax = x \tag{6.22}$$

Another mathematical skill worth developing is the ability to **understand the relevance of a theorem to the problem you are solving**.

Can you do so for Theorem 6.1? Here are some of the key features with regard to solving the current Markov chain, in which A is a regular stochastic matrix:

1. The fact that the sequence of vectors $X = (x_0, x_1, x_2, \ldots)$ converges to x means that, if you were to compute $x_1, x_2, \ldots$, these probability vectors that describe the fractions of households using AT&T, MCI, and Sprint at time k will indeed get closer and closer to some final probability vector, namely, the steady-state vector x.

2. The statement that *for all* probability vectors x_0, the sequence $X = (x_0, x_1, x_2, \ldots)$ converges to x means that, *regardless of the initial fractions of households using AT&T, MCI, and Sprint*, the system will, over time, approach the state described by the steady-state vector. In other words, the initial state of the system has no effect on the long-term behavior—a rather unexpected feature—although the initial state does affect how much time it takes, that is, how large k must be, for the state of the system to approach that of the steady-state vector.

3. The fact that x satisfies (6.22) means that there is no need to compute $x_1, x_2, \ldots$. Because you know that this sequence converges to the steady-state probability vector x, you need only find that vector by solving the following system of linear equations:

$$(A - I)x = 0 \tag{6.23}$$

$$ex = 1 \quad \text{(where } e = [1, 1, \ldots, 1]) \tag{6.24}$$

The group of equations in (6.23) comes directly from (6.22). The final equation in (6.24) ensures that the components of x sum to 1 and hence that x is a probability vector. (The components of x will be nonnegative automatically.)

The use of Theorem 6.1 is now illustrated on the current model in which

$$A = \begin{bmatrix} 0.8 & 0.2 & 0.2 \\ 0.1 & 0.7 & 0.2 \\ 0.1 & 0.1 & 0.6 \end{bmatrix}$$

Note that this stochastic matrix is regular because every element of A is strictly positive. To apply Theorem 6.1, solve the following system of linear equations from (6.23) and (6.24):

$$\begin{bmatrix} A - I \\ e \end{bmatrix} x = \begin{bmatrix} -0.2 & 0.2 & 0.2 \\ 0.1 & -0.3 & 0.2 \\ 0.1 & 0.1 & -0.4 \\ 1.0 & 1.0 & 1.0 \end{bmatrix} \begin{bmatrix} x_1 \\ x_2 \\ x_3 \end{bmatrix} = \begin{bmatrix} 0 \\ 0 \\ 0 \\ 1 \end{bmatrix}$$

Applying Gauss-Jordan elimination to the associated augmented matrix results in the following reduced row-echelon form:

$$\begin{bmatrix} 1 & 0 & 0 & 0.5 \\ 0 & 1 & 0 & 0.3 \\ 0 & 0 & 1 & 0.2 \\ 0 & 0 & 0 & 0.0 \end{bmatrix}$$

The steady-state vector is in the first three rows of the last column of this matrix. These values indicate that, in the long run, 50% of households use AT&T, 30% use MCI, and 20% use Sprint. Compare this approach with that of computing the values in Table 6.1.

6.3.2 Using Eigenvalues and Eigenvectors to Solve Dynamical Systems

Unfortunately, not all dynamical systems are Markov chains with a regular stochastic matrix, in which case, Theorem 6.1 is not applicable. For a general dynamical system, the long-term behavior generally depends on the initial state. It is for such systems that eigenvectors and eigenvalues are especially useful. To see one such example, suppose that you can control the initial state of a dynamical system and you choose the initial state to be an eigenvector, $\mathbf{x}_0$, corresponding to an eigenvalue λ of the matrix A in the linear difference equation for this system. Then it follows from the linear difference equation that

$$\mathbf{x}_1 = A\mathbf{x}_0 = \lambda\mathbf{x}_0$$
$$\mathbf{x}_2 = A\mathbf{x}_1 = \lambda\mathbf{x}_1 = \lambda(\lambda\mathbf{x}_0) = \lambda^2\mathbf{x}_0$$
$$\mathbf{x}_3 = A\mathbf{x}_2 = \lambda\mathbf{x}_2 = \lambda(\lambda^2\mathbf{x}_0) = \lambda^3\mathbf{x}_0$$
$$\vdots$$

As you are asked to prove in Exercise 7:

$$\mathbf{x}_k = \lambda^k\mathbf{x}_0, \quad \text{for } k = 0, 1, 2, \ldots \tag{6.25}$$

To simplify the following discussion, suppose that $\lambda > 0$. The closed-form expression for $\mathbf{x}_k$ in (6.25) allows the following analysis of the long-term behavior of $\mathbf{x}_k$, depending on whether $\lambda < 1$, $\lambda = 1$, or $\lambda > 1$, *without having to compute the sequence* $\mathbf{X} = (\mathbf{x}_0, \mathbf{x}_1, \mathbf{x}_2, \ldots)$ *explicitly.*

Analysis for the Case $0 < \lambda < 1$. If $0 < \lambda < 1$, then, as $k \to \infty$, λ^k gets closer to 0. Thus, from (6.25), $\mathbf{x}_k = \lambda^k\mathbf{x}_0$ gets closer to the zero vector, $\mathbf{0}$, as $k \to \infty$. In other words, if the system is started in the state described by an eigenvector $\mathbf{x}_0$ corresponding to an eigenvalue λ with $0 < \lambda < 1$, then the state of the system approaches the zero vector as $k \to \infty$, as shown in Figure 6.6(a).

Analysis for the Case $\lambda = 1$. If $\lambda = 1$, then, for each k, $\lambda^k = 1$. Thus, (6.25) says that for each k, $\mathbf{x}_k = \lambda^k\mathbf{x}_0 = \mathbf{x}_0$. In other words, if the system is started in the initial state described by an eigenvector $\mathbf{x}_0$ corresponding to the eigenvalue $\lambda = 1$, then the system remains in that state forever, as shown in Figure 6.6(b).

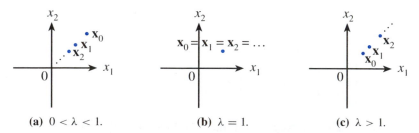

(a) $0 < \lambda < 1$. **(b)** $\lambda = 1$. **(c)** $\lambda > 1$.

Figure 6.6 *The Sequence* $\mathbf{X} = (\mathbf{x}_0, \mathbf{x}_1, \mathbf{x}_2, \ldots)$ *for a Dynamical System Starting with an Eigenvector* $\mathbf{x}_0$ *Corresponding to an Eigenvalue* $\lambda > 0$.

Analysis for the Case $\lambda > 1$. If $\lambda > 1$, then $\lambda^k \to \infty$ as $k \to \infty$. Consequently, from (6.25), the positive components of $\lambda^k \mathbf{x}_0$ approach $+\infty$. As a result, for any component of $\mathbf{x}_0$ that is positive, the corresponding component of $\mathbf{x}_k = \lambda^k \mathbf{x}_0$ approaches ∞ as $k \to \infty$. Likewise, for any component of $\mathbf{x}_0$ that is negative, the corresponding component of $\mathbf{x}_k = \lambda^k \mathbf{x}_0$ approaches $-\infty$ as $k \to \infty$. In other words, if the system is started in the state described by an eigenvector $\mathbf{x}_0$ corresponding to an eigenvalue $\lambda > 1$, then the system *diverges*. That is, one or more components of the vectors in the sequence $\mathbf{X} = (\mathbf{x}_0, \mathbf{x}_1, \mathbf{x}_2, \ldots)$ approaches $\pm\infty$ as $k \to \infty$. This case is shown in Figure 6.6(c).

The foregoing analysis illustrates one advantage to working with eigenvectors and eigenvalues of A. You will now see another approach for determining the long-term behavior of a dynamical system when you cannot control the initial state.

6.3.3 Using a Basis of Eigenvectors to Analyze the Long-Term Behavior

The closed-form analysis in Section 6.3.2 is based on the assumption that the initial state of the system, $\mathbf{x}_0$, is an eigenvector of A. A similar type of analysis is possible when $\mathbf{x}_0$ is not an eigenvector of A, *provided that you can find a basis for R^n consisting of n eigenvectors of A*. Such a basis is called an **eigenvector basis**.

So, suppose that $\mathbf{v}_1, \ldots, \mathbf{v}_n$ form an eigenvector basis in which each eigenvector $\mathbf{v}_i$ corresponds to an eigenvalue λ_i of A. As a result, every n-vector is a linear combination of these basis vectors. In particular, for the initial state $\mathbf{x}_0$, there are real numbers $t_1, \ldots, t_n$ such that

$$\mathbf{x}_0 = t_1 \mathbf{v}_1 + \cdots + t_n \mathbf{v}_n \tag{6.26}$$

To illustrate, recall, from Section 6.2.1, the dynamical system for population movement between the city and suburbs whose linear difference equation is described by the following matrix:

$$A = \begin{bmatrix} 0.95 & 0.03 \\ 0.05 & 0.97 \end{bmatrix}$$

Assume that the initial state, $\mathbf{x}_0$, is 40,000 people living in the city and 60,000 living in the suburbs. You can verify that the eigenvalues of A are $\lambda_1 = 1$ and $\lambda_2 = 0.92$ and that two corresponding eigenvectors are

$$\mathbf{v}_1 = \begin{bmatrix} 3 \\ 5 \end{bmatrix} \quad \text{and} \quad \mathbf{v}_2 = \begin{bmatrix} 1 \\ -1 \end{bmatrix}$$

It is not hard to check that $\mathbf{v}_1$ and $\mathbf{v}_2$ are linearly independent and thus form a basis for R^2. You can therefore express $\mathbf{x}_0$ as the following linear combination of $\mathbf{v}_1$ and $\mathbf{v}_2$:

$$\mathbf{x}_0 = \begin{bmatrix} 40000 \\ 60000 \end{bmatrix} = 12500 \begin{bmatrix} 3 \\ 5 \end{bmatrix} + 2500 \begin{bmatrix} 1 \\ -1 \end{bmatrix} \tag{6.27}$$

In general, once you are able to express $\mathbf{x}_0$ as a linear combination of the eigenvectors $\mathbf{v}_1, \ldots, \mathbf{v}_n$, as in (6.26), you can use the linear difference equation to compute the sequence of vectors $\mathbf{X} = (\mathbf{x}_0, \mathbf{x}_1, \mathbf{x}_2, \ldots)$, as follows:

$$\mathbf{x}_1 = A\mathbf{x}_0 = A\sum_{i=1}^{n} t_i\mathbf{v}_i = \sum_{i=1}^{n} t_i A\mathbf{v}_i = \sum_{i=1}^{n} t_i\lambda_i\mathbf{v}_i$$

$$\mathbf{x}_2 = A\mathbf{x}_1 = A\sum_{i=1}^{n} t_i\lambda_i\mathbf{v}_i = \sum_{i=1}^{n} t_i\lambda_i A\mathbf{v}_i = \sum_{i=1}^{n} t_i(\lambda_i)^2\mathbf{v}_i$$

$$\mathbf{x}_3 = A\mathbf{x}_2 = A\sum_{i=1}^{n} t_i(\lambda_i)^2\mathbf{v}_i = \sum_{i=1}^{n} t_i(\lambda_i)^2 A\mathbf{v}_i = \sum_{i=1}^{n} t_i(\lambda_i)^3\mathbf{v}_i$$

$$\vdots$$

As you are asked to prove in Exercise 8:

$$\mathbf{x}_k = t_1(\lambda_1)^k\mathbf{v}_1 + \cdots + t_n(\lambda_n)^k\mathbf{v}_n, \quad \text{for } k = 0, 1, 2, \ldots \tag{6.28}$$

A closed-form analysis is possible when all eigenvalues except one, say, λ_1, have absolute value < 1. In this case, for $i = 2, \ldots, n$, $(\lambda_i)^k$ approaches 0 as $k \to \infty$. Thus, from (6.28), for large values of k:

$$\mathbf{x}_k \approx t_1(\lambda_1)^k\mathbf{v}_1 \tag{6.29}$$

Also, using (6.29) and the linear difference equation,

$$\mathbf{x}_{k+1} = A\mathbf{x}_k \approx t_1(\lambda_1)^k A\mathbf{v}_1 = \lambda_1 t_1(\lambda_1)^k\mathbf{v}_1 \approx \lambda_1\mathbf{x}_k \tag{6.30}$$

To interpret this result in the context of a specific problem, consider the population migration example, in which

$$\lambda_1 = 1, \quad \lambda_2 = 0.92, \quad \mathbf{v}_1 = \begin{bmatrix} 3 \\ 5 \end{bmatrix}, \quad \mathbf{v}_2 = \begin{bmatrix} 1 \\ -1 \end{bmatrix}, \quad \text{and} \quad \mathbf{x}_0 = \begin{bmatrix} 40000 \\ 60000 \end{bmatrix}$$

so, $t_1 = 12500$ and $t_2 = 2500$ [see (6.27)]. Then (6.29) becomes:

$$\mathbf{x}_k \approx t_1(1)^k\mathbf{v}_1 = 12500\begin{bmatrix} 3 \\ 5 \end{bmatrix} = \begin{bmatrix} 37500 \\ 62500 \end{bmatrix} \tag{6.31}$$

The result in (6.31) indicates that, as k get large, there will be approximately 37,500 people living in the city and 62,500 people living in the suburbs. Also, from (6.30),

$$\mathbf{x}_{k+1} \approx \lambda_1\mathbf{x}_k = (1)\begin{bmatrix} 37500 \\ 62500 \end{bmatrix}$$

Determining When There Is an Eigenvector Basis

The foregoing analysis is based on the assumption that there is a basis for R^n consisting of n eigenvectors of A. Unfortunately, this is not the case for every $(n \times n)$ matrix A, as you are asked to show in Exercise 13. However, the following theorem shows that the eigenvectors corresponding to *distinct* eigenvalues of a matrix A are linearly independent.

THEOREM 6.2

If $\mathbf{v}_1, \ldots, \mathbf{v}_r$ are eigenvectors corresponding to distinct eigenvalues $\lambda_1, \ldots, \lambda_r$ of a matrix A, then $\mathbf{v}_1, \ldots, \mathbf{v}_r$ are linearly independent.

Proof.

The method for proving this theorem uses the following new technique.

When trying to prove that the statement *p implies q* is true, you assume that the hypothesis *p* is true. Your objective is to establish that *q* is true. With the **contradiction method**, you do so by ruling out the possibility that *q* is false. This is accomplished by assuming that *q* is false and then, using this assumption and the hypothesis *p*, obtaining a contradiction in the form of a statement that you know is false and cannot happen. You should consider using the contradiction method when the conclusion *q* contains the key words "no" or "not." (See Appendix A.9 for additional details.)

The contradiction method is illustrated now with the current theorem, whose conclusion can be reworded as "... then $\mathbf{v}_1, \ldots, \mathbf{v}_r$ are *not* linearly dependent." You should therefore assume that the hypothesis is true, that is, that $\mathbf{v}_1, \ldots, \mathbf{v}_r$ are eigenvectors corresponding to distinct eigenvalues $\lambda_1, \ldots, \lambda_r$ of an $(n \times n)$ matrix A. Also assume that the conclusion is not true, that is, that $\mathbf{v}_1, \ldots, \mathbf{v}_r$ *are* linearly dependent.

The objective now is to use these two assumptions to arrive at a statement contradicting something that you know is true. One of the disadvantages of the contradiction method is that you do not know, beforehand, what the contradiction will be. Finding a contradiction takes creativity, persistence, and sometimes luck. In the current example, the contradiction is that one of the vectors $\mathbf{v}_i$ is $\mathbf{0}$. This cannot happen because $\mathbf{v}_i$ is an eigenvector, which, by definition, is a nonzero vector. To reach this contradiction requires the following creative steps.

From the assumption that $\mathbf{v}_1, \ldots, \mathbf{v}_r$ are linearly dependent, find the first subscript p such that $\mathbf{v}_1, \ldots, \mathbf{v}_p$ are linearly independent, but $\mathbf{v}_1, \ldots, \mathbf{v}_p, \mathbf{v}_{p+1}$ are not. Then, $\mathbf{v}_{p+1}$ is a linear combination of the vectors $\mathbf{v}_1, \ldots, \mathbf{v}_p$, so, there are real numbers $t_1, \ldots, t_p$, such that

$$\mathbf{v}_{p+1} = t_1\mathbf{v}_1 + \cdots + t_p\mathbf{v}_p \tag{6.32}$$

Multiplying both sides of (6.32) on the left by A, you obtain:

$$A\mathbf{v}_{p+1} = t_1 A\mathbf{v}_1 + \cdots + t_p A\mathbf{v}_p \tag{6.33}$$

Using the fact that $A\mathbf{v}_i = \lambda_i\mathbf{v}_i$ for each i, (6.33) becomes

$$\lambda_{p+1}\mathbf{v}_{p+1} = t_1\lambda_1\mathbf{v}_1 + \cdots + t_p\lambda_p\mathbf{v}_p \tag{6.34}$$

Now multiply (6.32) through by λ_{p+1} and subtract the result from (6.34) to obtain

$$\mathbf{0} = t_1(\lambda_1 - \lambda_{p+1})\mathbf{v}_1 + \cdots + t_p(\lambda_p - \lambda_{p+1})\mathbf{v}_p \tag{6.35}$$

Because $\mathbf{v}_1, \ldots, \mathbf{v}_p$ are linearly independent, it must be the case that all the numbers $t_i(\lambda_i - \lambda_{p+1})$ in (6.35) are 0. However, because the eigenvalues are all distinct, as stated in the hypothesis of the theorem, it must also be the case that $t_1 = t_2 = \cdots = t_p = 0$. But then, (6.32) leads to the desired contradiction that $\mathbf{v}_{p+1} = \mathbf{0}$, thus completing the proof. ∎

Theorem 6.2 says that the eigenvectors corresponding to distinct eigenvalues of a matrix A are linearly independent. Thus, if the n eigenvalues of an $(n \times n)$ matrix A are distinct, any n corresponding eigenvectors form a basis for R^n. This is the case with the (2×2) matrix A in the migration problem, whose two distinct eigenvalues are 1 and 0.92.

6.3.4   A Visualization of a Dynamical System

You can gain further insight into the behavior of a dynamical system when the associated matrix, A, is (2×2). This is accomplished by plotting, for various initial vectors $\mathbf{x}_0$, the **trajectory**. That is, by plotting the sequence of points $\mathbf{X} = (\mathbf{x}_0, \mathbf{x}_1, \mathbf{x}_2, \ldots)$ obtained from the linear difference equation.

An Attractor of the System

To illustrate, consider a dynamical system whose associated A matrix is

$$A = \begin{bmatrix} 0.8 & 0.0 \\ 0.0 & 0.5 \end{bmatrix}$$

The eigenvalues of A are the diagonal elements 0.8 and 0.5, and two linearly independent eigenvectors are

$$\mathbf{v}_1 = \begin{bmatrix} 1 \\ 0 \end{bmatrix} \quad \text{and} \quad \mathbf{v}_2 = \begin{bmatrix} 0 \\ 1 \end{bmatrix}$$

For any initial state $\mathbf{x}_0$, there are real numbers t_1 and t_2 such that $\mathbf{x}_0 = t_1 \mathbf{v}_1 + t_2 \mathbf{v}_2$ and, according to the analysis in Section 6.3.3,

$$\mathbf{x}_k = t_1 (0.8)^k \mathbf{v}_1 + t_2 (0.5)^k \mathbf{v}_2, \quad \text{for } k = 0, 1, 2, \ldots \tag{6.36}$$

Because both eigenvalues are less than 1 in absolute value, from (6.36) you can see that $\mathbf{x}_k$ approaches $\mathbf{0}$ as $k \to \infty$, regardless of the value of $\mathbf{x}_0$. For this reason, the origin $\mathbf{0}$ is called an **attractor** of the dynamical system.

Each initial vector $\mathbf{x}_0$ gives rise to a sequence $\mathbf{X} = (\mathbf{x}_0, \mathbf{x}_1, \mathbf{x}_2, \ldots)$ that approaches $\mathbf{0}$. What is interesting is the specific way in which these sequences approach $\mathbf{0}$. Several such trajectories are shown in Figure 6.7 corresponding to different initial vectors

$$\mathbf{x}_0 = \begin{bmatrix} t_1 \\ t_2 \end{bmatrix}$$

in which $t_1 = \pm 3$ and/or $t_2 = \pm 3$. For each initial vector, the points in the sequence $\mathbf{X} = (\mathbf{x}_0, \mathbf{x}_1, \mathbf{x}_2, \ldots)$ are connected together to show the corresponding trajectory more easily.

A Repellor of the System

To illustrate another type of trajectory, consider a dynamical system whose associated A matrix is

$$A = \begin{bmatrix} 1.5 & 0.0 \\ 0.0 & 1.8 \end{bmatrix}$$

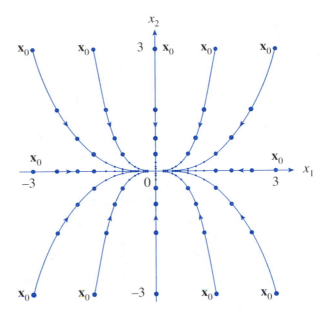

Figure 6.7 *Trajectories When the Origin is an Attractor.*

The eigenvalues of A are the diagonal elements 1.5 and 1.8, and two linearly independent eigenvectors are

$$\mathbf{v}_1 = \begin{bmatrix} 1 \\ 0 \end{bmatrix} \quad \text{and} \quad \mathbf{v}_2 = \begin{bmatrix} 0 \\ 1 \end{bmatrix}.$$

For any initial state $\mathbf{x}_0$, there are real numbers t_1 and t_2 such that $\mathbf{x}_0 = t_1\mathbf{v}_1 + t_2\mathbf{v}_2$ and, according to the analysis in Section 6.3.3,

$$\mathbf{x}_k = t_1(1.5)^k\mathbf{v}_1 + t_2(1.8)^k\mathbf{v}_2, \quad \text{for } k = 0, 1, 2, \dots \tag{6.37}$$

As long as either t_1 or t_2 is not zero, that is, as long as the initial state of the system is not the origin, the sequence $\mathbf{X} = (\mathbf{x}_0, \mathbf{x}_1, \mathbf{x}_2, \dots)$ diverges because one or more components of $\mathbf{x}_k$ approaches $\pm\infty$. For this reason, the origin is called a **repellor**. Several trajectories from (6.37) are shown in Figure 6.8.

A Saddle Point of the System

To illustrate yet another type of trajectory, consider a dynamical system whose associated A matrix is

$$A = \begin{bmatrix} 1.5 & 0.0 \\ 0.0 & 0.5 \end{bmatrix}$$

The eigenvalues of A are the diagonal elements 1.5 and 0.5, and two linearly independent

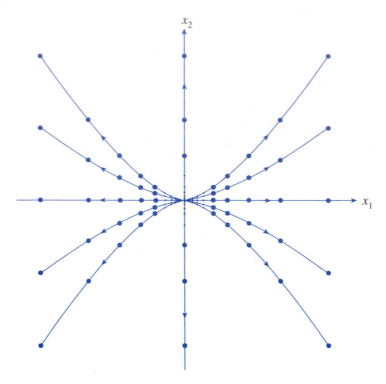

Figure 6.8 *Trajectories When the Origin is a Repellor.*

eigenvectors are

$$\mathbf{v}_1 = \begin{bmatrix} 1 \\ 0 \end{bmatrix} \quad \text{and} \quad \mathbf{v}_2 = \begin{bmatrix} 0 \\ 1 \end{bmatrix}$$

For any initial state $\mathbf{x}_0$, there are real numbers t_1 and t_2 such that $\mathbf{x}_0 = t_1\mathbf{v}_1 + t_2\mathbf{v}_2$ and, according to the analysis in Section 6.3.3,

$$\mathbf{x}_k = t_1(1.5)^k\mathbf{v}_1 + t_2(0.5)^k\mathbf{v}_2, \quad \text{for } k = 0, 1, 2, \dots \tag{6.38}$$

The analysis of the behavior of the system in this case depends on the values of t_1 and t_2, that is, on the initial state $\mathbf{x}_0$.

If $t_1 = t_2 = 0$, then (6.38) indicates that $\mathbf{x}_k = \mathbf{0}$, for $k = 0, 1, 2, \dots$. In other words, if the system starts at the origin, then the system remains at the origin forever.

If $t_1 \neq 0$ and $t_2 = 0$, then (6.38) becomes

$$\mathbf{x}_k = t_1(1.5)^k\mathbf{v}_1 = t_1(1.5)^k \begin{bmatrix} 1 \\ 0 \end{bmatrix} = \begin{bmatrix} t_1(1.5)^k \\ 0 \end{bmatrix}, \quad \text{for } k = 0, 1, 2, \dots$$

Now $t_1 \neq 0$, so, as $k \to \infty$, $t_1(1.5)^k$ approaches ∞ if $t_1 > 0$ and $t_1(1.5)^k$ approaches $-\infty$ if $t_1 < 0$. In other words, as shown in Figure 6.9, if the system starts on the x_1-axis not at the origin ($t_1 \neq 0$ and $t_2 = 0$), then the system stays on the x_1-axis and approaches $\pm\infty$.

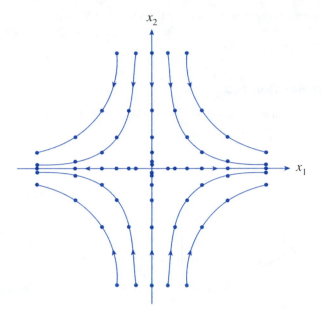

Figure 6.9 *Trajectories When the Origin is a Saddle Point.*

In contrast, if $t_1 = 0$ and $t_2 \neq 0$, then (6.38) becomes

$$\mathbf{x}_k = t_2 (0.5)^k \mathbf{v}_2 = t_2 (0.5)^k \begin{bmatrix} 0 \\ 1 \end{bmatrix} = \begin{bmatrix} 0 \\ t_2 (0.5)^k \end{bmatrix}, \qquad \text{for } k = 0, 1, 2, \dots$$

Now $t_2 \neq 0$, so $t_2 (0.5)^k$ approaches 0 as $k \to \infty$. In other words, as shown in Figure 6.9, if the system starts on the x_2-axis not at the origin ($t_1 = 0$ and $t_2 \neq 0$), then the system stays on the x_2-axis and approaches the origin.

Finally, if $t_1 \neq 0$ and $t_2 \neq 0$, then (6.38) is

$$\mathbf{x}_k = \begin{bmatrix} t_1 (1.5)^k \\ t_2 (0.5)^k \end{bmatrix}$$

Now $t_1 (1.5)^k$ approaches $\pm\infty$ as $k \to \infty$, so the first component of $\mathbf{x}_k$ diverges. In contrast, $t_2 (0.5)^k$ approaches 0 as $k \to \infty$, so the second component of $\mathbf{x}_k$ approaches 0 as $k \to \infty$. In other words, if the system starts at an initial point $\mathbf{x}_0$ that is not on the axes, then the first component of $\mathbf{x}_k$ diverges to $\pm\infty$ and the second component approaches 0, as shown in Figure 6.9. In this case, the origin is said to be a **saddle point**.

The trajectories in Figure 6.7 through Figure 6.9 are based on a diagonal matrix, A, that is, a matrix in which all elements not on the diagonal are 0. When a (2×2) matrix A does not have this property, the material in Section 6.4 is needed to provide a visualization of the trajectories.

In spite of the closed-form analysis presented in this section, it may be the case that the only way to analyze a dynamical system is to use the initial state, $\mathbf{x}_0$, and the linear difference equation to compute the sequence $X = (x_0, x_1, x_2, \ldots)$ explicitly. Even in this case, eigenvectors and eigenvalues can reduce greatly the computational effort involved, as explained in the next section.

Exercises for Section 6.3

1. Determine if the following stochastic matrix is regular:

$$\begin{bmatrix} 0.7 & 0 \\ 0.3 & 1 \end{bmatrix}$$

2. Determine if the following stochastic matrix is regular:

$$\begin{bmatrix} 0.7 & 0.2 & 0.0 \\ 0.3 & 0.5 & 0.6 \\ 0.0 & 0.5 & 0.4 \end{bmatrix}$$

3. Explain what Theorem 6.1 in Section 6.3.1 is saying about the eigenvalues and eigenvectors of the matrix A.

4. A student said that Theorem 6.1 in Section 6.3.1 is not very interesting because $x = 0$ always satisfies $Ax = x$. Explain what is wrong with this observation.

5. Find the steady-state vector for the Markov chain in Exercise 1 of Section 6.2 and explain the meaning and use of that vector in the context of that problem.

6. Find the steady-state vector for the Markov chain in Exercise 2 of Section 6.2 and explain the meaning and use of that vector in the context of that problem.

7. Consider a dynamical system whose linear difference equation is $x_{k+1} = Ax_k$. Use induction to prove that if the initial state of the system is an eigenvector x_0 corresponding to an eigenvalue λ of A, then, for each $k = 0, 1, 2, \ldots$, $x_k = \lambda^k x_0$.

8. Consider a dynamical system whose linear difference equation is $x_{k+1} = Ax_k$ and let $v_1, \ldots, v_n$ be a basis for R^n in which each v_i is an eigenvector corresponding to an eigenvalue λ_i of A. Also, let $t_1, \ldots, t_n$ be real numbers for which $x_0 = t_1 v_1 + \cdots + t_n v_n$. Use induction to prove that for each $k = 0, 1, 2, \ldots$, $x_k = t_1 (\lambda_1)^k v_1 + \cdots + t_n (\lambda_n)^k v_n$.

9. Suppose that the initial state of a dynamical system is an eigenvector x_0 corresponding to an eigenvalue $\lambda < 0$ for the matrix A in the linear difference equation. Perform a closed-form analysis on the sequence $X = (x_0, x_1, x_2, \ldots)$, as in Section 6.3.2. Include a figure similar to Figure 6.6.

10. Let v_1 and v_2 be two linearly independent eigenvectors corresponding to the two eigenvalues λ_1 and λ_2 of the (2×2) matrix A for the linear difference equation of a dynamical system. Suppose that t_1 and t_2 are real numbers for which the initial state $x_0 = t_1 v_1 + t_2 v_2$. From Exercise 8, you know that

$$x_k = t_1 (\lambda_1)^k v_1 + t_2 (\lambda_2)^k v_2$$

Perform an analysis of what can possibly happen to (x_k) as $k \to \infty$, assuming that $\lambda_2 > 1$. (Hint: Begin by considering what happens when $t_2 \neq 0$ and then when $t_2 = 0$.)

11. For each of the following matrices in the linear difference equation of a dynamical system, determine if the origin is an attractor, a repellor, or a saddle point.

    (a) $A = \begin{bmatrix} 1.1 & 0.0 \\ 0.0 & 0.5 \end{bmatrix}$ 　　(b) $A = \begin{bmatrix} 1.1 & 0.0 \\ 0.0 & 2.5 \end{bmatrix}$ 　　(c) $A = \begin{bmatrix} 0.1 & 0.0 \\ 0.0 & 0.5 \end{bmatrix}$

12. Analyze the trajectories of a dynamical system whose linear difference equation is described by the following matrix:

    $$A = \begin{bmatrix} 1 & 0.0 \\ 0 & 0.5 \end{bmatrix}$$

    Draw a graph that shows several sample trajectories.

13. Show that the eigenvectors of a matrix A do not always constitute a basis for R^n by creating a (2×2) matrix A that has only one linearly independent eigenvector.

14. Show that even if a matrix does not have distinct eigenvalues, it may still be possible to find a basis of eigenvectors by constructing a (2×2) matrix whose two eigenvalues are the same but for which there are two linearly independent eigenvectors.

15. Construct a (2×2) matrix that has two distinct eigenvalues but for which there is only one linearly independent eigenvector, or explain why you cannot do so.

16. Use the contradiction method to show that if each component of the n-vector $\mathbf{x} = (x_1, \ldots, x_n)$ is ≥ 0 and $x_1 + \cdots + x_n = 0$, then $\mathbf{x} = \mathbf{0}$. State clearly what you are assuming and what your contradiction is.

6.4 Diagonalization

In the event that the closed-form analysis in Section 6.3 is not applicable, one way to solve a discrete dynamical system is to use the initial state, $\mathbf{x}_0$, and the linear difference equation $\mathbf{x}_{k+1} = A\mathbf{x}_k$ to compute the sequence $\mathbf{X} = (\mathbf{x}_0, \mathbf{x}_1, \mathbf{x}_2, \ldots)$ explicitly. To illustrate, recall the dynamical system in which the two components of

$$\mathbf{x}_k = \begin{bmatrix} a_k \\ b_k \end{bmatrix}$$

represent the number of people living in the city and in the suburbs at the beginning of year k. The associated linear difference equation is

$$\mathbf{x}_{k+1} = A\mathbf{x}_k = \begin{bmatrix} 0.95 & 0.03 \\ 0.05 & 0.97 \end{bmatrix} \begin{bmatrix} a_k \\ b_k \end{bmatrix}$$

Assuming that there are 40,000 people living in the city and 60,000 living in the suburbs at the beginning of year 0, Table 6.2 shows the number of people living in the city and suburbs at the beginning of years 10, 20, 30, 40, and 50, obtained by computing $\mathbf{x}_1, \ldots, \mathbf{x}_{50}$.

　　Here you see again the two drawbacks to this approach, namely, (1) the effort needed to compute $\mathbf{x}_k$ for large values of k and (2) that you might not be able to "see what is happening" to the sequence $\mathbf{X} = (\mathbf{x}_0, \mathbf{x}_1, \mathbf{x}_2, \ldots)$. In this section, you will learn how and when eigenvalues and eigenvectors can reduce this computational effort significantly.

	$\mathbf{x}_0$	$\mathbf{x}_{10}$	$\mathbf{x}_{20}$	$\mathbf{x}_{30}$	$\mathbf{x}_{40}$	$\mathbf{x}_{50}$
City	40000	38586	37972	37705	37589	37539
Suburbs	60000	61414	62028	62295	62411	62461

Table 6.2 *Computing* $\mathbf{x}_{10}, \mathbf{x}_{20}, \mathbf{x}_{30}, \mathbf{x}_{40}, \mathbf{x}_{50}$ *for the Population-Migration Problem.*

6.4.1 The Advantage of a Diagonal Matrix

Recall, from Section 2.4.2, that a diagonal matrix is an $(n \times n)$ matrix D in which all elements, except possibly the diagonal elements, D_{ii}, are 0. In the event that the matrix A in the linear difference equation of a dynamical system is a diagonal matrix, the effort required to compute the sequence $\mathbf{X} = (\mathbf{x}_0, \mathbf{x}_1, \mathbf{x}_2, \ldots)$ is reduced substantially because, as you are asked to show in Exercise 7,

$$
\mathbf{x}_k = A^k \mathbf{x}_0 = \begin{bmatrix} (A_{11})^k & 0 & \cdots & 0 \\ 0 & (A_{22})^k & \cdots & 0 \\ \vdots & \vdots & \ddots & \vdots \\ 0 & 0 & \cdots & (A_{nn})^k \end{bmatrix} \mathbf{x}_0 \tag{6.39}
$$

In other words, when A is a diagonal matrix, you can find the value of component i of $\mathbf{x}_k$ by multiplying $(A_{ii})^k$ by component i of $\mathbf{x}_0$.

To illustrate, consider a dynamical system in which

$$
\mathbf{x}_0 = \begin{bmatrix} 12500 \\ 2500 \end{bmatrix} \quad \text{and} \quad A = \begin{bmatrix} 1 & 0 \\ 0 & 0.92 \end{bmatrix}
$$

Then, to compute $\mathbf{x}_{75}$, for example, you have from (6.39) that

$$
\mathbf{x}_{75} = A^{75}\mathbf{x}_0 = \begin{bmatrix} (1)^{75} & 0 \\ 0 & (0.92)^{75} \end{bmatrix} \begin{bmatrix} 12500 \\ 2500 \end{bmatrix} = \begin{bmatrix} 12500 \\ 5 \end{bmatrix} \tag{6.40}
$$

Unfortunately, the matrix A in a dynamical system need not be—and generally is not—diagonal. In such cases, mathematicians have discovered that, under certain circumstances, it is possible to use the eigenvalues and eigenvectors of A to rewrite A in a form that has the computational advantages of a diagonal matrix, as is now explained.

6.4.2 A Diagonalizable Matrix

The benefits of a diagonal matrix in the foregoing computation of the sequence $\mathbf{x}_k$ are still realized when A is *similar* to a diagonal matrix D. Recall, from Definition 5.5 in Section 5.4.3, that A is similar to D if and only if there is an invertible $(n \times n)$ matrix P such that

$$
A = PDP^{-1} \tag{6.41}
$$

In the event that there is a diagonal matrix D such that A is similar to D, A is said to be **diagonalizable**.

A diagonalizable matrix has advantages comparable to those of a diagonal matrix when computing the sequence $\mathbf{X} = (\mathbf{x}_0, \mathbf{x}_1, \mathbf{x}_2, \ldots)$ for a dynamical system because, from (6.41),

$$\mathbf{x}_1 = A\mathbf{x}_0 = (PDP^{-1})\mathbf{x}_0$$

$$\mathbf{x}_2 = A\mathbf{x}_1 = (PD\underbrace{P^{-1})(P}_{I}DP^{-1}\mathbf{x}_0) = (PD^2P^{-1})\mathbf{x}_0$$

$$\mathbf{x}_3 = A\mathbf{x}_2 = (PD\underbrace{P^{-1})(P}_{I}D^2P^{-1})\mathbf{x}_0 = (PD^3P^{-1})\mathbf{x}_0$$

$$\vdots$$

As you are asked to show in Exercise 8,

$$\mathbf{x}_k = (PD^kP^{-1})\mathbf{x}_0, \quad \text{for } k = 0, 1, 2, \ldots \tag{6.42}$$

The use of (6.42) is now illustrated with the dynamical system for representing population migration. The associated matrix A is diagonalizable because, as you can verify,

$$A = \underbrace{\begin{bmatrix} 0.95 & 0.03 \\ 0.05 & 0.97 \end{bmatrix}}_{} = \underbrace{\begin{bmatrix} 3 & 1 \\ 5 & -1 \end{bmatrix}}_{P} \underbrace{\begin{bmatrix} 1 & 0 \\ 0 & 0.92 \end{bmatrix}}_{D} \underbrace{\begin{bmatrix} 0.125 & 0.125 \\ 0.625 & -0.375 \end{bmatrix}}_{P^{-1}}$$

(You will learn how to find such a matrix P in Section 6.4.3.) Therefore, you can compute $\mathbf{x}_k$, for any desired value of k, using (6.42). For example, with the initial state in which there are 40,000 people living in the city and 60,000 in the suburbs you have that

$$\mathbf{x}_{50} = (PD^{50}P^{-1})\mathbf{x}_0 = P(D^{50}(P^{-1}\mathbf{x}_0))$$

$$= \begin{bmatrix} 3 & 1 \\ 5 & -1 \end{bmatrix} \left(\begin{bmatrix} 1 & 0 \\ 0 & 0.92 \end{bmatrix}^{50} \left(\begin{bmatrix} 0.125 & 0.125 \\ 0.625 & -0.375 \end{bmatrix} \begin{bmatrix} 40000 \\ 60000 \end{bmatrix} \right) \right)$$

$$= \begin{bmatrix} 3 & 1 \\ 5 & -1 \end{bmatrix} \left(\begin{bmatrix} (1)^{50} & 0 \\ 0 & (0.92)^{50} \end{bmatrix} \begin{bmatrix} 12500 \\ 2500 \end{bmatrix} \right) = \begin{bmatrix} 37539 \\ 62461 \end{bmatrix}$$

Observe that this result agrees with that in the final column of Table 6.2 yet requires substantially less computational effort to obtain.

You can use (6.42) to compute $\mathbf{x}_k$ for larger values of k and see what is happening to the sequence $\mathbf{X} = (\mathbf{x}_0, \mathbf{x}_1, \mathbf{x}_2, \ldots)$. For example, $\mathbf{x}_{75} = (PD^{75}P^{-1})\mathbf{x}_0 = P(D^{75}(P^{-1}\mathbf{x}_0))$.

The value of $(D^{75}(P^{-1}\mathbf{x}_0))$ is given in (6.40), so

$$\mathbf{x}_{75} = P(D^{75}(P^{-1}\mathbf{x}_0)) = \begin{bmatrix} 3 & 1 \\ 5 & -1 \end{bmatrix} \begin{bmatrix} 12500 \\ 5 \end{bmatrix} = \begin{bmatrix} 37505 \\ 62495 \end{bmatrix}$$

(The initial vector in (6.40) is equal to $P^{-1}\mathbf{x}_0$ in the current example.)

Now you can be more confident that, in the long run, there will be about 37,500 people living in the city and 62,500 people living in the suburbs. In fact, you might recall that the matrix A in this problem is a regular stochastic matrix. You can therefore use Theorem 6.1 in Section 6.3.1, which states that there is a steady-state probability vector associated with A whose value is computed by solving the following system of linear equations:

$$(A - I)\mathbf{x} = \mathbf{0}$$

$$\mathbf{ex} = 1 \quad \text{where } \mathbf{e} = (1, \ldots, 1)$$

Doing so for the matrix A in this example yields the steady-state vector $\mathbf{x} = (0.375, 0.625)$. Because there are a total of 100,000 people living in the city and suburbs, this steady-state vectors indicates that, in the long run, there will be $0.375(100000) = 37,500$ people living in the city and $0.625(100000) = 62,500$ people living in the suburbs. These figures confirm the numerical computations presented in this section.

6.4.3 Diagonalizing a Matrix

You now know the computational value of diagonalizing A when solving a dynamical system numerically. The natural question to ask is, "How do you determine if a matrix A is diagonalizable and, if A is diagonalizable, how do you find P and D?" It is here that the eigenvalues and eigenvectors of A play a role. This use of eigenvalues and eigenvectors is shown in the following theorem.

THEOREM 6.3

An $(n \times n)$ matrix A is diagonalizable if and only if A has n linearly independent eigenvectors. If A is diagonalizable with $A = PDP^{-1}$, where D is a diagonal matrix, then the diagonal elements of D are the n eigenvalues of A and the columns of P are n linearly independent eigenvectors of A corresponding to the n eigenvalues.

Proof.

Assume first that A is diagonalizable, then, by definition, there is a diagonal matrix D and an invertible matrix P such that $A = PDP^{-1}$, or, equivalently, after multiplying both sides on the right by P,

$$AP = PD \tag{6.43}$$

It is necessary to show that A has n linearly independent eigenvectors. This is accomplished by showing that each of the n columns of P, which are linearly independent because P is invertible, is an eigenvector of A. To that end, use the choose method to choose a column j of P. Let that column of P be denoted by the vector $\mathbf{v}_j$ and let the diagonal elements of D be $\lambda_1, \ldots, \lambda_n$ (which, as yet, are not known to be eigenvalues of A). Then column j of the left side of (6.43) is

$$(AP)_{*j} = AP_{*j} = A\mathbf{v}_j \tag{6.44}$$

Column j of the right side of (6.43) is

$$(PD)_{*j} = PD_{*j} = [\mathbf{v}_1 \cdots \mathbf{v}_j \cdots \mathbf{v}_n] \begin{bmatrix} 0 \\ \vdots \\ \lambda_j \\ \vdots \\ 0 \end{bmatrix} = \lambda_j \mathbf{v}_j \tag{6.45}$$

Equating the right sides of (6.44) and (6.45) yields $A\mathbf{v}_j = \lambda_j \mathbf{v}_j$. Noting that $\mathbf{v}_j \neq \mathbf{0}$, you have the desired conclusion that $\mathbf{v}_j$ is an eigenvector of A corresponding to the eigenvalue λ_j. This also establishes the second sentence of the statement of the theorem.

For the converse, suppose A has n linearly independent eigenvectors, say, $\mathbf{v}_1, \ldots, \mathbf{v}_n$, corresponding to the n eigenvalues $\lambda_1, \ldots, \lambda_n$. To show that A is diagonalizable, by definition, it is necessary to show that there is a diagonal matrix D and an invertible matrix P such that $A = PDP^{-1}$. This is accomplished by constructing D as the diagonal matrix whose diagonal elements are the eigenvalues $\lambda_1, \ldots, \lambda_n$. Also, define column j of P to be $\mathbf{v}_j$. Now P is invertible because the columns of P are the linearly independent vectors $\mathbf{v}_1, \ldots, \mathbf{v}_n$. It remains to show that $A = PDP^{-1}$, or, equivalently, that $AP = PD$. To show that these two matrices are equal reduces to showing that column j of AP is equal to column j of PD. But column j of AP is given in (6.44) and column j of PD is given in (6.45). The fact that $A\mathbf{v}_j = \lambda_j \mathbf{v}_j$ follows from the fact that $\mathbf{v}_j$ is an eigenvector of A corresponding to the eigenvalue λ_j. The proof is now complete. ■

Theorem 6.3 says that A is diagonalizable if and only if there are n eigenvectors of A that form a basis for R^n. To use Theorem 6.3, perform the following four steps.

Steps for Attempting to Diagonalize a Matrix A

Step 1. Find the eigenvalues of A by setting the characteristic polynomial equal to 0 and solving. Solutions to this equation are the eigenvalues $\lambda_1, \ldots, \lambda_n$.

Step 2. Attempt to find n linearly independent eigenvectors of A, say, $\mathbf{v}_1, \ldots, \mathbf{v}_n$. If successful, go to Step 3; otherwise, stop. If you can somehow determine that A does not have n linearly independent eigenvectors then, by Theorem 6.3, A is not diagonalizable.

Step 3. Use the linearly independent vectors found in Step 2 to form the columns of the invertible matrix P. The order is unimportant, thus, for example, $P = [\mathbf{v}_1, \ldots, \mathbf{v}_n]$.

Step 4. Construct the diagonal matrix D using the eigenvalues from Step 1 as the diagonal elements of D. Be sure that the eigenvalue D_{jj} corresponds to the eigenvector in column j of P.

The foregoing steps for diagonalizing a matrix are now illustrated with a specific numerical example.

EXAMPLE 6.6 Diagonalizing a Diagonalizable Matrix

Perform these steps to diagonalize the matrix

$$A = \begin{bmatrix} -4 & 2 & -10 \\ -5 & 3 & -10 \\ 1 & -1 & 3 \end{bmatrix}$$

Step 1. Compute the eigenvalues of A by setting the characteristic polynomial, $\det(A - \lambda I)$, equal to 0 and solving for λ. Doing so by hand is time consuming in this example, however, using appropriate technology provides the characteristic polynomial $\det(A - \lambda I) = -(\lambda^3 - 2\lambda^2 - 5\lambda + 6) = -[(\lambda + 2)(\lambda - 3)(\lambda - 1)]$. Setting this polynomial equal to 0 and solving yields $\lambda_1 = -2$, $\lambda_2 = 3$, and $\lambda_3 = 1$.

Step 2. Because the three eigenvalues found in Step 1 are distinct, by Theorem 6.2 in Section 6.3.3, you know that any three corresponding eigenvectors are linearly independent. Three such eigenvectors are

$$\mathbf{v}_1 = \begin{bmatrix} 1 \\ 1 \\ 0 \end{bmatrix}, \quad \mathbf{v}_2 = \begin{bmatrix} 2 \\ 2 \\ -1 \end{bmatrix}, \quad \text{and} \quad \mathbf{v}_3 = \begin{bmatrix} -2 \\ 0 \\ 1 \end{bmatrix}$$

Step 3. Use the vectors in Step 2 to form the matrix P:

$$P = [\mathbf{v}_1 \quad \mathbf{v}_2 \quad \mathbf{v}_3] = \begin{bmatrix} 1 & 2 & -2 \\ 1 & 2 & 0 \\ 0 & -1 & 1 \end{bmatrix}$$

Step 4. Use the eigenvalues in Step 1 to form the diagonal matrix D so that the eigenvector in column j of P corresponds to the eigenvalue D_{jj}:

$$D = \begin{bmatrix} -2 & 0 & 0 \\ 0 & 3 & 0 \\ 0 & 0 & 1 \end{bmatrix}$$

You can verify that $A = PDP^{-1}$, or equivalently, that $AP = PD$.

EXAMPLE 6.7 A Matrix that Is Not Diagonalizable

In trying to diagonalize the matrix

$$A = \begin{bmatrix} 3 & 1 \\ 1 & -1 \end{bmatrix}$$

you find that the eigenvalues of A are $\lambda_1 = 2$ and $\lambda_2 = 2$. As you can verify, one eigenvector corresponding to the eigenvalue 2 is

$$\mathbf{v}_1 = \begin{bmatrix} 1 \\ 1 \end{bmatrix}$$

All other eigenvectors are a multiple of $\mathbf{v}_1$, so A does not have two linearly independent eigenvectors. Hence, by Theorem 6.3, A is not diagonalizable.

The fact that the eigenvalues of the matrix A in Example 6.6 are distinct ensures, by Theorem 6.2, that any three corresponding eigenvectors are linearly independent and hence, by Theorem 6.3, that A is diagonalizable. In contrast, the eigenvalues of the matrix A in Example 6.7 are not distinct. In this example, it is not possible to find the necessary number of linearly independent eigenvectors. However, that need not always be the case, as discussed next.

Diagonalizing Matrices Whose Eigenvalues Are Not Distinct

You can diagonalize a matrix A even when the eigenvalues are not distinct, *provided that you can find n linearly independent eigenvectors*. The approach to doing so is to find a basis of linearly independent vectors for each eigenspace. If, in total, there are n eigenvectors, they are all linearly independent (proof omitted) so, by Theorem 6.3, A is diagonalizable.

To illustrate this favorable situation, consider the matrix

$$A = \begin{bmatrix} 8 & 10 & -10 \\ 0 & -2 & 0 \\ 5 & 5 & -7 \end{bmatrix}$$

To diagonalize A, first find the eigenvalues of A which, in this case, are $\lambda_1 = -2, \lambda_2 = -2,$ $\lambda_3 = 3$.

Had the three eigenvalues been distinct, you could conclude from Theorem 6.2 that any three corresponding eigenvectors are linearly independent and hence, by Theorem 6.3, that A is diagonalizable. The repeated eigenvalue -2, however, makes it necessary to do more work to determine if there are three linearly independent eigenvectors. This is accomplished by using the technique in Example 6.5 in Section 6.1.2 to find the following bases for the two eigenspaces:

$$\text{Basis for the eigenspace for } \lambda = -2: \quad \mathbf{v}_1 = \begin{bmatrix} -1 \\ 1 \\ 0 \end{bmatrix} \quad \text{and} \quad \mathbf{v}_2 = \begin{bmatrix} 0 \\ 1 \\ 1 \end{bmatrix}$$

$$\text{Basis for the eigenspace for } \lambda = 3: \quad \mathbf{v}_3 = \begin{bmatrix} 2 \\ 0 \\ 1 \end{bmatrix}$$

You can verify that $\mathbf{v}_1$, $\mathbf{v}_2$, and $\mathbf{v}_3$ are linearly independent and thus, A is diagonalizable. Specifically, the columns of the invertible matrix P are the eigenvectors $\mathbf{v}_1$, $\mathbf{v}_2$, and $\mathbf{v}_3$:

$$P = [\mathbf{v}_1 \quad \mathbf{v}_2 \quad \mathbf{v}_3] = \begin{bmatrix} -1 & 0 & 2 \\ 1 & 1 & 0 \\ 0 & 1 & 1 \end{bmatrix}$$

The eigenvalues $\lambda_1 = -2$, $\lambda_2 = -2$, and $\lambda_3 = 3$, in that order, give rise to the following diagonal matrix D:

$$D = \begin{bmatrix} -2 & 0 & 0 \\ 0 & -2 & 0 \\ 0 & 0 & 3 \end{bmatrix}$$

You can verify that A is diagonalizable by checking that $AP = PD$.

In this section, you have seen the role of eigenvalues and eigenvectors in solving a dynamical system and in diagonalizing a matrix. You knowledge of the material in this chapter is brought together in Section 6.5.

Exercises for Section 6.4

1. For the matrices

$$A = \begin{bmatrix} 5 & -6 \\ 2 & -4 \end{bmatrix}, \quad D = \begin{bmatrix} 2 & 0 \\ 0 & -1 \end{bmatrix}, \quad \text{and} \quad P = \begin{bmatrix} 2 & 1 \\ 1 & 1 \end{bmatrix}$$

a student said that A is not diagonalizable because $A \neq PDP^{-1}$. Do you agree with this statement? Why or why not? Explain.

2. For the following matrices, verify that $A = PDP^{-1}$ by computing AP and AD. Can you conclude that A is diagonalizable? Why or why not? Explain.

$$A = \begin{bmatrix} 3 & 0 \\ 2 & -1 \end{bmatrix}, \quad D = \begin{bmatrix} 3 & 2 \\ 0 & -1 \end{bmatrix}, \quad \text{and} \quad P = \begin{bmatrix} 2 & 1 \\ 1 & 1 \end{bmatrix}$$

3. For the following matrices D and P, can you conclude that the given matrix A is diagonalizable? Why or why not? Explain.

$$A = \begin{bmatrix} 8 & 5 \\ -10 & -7 \end{bmatrix}, \quad D = \begin{bmatrix} -2 & 0 \\ 0 & 3 \end{bmatrix}, \quad \text{and} \quad P = \begin{bmatrix} 1 & -1 \\ -2 & 1 \end{bmatrix}$$

4. For the following matrices D and P, can you conclude that the given matrix A is diagonalizable? Why or why not? Explain.

$$A = \begin{bmatrix} 2 & 2 \\ 0 & 0 \end{bmatrix}, \quad D = \begin{bmatrix} 2 & 0 \\ 0 & 2 \end{bmatrix}, \quad \text{and} \quad P = \begin{bmatrix} 1 & -1 \\ 1 & -1 \end{bmatrix}$$

5. Verify that the matrices A, P, and D in Example 6.6 satisfy $A = PDP^{-1}$ by computing AP and PD.

6. The eigenvalues of a (4×4) matrix A are $\lambda_1 = 2$, $\lambda_2 = -3$, $\lambda_3 = 4$, and $\lambda_4 = 2$. Is A diagonalizable? Why or why not? Explain?

7. Suppose that D is an $(n \times n)$ diagonal matrix with diagonal elements $D_{11}, \ldots, D_{nn}$ and $\mathbf{x}$ is an n-vector. Use induction to prove that for each $k = 1, 2, \ldots$,

$$D^k \mathbf{x} = \begin{bmatrix} (D_{11})^k x_1 \\ \vdots \\ (D_{nn})^k x_n \end{bmatrix}$$

8. Suppose that the matrix A in a linear difference equation of a dynamical system is diagonalizable and so $A = PDP^{-1}$ for some diagonal matrix D and invertible matrix P. Let $\mathbf{x}_0$ be the initial state of the system and $\mathbf{X} = (\mathbf{x}_0, \mathbf{x}_1, \mathbf{x}_2, \ldots)$ be the sequence obtained from the linear difference equation, that is, $\mathbf{x}_{k+1} = A\mathbf{x}_k$. Use induction to prove that for each $k = 1, 2, \ldots$, $\mathbf{x}_k = (PD^k P^{-1})\mathbf{x}_0$.

9. Use the numerical results in Section 6.1.1 to find the matrices P and D in the diagonalization of the matrix

$$A = \begin{bmatrix} 0 & 2 \\ 3 & -1 \end{bmatrix}$$

10. Suppose that in working with a particular dynamical system, the following diagonalization of the matrix A in the linear difference equation was found:

$$D = \begin{bmatrix} -3 & 0 \\ 0 & 2 \end{bmatrix} \quad \text{and} \quad P = \begin{bmatrix} 2 & 1 \\ 1 & 1 \end{bmatrix}$$

(a) What are the eigenvalues of A?
(b) Identify two eigenvectors of A.
(c) Use P and D to compute $\mathbf{x}_3$ if
(d) What is A?

$$\mathbf{x}_0 = \begin{bmatrix} 1 \\ 2 \end{bmatrix}$$

11. Use the following three linearly independent eigenvectors $\mathbf{v}_1$, $\mathbf{v}_2$, and $\mathbf{v}_3$ to diagonalize the given matrix A by identifying P and D:

$$\mathbf{v}_1 = \begin{bmatrix} -10 \\ 2 \\ 5 \end{bmatrix}, \quad \mathbf{v}_2 = \begin{bmatrix} 0 \\ 1 \\ 0 \end{bmatrix}, \quad \mathbf{v}_3 = \begin{bmatrix} 0 \\ 0 \\ 1 \end{bmatrix}, \quad A = \begin{bmatrix} 2 & 0 & 0 \\ -1 & -3 & 0 \\ 1 & 0 & 4 \end{bmatrix}$$

12. Use the following three linearly independent eigenvectors v_1, v_2, and v_3 to diagonalize the given matrix A by identifying P and D:

$$v_1 = \begin{bmatrix} -1 \\ 2 \\ 2 \end{bmatrix}, \quad v_2 = \begin{bmatrix} 2 \\ -7 \\ 0 \end{bmatrix}, \quad v_3 = \begin{bmatrix} 1 \\ 0 \\ 0 \end{bmatrix}, \quad A = \begin{bmatrix} 4 & 2 & -1 \\ 0 & -3 & 5 \\ 0 & 0 & 2 \end{bmatrix}$$

13. Consider the matrix

$$A = \begin{bmatrix} 4 & 2 & -8 \\ 0 & 6 & 6 \\ 0 & 0 & 4 \end{bmatrix}$$

   (a) Find a basis for the eigenspace corresponding to the eigenvalue 4.
   (b) Find a basis for the eigenspace corresponding to the eigenvalue 6.
   (c) On the basis of your results in parts (a) and (b), explain why A cannot be diagonalized.

14. Diagonalize the matrix

$$A = \begin{bmatrix} 1 & -8 & -2 \\ 0 & 3 & 0 \\ 0 & 0 & 3 \end{bmatrix}$$

   by performing the following steps.
   (a) Find a basis for the eigenspace corresponding to the eigenvalue $\lambda = 1$.
   (b) Find a basis for the eigenspace corresponding to the eigenvalue $\lambda = 3$.
   (c) Use the three linearly independent eigenvectors from parts (a) and (b) to diagonalize A by identifying P and D.

15. Consider a problem involving an $(n \times n)$ matrix A in which you need to compute

$$\sum_{k=1}^{N} A^k$$

   Show how the computational effort is reduced if A is a diagonal matrix.

6.5 Problem Solving with Eigenvalues and Eigenvectors

Your knowledge of eigenvalues and eigenvectors is unified by solving the problem of the National Wildlife Foundation presented at the beginning of this chapter. One of the objectives is to show you how to interpret and explain mathematical statements in the context of a specific problem.

6.5.1 The Problem of the National Wildlife Foundation

The National Wildlife Foundation is trying to determine the importance of the rabbit population to that of the timber wolves, whose primary diet in Yellowstone National Park is the rabbit. Given the number of wolves at the beginning of year k, say, W_k, and the number of hundreds of rabbits at the beginning of that year, say, R_k, historical data indicate that the number of wolves at the beginning of the next year, W_{k+1}, and the number of hundreds of rabbits at the beginning of the next year, R_{k+1}, depend on W_k and R_k according to the following relationships:

$$W_{k+1} = 0.72W_k + 0.2R_k \tag{6.46}$$
$$R_{k+1} = -0.12W_k + 1.1R_k \tag{6.47}$$

As a consultant, you have been asked to determine what will happen to the wolf and rabbit populations over a long period of time, assuming that nothing happens to change the foregoing equations.

You can interpret the coefficient 0.72 in (6.46) as the fraction of wolves that would survive in a year if there were no rabbits ($R_k = 0$). Likewise, you can interpret the coefficient 1.1 in (6.47) to mean that, if there were no wolves ($W_k = 0$), then there would be 1.1 times as many rabbits at the beginning of the next year as there are at the beginning of this year. The coefficient 0.2 in (6.46) is the rate at which the rabbit population contributes to an increase in the wolf population from year to year. The coefficient -0.12 in (6.47) is the rate at which the wolf population contributes to a decrease in the rabbit population from year to year.

In any event, this problem involves the study of the long-term behavior of a discrete dynamical system in which the state of the system at the beginning of year k is the number of wolves and hundreds of rabbits at that point in time. Specifically, for $k = 0, 1, 2, \ldots$, let

$$\mathbf{x}_k = \begin{bmatrix} W_k \\ R_k \end{bmatrix} = \begin{bmatrix} \text{number of wolves at the start of year } k \\ \text{number of hundreds of rabbits at the start of year } k \end{bmatrix}$$

The matrix A in the linear difference equation $\mathbf{x}_{k+1} = A\mathbf{x}_k$ that describes how this system evolves from one year to the next consists of the coefficients on the right sides of (6.46) and (6.47). Specifically,

$$A = \begin{bmatrix} 0.72 & 0.2 \\ -0.12 & 1.1 \end{bmatrix} \tag{6.48}$$

6.5.2 Problem Solution

From the material in Section 6.3, you know that the eigenvalues and eigenvectors of A are used to analyze the long-term behavior of this system. The process of doing so is now described.

Finding the Eigenvalues and Eigenvectors

To find the eigenvalues of A, set the characteristic polynomial, namely, $\det(A - \lambda I)$, equal to zero and solve for λ:

$$\det(A - \lambda I) = \begin{bmatrix} 0.72 - \lambda & 0.2 \\ -0.12 & 1.1 - \lambda \end{bmatrix} = (0.72 - \lambda)(1.1 - \lambda) + 0.2(0.12) = 0$$

Using suitable technology, you obtain the following two eigenvalues:

$$\lambda_1 = 0.8 \quad \text{and} \quad \lambda_2 = 1.02 \tag{6.49}$$

Using the technique in Section 6.1.2 or appropriate technology, you can also identify the following two eigenvectors associated with these two eigenvalues:

$$\mathbf{v}_1 = \begin{bmatrix} 5 \\ 2 \end{bmatrix} \quad \text{and} \quad \mathbf{v}_2 = \begin{bmatrix} 2 \\ 3 \end{bmatrix} \tag{6.50}$$

Analyzing the Long-Term Behavior

Recognizing that the two eigenvalues $\lambda_1 = 0.8$ and $\lambda_2 = 1.02$ are distinct, you can conclude from Theorem 6.2 in Section 6.3.3, that the two eigenvectors in (6.50) are linearly independent. Thus, you can express the initial state $\mathbf{x}_0$ as a linear combination of the two eigenvectors, that is, there are real number t_1 and t_2 such that

$$\mathbf{x}_0 = t_1 \mathbf{v}_1 + t_2 \mathbf{v}_2 \tag{6.51}$$

But what *is* the initial state of this system?

You should now realize, if you have not already done so, that the initial state of this system is not given. Rather than report that you are lacking necessary information, you can provide some useful information—even without knowing the initial state—by performing the type of analysis in Section 6.3.3. In particular, whatever the value of $\mathbf{x}_0$, you have the following closed-form expression for the state of the system at the beginning of each year k:

$$\mathbf{x}_k = t_1 (\lambda_1)^k \mathbf{v}_1 + t_2 (\lambda_2)^k \mathbf{v}_2 = t_1 (0.8)^k \mathbf{v}_1 + t_2 (1.02)^k \mathbf{v}_2 \tag{6.52}$$

Because $(0.8)^k$ approaches 0 as $k \to \infty$, from (6.52) and (6.50) you have that, for large values of k,

$$\mathbf{x}_k \approx t_2 (1.02)^k \mathbf{v}_2 = t_2 (1.02)^k \begin{bmatrix} 2 \\ 3 \end{bmatrix} \tag{6.53}$$

At this point, the analysis depends on the value of t_2.

Analysis for the Case $t_2 = 0$. If $t_2 = 0$, then (6.53) indicates that the system approaches the zero vector. What does this mean in the context of this problem? The answer is that, starting in an initial state in which $t_2 = 0$ results in both the wolves and rabbits eventually dying out. So, how can it happen that the initial state has $t_2 = 0$?

To answer this question, use (6.50) and (6.51) to conclude that when $t_2 = 0$, the initial state of the system is

$$\mathbf{x}_0 = t_1 \mathbf{v}_1 = t_1 \begin{bmatrix} 5 \\ 2 \end{bmatrix}$$

In other words, if, in the initial state, there are 5 wolves to every 2 hundred rabbits, or equivalently, 1 wolf to every 40 rabbits, both the wolf and rabbit populations will eventually die out. In fact, from (6.52), with $t_2 = 0$, each population decreases at the yearly rate of the eigenvalue $\lambda_1 = 0.8$.

Analysis for the Case $t_2 \neq 0$. When $t_2 \neq 0$, from (6.53) you can say that the wolf and rabbit populations eventually grow each year, and, if nothing changes, they will increase indefinitely because $(1.02)^k$ approaches ∞ as $k \to \infty$. You can provide additional information on the rate of growth of these two populations because, from (6.53), you have that

$$\mathbf{x}_{k+1} \approx t_2 (1.02)^{k+1} \mathbf{v}_2 = (1.02) t_2 (1.02)^k \mathbf{v}_2 \approx 1.02 \mathbf{x}_k$$

That is, for large values of k, the populations of wolves and rabbits at the beginning of the following year is approximately 1.02 times those populations at the beginning of the current year. In summary, if the initial number of wolves and rabbits is not in the ratio of 5 to 2 hundred (that is, if $t_2 \neq 0$), then, in the long run, both populations increase indefinitely at the approximate rate of 2% per year.

Because the matrix A is (2×2), you can supplement your presentation to the National Wildlife Foundation by including a graph that shows trajectories for various initial states.

6.5.3 Further Discussion and Analysis

During your presentation, one of the members of the board said that the recent increase in birds of prey has increased competition not only for the rabbits, but also for other food sources for the wolves. She expressed concern that, although the wolves previously accounted for an average decrease of about 12 rabbits per hundred, as indicated by the value -0.12 in (6.47), the wolves might now account for an average decrease of 13, 14, 15, or more rabbits per hundred each year. She asked how this change would affect the long-term balance of the rabbits and wolves.

You could of course use trial-and-error to change the value of A_{21} from its current value of -0.12 to each new value and repeat the analysis in Section 6.5.2. You are asked to pursue this approach in Exercise 5, however, doing so for each new value of A_{21} is time consuming. A better alternative is to perform a mathematical analysis, as follows.

Using Mathematics to Obtain the Solution

The approach is to think of the number A_{21} as a variable, say, c. Thus, the matrix for the linear difference equation becomes:

$$A = \begin{bmatrix} 0.72 & 0.2 \\ c & 1.1 \end{bmatrix} \tag{6.54}$$

You want to determine how the value of c affects the long-term behavior of the system.

In this regard, you know that the eigenvalues of A in (6.54) are important to this analysis. When $c = -0.12$, the two eigenvalues are 0.8 and 1.02, as found in (6.49) in Section 6.5.2. Once you find the eigenvalues of A in (6.54) and two corresponding eigenvectors $\mathbf{v}_1$ and $\mathbf{v}_2$, you can write, as in (6.52) in Section 6.5.2,

$$\mathbf{x}_k = t_1(\lambda_1)^k \mathbf{v}_1 + t_2(\lambda_2)^k \mathbf{v}_2 \tag{6.55}$$

The key concern is the case when both eigenvalues of A are less than 1. For then, from (6.55), $\mathbf{x}_k$ approaches $\mathbf{0}$ as $k \to \infty$. That is, for any initial state, the system approaches the zero vector which, in this case, means that the wolf and rabbit populations both die out. The question, therefore, is, "What values of c cause both eigenvalues of A in (6.54) to be less than 1?"

To answer this question, construct the characteristic equation of A:

$$\det(A - \lambda I) = \det \begin{bmatrix} 0.72 - \lambda & 0.2 \\ c & 1.1 - \lambda \end{bmatrix}$$

$$= (0.72 - \lambda)(1.1 - \lambda) - 0.2c = 0 \tag{6.56}$$

The desired value for c is obtained by replacing λ with the critical value of 1 in (6.56) and solving for c to obtain:

$$c = \frac{(0.72 - 1)(1.1 - 1)}{0.2} = -0.14$$

By substituting $c = -0.14$ in (6.56) and solving for λ by hand, or using appropriate technology, you find that the two eigenvalues of A are $\lambda_1 = 0.82$ and $\lambda_2 = 1$. Any further decrease in the value of c below -0.14 results in both eigenvalues of A being less than 1. For example, if $c = -0.15$, then the two eigenvalues of A are $\lambda_1 = 0.8319$ and $\lambda_2 = 0.9881$. In summary, if the wolves account for an average decrease of more than 14 rabbits per hundred each year, then, in the long run, both the wolf and rabbit populations become extinct.

Limitations of the Model

It is also important to note that the model described by

$$W_{k+1} = 0.72W_k + 0.2R_k \tag{6.57}$$
$$R_{k+1} = -0.12W_k + 1.1R_k \tag{6.58}$$

has some limitations. For example, suppose that $W_0 = 0$ and $R_0 > 0$, meaning that there are no wolves initially but there are rabbits. If (6.57) were valid in this case, then you would be able to conclude that $W_1 = 0.2R_0 > 0$, meaning that there are wolves at the beginning of year 1. It is clearly not possible for there to be wolves at the beginning of year 1 when there are no wolves at the beginning of year 0. In other words, equations (6.57) and (6.58) are not valid for all values of W_0 and R_0.

As another example of the limitation of (6.57) and (6.58), suppose that the ratio of the initial wolf to rabbit populations satisfies the following relationship:

$$\frac{W_0}{R_0} > \frac{110}{12}$$

Then, $12W_0 > 110R_0$, or equivalently, by subtracting $12W_0$ from both sides and dividing through by 100,

$$-0.12W_0 + 1.1R_0 < 0 \tag{6.59}$$

If (6.58) were valid in this case, you would then be able to conclude from (6.59) that $R_1 = -0.12W_0 + 1.1R_0 < 0$, which is clearly impossible.

You now know that (6.57) and (6.58) are valid only for certain values of W_0 and R_0. You should accept conclusions from this model only when the initial conditions are such that (6.57) and (6.58) are valid.

Exercises for Section 6.5

PROJECT 6.1: The Problem of the National Wildlife Foundation

Consider the dynamical system for monitoring the yearly rabbit and wolf populations, in which

$$\mathbf{x}_k = \begin{bmatrix} W_k \\ R_k \end{bmatrix} = \begin{bmatrix} \text{number of wolves at the start of year } k \\ \text{number of hundreds of rabbits at the start of year } k \end{bmatrix}$$

The linear difference equation that describes how these populations change from one year to the next is $\mathbf{x}_{k+1} = A\mathbf{x}_k$, in which

$$A = \begin{bmatrix} 0.72 & 0.2 \\ -0.12 & 1.1 \end{bmatrix}$$

Use a graphing calculator, MATLAB, or other software package, as indicated by your instructor, to perform the computations in Exercise 1 through Exercise 6. Whenever possible, use matrix and vector operations available on your system. Write the answer and the sequence of operations you performed to obtain the answer.

1. Verify that $\lambda_1 = 0.8$ and $\lambda_2 = 1.02$ are eigenvalues of A and that the following are eigenvectors corresponding to these two eigenvalues:

$$\mathbf{v}_1 = \begin{bmatrix} 5 \\ 2 \end{bmatrix} \quad \text{and} \quad \mathbf{v}_2 = \begin{bmatrix} 2 \\ 3 \end{bmatrix}$$

2. Use the linear difference equation to compute and plot the trajectory of how many wolves and rabbits there will be at the beginning of each of the next fifteen years when there are initially 100 wolves and 4000 rabbits. Expain how these results support some of the mathematical analysis in Section 6.5.2.

3. Use the linear difference equation to compute how many wolves and rabbits there will be at the beginning of each of the next five years when there are initially 100 wolves and 8000 rabbits. Expain why the fact that the wolf and rabbit populations are decreasing in each of these years does not contradict the following statement in Section 6.5.2: "In other words, if the initial number of wolves and rabbits is not in the ratio of 5 to 2 hundred (that is, if $t_2 \neq 0$), then, in the long run, both populations increase indefinitely at the approximate rate of 2% per year." Justify your explanation with appropriate computational results and a plot of the trajectory.

4. According to the analysis in Section 6.5.3, when the value of A_{21} is -0.14, the eigenvalues of A are $\lambda_1 = 0.82$ and $\lambda_2 = 1$.
   (a) Find two eigenvectors $\mathbf{v}_1$ and $\mathbf{v}_2$ corresponding to the eigenvalues λ_1 and λ_2.
   (b) Find the values of t_1 and t_2 so that the initial state

$$\mathbf{x}_0 = \begin{bmatrix} 100 \\ 80 \end{bmatrix}$$

   satisfies $\mathbf{x}_0 = t_1 \mathbf{v}_1 + t_2 \mathbf{v}_2$.
   (c) Use an appropriately modified expression as in (6.52) to compute $\mathbf{x}_{15}$.

5. Suppose that there are 100 wolves and 8000 rabbits initially. Compute and display, on three different graphs, the three 15-year trajectories corresponding to the following three values of A_{21}: $-0.13, -0.14, -0.15$. Explain how these computational results support the mathematical analysis in Section 6.5.3 and also in part (c) of the previous exercise.

6. Verify the analysis in Section 6.5.3 by performing the following steps using your technology.
   (a) Create the following *symbolic* matrix in terms of the variables λ and c:

$$A = \begin{bmatrix} 0.72 - \lambda & 0.2 \\ c & 1.1 - \lambda \end{bmatrix}$$

   (b) Set the determinant of A in part (a) equal to 0 and solve for c in terms of λ.
   (c) Set the value of $\lambda = 1$ in part (b) and obtain the desired value for c.

PROJECT 6.2: Studying A Variety of Dynamical Systems
Use a graphing calculator, MATLAB, or other software package, as indicated by your instructor, to perform the computations in Exercise 7 through Exercise 10. Whenever possible, use matrix and vector operations available on your system. Write the answer and the sequence of operations you performed to obtain the answer.

7. For the dynamical system in Exercise 3 in Section 6.2, what share of the market is expected to drink Coke, Pepsi, and other brands of soft drinks in the long run?

8. For the dynamical system in Exercise 4 in Section 6.2, if the annual production of Germany is currently 75 billion dollars, that of Japan is 50 billion dollars, and that of the U.S. is 100 billion dollars, what will these figures be in the long run?

9. Perform the following instructions pertaining to the dynamical system in Exercise 5 in Section 6.2.
   (a) Find the three eigenvalues λ_1, λ_2, and λ_3 and three corresponding eigenvectors $\mathbf{v}_1$, $\mathbf{v}_2$, and $\mathbf{v}_3$ of the matrix A in the linear difference equation.
   (b) Diagonalize A by creating a diagonal matrix D and an invertible matrix P so that $A = PDP^{-1}$.
   (c) Use the diagonalization in part (b) to compute the production, inventory, and demand at the end of the planning horizon, assuming there are 4000 tons of steel in inventory to begin with and that 10,000 tons of steel were produced in the preceding month.
   (d) Confirm the results in part (c) by finding real numbers t_1, t_2, and t_3 so that the initial

state of the system in part (c) satisfies $\mathbf{x}_0 = t_1\mathbf{v}_1 + t_2\mathbf{v}_2 + t_3\mathbf{v}_3$. Then compute the desired value of $\mathbf{x}_k$, as follows:

$$\mathbf{x}_k = (\lambda_1)^k t_1\mathbf{v}_1 + (\lambda_2)^k t_2\mathbf{v}_2 + (\lambda_3)^k t_3\mathbf{v}_3$$

10. Perform the following instructions pertaining to the dynamical system in Exercise 6 in Section 6.2.
    (a) Use the linear difference equation to determine the number of baby, preadult, and adult female condors there are at the beginning of years 10, 20, 30, 40, and 50.
    (b) Confirm the fact that the condor population is becoming extinct by showing that the absolute value of each of the three eigenvalues of the matrix A in the linear difference equation is less than 1. (Note: Two of these eigenvalues are complex numbers. The absolute value of the complex number $a + bi$ is $\sqrt{a^2 + b^2}$.)
    (c) Use trial and error to determine what the survival rate of babies to preadults must be in order for the real eigenvalue of A to exceed 1, and hence for the condors to survive.

Chapter Summary

An eigenvector of an $(n \times n)$ matrix A is a nonzero vector $\mathbf{x}$ for which there is a scalar λ such that $A\mathbf{x} = \lambda\mathbf{x}$. An eigenvalue is a scalar for which there is a nonzero n-vector $\mathbf{x}$ such that $A\mathbf{x} = \lambda\mathbf{x}$. Eigenvectors and eigenvalues are often used to reduce the computational effort needed to solve a problem. The efficiency is a result of the fact that, when $\mathbf{x}$ is an eigenvector corresponding to an eigenvalue λ, you can compute $A\mathbf{x}$ without having to perform matrix multiplication. Rather, you need only compute the value of $\lambda\mathbf{x}$.

A given nonzero n-vector $\mathbf{x}$ is an eigenvector of A if and only if the there is a value of λ such that $A\mathbf{x} = \lambda\mathbf{x}$. To check if a given value of λ is an eigenvalue of A, use Gauss-Jordan elimination to see if there is a nonzero n-vector $\mathbf{x}$ that satisfies the system of linear equations $(A - \lambda I)\mathbf{x} = \mathbf{0}$. If so, then λ is an eigenvalue of A, otherwise λ is not.

One approach to finding all the eigenvalues of A is to solve the characteristic equation $\det(A - \lambda I) = 0$. Doing so requires finding the n (not necessarily distinct) roots of the characteristic polynomial $\det(A - \lambda I)$, some of which may be complex numbers. When it is not possible to find these n roots in closed-form, numerical methods are used to approximate the eigenvalues.

You cannot list all the eigenvectors of a matrix because there are an infinite number of them. However, if you can find distinct eigenvalues $\lambda_1, \ldots, \lambda_r$, then, for each λ_i, you can find a basis of the corresponding eigenspace $\{n\text{-vectors } \mathbf{x} : A\mathbf{x} = \lambda_i\mathbf{x}\}$.

Eigenvalues and eigenvectors are used to study discrete dynamical systems in which the n-vectors $\mathbf{x}_k$ represent the state of the system at time k, for $k = 0, 1, 2, \ldots$. An $(n \times n)$ matrix, A, describes how the system changes from one time period to the next through the linear difference equation

$$\mathbf{x}_{k+1} = A\mathbf{x}_k \tag{6.60}$$

Often, the objective of such problems is to determine the long-term behavior of the system. That is, given the initial state, $\mathbf{x}_0$, you want to know what happens to $\mathbf{x}_k$ as $k \to \infty$. When the dynamical system is a Markov chain whose A matrix is stochastic, as indicated by the fact that the elements of A are nonnegative and each column of A sums to 1, and regular, there is a steady-state probability vector that controls the long-term behavior of the system, regardless of the initial state.

Eigenvalues and eigenvectors also play a significant role in determining the long-term behavior of other dynamical systems. For example, suppose you can find a basis for R^n consisting of n eigenvectors $\mathbf{v}_1, \ldots, \mathbf{v}_n$ corresponding to the eigenvalues $\lambda_1, \ldots, \lambda_n$ of the matrix A in the linear difference equation. In this case, there are real numbers $t_1, \ldots, t_n$ such that the initial state $\mathbf{x}_0 = t_1 \mathbf{v}_1 + \cdots + t_n \mathbf{v}_n$. It then follows from the linear difference equation that

$$\mathbf{x}_k = t_1 (\lambda_1)^k \mathbf{v}_1 + \cdots + t_n (\lambda_n)^k \mathbf{v}_n, \quad \text{for } k = 0, 1, 2, \ldots \tag{6.61}$$

You can now use (6.61) to perform an analysis of what happens to $(\mathbf{x}_k)$ as $k \to \infty$ with respect to the values of $t_1, \ldots, t_n$ and $\lambda_1, \ldots, \lambda_n$.

In the event that this analysis is unsuccessful, you can always compute the sequence $\mathbf{X} = (\mathbf{x}_0, \mathbf{x}_1, \mathbf{x}_2, \ldots)$ explicitly by using the linear difference equation $\mathbf{x}_{k+1} = A\mathbf{x}_k$. The computational effort is reduced significantly when A has n linearly independent eigenvectors $\mathbf{v}_1, \ldots, \mathbf{v}_n$ corresponding to the n eigenvalues $\lambda_1, \ldots, \lambda_n$. In this case, A is diagonalizable. Letting P be the matrix in which column j is $\mathbf{v}_j$ and D be the diagonal matrix in which $D_{jj} = \lambda_j$, it follows that $A = PDP^{-1}$, and so,

$$\mathbf{x}_k = (PD^k P^{-1})\mathbf{x}_0, \quad \text{for } k = 0, 1, 2, \ldots$$

One condition under which you can diagonalize A is when the n eigenvalues of A are all distinct. In this case, any corresponding set of n eigenvectors is linearly independent. When the eigenvalues are not distinct, A is still diagonalizable if you can find n linearly independent eigenvectors. To do so, find bases for the eigenspaces corresponding to the distinct eigenvectors. If these bases provide n vectors in total, then A is diagonalizable.

Mathematical Thinking Processes

In addition to the specific material on eigenvalues and eigenvectors, you have learned the following general mathematical thinking processes.

1. Imposing Solvability Conditions   When you cannot find the solution to a challenging problem, try to find conditions on the problem data under which you *can* solve the problem, either by obtaining a closed-form or a numerical-method solution.

2. Interpreting Mathematical Statements   When reading mathematical statements pertaining to a specific problem, interpret the meaning of those statements in the context of the problem. Likewise, explain the relevance of any mathematical results you obtain.

3. Contradiction   When attempting to prove an implication *p implies q*, consider the contradiction method when the statement *q* contains the keyword *no* or *not*. With this method, you use the assumption that the hypothesis *p* is true and that the conclusion *q* is not true to reach a contradiction to some statement that you know is true. In so doing, you rule out the possibility that *q* is false and thus can conclude that *q*, and hence *p implies q*, is true.

Chapter 7

Orthogonality and Inner Product Spaces

As you learned in Chapter 1, the dot product is an operation that combines two n-vectors to produce a real number. This operation arises in many applications and also provides a useful geometric relationship between the two n-vectors. For example, as stated in Section 1.2.4, two n-vectors are perpendicular, or *orthogonal*, if their dot product is 0. Generalizations of orthogonality and their role in solving problems, such as the following, are the topic of this chapter.

The Production Problem of Philadelphia Paints

Philadelphia Paints sells a variety of different paints. One of their newer products is a latex paint that is produced on a weekly basis. In reviewing available production costs for this paint, the CEO (Chief Executive Officer) noticed that the cost per gallon decreases as the number of gallons produced increases, but only up to a certain point. After this point, the cost begins to rise. She has asked you, the Manager of the Production Department, to use the cost data from the Accounting Department to determine the optimal number of gallons to produce in each batch.

7.1 Orthogonality in R^n

In this section, the concept of orthogonality of two n-vectors is generalized to a group of n-vectors. A review of the dot product and other related operations provides the basis for the subsequent generalization.

7.1.1 A Review of the Dot Product and Related Operations on n-Vectors

Recall, from Section 1.2.4, that the *dot product* of the n-vectors $\mathbf{u} = (u_1, \ldots, u_n)$ and $\mathbf{v} = (v_1, \ldots, v_n)$ is defined as follows:

$$\mathbf{u} \cdot \mathbf{v} = \sum_{i=1}^{n} u_i v_i = u_1 v_1 + \cdots + u_n v_n \quad \text{(dot product)} \tag{7.1}$$

EXAMPLE 7.1 Computing the Dot Product of Two Vectors
The following table shows how to compute the dot product for three different pairs of vectors:

$\mathbf{u}$	$\mathbf{v}$	$u_1 v_1 + \cdots + u_n v_n = \mathbf{u} \cdot \mathbf{v}$
$(1, 2)$	$(0, 3)$	$1(0) + 2(3) = 6$
$(1, 2)$	$(0, -3)$	$1(0) + 2(-3) = -6$
$(1, 2, 0)$	$(-4, 2, -2)$	$1(-4) + 2(2) + 0(-2) = 0$

Numerous uses of the dot product in applications and problem solving are presented beginning in Section 1.2.4 and throughout the rest of this book and are therefore not repeated here.

Recall also that the *length*, also called the *norm*, of an n-vector $\mathbf{u} = (u_1, \ldots, u_n)$ is

$$\|\mathbf{u}\| = \sqrt{u_1^2 + \cdots + u_n^2} \quad \text{(length of } \mathbf{u}) \tag{7.2}$$

EXAMPLE 7.2 Computing the Length of a Vector
If $\mathbf{u} = (1, -2, 3, 0)$, then

$$\|\mathbf{u}\| = \sqrt{1^2 + (-2)^2 + 3^2 + 0^2} = \sqrt{14}$$

One additional use of the norm is to compute the distance, $d(\mathbf{u}, \mathbf{v})$, between two n-vectors $\mathbf{u} = (u_1, \ldots, u_n)$ and $\mathbf{v} = (v_1, \ldots, v_n)$:

$$d(\mathbf{u}, \mathbf{v}) = \|\mathbf{u} - \mathbf{v}\| = \sqrt{(u_1 - v_1)^2 + \cdots + (u_n - v_n)^2} \tag{7.3}$$

EXAMPLE 7.3 Computing the Distance Between Two Vectors

The distance between the two vectors

$$\mathbf{u} = (4, -1, 2) \quad \text{and} \quad \mathbf{v} = (1, 3, 2)$$

is

$$d(\mathbf{u}, \mathbf{v}) = \sqrt{(4-1)^2 + (-1-3)^2 + (2-2)^2} = \sqrt{9 + 16 + 0} = 5$$

A **unit vector** is an n-vector whose length is 1. For example, $\mathbf{u} = (1/\sqrt{2}, 1/\sqrt{2})$ is a unit vector because $\|\mathbf{u}\| = 1$. Given a nonzero vector $\mathbf{v}$, it is possible to create a unit vector that points in the same direction as $\mathbf{v}$. This process of **normalizing** or **scaling** requires multiplying each component of $\mathbf{v}$ by 1 over the norm of $\mathbf{v}$. Specifically, to normalize a nonzero vector $\mathbf{v}$, compute

$$\mathbf{u} = \frac{1}{\|\mathbf{v}\|}\mathbf{v}$$

Now $\mathbf{u}$ points in the same direction as $\mathbf{v}$ and is a unit vector because

$$\|\mathbf{u}\| = \left\|\frac{1}{\|\mathbf{v}\|}\mathbf{v}\right\| = \frac{1}{\|\mathbf{v}\|}\|\mathbf{v}\| = 1$$

EXAMPLE 7.4 Normalizing a Vector

To normalize the vector

$$\mathbf{v} = (4, -3)$$

whose norm is 5, compute

$$\mathbf{u} = \frac{1}{\|\mathbf{v}\|}\mathbf{v} = \frac{1}{5}(4, -3) = \left(\frac{4}{5}, -\frac{3}{5}\right)$$

The vector $\mathbf{u}$ points in the same direction as $\mathbf{v}$ but has length 1, as shown in Figure 7.1.

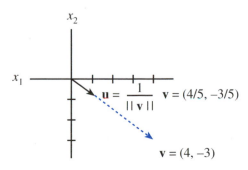

Figure 7.1 *Normalizing the Vector* $\mathbf{v} = (4, -3)$.

Various properties of the dot product and its relation to the length of an n-vector are given in Theorem 1.3 in Section 1.2.5, repeated here for easy reference.

THEOREM 7.1

Suppose that $\mathbf{u}$, $\mathbf{v}$, and $\mathbf{w}$ are n-vectors and that t is a real number. Then the following properties hold:

(a) $\mathbf{u} \cdot \mathbf{u} \geq 0$.

(b) $\mathbf{u} \cdot \mathbf{u} = 0$ if and only if $\mathbf{u} = \mathbf{0}$.

(c) $\|t\mathbf{u}\| = |t|\|\mathbf{u}\|$.

(d) $\mathbf{u} \cdot \mathbf{u} = \|\mathbf{u}\|^2$, that is, $\|\mathbf{u}\| = \sqrt{\mathbf{u} \cdot \mathbf{u}}$.

(e) $\mathbf{u} \cdot \mathbf{v} = \mathbf{v} \cdot \mathbf{u}$.

(f) $\mathbf{u} \cdot (\mathbf{v} + \mathbf{w}) = (\mathbf{u} \cdot \mathbf{v}) + (\mathbf{u} \cdot \mathbf{w})$.

(g) $t(\mathbf{u} \cdot \mathbf{v}) = (t\mathbf{u}) \cdot \mathbf{v} = \mathbf{u} \cdot (t\mathbf{v})$.

7.1.2 An Orthogonal Set of n-Vectors

Two n-vectors are orthogonal if their dot product is 0. A generalization to a collection of n-vectors is obtained by requiring that each pair of distinct vectors in the collection be orthogonal, as stated in the following definition.

DEFINITION 7.1

A set of n-vectors $\{\mathbf{v}_1, \ldots, \mathbf{v}_k\}$ is an **orthogonal set** if and only if

$$\mathbf{v}_i \cdot \mathbf{v}_j = 0, \quad \text{for all } i, j = 1, \ldots, k \text{ with } i \neq j$$

The three standard unit vectors in R^3, shown in Figure 7.2(a), form an orthogonal set. So do the vectors in Figure 7.2(b), as established in the following example.

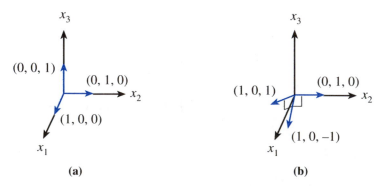

(a)   **(b)**

Figure 7.2 *Sets of Orthogonal Vectors.*

EXAMPLE 7.5 An Orthogonal Set of *n*-Vectors

The set $\{\mathbf{v}_1, \mathbf{v}_2, \mathbf{v}_3\}$ in which

$$\mathbf{v}_1 = (1, 0, -1), \quad \mathbf{v}_2 = (0, 1, 0), \quad \text{and} \quad \mathbf{v}_3 = (1, 0, 1)$$

is an orthogonal set because

$$\mathbf{v}_1 \cdot \mathbf{v}_2 = (1, 0, -1) \cdot (0, 1, 0) = 0$$
$$\mathbf{v}_1 \cdot \mathbf{v}_3 = (1, 0, -1) \cdot (1, 0, 1) = 0$$
$$\mathbf{v}_2 \cdot \mathbf{v}_3 = (0, 1, 0) \cdot (1, 0, 1) = 0$$

Orthogonal Bases

There are often computational advantages to working with an orthogonal set of vectors. For example, determining if an arbitrary collection of *n*-vectors is linear independent usually involves substantial effort. When the vectors are nonzero and orthogonal, however, no work is needed because, as the next theorem shows, those vectors are always linearly independent.

THEOREM 7.2

If $\{\mathbf{v}_1, \ldots, \mathbf{v}_k\}$ is an orthogonal set of nonzero *n*-vectors, then these *n*-vectors are linearly independent.

Proof.

As usual, to show that $\mathbf{v}_1, \ldots, \mathbf{v}_k$ are linearly independent, let $t_1, \ldots, t_k$ be real numbers such that

$$t_1\mathbf{v}_1 + \cdots + t_k\mathbf{v}_k = \mathbf{0} \tag{7.4}$$

According to the definition, it is necessary to show that for each $i = 1, \ldots, k, t_i = 0$.

The key words *for each* suggest choosing an arbitrary integer i with $1 \leq i \leq k$ and showing that $t_i = 0$. To do so, dot product both sides of (7.4) by $\mathbf{v}_i$ to obtain

$$t_1(\mathbf{v}_1 \cdot \mathbf{v}_i) + \cdots + t_i(\mathbf{v}_i \cdot \mathbf{v}_i) + \cdots + t_k(\mathbf{v}_k \cdot \mathbf{v}_i) = \mathbf{0} \cdot \mathbf{v}_i \tag{7.5}$$

The right side of (7.5) is 0 and, because $\{\mathbf{v}_1, \ldots, \mathbf{v}_k\}$ is an orthogonal set, each term on the left of (7.5), except for the term $t_i(\mathbf{v}_i \cdot \mathbf{v}_i)$, is also 0, so (7.5) becomes

$$t_i(\mathbf{v}_i \cdot \mathbf{v}_i) = 0 \tag{7.6}$$

The desired result that $t_i = 0$ follows on dividing (7.6) through by the nonzero number $\mathbf{v}_i \cdot \mathbf{v}_i$ (because $\mathbf{v}_i \neq \mathbf{0}$), thus completing the proof. ∎

Thus, for example, the three orthogonal vectors in Example 7.5 are linearly independent and hence form an **orthogonal basis** for R^3, meaning a basis whose vectors are an orthogonal set.

As another example of the computational advantage of working with a collection of orthogonal *n*-vectors, recall that you are free to choose any basis that is advantageous to solving your problem. For a chosen $(n \times n)$ basis matrix B for R^n, you can find the components of an n-vector $\mathbf{v}$ relative to B, as follows (see Section 5.4.1):

$$[\mathbf{v}]_B = B^{-1}\mathbf{v} \tag{7.7}$$

The computation in (7.7) involves substantial effort when the matrix B is large. If the columns of B form an orthogonal set, then the effort is simplified significantly, as shown in the following theorem.

THEOREM 7.3

Let $\{\mathbf{v}_1, \ldots, \mathbf{v}_k\}$ be an orthogonal basis for a subspace W of R^n. Then for any $\mathbf{v}$ in W, the unique real numbers $t_1, \ldots, t_k$ for which

$$\mathbf{v} = t_1\mathbf{v}_1 + \cdots + t_k\mathbf{v}_k$$

are

$$t_i = \frac{\mathbf{v} \cdot \mathbf{v}_i}{\mathbf{v}_i \cdot \mathbf{v}_i}, \quad \text{for } i = 1, \ldots, k \tag{7.8}$$

Proof.
Because $\{\mathbf{v}_1, \ldots, \mathbf{v}_k\}$ is a basis for W, there are unique real numbers $t_1, \ldots, t_k$ such that

$$\mathbf{v} = t_1\mathbf{v}_1 + \cdots + t_k\mathbf{v}_k \tag{7.9}$$

It remains to show that for each $i = 1, \ldots, k$, t_i is given by (7.8). Because of the key words *for each*, choose an integer i with $1 \le i \le k$ and show that t_i satisfies (7.8). For this chosen value of i, dot product both sides of (7.9) by $\mathbf{v}_i$ to obtain

$$\mathbf{v} \cdot \mathbf{v}_i = t_1(\mathbf{v}_1 \cdot \mathbf{v}_i) + \cdots + t_i(\mathbf{v}_i \cdot \mathbf{v}_i) + \cdots + t_k(\mathbf{v}_k \cdot \mathbf{v}_i) \tag{7.10}$$

Because $\{\mathbf{v}_1, \ldots, \mathbf{v}_k\}$ is an orthogonal basis, each term on the right side of (7.10) is 0, except for the term $t_i(\mathbf{v}_i \cdot \mathbf{v}_i)$, so (7.10) reduces to

$$\mathbf{v} \cdot \mathbf{v}_i = t_i(\mathbf{v}_i \cdot \mathbf{v}_i)$$

You obtain (7.8) after dividing both sides of the foregoing equality by the nonzero number $\mathbf{v}_i \cdot \mathbf{v}_i$ (because $\mathbf{v}_i \neq \mathbf{0}$), thus completing the proof. ∎

Theorem 7.3 says that if you have an orthogonal basis for a subspace W of R^n, then, to find the components of an n-vector $\mathbf{v} \in W$ relative to this basis, you need only compute the real numbers $t_1, \ldots, t_k$ using (7.8). This process is illustrated in the following example.

EXAMPLE 7.6 Finding the Components of a Vector Relative to an Orthogonal Basis
To find the components of the vector

$$\mathbf{v} = (3, -2, -1)$$

relative to the orthogonal basis $B = \{\mathbf{v}_1, \mathbf{v}_2, \mathbf{v}_3\}$ in Example 7.5, use (7.8) to compute

$$t_1 = \frac{\mathbf{v} \cdot \mathbf{v}_1}{\mathbf{v}_1 \cdot \mathbf{v}_1} = \frac{(3, -2, -1) \cdot (1, 0, -1)}{(1, 0, -1) \cdot (1, 0, -1)} = \frac{4}{2} = 2$$

$$t_2 = \frac{\mathbf{v} \cdot \mathbf{v}_2}{\mathbf{v}_2 \cdot \mathbf{v}_2} = \frac{(3, -2, -1) \cdot (0, 1, 0)}{(0, 1, 0) \cdot (0, 1, 0)} = \frac{-2}{1} = -2$$

$$t_3 = \frac{\mathbf{v} \cdot \mathbf{v}_3}{\mathbf{v}_3 \cdot \mathbf{v}_3} = \frac{(3, -2, -1) \cdot (1, 0, 1)}{(1, 0, 1) \cdot (1, 0, 1)} = \frac{2}{2} = 1$$

Thus, the components of $\mathbf{v}$ relative to this basis are given by

$$[\mathbf{v}]_B = \begin{bmatrix} 2 \\ -2 \\ 1 \end{bmatrix}$$

Orthonormal Bases

The effort involved in computing the components of an n-vector in a subspace W relative to a given orthogonal basis is simplified even further if each n-vector in the basis has length 1. The reason for this is that, in this case, each of the denominators $\mathbf{v}_i \cdot \mathbf{v}_i$ in (7.8) is 1 and therefore that computation and the subsequent division need not be performed. Such a basis is called an **orthonormal basis**. The standard basis $\{\mathbf{e}_1, \ldots, \mathbf{e}_n\}$ for R^n is an orthonormal basis, as is the basis in the following example.

EXAMPLE 7.7 An Orthonormal Basis for R^3
The vectors

$$\mathbf{v}_1 = (1/\sqrt{2}, 0, -1/\sqrt{2}), \quad \mathbf{v}_2 = (0, 1, 0), \quad \text{and} \quad \mathbf{v}_3 = (1/\sqrt{2}, 0, 1/\sqrt{2})$$

obtained by normalizing the vectors in the orthogonal basis in Example 7.5 each have length 1 and thus form an orthonormal basis for R^3.

7.1.3 Orthogonal Complements

You know what it means for a given n-vector $\mathbf{v}$ to be orthogonal to another n-vector. A generalization of this concept that is useful in problem solving is when $\mathbf{v}$ is orthogonal to every vector in a subspace W of R^n, as formalized in the following definition.

DEFINITION 7.2

An n-vector $\mathbf{v}$ is **orthogonal to a subspace** W of R^n if and only if $\mathbf{v}$ is orthogonal to every vector in W. The **orthogonal complement of** W, written $W^\perp$ and pronounced "W perp," is the set of all vectors that are orthogonal to W, that is,

$$W^\perp = \{\mathbf{v} \in R^n : \text{for all } \mathbf{w} \in W, \mathbf{v} \cdot \mathbf{w} = 0\}$$

To illustrate the concepts in Definition 7.2, look at the vector $\mathbf{y}$ on the line L perpendicular to the plane P in Figure 7.3. Now $\mathbf{y}$ is orthogonal to the subspace P because $\mathbf{y}$ is orthogonal to every vector in P. Note also that both L and P are subspaces of R^3. In fact,

L is $P^\perp$ because L is the set of vectors that are orthogonal to all vectors in P. Likewise, P is $L^\perp$ because P is the set of vectors that are orthogonal to every vector in L. That is,

$$L = P^\perp \quad \text{and} \quad P = L^\perp$$

Two other useful properties of the orthogonal complement of a subspace are provided in the following theorem.

THEOREM 7.4

For a subspace W of R^n,

(a) $W^\perp$ is a subspace of R^n.

(b) Let $\mathbf{w}_1, \ldots, \mathbf{w}_k$ be vectors that span W. A vector $\mathbf{v}$ is in $W^\perp$ if and only if $\mathbf{v}$ is orthogonal to each vector $\mathbf{w}_i$, for $i = 1, \ldots, k$.

Proof.
The key question associated with (a) is, "How can I show that a set (namely, $W^\perp$) is a subspace of R^n?" Noting that $W^\perp$ is not empty because $\mathbf{0} \in W^\perp$, according to Theorem 4.2 in Section 4.2.2, it suffices to show both of the following:

1. For all $\mathbf{u}, \mathbf{v} \in W^\perp, \mathbf{u} + \mathbf{v} \in W^\perp$.
2. For all real numbers t and vectors $\mathbf{v} \in W^\perp, t\mathbf{v} \in W^\perp$.

To verify (1), the key words *for all* suggest using the choose method to choose n-vectors $\mathbf{u}, \mathbf{v} \in W^\perp$, for which it must be shown that $\mathbf{u} + \mathbf{v} \in W^\perp$. By Definition 7.2, to show that the vector $\mathbf{u} + \mathbf{v}$ is in $W^\perp$, it is necessary to show that

for all n-vectors $\mathbf{w} \in W, \mathbf{w} \cdot (\mathbf{u} + \mathbf{v}) = 0$

Once again, the key words *for all* suggest using the choose method to choose an n-vector $\mathbf{w} \in W$, for which you must show that

$$\mathbf{w} \cdot (\mathbf{u} + \mathbf{v}) = (\mathbf{w} \cdot \mathbf{u}) + (\mathbf{w} \cdot \mathbf{v}) = 0 \tag{7.11}$$

But, using the fact that $\mathbf{u}, \mathbf{v} \in W^\perp$, you know by Definition 7.2 that $\mathbf{u}$ is orthogonal to every vector in W. In particular, by specialization, $\mathbf{w} \cdot \mathbf{u} = \mathbf{w} \cdot \mathbf{v} = 0$. Thus, (7.11) is true and the

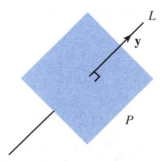

Figure 7.3 *The Orthogonal Complement of a Subspace.*

proof of (1) is complete.

To verify (2), again use the choose method to choose a real number t and a vector $\mathbf{v} \in W^{\perp}$, for which you must show that $t\mathbf{v} \in W^{\perp}$, that is, that

for all $\mathbf{w} \in W$, $(t\mathbf{v}) \cdot \mathbf{w} = 0$

The choose method is used again to choose a vector $\mathbf{w} \in W$. It then follows that $(t\mathbf{v}) \cdot \mathbf{w} = t(\mathbf{v} \cdot \mathbf{w}) = 0$ from the fact that $\mathbf{v} \in W^{\perp}$ and hence $\mathbf{v} \cdot \mathbf{w} = 0$. This completes part (a). The proof of part (b) is left to Exercise 11. ■

Now that you know some of the advantages of working with orthogonal and orthonormal bases, the issue is how to find such bases. This is the topic of the next section.

Exercises for Section 7.1

1. Perform the following operations on the vectors $\mathbf{u} = (-1, 2, 2)$ and $\mathbf{v} = (0, -3, 3)$.
   (a) Compute $\mathbf{u} \cdot \mathbf{v}$ and use this value to describe the geometric relationship of $\mathbf{u}$ and $\mathbf{v}$.
   (b) Find the distance from $\mathbf{u}$ to $\mathbf{v}$.
   (c) Normalize $\mathbf{u}$.

2. Repeat the previous exercise for the vectors $\mathbf{u} = (3, 0, -4)$ and $\mathbf{v} = (4, -1, 3)$.

3. Do the following vectors constitute an orthogonal set?

$$\mathbf{v}_1 = (-2, 1, 0), \quad \mathbf{v}_2 = (2, 4, 0), \quad \text{and} \quad \mathbf{v}_3 = (0, 0, -3)$$

4. Do the following vectors constitute an orthogonal set?

$$\mathbf{v}_1 = (1, 0, -5), \quad \mathbf{v}_2 = (-10, 1, -2), \quad \text{and} \quad \mathbf{v}_3 = (1, 10, 0)$$

5. Are the vectors in Exercise 3 a basis for R^3? Why or why not? Explain.

6. A student said that if the vectors in Exercise 4 are not orthogonal, then they do not form a basis for R^3. Do you agree? Why or why not?

7. Use Theorem 7.3 in Section 7.1.2 to find the components of the vector $\mathbf{v} = (1, 2, -9)$ relative to the basis in Exercise 3.

8. Use Theorem 7.3 in Section 7.1.2 to find the components of the vector $\mathbf{v} = (10, -3, 8)$ relative to the following basis for the subspace W spanned by these two vectors:

$$\mathbf{v}_1 = (-2, 2, -5) \quad \text{and} \quad \mathbf{v}_2 = (6, 1, -2)$$

9. Consider the subspace W of R^2 defined by the line $y = 2x$.
   (a) Show that the vector $\mathbf{v} = (-2, 1)$ is orthogonal to W.
   (b) Find the orthogonal complement, $W^{\perp}$, of W and draw an appropriate figure.

10. Consider the subspace W of R^3 defined by the plane $ax + by + cz = 0$, in which a, b, and c are given real numbers.
    (a) Show that the vector $\mathbf{v} = (a, b, c)$ is orthogonal to W.
    (b) Find the orthogonal complement, $W^{\perp}$, of W. Describe $W^{\perp}$ and how it is related to W geometrically.

11. Complete Theorem 7.4 in Section 7.1.3 by proving that for vectors $\mathbf{w}_1, \ldots, \mathbf{w}_k$ that span a subspace W of R^n, a vector $\mathbf{v}$ is in $W^{\perp}$ if and only if $\mathbf{v}$ is orthogonal to each

vector $\mathbf{w}_i$, for $i = 1, \ldots, k$.

12. Suppose that W is a subspace of R^n and that you need to show that a vector $\mathbf{v}$ is in $W^\perp$. Explain why Theorem 7.4(b) is useful. (Hint: Look at the definition of $W^\perp$.)

13. Prove that if $\mathbf{v}$ is an n-vector that belongs to both a subspace W of R^n and also to $W^\perp$, then $\mathbf{v} = \mathbf{0}$.

14. A student said that the following three vectors:

$$\mathbf{v}_1 = (1, 0, -5), \quad \mathbf{v}_2 = (-10, 1, -2), \quad \text{and} \quad \mathbf{v}_3 = (0, 0, 0)$$

form an orthogonal set and, therefore, by Theorem 7.2 in Section 7.1.2 form a basis for R^3. Do you agree? Why or why not? Explain.

7.2 Orthogonal Projections and the Gram-Schmidt Process

In this section, a numerical method for finding an orthogonal basis for a subspace of R^n is presented. This method requires a generalization of orthogonal projections, presented in Section 1.4.2 and repeated here.

7.2.1 The Orthogonal Projection of One Vector onto Another Vector

The problem of finding an orthogonal projection of one n-vector onto a second n-vector is stated as follows.

> **PROBLEM 7.1 The Orthogonal Projection of One Vector onto Another Vector**
> Given an n-vector $\mathbf{v}$ and a nonzero n-vector $\mathbf{d}$, find two n-vectors $\mathbf{p}$ and $\mathbf{q}$ with $\mathbf{p}$ pointing in the same direction as $\mathbf{d}$ and $\mathbf{q}$ orthogonal to $\mathbf{d}$, as shown in Figure 7.4, so that
>
> $$\mathbf{v} = \mathbf{p} + \mathbf{q}$$
>
> The vector $\mathbf{p}$ is the *orthogonal projection of* $\mathbf{v}$ *onto* $\mathbf{d}$ and $\mathbf{q}$ is the *component of* $\mathbf{v}$ *orthogonal to* $\mathbf{d}$.

The solution to this problem, presented in Section 1.4.2, is based on the observations that (1) $\mathbf{p}$ points in the same direction as $\mathbf{d}$ and (2) $\mathbf{v}$ is the sum of $\mathbf{p}$ and $\mathbf{q}$. Therefore,

$$\mathbf{p} = t\mathbf{d}, \quad \text{for some real number } t \text{ and} \tag{7.12}$$

$$\mathbf{v} = \mathbf{p} + \mathbf{q} \tag{7.13}$$

To find the value of t in (7.12), dot product both sides of (7.13) by $\mathbf{d}$ to obtain

$$\mathbf{v} \cdot \mathbf{d} = \mathbf{p} \cdot \mathbf{d} + \mathbf{q} \cdot \mathbf{d} \tag{7.14}$$

Because $\mathbf{q}$ and $\mathbf{d}$ are orthogonal, $\mathbf{q} \cdot \mathbf{d} = 0$. Substituting this and $\mathbf{p} = t\mathbf{d}$ from (7.12) in (7.14) yields

$$\mathbf{v} \cdot \mathbf{d} = t(\mathbf{d} \cdot \mathbf{d})$$

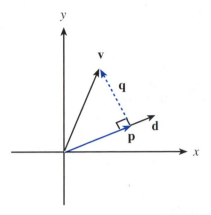

Figure 7.4 *The Orthogonal Projection of One Vector onto Another Vector.*

Dividing both sides of the foregoing equality by the nonzero number $\mathbf{d} \cdot \mathbf{d}$ (because $\mathbf{d} \neq \mathbf{0}$), you have that

$$t = \frac{\mathbf{v} \cdot \mathbf{d}}{\mathbf{d} \cdot \mathbf{d}}$$

Thus, from (7.12),

$$\mathbf{p} = t\mathbf{d} = \frac{\mathbf{v} \cdot \mathbf{d}}{\mathbf{d} \cdot \mathbf{d}}\mathbf{d}$$

Finally, substituting this expression for $\mathbf{p}$ in (7.13) and solving for $\mathbf{q}$ yields the following solution to Problem 7.1.

SOLUTION to Problem 7.1
Given two n-vectors $\mathbf{v}$ and $\mathbf{d}$ with $\mathbf{d} \neq \mathbf{0}$, the orthogonal projection of $\mathbf{v}$ onto $\mathbf{d}$ is

$$\mathbf{p} = \frac{\mathbf{v} \cdot \mathbf{d}}{\mathbf{d} \cdot \mathbf{d}}\mathbf{d} \quad \text{(orthogonal projection of } \mathbf{v} \text{ onto } \mathbf{d}\text{)}$$

The component of $\mathbf{v}$ orthogonal to $\mathbf{d}$ is

$$\mathbf{q} = \mathbf{v} - \mathbf{p} = \mathbf{v} - \frac{\mathbf{v} \cdot \mathbf{d}}{\mathbf{d} \cdot \mathbf{d}}\mathbf{d} \quad \text{(component of } \mathbf{v} \text{ orthogonal to } \mathbf{d}\text{)}$$

EXAMPLE 7.8 Finding the Orthogonal Projection of One Vector onto Another Vector
For the vectors

$$\mathbf{v} = (1, 3) \quad \text{and} \quad \mathbf{d} = (4, 2)$$

you have

p = orthogonal projection of **v** onto **d**

$$= \frac{\mathbf{v} \cdot \mathbf{d}}{\mathbf{d} \cdot \mathbf{d}} \mathbf{d} = \frac{(1, 3) \cdot (4, 2)}{(4, 2) \cdot (4, 2)} (4, 2) = \left(\frac{10}{20}\right) (4, 2) = (2, 1)$$

q = component of **v** orthogonal to **d**

$$= \mathbf{v} - \mathbf{p} = (1, 3) - (2, 1) = (-1, 2)$$

7.2.2 The Orthogonal Projection of a Vector onto a Subspace

A sequential generalization of projecting a given n-vector **v** onto another given n-vector **d**, as shown in Figure 7.5(a), is now presented. One generalization is obtained by replacing the vector **d** with a line through the origin in the direction of **d**, as shown in Figure 7.5(b). In other words, you are now projecting the n-vector **v** onto the set of n-vectors constituting a line in a given direction **d**. You are asked to show in Exercise 9 that the solution to this problem is the same as the solution to Problem 7.1.

The next generalization is obtained by projecting a given n-vector **v** onto a plane, rather than a line, as shown in Figure 7.5(c). In particular, you want to find a vector **p** in the plane and a vector **q** orthogonal to the plane so that

$$\mathbf{v} = \mathbf{p} + \mathbf{q}$$

A final generalization of this problem follows. Recall that, for a subspace W of R^n, $W^\perp$ is the set of n-vectors that are orthogonal to every vector in W.

> **PROBLEM 7.2 The Orthogonal Decomposition of a Vector**
> Given an n-vector **v** and a subspace W of R^n, find two n-vectors **p** and **q** with $\mathbf{p} \in W$ and $\mathbf{q} \in W^\perp$ such that
>
> $$\mathbf{v} = \mathbf{p} + \mathbf{q}$$

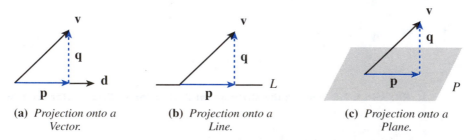

(**a**) *Projection onto a Vector.*   (**b**) *Projection onto a Line.*   (**c**) *Projection onto a Plane.*

Figure 7.5 *The Orthogonal Projection of a Vector onto (a) a Vector, (b) a Line, and (c) a Plane.*

Writing **v** in this way is called the **orthogonal decomposition** of **v** into **p** and **q**. The vector **p** is the orthogonal projection of **v** onto the subspace W and is often written as

$$\text{proj}_W \mathbf{v}$$

The vector **q** is the component of **v** orthogonal to W.

To find closed-form expressions for **p** and **q** in Problem 7.2, you need to have an orthogonal basis $\mathbf{w}_1, \ldots, \mathbf{w}_k$ for W. In this case, the solution to Problem 7.2 is obtained by applying the following theorem.

THEOREM 7.5

Let $\mathbf{w}_1, \ldots, \mathbf{w}_k$ be an orthogonal basis for a subspace W of R^n. Then for every n-vector **v**, there are unique n-vectors $\mathbf{p} \in W$ and $\mathbf{q} \in W^\perp$ such that

$$\mathbf{v} = \mathbf{p} + \mathbf{q} \tag{7.15}$$

In particular,

$$\mathbf{p} = \frac{\mathbf{v} \cdot \mathbf{w}_1}{\mathbf{w}_1 \cdot \mathbf{w}_1} \mathbf{w}_1 + \cdots + \frac{\mathbf{v} \cdot \mathbf{w}_k}{\mathbf{w}_k \cdot \mathbf{w}_k} \mathbf{w}_k \quad \text{and} \quad \mathbf{q} = \mathbf{v} - \mathbf{p} \tag{7.16}$$

Proof.

The key words *for every* in the conclusion suggest using the choose method to choose an n-vector **v**, for which you must show that there are unique n-vectors **p** and **q** such that (7.15) holds. According to the uniqueness method, the first step is to use the construction method to produce **p** and **q**. The construction of these vectors is given in (7.16) so, it remains to show that for these values of **p** and **q**, $\mathbf{p} \in W$, $\mathbf{q} \in W^\perp$, and $\mathbf{v} = \mathbf{p} + \mathbf{q}$.

From (7.16), you can see that $\mathbf{v} = \mathbf{p} + \mathbf{q}$, so you need only verify the first two properties. But **p**, as defined in (7.16), is in W because **p** is a linear combination of the vectors $\mathbf{w}_1, \ldots, \mathbf{w}_k$, each of which is in the subspace W. To show that $\mathbf{q} \in W^\perp$, by definition, you must show that **q** is orthogonal to every vector in W. By Theorem 7.4(b) in Section 7.1.3, it suffices to show that **q** is orthogonal to each of the basis vectors for W. That is, you must show that

for each $i = 1, \ldots, k$, $\mathbf{q} \cdot \mathbf{w}_i = 0$

Recognizing the key words *for each*, you should use the choose method to choose an integer i with $1 \le i \le k$ and show that

$$\mathbf{q} \cdot \mathbf{w}_i = 0$$

To that end, using (7.16) and the fact that $\mathbf{w}_1, \ldots, \mathbf{w}_k$ are an orthogonal basis for W, it follows that

$$\mathbf{q} \cdot \mathbf{w}_i = (\mathbf{v} - \mathbf{p}) \cdot \mathbf{w}_i = \mathbf{v} \cdot \mathbf{w}_i - \mathbf{p} \cdot \mathbf{w}_i$$

$$= \mathbf{v} \cdot \mathbf{w}_i - \frac{\mathbf{v} \cdot \mathbf{w}_1}{\mathbf{w}_1 \cdot \mathbf{w}_1}(\mathbf{w}_1 \cdot \mathbf{w}_i) - \cdots - \frac{\mathbf{v} \cdot \mathbf{w}_i}{\mathbf{w}_i \cdot \mathbf{w}_i}(\mathbf{w}_i \cdot \mathbf{w}_i) - \cdots - \frac{\mathbf{v} \cdot \mathbf{w}_k}{\mathbf{w}_k \cdot \mathbf{w}_k}(\mathbf{w}_k \cdot \mathbf{w}_i)$$

$$= \mathbf{v} \cdot \mathbf{w}_i - \mathbf{v} \cdot \mathbf{w}_i = 0$$

To complete the proof, you must establish the uniqueness of $\mathbf{p}$ and $\mathbf{q}$. To that end, suppose that $\mathbf{a} \in W$ and $\mathbf{b} \in W^\perp$ also satisfy

$$\mathbf{v} = \mathbf{a} + \mathbf{b} \tag{7.17}$$

You must show that $\mathbf{a} = \mathbf{p}$ and $\mathbf{b} = \mathbf{q}$. To that end, from (7.15) and (7.17), you have that

$$\mathbf{a} + \mathbf{b} = \mathbf{p} + \mathbf{q}$$

or, equivalently, that

$$\mathbf{p} - \mathbf{a} = \mathbf{b} - \mathbf{q} \tag{7.18}$$

Now both $\mathbf{p}$ and $\mathbf{a}$ are in the subspace W, so $\mathbf{p} - \mathbf{a} \in W$. Likewise, $\mathbf{b}$ and $\mathbf{q}$ are in the subspace $W^\perp$, so $\mathbf{b} - \mathbf{q} \in W^\perp$. So, from (7.18), $\mathbf{p} - \mathbf{a}$ is in both W and $W^\perp$. Thus, $(\mathbf{p} - \mathbf{a}) \cdot (\mathbf{p} - \mathbf{a}) = 0$, which means that $\mathbf{p} - \mathbf{a} = \mathbf{0}$, that is, $\mathbf{p} = \mathbf{a}$. Then, from (7.18), it follows that $\mathbf{b} = \mathbf{q}$, thus completing the proof. ∎

The fact that $\mathbf{p}$ and $\mathbf{q}$ are unique means that this decomposition of $\mathbf{v}$ does not depend on the choice of the orthogonal basis for W. That is, any orthogonal basis produces the same values for $\mathbf{p}$ and $\mathbf{q}$. By using orthogonal projections, it is now possible to present a method for finding an orthogonal basis.

7.2.3 The Gram-Schmidt Process for Finding an Orthogonal Basis

In Section 7.1, you saw several reasons why working with an orthogonal or orthonormal basis is desirable. One numerical method for finding such a basis for a subspace W of R^n is presented now. The algorithm starts with any basis for W, say, $\{\mathbf{x}_1, \ldots, \mathbf{x}_k\}$, and sequentially creates an orthogonal basis $\{\mathbf{v}_1, \ldots, \mathbf{v}_k\}$. In particular, the algorithm starts with $\mathbf{v}_1 = \mathbf{x}_1$ and then uses $\mathbf{x}_1$, $\mathbf{x}_2$, and $\mathbf{v}_1$ to find the next orthogonal vector, $\mathbf{v}_2$. The procedure then uses $\mathbf{x}_1$, $\mathbf{x}_2$, and $\mathbf{x}_3$ together with $\mathbf{v}_1$ and $\mathbf{v}_2$ to find the next orthogonal vector, $\mathbf{v}_3$, and so on.

Finding an Orthogonal Basis for a Two-Dimensional Subspace

You can visualize how this orthogonal basis is constructed by considering the subspace represented by a plane in R^3 with a basis consisting of the two linearly independent vectors $\mathbf{x}_1$ and $\mathbf{x}_2$, as shown in Figure 7.6. The first vector of the orthogonal basis is

$$\mathbf{v}_1 = \mathbf{x}_1$$

The second vector, $\mathbf{v}_2$, must be orthogonal to $\mathbf{v}_1$. From your knowledge of orthogonal projections, you know that one such vector is the component of $\mathbf{x}_2$ orthogonal to $\mathbf{x}_1$. In the context of the notation of the solution to Problem 7.1 in Section 7.2.1, you can construct $\mathbf{v}_2 = \mathbf{q}$, given $\mathbf{v} = \mathbf{x}_2$ and $\mathbf{d} = \mathbf{v}_1$, as follows:

$$\mathbf{v}_2 = \mathbf{x}_2 - \frac{\mathbf{x}_2 \cdot \mathbf{v}_1}{\mathbf{v}_1 \cdot \mathbf{v}_1} \mathbf{v}_1$$

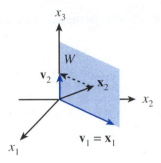

Figure 7.6 *Constructing an Orthogonal Basis for a Plane in R^3.*

Equivalently stated, letting W_1 be the subspace spanned by $\mathbf{x}_1$,

$$\mathbf{v}_2 = \mathbf{x}_2 - \text{proj}_{W_1} \mathbf{x}_2$$

This vector $\mathbf{v}_2$ is orthogonal to $\mathbf{v}_1$, as shown in Figure 7.6.

EXAMPLE 7.9 Finding an Orthogonal Basis for a Two-Dimensional Subspace of R^3
Given the basis

$$\mathbf{x}_1 = (1, 0, -1) \quad \text{and} \quad \mathbf{x}_2 = (1, -2, 3)$$

for the subspace W spanned by these two vectors, an orthogonal basis for W is

$$\mathbf{v}_1 = \mathbf{x}_1 = (1, 0, -1)$$

$$\mathbf{v}_2 = \mathbf{x}_2 - \frac{\mathbf{x}_2 \cdot \mathbf{v}_1}{\mathbf{v}_1 \cdot \mathbf{v}_1} \mathbf{v}_1 = (1, -2, 3) - \left(\frac{-2}{2}\right)(1, 0, -1) = (2, -2, 2)$$

Finding an Orthogonal Basis for a Three-Dimensional Subspace

The generalization of this process is understood by seeing how an orthogonal basis $\{\mathbf{v}_1, \mathbf{v}_2, \mathbf{v}_3\}$ for R^3 is constructed starting with any basis $\{\mathbf{x}_1, \mathbf{x}_2, \mathbf{x}_3\}$. Of course,

$$\mathbf{v}_1 = \mathbf{x}_1$$

$$\mathbf{v}_2 = \mathbf{x}_2 - \text{proj}_{W_1} \mathbf{x}_2 = \mathbf{x}_2 - \frac{\mathbf{x}_2 \cdot \mathbf{v}_1}{\mathbf{v}_1 \cdot \mathbf{v}_1} \mathbf{v}_1$$

It remains to find a vector $\mathbf{v}_3$ orthogonal to both $\mathbf{v}_1$ and $\mathbf{v}_2$. Such a vector is created by letting W_2 be the subspace spanned by $\{\mathbf{x}_1, \mathbf{x}_2\}$ and realizing that $\{\mathbf{v}_1, \mathbf{v}_2\}$ is an orthogonal basis for W_2. Then, using Theorem 7.5,

$$\mathbf{v}_3 = \mathbf{x}_3 - \text{proj}_{W_2} \mathbf{x}_3 = \mathbf{x}_3 - \frac{\mathbf{x}_3 \cdot \mathbf{v}_1}{\mathbf{v}_1 \cdot \mathbf{v}_1} \mathbf{v}_1 - \frac{\mathbf{x}_3 \cdot \mathbf{v}_2}{\mathbf{v}_2 \cdot \mathbf{v}_2} \mathbf{v}_2$$

EXAMPLE 7.10 Finding an Orthogonal Basis for R^3

Given the basis

$$\mathbf{x}_1 = (1, 0, -1), \quad \mathbf{x}_2 = (1, -2, 3), \quad \text{and} \quad \mathbf{x}_3 = (-2, 4, 0)$$

for R^3, an orthogonal basis for R^3 is

$$\mathbf{v}_1 = \mathbf{x}_1 = (1, 0, -1) \quad \text{(see Example 7.9)}$$
$$\mathbf{v}_2 = \mathbf{x}_2 - \frac{\mathbf{x}_2 \cdot \mathbf{v}_1}{\mathbf{v}_1 \cdot \mathbf{v}_1}\mathbf{v}_1 = (2, -2, 2) \quad \text{(see Example 7.9)}$$

$$\mathbf{v}_3 = \mathbf{x}_3 - \frac{\mathbf{x}_3 \cdot \mathbf{v}_1}{\mathbf{v}_1 \cdot \mathbf{v}_1}\mathbf{v}_1 - \frac{\mathbf{x}_3 \cdot \mathbf{v}_2}{\mathbf{v}_2 \cdot \mathbf{v}_2}\mathbf{v}_2$$

$$= (-2, 4, 0) - \left(\frac{-2}{2}\right)(1, 0, -1) - \left(\frac{-12}{12}\right)(2, -2, 2) = (1, 2, 1)$$

The Gram-Schmidt Process

The generalization of this step-by-step approach is the following **Gram-Schmidt process**:

The Gram-Schmidt Process

Given a basis $\{\mathbf{x}_1, \ldots, \mathbf{x}_k\}$ for a subspace W of R^n, an orthogonal basis $\{\mathbf{v}_1, \ldots, \mathbf{v}_k\}$ for W is obtained by computing:

$$\mathbf{v}_1 = \mathbf{x}_1$$

$$\mathbf{v}_2 = \mathbf{x}_2 - \frac{\mathbf{x}_2 \cdot \mathbf{v}_1}{\mathbf{v}_1 \cdot \mathbf{v}_1}\mathbf{v}_1$$

$$\mathbf{v}_3 = \mathbf{x}_3 - \frac{\mathbf{x}_3 \cdot \mathbf{v}_1}{\mathbf{v}_1 \cdot \mathbf{v}_1}\mathbf{v}_1 - \frac{\mathbf{x}_3 \cdot \mathbf{v}_2}{\mathbf{v}_2 \cdot \mathbf{v}_2}\mathbf{v}_2$$

$$\vdots$$

$$\mathbf{v}_k = \mathbf{x}_k - \frac{\mathbf{x}_k \cdot \mathbf{v}_1}{\mathbf{v}_1 \cdot \mathbf{v}_1}\mathbf{v}_1 - \frac{\mathbf{x}_k \cdot \mathbf{v}_2}{\mathbf{v}_2 \cdot \mathbf{v}_2}\mathbf{v}_2 - \cdots - \frac{\mathbf{x}_k \cdot \mathbf{v}_{k-1}}{\mathbf{v}_{k-1} \cdot \mathbf{v}_{k-1}}\mathbf{v}_{k-1}$$

The fact that this algorithm does produce an orthogonal basis for W is proved now.

THEOREM 7.6

If $\{\mathbf{x}_1, \ldots, \mathbf{x}_k\}$ is a basis for a subspace W of R^n, $\{\mathbf{v}_1, \ldots, \mathbf{v}_k\}$ are the vectors produced by the Gram-Schmidt process, and $W_i = \text{span} \{\mathbf{x}_1, \ldots, \mathbf{x}_i\}$, then, for each $i = 1, \ldots, k$,

(a) $W_i = \text{span} \{\mathbf{v}_1, \ldots, \mathbf{v}_i\}$.

(b) $\{\mathbf{v}_1, \ldots, \mathbf{v}_i\}$ is an orthogonal basis for W_i.

Proof.

The conclusion of this theorem requires showing that two statements (a) and (b) are true for each value of the integer $i = 1, \ldots, k$. It is therefore appropriate to use induction on the number of vectors, i, constructed by the algorithm.

Accordingly, you must first show that the conclusion is true when $i = 1$, which is indeed the case. Next you assume the following induction hypotheses:

(a) $W_i = \text{span } \{\mathbf{v}_1, \ldots, \mathbf{v}_i\}$.

(b) The set $\{\mathbf{v}_1, \ldots, \mathbf{v}_i\}$ produced by the algorithm is an orthogonal basis for W_i.

It remains to show that if the vector $\mathbf{v}_{i+1}$ is produced by the algorithm, then $W_{i+1} = \text{span } \{\mathbf{v}_1, \ldots, \mathbf{v}_{i+1}\}$ and $\{\mathbf{v}_1, \ldots, \mathbf{v}_{i+1}\}$ is an orthogonal basis for W_{i+1}.

The latter statement is proved first. To that end, it is shown that:

1. $\mathbf{v}_1, \ldots, \mathbf{v}_{i+1}$ are in W_{i+1}.
2. $\mathbf{v}_1, \ldots, \mathbf{v}_{i+1}$ are orthogonal.
3. $\mathbf{v}_1, \ldots, \mathbf{v}_{i+1}$ are linearly independent.
4. $\mathbf{v}_1, \ldots, \mathbf{v}_{i+1}$ are a basis for W_{i+1}.

To see (1), note first that

$$\mathbf{v}_1, \ldots, \mathbf{v}_i \in \text{span}\{\mathbf{v}_1, \ldots, \mathbf{v}_i\} \quad \text{(definition of span)}$$
$$= W_i \quad \text{[induction hypothesis for part (a)]}$$
$$\subseteq W_{i+1} \quad (W_{i+1} \text{ contains all of } W_i)$$

It remains to show that $\mathbf{v}_{i+1} \in W_{i+1}$. From the algorithm, $\mathbf{v}_{i+1}$ is defined as follows:

$$\mathbf{v}_{i+1} = \mathbf{x}_{i+1} - \text{proj}_{W_i} \mathbf{x}_{i+1} \tag{7.19}$$

Now $\mathbf{x}_{i+1} \in W_{i+1}$ and $\text{proj}_{W_i} \mathbf{x}_{i+1} \in W_i \subseteq W_{i+1}$, so $\mathbf{v}_{i+1} \in W_{i+1}$. This establishes (1).

To see that (2) is true, note that part (b) of the induction hypothesis states that $\{\mathbf{v}_1, \ldots, \mathbf{v}_i\}$ is an orthogonal basis for W_i, so $\mathbf{v}_1, \ldots, \mathbf{v}_i$ are orthogonal. The fact that $\mathbf{v}_{i+1}$ defined by (7.19) is orthogonal to each of the vectors $\mathbf{v}_1, \ldots, \mathbf{v}_i$ follows from the fact that $\mathbf{v}_{i+1} \in (W_i)^{\perp}$ (see Theorem 7.5 in Section 7.2.2) and so $\mathbf{v}_{i+1}$ is orthogonal to each vector in W_i and, in particular, to $\mathbf{v}_1, \ldots, \mathbf{v}_i$. This shows that (2) is true.

You can use Theorem 7.2 in Section 7.1.2 to conclude that $\mathbf{v}_1, \ldots, \mathbf{v}_{i+1}$ are linearly independent. To do so, you need only be sure that these vectors are nonzero. The fact that $\mathbf{v}_1, \ldots, \mathbf{v}_i$ are nonzero follows from part (b) of the induction hypothesis because these vectors constitute a basis for W_i and no vector in a basis is the zero vector. Also, $\mathbf{v}_{i+1} \neq \mathbf{0}$ because $\mathbf{x}_{i+1}$ is not in $W_i = \text{span}\{\mathbf{v}_1, \ldots, \mathbf{v}_i\}$. This now establishes (3).

To see that $\mathbf{v}_1, \ldots, \mathbf{v}_{i+1}$ is a basis for W_{i+1}, note that $\dim(W_{i+1}) = i + 1$ because $\mathbf{v}_1, \ldots, \mathbf{v}_{i+1}$ are $i + 1$ linearly independent vectors in W_{i+1}. Thus, by Theorem 4.9 in Section 4.4.1, these $i + 1$ linearly independent vectors are a basis for W_{i+1}. Finally, because $\mathbf{v}_1, \ldots, \mathbf{v}_{i+1}$ is a basis for W_{i+1}, these vectors span W_{i+1}. The proof is now complete. ∎

In this section, you have seen how orthogonal projections are used to find an orthogonal basis. In the next section, you will see how orthogonality is used in solving an important and practical set of problems.

Exercises for Section 7.2

1. Find the projection of the vector $\mathbf{x} = (-1, 3)$ onto the vector $\mathbf{y} = (-4, 2)$ and the component of $\mathbf{x}$ orthogonal to $\mathbf{y}$. Draw an appropriate figure.

2. Find the projection of the vector $\mathbf{y} = (1, 0, -2)$ onto the vector $\mathbf{x} = (-3, 1, 0)$ and the component of $\mathbf{y}$ orthogonal to $\mathbf{x}$.

3. Use the following orthogonal basis to find the projection of the vector $\mathbf{v} = (2, 1, -3)$ onto the space W spanned by these two vectors:

$$\mathbf{v}_1 = (1, 0, -1) \quad \text{and} \quad \mathbf{v}_2 = (1, 0, 1)$$

4. Show that the projection of the vector $\mathbf{v}$ onto the subspace W in Exercise 3 is the same as the projection obtained in Exercise 3 if the following orthogonal basis for W is used:

$$\mathbf{v}_1 = (2, 0, -3) \quad \text{and} \quad \mathbf{v}_2 = (6, 0, 4)$$

5. Use the Gram-Schmidt process to find an orthogonal basis for R^2 using the following basis:

$$\mathbf{x}_1 = (-3, 1) \quad \text{and} \quad \mathbf{x}_2 = (0, -1)$$

Draw a figure showing the original basis and the orthogonal basis.

6. Use the Gram-Schmidt process to find an orthogonal basis for the subspace W spanned by the following two vectors:

$$\mathbf{x}_1 = (-2, 0, 1) \quad \text{and} \quad \mathbf{x}_2 = (0, -3, 2)$$

7. Use the Gram-Schmidt process to find an orthogonal basis for the subspace W spanned by the following two vectors:

$$\mathbf{x}_1 = (1, 0, -3, 0) \quad \text{and} \quad \mathbf{x}_2 = (0, -4, 0, 2)$$

8. Find the components of $\mathbf{v} = (2, 4, -6, 2)$ relative to the orthogonal basis you found in Exercise 7.

9. Consider the problem of finding the projection of an n-vector $\mathbf{v}$ onto a line through the origin in the direction of the n-vector $\mathbf{x}$. Prove that for any vector $\mathbf{d}$ on the line, the projection of $\mathbf{v}$ onto $\mathbf{d}$ results in the same vector.

10. Suppose that W is the subspace of R^n spanned by the vectors $\mathbf{v}_1, \ldots, \mathbf{v}_k$. Prove that if $\mathbf{v}$ is an n-vector that is not in W, then the component of $\mathbf{v}$ orthogonal to W is not the zero vector. (Hint: Use the contrapositive method because the word *not* appears in the conclusion.)

11. Suppose that $\mathbf{v}$ is an n-vector on a line L through the origin in the direction $\mathbf{d}$.
    (a) Prove that the projection of $\mathbf{v}$ onto L is $\mathbf{v}$.
    (b) Draw an appropriate figure in two dimensions to illustrate part (a).
    (c) State a generalization of part (a) to a subspace W of R^n.

12. Suppose that the n-vector $\mathbf{v}$ is perpendicular to a line L through the origin in the direction $\mathbf{d}$.
    (a) Prove that the projection of $\mathbf{v}$ onto L is $\mathbf{0}$.
    (b) Draw an appropriate figure in two dimensions to illustrate part (a).
    (c) State a generalization of part (a) to a subspace W of R^n.

7.3 Applications to Regression Models

A problem that arises in business, economics, statistics, engineering, and the natural sciences is to predict the value of a quantity of interest. What will be the sales of a particular soft drink next year? How much gasoline is needed for the upcoming summer months? How much winter wheat can be expected from the next harvest? You can think of each of these unknown quantities as a variable, say, y. One approach to determining a value for y is to use historical data in the belief that previous values of y provide a reliable forecast of future values. Another approach—the one used here—is to identify another variable, say, x, on which the value of y is assumed to depend according to a functional relationship of the following form:

$$y = f(x)$$

Once the function $f(x)$ is determined, you can predict a value for y by identifying an appropriate value for x and computing $y = f(x)$. An underlying assumption, therefore, is that you are able to obtain a reliable value for x on which to base the prediction. This section explains the steps involved in developing these types of **regression models** and how historical data and orthogonal projections are used to obtain their solutions.

7.3.1 General Steps to Developing and Using Regression Models

To illustrate the steps involved in building and using a regression model, consider the problem faced by Connie Baker. Connie is looking to buy a house in the near future and needs to know what the monthly mortgage payment, y, per \$1000 of principal borrowed is going to be on a 15-year loan.

Identifying the Dependent and Independent Variables

The first step is to identify a variable x on which the value of the desired variable y depends. For this reason, y is called the **dependent variable** and x is called the **independent variable**. For Connie's problem, you know that the monthly mortgage payment depends on the interest rate at the time of the loan, so you might define the following dependent variable y and independent variable x:

y = the monthly mortgage payment per \$1000 of principal borrowed

x = the annual interest rate of the loan, as a percentage

Identifying a Relationship Between the Dependent and the Independent Variables

Although you might not know the exact manner in which y depends on x, to apply the techniques in this section, you must have an idea of the *form* of the relationship between x and y. For example, if you believe that y depends on x in a linear way, then you can represent this relationship mathematically as

$$y = mx + b \qquad (7.20)$$

where m is the slope (that is, the change in y per unit of increase in x) and b is the y-intercept (that is, the value of y when $x = 0$). Alternatively, if you believe a quadratic relationship is more appropriate, then you can represent this relationship mathematically as

$$y = ax^2 + bx + c$$

You should choose the form of relationship based on your knowledge of the specific application, on historical data, or by other means available to you. For the problem of Connie Baker, it is reasonable to assume that the monthly mortgage payment, y, depends on the interest rate, x, according to the linear relationship (7.20).

Although it is necessary to specify the general form of the relationship between y and x, you need not know the specific form. For example, the general form of a linear relationship is given in (7.20) but the specific form, that is, the specific values of the *parameters m* and b, are not known.

Finding Values for the Parameters

A major task in solving the problem is to determine appropriate values for the unknown parameters. Historical data and orthogonal projections are used to do so in Section 7.3.2 and Section 7.3.3. Once the values of the parameters are obtained, you can then use the specific functional relationship between y and x to predict a value for y, as explained next.

Using a Regression Model

To illustrate the prediction process once the values of the parameters are determined, recall the problem of Connie Baker. Suppose you determine that the monthly mortgage payment, y, depends on the interest rate, x, according to the following specific linear relationship:

$$y = 0.567x + 5.03 \tag{7.21}$$

Connie can now use (7.21) to predict her monthly mortgage payment per $1000 of loan at a specific interest rate x. For example, if $x = 10$ percent, then, from (7.21), the monthly payment per $1000 borrowed is

$$y = 0.567(10) + 5.03 = 10.70$$

That is, at an interest rate of 10%, the monthly mortgage payment per $1000 of loan is $10.70.

It is also useful to interpret the meaning of the slope 0.567 in (7.21) in the context of this problem. This number indicates the increase in the monthly payment, y, per unit of increase in the interest rate, x. Thus, each increase in the interest rate of one percentage point results in an increase in the monthly payments of $0.567 per $1000 borrowed.

The intercept of 5.03 in (7.21) is the monthly payment when the value of the interest rate is 0. Because an interest rate of 0 percent does not happen, the intercept 5.03 has no useful meaning in this problem.

When predicting the value of the dependent variable, keep the following issues in mind:

1. The functional relationship you assume between y and x might only be valid over a certain range of values for x. In such cases, it is not appropriate to use the model to predict a value for y unless the value for x is within this range.

2. The specific values you obtain for the parameters are only estimates of their true values.

For example, the values 0.567 and 5.03 in (7.21) for the problem of Connie Baker are estimates of the actual values of the parameters m and b in the equation

$$y = mx + b$$

Thus, when you predict that an interest rate of $x = 10$ percent results in a monthly mortgage payment of $y = 0.567(10) + 5.03 = 10.70$ dollars, you should realize that this prediction is an estimate of the monthly payment. The actual payment may differ from this estimate for a variety of reasons. Statistical methods are used to quantify the degree of certainty you can have in the prediction. These techniques are outside the scope of this book.

7.3.2 Simple Linear Regression and General Least-Squares Problems

In this section, a **linear regression model** is studied, that is, a regression model in which it is assumed that the following linear relationship holds between the dependent variable y and the independent variable x:

$$y = mx + b \tag{7.22}$$

The line described by (7.22) is called the **regression line**. The next question is how to estimate the parameters m and b.

Using Historical Data to Estimate the Parameters

One typical approach to estimating the values of the parameters is to use existing data that provide ordered pairs of values for the variables x and y. That is, you collect k pairs of data $(x_1, y_1), \ldots, (x_k, y_k)$. To illustrate, suppose that Connie Baker has obtained the monthly payments shown in Table 7.1 for interest rates of 6, 7, 8, and 9 percent. The objective is to use these data to determine values for the parameters m and b in the regression line in (7.22). To find the values for m and b, you would ideally like the four data points to lie on the same line so that the line correctly predicts the monthly payments for the given interest rates of $x = 6, 7, 8$, and 9 percent. However, all data points may not lie on the same line because the values are inaccurate or because the relation between the monthly payment and the interest rate is not linear. As a result, you may not be able to choose values for m and b so that all data points lie on the same line. The four data points are plotted on the graph in Figure 7.7. Different lines corresponding to different choices of m and b are also shown. Because none of the lines fits the data exactly, how do you identify the "best" line?

The Measure of Least-Squared Error

What is needed is a way to measure how well a given line fits the data. One such measure is a nonnegative number with the property that, the smaller the number, the better the line

Interest rate (%)	6	7	8	9
Monthly payments ($ / $1000 of loan)	8.44	8.99	9.56	10.14

Table 7.1 *Monthly Mortgage Payments for Various Interest Rates.*

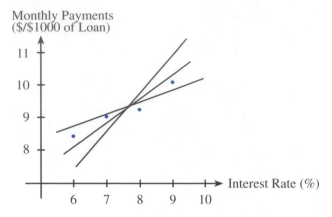

Figure 7.7 *Finding a Line to Fit the Data.*

fits the data. To solve this problem, you need to determine values for m and b that provide the smallest value of this measure.

To illustrate, suppose you have specific values for m and b, say, $m = 0.50$ and $b = 5.50$, so the equation of the line under consideration is

$$y = 0.50x + 5.50 \tag{7.23}$$

For a given value of the interest rate x, you can use (7.23) to predict the value for the monthly payment per \$1000 borrowed, y. In particular, you can predict the monthly payment for the given data values $x = 6, 7, 8,$ and 9 percent and compare those values with the actual monthly payments given in the data. These computations and comparisons are shown in the following table using (7.23):

Interest rate (x)	6	7	8	9
Predicted payments ($0.50x + 5.50$)	8.50	9.00	9.50	10.00
Actual payments	8.44	8.99	9.56	10.14

From the values in the foregoing table, you can see that the predicted payments are different from the actual payments—sometimes more and sometimes less. For a particular interest rate x in the data, the amount by which the predicted payment differs from the actual payment provides a measure of how well the line fits this one data point. For example, for an interest rate of $x = 6$ percent, you have:

predicted payment − actual payment = $8.50 - 8.44 = 0.06$

Likewise, for an interest rate of $x = 9$ percent,

predicted payment − actual payment = $10.00 - 10.14 = -0.14$

To compensate for the fact that sometimes the predicted payment is larger and sometimes smaller than the actual payment, an appropriate measure is the *square* of the difference between these two numbers. So, for a given value of the interest rate x in the data and its corresponding actual payment, say, a, a measure of the error incurred by using the predicted value of $y = mx + b$ instead of the actual payment a is

$$\text{squared error for interest rate } x = (y - a)^2 = (mx + b - a)^2 \tag{7.24}$$

Note that (7.24) provides a measure of the error for one value of x from the data. You can compute a similar measure of error for each value of x in the data. The desired measure of goodness for the line $y = mx + b$ is then taken to be the sum of the individual squared errors of each of the values for x in the data. That is, given values for m and b, and using the $k = 4$ data values $x_1 = 6$, $x_2 = 7$, $x_3 = 8$, and $x_4 = 9$ together with the four corresponding actual monthly payments of $y_1 = 8.44$, $y_2 = 8.99$, $y_3 = 9.56$, and $y_4 = 10.14$, the measure of goodness for the line $y = mx + b$ is the following **sum of squared errors** (SSE):

$$\text{SSE} = \sum_{i=1}^{k}(mx_i + b - y_i)^2 \tag{7.25}$$

$$= (mx_1 + b - y_1)^2 + (mx_2 + b - y_2)^2 +$$
$$(mx_3 + b - y_3)^2 + (mx_4 + b - y_4)^2$$

$$= (6m + b - 8.44)^2 + (7m + b - 8.99)^2 +$$
$$(8m + b - 9.56)^2 + (9m + b - 10.14)^2$$

This problem of using known data to determine the parameters in a linear relationship between a dependent variable y and an independent variable x arises in many applications. Describing this problem in a more general way using matrix-vector notation provides the basis for the subsequent solution.

The General Least-Squares Problem

When two variables x and y are believed to be related in a linear way, the objective is to determine the slope m and intercept b of the line

$$y = mx + b \tag{7.26}$$

This is accomplished by obtaining k actual values for x and their corresponding values for y, say, $(x_1, y_1), \ldots, (x_k, y_k)$. These data are used to find values for m and b that make the line in (7.26) fit these data most accurately, using the measure of least-squared error in (7.25). That is, for given values of m and b, you can use (7.26) to predict the y-value at each of the data values $x_1, \ldots, x_k$ by computing the following k-vector **p** of predicted values:

$$\mathbf{p} = \begin{bmatrix} p_1 \\ \vdots \\ p_k \end{bmatrix} = \begin{bmatrix} mx_1 + b \\ \vdots \\ mx_k + b \end{bmatrix} = \begin{bmatrix} x_1 & 1 \\ \vdots & \vdots \\ x_k & 1 \end{bmatrix} \begin{bmatrix} m \\ b \end{bmatrix} \tag{7.27}$$

Using matrix-vector notation, let A be the $(k \times 2)$ matrix in (7.27)—hereafter called the **design matrix**—whose first column contains the k data values $x_1, \ldots, x_k$ and whose second column contains all 1's. Let **z** be the 2-vector of the unknown parameters m and b, called the **parameter vector**. Then, from (7.27), the predicted values of y at the k data points $x_1, \ldots, x_k$ are the components of the column vector $\mathbf{p} = A\mathbf{z}$.

You can then compare the predicted values, $\mathbf{p}$, to the actual values of y from the data, denoted by the column vector $\mathbf{y} \in R^{k \times 1}$. Using the measure of SSE, you can compute how well the line $y = mx + b$ fits the data in matrix-vector notation, as follows:

$$\text{SSE} = (p_1 - y_1)^2 + \cdots + (p_k - y_k)^2 = (\mathbf{p} - \mathbf{y})'(\mathbf{p} - \mathbf{y})$$
$$= (A\mathbf{z} - \mathbf{y})'(A\mathbf{z} - \mathbf{y}) = \|A\mathbf{z} - \mathbf{y}\|^2$$

In summary, the **linear-regression problem** is the following.

PROBLEM 7.3 The Linear Regression Problem

Given k pairs of data values stored in the column vectors $\mathbf{x}$ and $\mathbf{y}$, find values for the variables

$$\mathbf{z} = \begin{bmatrix} m \\ b \end{bmatrix}$$

that represent the slope and intercept of the line $y = mx + b$ so as to

minimize $\|A\mathbf{z} - \mathbf{y}\|^2$

where

$$A = \begin{bmatrix} x_1 & 1 \\ \vdots & \vdots \\ x_k & 1 \end{bmatrix}$$

You know that abstraction is the process of thinking of a concept less in terms of specific items and more in terms of general objects. This technique is used in Section 4.1 to create the axiomatic system of a vector space. You can also apply abstraction to Problem 7.3. To do so, think of A as any $(k \times n)$ matrix, rather than the specific one in Problem 7.3. You then obtain the following more general **least-squares problem**.

PROBLEM 7.4 The General Least-Squares Problem

For a $(k \times n)$ matrix A and a column vector $\mathbf{y} \in R^{k \times 1}$, find a column vector $\mathbf{z} \in R^{n \times 1}$ to

minimize $\|A\mathbf{z} - \mathbf{y}\|^2$

You will now see how orthogonal projections are used to obtain the desired values for $\mathbf{z}$.

7.3.3 Using Orthogonal Projections to Solve Least-Squares Problems

If you could find an n-vector $\mathbf{z}$ such that

$$A\mathbf{z} = \mathbf{y} \tag{7.28}$$

then $\mathbf{z}$ is the solution to Problem 7.4 because the value of $\|A\mathbf{z} - \mathbf{y}\|^2$ would be 0 and you cannot make the value of this nonnegative quantity smaller. When there is no $\mathbf{z}$ satisfying

(7.28), the next best approach is to find an n-vector $\mathbf{z}$ for which $A\mathbf{z}$ is as close, in terms of the sum of squared error, to the given k-vector $\mathbf{y}$ as possible.

To visualize such a solution to the least-squares problem, note first that to minimize $\|A\mathbf{z} - \mathbf{y}\|^2$, you can minimize $\|A\mathbf{z} - \mathbf{y}\|$ itself. Thus, if you want to solve Problem 7.4, you need to find an n-vector $\mathbf{z}$ such that $A\mathbf{z}$ is as close as possible to $\mathbf{y}$, as shown in Figure 7.8. However, any n-vector $\mathbf{z}$ you choose results in $A\mathbf{z}$ being in the column space of A, denoted by Col A (see Section 4.4.2). Hence, you need to find the closest point in Col A to $\mathbf{y}$. How you do so is based on the following property of the orthogonal projection.

THEOREM 7.7

If $\mathbf{y}$ is a given k-vector and W is a subspace of R^k, then the orthogonal projection $\mathbf{p}$ of $\mathbf{y}$ onto W is the closest point in W to $\mathbf{y}$, that is,

$$\|\mathbf{y} - \mathbf{p}\| \leq \|\mathbf{y} - \mathbf{w}\|, \text{ for all } \mathbf{w} \in W$$

Proof.

You must show that for $\mathbf{p} = \text{proj}_W \mathbf{y}$,

$$\|\mathbf{y} - \mathbf{p}\| \leq \|\mathbf{y} - \mathbf{w}\|, \quad \text{for all } \mathbf{w} \in W$$

Recognize the key words *for all* and use the choose method to choose an n-vector $\mathbf{w} \in W$. You must show that

$$\|\mathbf{y} - \mathbf{p}\| \leq \|\mathbf{y} - \mathbf{w}\|$$

Now $\mathbf{w} \in W$ and $\mathbf{p} \in W$, so $\mathbf{w} - \mathbf{p} \in W$. By Theorem 7.5 in Section 7.2.2, $\mathbf{y} - \mathbf{p}$ is orthogonal to $\mathbf{w} - \mathbf{p}$, as shown in Figure 7.9. The desired inequality is derived from the dotted right triangle in Figure 7.9 because:

$$\|\mathbf{y} - \mathbf{w}\|^2 = \|\mathbf{y} - \mathbf{p}\|^2 + \|\mathbf{w} - \mathbf{p}\|^2 \quad \text{(Pythagorean theorem)}$$
$$\geq \|\mathbf{y} - \mathbf{p}\|^2$$

The result follows on taking the positive square root of both sides of the foregoing inequality. ■

You should now explain how Theorem 7.7 provides a solution to the least-squares problem. The first step in doing so is to match the notation of Theorem 7.7 to that of Problem 7.4, as follows:

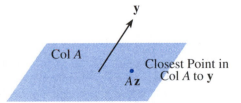

Figure 7.8 *A Geometric Solution to the Least-Squares Problem.*

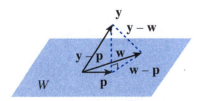

Figure 7.9 *Finding the Point in W Closest to* $\mathbf{y}$.

1. The vector $\mathbf{y}$ in Theorem 7.7 is the vector $\mathbf{y}$ in Problem 7.4.

2. The subspace W in Theorem 7.7 is Col A in Problem 7.4.

With this matching of notation, you can use Theorem 7.7 to obtain the following solution, $\mathbf{z}$, to the least-squares problem:

(a) Compute $\mathbf{p} =$ the projection of $\mathbf{y}$ onto Col A.

(b) Solve the (consistent) system of linear equations $A\mathbf{z} = \mathbf{p}$ for $\mathbf{z}$.

As you know, one of the uses of mathematics is to reduce the amount of computational effort needed to solve a problem. You can do so in this case by noting that $\mathbf{p}$, as computed in (a), has the property that $\mathbf{p} - \mathbf{y}$ is orthogonal to Col A and, in particular, to each column j of A. That is, $(A_{*j})^t(\mathbf{p} - \mathbf{y}) = 0$, or, in matrix-vector notation:

$$A^t(\mathbf{p} - \mathbf{y}) = \mathbf{0}$$

Using the fact that $\mathbf{p} = A\mathbf{z}$ from (b), you have that

$$A^t(\mathbf{p} - \mathbf{y}) = A^t\mathbf{p} - A^t\mathbf{y} = (A^tA)\mathbf{z} - A^t\mathbf{y} = \mathbf{0}$$

Equivalently stated, to solve the least-squares problems, you want to find an n-vector $\mathbf{z}$ that satisfies the following **normal equations**:

$$(A^tA)\mathbf{z} = A^t\mathbf{y} \tag{7.29}$$

In summary, the following is the solution to the least-squares problem.

> **SOLUTION to Problem 7.4**
>
> To solve the least-squares problem, use the given $(k \times n)$ matrix A and column vector $\mathbf{y} \in R^{k \times 1}$ to solve the following $(n \times n)$ system of linear equations for $\mathbf{z}$:
>
> $$(A^tA)\mathbf{z} = A^t\mathbf{y}$$

If the $(n \times n)$ matrix A^tA is invertible, then the solution to Problem 7.4 is:

$$\mathbf{z} = (A^tA)^{-1}A^t\mathbf{y}$$

A condition that ensures A^tA is invertible is given in the following theorem.

> **THEOREM 7.8**
>
> Let A be a $(k \times n)$ matrix. The $(n \times n)$ matrix A^tA is invertible if and only if the columns of A are linearly independent.

Proof.

The key words *if and only if* mean that two proofs are necessary. So, assume first that A^tA is invertible. To show that the columns of A are linearly independent, by definition, it must be shown that for all n-vectors $\mathbf{x}$ with $A\mathbf{x} = \mathbf{0}$, $\mathbf{x} = \mathbf{0}$. Use the choose method to choose an n-vector $\mathbf{x}$ with $A\mathbf{x} = \mathbf{0}$. Multiplying both sides by A^t, you have that

$$A^t(A\mathbf{x}) = (A^tA)\mathbf{x} = \mathbf{0} \tag{7.30}$$

From the assumption that $A'A$ is invertible, you can multiply both sides of (7.30) by $(A'A)^{-1}$ to conclude that $\mathbf{x} = \mathbf{0}$, and so the columns of A are linearly independent.

It remains to show that if the columns of A are linearly independent, then $A'A$ is invertible. To show that $A'A$ is invertible, it is shown that the columns of $A'A$ are linearly independent, that is, for all n-vectors $\mathbf{x}$ with $(A'A)\mathbf{x} = \mathbf{0}$, $\mathbf{x} = \mathbf{0}$. So, use the choose method to choose an n-vector $\mathbf{x}$ with $(A'A)\mathbf{x} = \mathbf{0}$. Multiplying through by the row vector $\mathbf{x}'$ yields

$$x' A' A \mathbf{x} = \mathbf{0} \tag{7.31}$$

Letting $\mathbf{y} = A\mathbf{x}$ and noting that $\mathbf{y}' = \mathbf{x}'A'$, you can write (7.31) as follows:

$$\mathbf{y}'\mathbf{y} = \mathbf{0}$$

The only way this can happen is if $\mathbf{y} = \mathbf{0}$, that is,

$$A\mathbf{x} = \mathbf{0} \tag{7.32}$$

The desired result that $\mathbf{x} = \mathbf{0}$ follows from the assumption that the columns of A are linearly independent.

The proof is now complete because it has been shown that the matrix $A'A$ is invertible if and only if the columns of A are linearly independent. ∎

For the problem of Connie Baker,

$$A = \begin{bmatrix} 6 & 1 \\ 7 & 1 \\ 8 & 1 \\ 9 & 1 \end{bmatrix} \quad \text{and} \quad \mathbf{y} = \begin{bmatrix} 8.44 \\ 8.99 \\ 9.56 \\ 10.14 \end{bmatrix}$$

so

$$A'A = \begin{bmatrix} 6 & 7 & 8 & 9 \\ 1 & 1 & 1 & 1 \end{bmatrix} \begin{bmatrix} 6 & 1 \\ 7 & 1 \\ 8 & 1 \\ 9 & 1 \end{bmatrix} = \begin{bmatrix} 230 & 30 \\ 30 & 4 \end{bmatrix} \quad \text{and} \quad A'\mathbf{y} = \begin{bmatrix} 281.31 \\ 37.13 \end{bmatrix}$$

Having made these computations, you can now solve the normal equations $(A'A)\mathbf{z} = A'\mathbf{y}$ for $\mathbf{z}$ by finding $(A'A)^{-1}$ because the columns of A are linearly independent, so, by Theorem 7.8, $A'A$ is invertible. Thus,

$$\mathbf{z} = (A'A)^{-1}A'\mathbf{y} = \begin{bmatrix} 0.2 & -1.5 \\ -1.5 & 11.5 \end{bmatrix} \begin{bmatrix} 281.31 \\ 37.13 \end{bmatrix} = \begin{bmatrix} 0.567 \\ 5.030 \end{bmatrix} = \begin{bmatrix} m \\ b \end{bmatrix}$$

In other words, these results indicate that the following line best fits the data according to the measure of least-squared error:

$$y = 0.567x + 5.03$$

The interpretation of the slope and intercept and the use of this line for making predictions was described in Section 7.3.1 and is not repeated here.

Exercises for Section 7.3

Measuring Devices is developing a device for determining the distance from the device to an object moving in a straight line at a constant speed, as a function of time. Complete Exercise 1 to Exercise 5 using the following data that show the distance of a moving power boat from the measuring device at various points in time:

Time (in seconds)	0	2	4	7	10	12
Distance (in feet)	5000	5075	5175	5300	5400	5500

1. Define the dependent variable, y, and the independent variable, x. Include units.

2. Suppose the relation $y = mx + b$ holds. What are the units of m and b? Interpret the meaning of m and b in the context of this problem.

3. What is the design matrix for this problem?

4. Use the design matrix in the previous exercise to write the normal equations for finding the slope and intercept of the line that best fits the data.

5. Use a calculator, software package, or other means to solve the normal equations in the previous exercise. Write the equation of the resulting line. How far is the object from the measuring device after 30 seconds?

The Agricultural Department of Kansas needs to estimate the annual winter-wheat harvest for the state. The director knows that the yield depends on the rainfall during the previous summer. Use the following data to complete Exercise 6 to Exercise 10:

Year	1991	1992	1993	1994	1995	1996	1997
Rainfall (in inches)	6.0	7.0	8.5	7.0	5.5	7.5	6.5
Yield (in 000's of bushels)	550	600	750	625	500	650	600

6. Define the dependent variable, y, and the independent variable, x. Include units.

7. Suppose the relation $y = mx + b$ holds. What are the units of m and b? Interpret the meaning of m and b in the context of this problem.

8. What is the design matrix for this problem?

9. Use the design matrix in the previous exercise to write the normal equations for finding the slope and intercept of the line that best fits the data.

10. Use a calculator, software package, or other means to solve the normal equations in the previous exercise. Write the equation of the resulting line. How much wheat can the state expect if there are 8 inches of rain in the previous summer?

Reread the problem of Philadelphia Paints presented at the beginning of this chapter. Then complete Exercise 11 through Exercise 13.

11. Define the dependent variable, y, and the independent variable, x. Include units.

12. By drawing a graph of the following data for this problem, postulate a functional relationship between y and x that conforms to the observation of the CEO, namely, that the cost per gallon of this paint decreases as the number of gallons produced increases,

but only up to a certain point, after which the cost begins to rise again. What are the parameters?

Thousands of gallons produced	1	2	3	4	4.5	5
Cost-per-gallon (in $)	6.00	5.00	5.50	4.75	5.50	4.00

Thousands of gallons produced	5	6.5	7.5	8	9	10
Cost-per-gallon (in $)	5.25	4.25	5.25	4.50	5.75	5.50

13. Explain how to use this model once you obtain values for the parameters.

Scientists need to know how many grams, g, of a radioactive material are left at various points in time, denoted by the independent variable t. Complete Exercise 14 through Exercise 18, assuming that the following relationship holds between g and t:

$$g = g_0 e^{-\alpha t} \tag{7.33}$$

14. What are the units of the parameter g_0? Interpret the meaning of g_0 in the context of this problem.

15. The relationship between g and t in (7.33) is not linear. By taking the natural logarithm of both sides of (7.33) and defining new variables, y and x, appropriately, as well as new parameters, m and b, show that you obtain the following linear relationship:

$$y = mx + b \tag{7.34}$$

16. Explain how to use data $(t_1, g_1), \ldots, (t_k, g_k)$ for the original variables t and g to create data $(x_1, y_1), \ldots, (x_k, y_k)$ for the new variables x and y in the previous exercise.

17. Assume you solve the normal equations associated with (7.34) and obtain estimates for m and b. How do you use those values to get estimates for the original parameters α and g_0?

18. Explain how to use the model in (7.33) once you obtain values for α and g_0.

For Exercise 19 through Exercise 22, consider using data $(x_1, y_1), \ldots, (x_k, y_k)$ to determine the parameters m and b for the following regression line:

$$y = mx + b$$

Recall the following notations:

$$A = \begin{bmatrix} x_1 & 1 \\ \vdots & \vdots \\ x_k & 1 \end{bmatrix}, \quad y = \begin{bmatrix} y_1 \\ \vdots \\ y_k \end{bmatrix}, \quad \text{and} \quad z = \begin{bmatrix} m \\ b \end{bmatrix}$$

19. By computing $A^t A$ and $A^t y$ explicitly, rewrite the normal equations $(A^t A)z = A^t y$ as the following system of two linear equations in the two unknowns m and b by filling

in the parentheses with appropriate expressions in terms of the data:

$$(\quad)m + (\quad)b = (\quad)$$
$$(\quad)m + (\quad)b = (\quad)$$

20. Find the inverse of the (2×2) matrix in the previous exercise and solve the two linear equations to obtain the values for m and b. Express your answer in terms of the following notations (which are typically used in statistics books):

$$\sum x = \sum_{i=1}^{k} x_i, \quad \sum x^2 = \sum_{i=1}^{k} x_i^2, \quad \sum y = \sum_{i=1}^{k} y_i, \quad \text{and} \quad \sum xy = \sum_{i=1}^{k} x_i y_i$$

21. Assume now that the value of y depends on the values of two independent variables, u and v, according to the following relation:

$$y = \beta_0 + \beta_1 u + \beta_2 v \tag{7.35}$$

Suppose you collect data in the form of k values for y, u, and v, say, $(u_1, v_1, y_1), \ldots, (u_k, v_k, y_k)$.

   (a) Follow the approach in Section 7.3.2 to derive a design matrix, A, and a parameter vector, $\mathbf{z}$, for determining the values of β_0, β_1, and β_2 that make the equation in (7.35) best fit the data, according to the measure of sum of squared errors. What are the dimensions of A and $\mathbf{z}$ and what are their elements?

   (b) How do you use the design matrix A to find the values of the parameter vector $\mathbf{z}$?

22. Repeat the previous exercise when y depends on the values of n independent variables, say, $u_1, \ldots, u_n$, according to the following relationship:

$$y = \beta_0 + \beta_1 u_1 + \cdots + \beta_n u_n$$

7.4 Inner Product Spaces

You have seen important uses of the dot product of two n-vectors $\mathbf{u}$ and $\mathbf{v}$:

$$\mathbf{u} \cdot \mathbf{v} = u_1 v_1 + \cdots + u_n v_n \quad \text{(dot product for n-vectors)} \tag{7.36}$$

In this section, generalization and abstraction are used to extend the concept of the dot product to two vectors $\mathbf{u}$ and $\mathbf{v}$ belonging to a vector space, thus enabling you to solve problems in this more general setting.

7.4.1 The Axiomatic System of an Inner Product Space

When $\mathbf{u}$ and $\mathbf{v}$ are vectors belonging to any vector space V, (7.36) results in a syntax error because the real numbers $u_1, \ldots, u_n$ and $v_1, \ldots, v_n$ are not defined. The only mathematical objects in V are vectors; real numbers are not elements of V. One way to avoid the syntax error is not to specify the right side of (7.36). That is, think of the left side of (7.36) as an operation that combines $\mathbf{u}$ and $\mathbf{v}$ to produce a real number whose value is not specified. To differentiate this general operation from the specific one in (7.36) that applies only to n-vectors, the general operation is called an *inner product* and is written as follows:

$$\langle \mathbf{u}, \mathbf{v} \rangle \quad \text{(inner product)} \tag{7.37}$$

The formula in (7.36) is a special case of (7.37) that applies to n-vectors.

You now have an abstract system consisting of a vector space V and the operation defined in (7.37) that combines two vectors in V to produce the real number $\langle \mathbf{u}, \mathbf{v} \rangle$. Axioms are needed to restore desired properties of the dot product that are lost through abstraction. For example, when $\mathbf{u}$ and $\mathbf{v}$ are n-vectors, you know that $\mathbf{u} \cdot \mathbf{v} = \mathbf{v} \cdot \mathbf{u}$. If you want to restore this property in the abstract system, then you can include the following axiom:

$$\langle \mathbf{u}, \mathbf{v} \rangle = \langle \mathbf{v}, \mathbf{u} \rangle$$

You must decide what properties to include as axioms. How you do so depends on the specific examples you are working with, the properties you feel are important, the kinds of results you eventually want to obtain about the abstract system, and more. One such set of axioms is presented in the following axiomatic system.

DEFINITION 7.3

An operation on a vector space V that associates to each pair of vectors $\mathbf{u}, \mathbf{v} \in V$, the real number $\langle \mathbf{u}, \mathbf{v} \rangle$ is an **inner product** on V if and only if the following properties hold for all vectors $\mathbf{u}, \mathbf{v}, \mathbf{w} \in V$ and for all real numbers t:

(1) $\langle \mathbf{u}, \mathbf{v} \rangle = \langle \mathbf{v}, \mathbf{u} \rangle$.

(2) $\langle \mathbf{u} + \mathbf{v}, \mathbf{w} \rangle = \langle \mathbf{u}, \mathbf{w} \rangle + \langle \mathbf{v}, \mathbf{w} \rangle$.

(3) $\langle t\mathbf{u}, \mathbf{v} \rangle = t \langle \mathbf{u}, \mathbf{v} \rangle$.

(4) $\langle \mathbf{u}, \mathbf{u} \rangle \geq 0$; and also, $\langle \mathbf{u}, \mathbf{u} \rangle = 0$ if and only if $\mathbf{u} = \mathbf{0}$.

A vector space with an inner product is an **inner product space**.

Note that the symbol "+" on the left side of (2) in Definition 7.3 combines two vectors and the symbol "+" on the right side combines two real numbers. Likewise, the expression $t\mathbf{u}$ on the left side of (3) in Definition 7.3 combines the real number t with the vector $\mathbf{u}$ to produce the vector $t\mathbf{u}$. In contrast, the expression $t \langle \mathbf{u}, \mathbf{v} \rangle$ is the multiplication of the real number t by the real number $\langle \mathbf{u}, \mathbf{v} \rangle$.

From Theorem 1.3 in Section 1.2.5, the operation of the dot product of two n-vectors in (7.36), is an inner product on R^n, so R^n together with this operation is an inner product space. Two other examples of inner product spaces that arise in physics, statistics, and engineering follow. In each case, the four axioms in Definition 7.3 are verified.

EXAMPLE 7.11 An Inner Product Space on R^2

For 2-vectors $\mathbf{u} = (u_1, u_2)$ and $\mathbf{v} = (v_1, v_2)$, the following operation is an inner product on R^2:

$$\langle \mathbf{u}, \mathbf{v} \rangle = 2u_1 v_1 + 5u_2 v_2$$

You can see that axiom (1) holds because

$$\langle \mathbf{u}, \mathbf{v} \rangle = 2u_1 v_1 + 5u_2 v_2 = 2v_1 u_1 + 5v_2 u_2 = \langle \mathbf{v}, \mathbf{u} \rangle$$

For axiom (2), let $\mathbf{w} = (w_1, w_2)$, then

$$\begin{aligned}
\langle \mathbf{u} + \mathbf{v}, \mathbf{w} \rangle &= 2(u_1 + v_1)w_1 + 5(u_2 + v_2)w_2 \\
&= (2u_1 w_1 + 5u_2 w_2) + (2v_1 w_1 + 5v_2 w_2) \\
&= \langle \mathbf{u}, \mathbf{w} \rangle + \langle \mathbf{v}, \mathbf{w} \rangle
\end{aligned}$$

For axiom (3), let t be a real number, then

$$\begin{aligned}
\langle t\mathbf{u}, \mathbf{v} \rangle &= 2(tu_1)v_1 + 5(tu_2)v_2 \\
&= t(2u_1 v_1 + 5u_2 v_2) \\
&= t\langle \mathbf{u}, \mathbf{v} \rangle
\end{aligned}$$

For axiom (4), note that

$$\begin{aligned}
\langle \mathbf{u}, \mathbf{u} \rangle &= 2u_1 u_1 + 5u_2 u_2 \\
&= 2u_1^2 + 5u_2^2 \\
&\geq 0
\end{aligned}$$

Thus, if $\mathbf{u} = \mathbf{0}$, then $u_1 = 0$ and $u_2 = 0$, so $\langle \mathbf{u}, \mathbf{u} \rangle = 0$. Finally, if $\langle \mathbf{u}, \mathbf{u} \rangle = 0$, then $2u_1^2 + 5u_2^2 = 0$, so it must be that $u_1 = u_2 = 0$, hence $\mathbf{u} = \mathbf{0}$.

A sequential generalization of Example 7.11 arises by first replacing the numbers 2 and 5 with any positive real numbers a and b, and then by replacing $\mathbf{u}$ and $\mathbf{v}$ with n-vectors. You are asked to show in Exercise 1 and Exercise 2 that these generalizations do in fact result in inner product spaces. Such an inner product arises in a generalization of the least-squares problem introduced in Section 7.3.

The next example deals with polynomials of degree n, which are written either as $\mathbf{p}$ or as $\mathbf{p}(x)$.

EXAMPLE 7.12 The Inner Product of Two Polynomials

For polynomials $\mathbf{p}, \mathbf{q} \in P^n$, the vector space of all polynomials of degree n or less, let $(x_0, x_1, \ldots, x_n)$ be $n + 1$ distinct real numbers and define

$$\langle \mathbf{p}, \mathbf{q} \rangle = \mathbf{p}(x_0)\mathbf{q}(x_0) + \cdots + \mathbf{p}(x_n)\mathbf{q}(x_n)$$

You are asked to verify in Exercise 3 that axioms (1) - (3) in Definition 7.3 hold. To see that axiom (4) holds, note that

$$\langle \mathbf{p}, \mathbf{p} \rangle = [\mathbf{p}(x_0)]^2 + \cdots + [\mathbf{p}(x_n)]^2 \geq 0$$

Thus, if $\mathbf{p} = \mathbf{0}$, then $\mathbf{p}(x_0) = \cdots = \mathbf{p}(x_n) = 0$, so $\langle \mathbf{p}, \mathbf{p} \rangle = 0$. Finally, if $\langle \mathbf{p}, \mathbf{p} \rangle = 0$, then $[\mathbf{p}(x_0)]^2 + \cdots + [\mathbf{p}(x_n)]^2 = 0$, so it must be that $\mathbf{p}(x_0) = \cdots = \mathbf{p}(x_n) = 0$. But the only way that the polynomial $\mathbf{p}$ of degree n or less can have $n + 1$ roots is if $\mathbf{p}$ is the zero polynomial, that is, $\mathbf{p} = \mathbf{0}$.

7.4.2 Length, Distance, and Orthogonality

An important geometric property of a vector $\mathbf{v} = (v_1, \ldots, v_n) \in R^n$ is the length, or norm, of $\mathbf{v}$, defined as follows:

$$\|\mathbf{v}\| = \sqrt{v_1^2 + \cdots + v_n^2} \quad \text{(norm of an } n\text{-vector)} \tag{7.38}$$

Such a property for a vector $\mathbf{v}$ in an inner product space V is often useful in problem solving, however, the right side of (7.38) is undefined when $\mathbf{v}$ is a vector in V, thus resulting in a syntax error.

One way to overcome this error is to express the length of an n-vector in terms of the dot product, as follows

$$\|\mathbf{v}\| = \sqrt{\mathbf{v} \cdot \mathbf{v}} \quad \text{(expressing the norm as a dot product)} \tag{7.39}$$

Now, to generalize (7.39) to a vector $\mathbf{v}$ in an inner product space V, replace the dot product with the inner product in V, which results in the following **length** or **norm** of $\mathbf{v}$:

$$\|\mathbf{v}\| = \sqrt{\langle \mathbf{v}, \mathbf{v} \rangle} \quad \text{(expressing the norm as an inner product)} \tag{7.40}$$

A **unit vector** in an inner product space is a vector whose length, as defined by (7.40), is 1.

You can also define the **distance between two vectors u** and **v** in an inner product space V to be $\|\mathbf{u} - \mathbf{v}\|$. Also, $\mathbf{u}$ and $\mathbf{v}$ are **orthogonal** if and only if $\langle \mathbf{u}, \mathbf{v} \rangle = 0$.

EXAMPLE 7.13 Length in an Inner Product Space

For the polynomials

$$\mathbf{p} = \mathbf{p}(x) = \frac{1}{\sqrt{2}}x^2 \quad \text{and} \quad \mathbf{q} = \mathbf{q}(x) = x^2 + 3x - 1$$

and the points $x_0 = -1$, $x_1 = 0$, and $x_2 = 1$, you have that

$$\mathbf{p}(x_0) = \mathbf{p}(-1) = \frac{1}{\sqrt{2}}, \quad \mathbf{q}(x_0) = \mathbf{q}(-1) = -3$$

$$\mathbf{p}(x_1) = \mathbf{p}(0) = 0, \quad \mathbf{q}(x_1) = \mathbf{q}(0) = -1$$

$$\mathbf{p}(x_2) = \mathbf{p}(1) = \frac{1}{\sqrt{2}}, \quad \mathbf{q}(x_2) = \mathbf{q}(1) = 3$$

Thus, the inner product in Example 7.12 results in

$$\langle \mathbf{p}, \mathbf{q} \rangle = \mathbf{p}(x_0)\mathbf{q}(x_0) + \mathbf{p}(x_1)\mathbf{q}(x_1) + \mathbf{p}(x_2)\mathbf{q}(x_2)$$

$$= \mathbf{p}(-1)\mathbf{q}(-1) + \mathbf{p}(0)\mathbf{q}(0) + \mathbf{p}(1)\mathbf{q}(1)$$

$$= \frac{1}{\sqrt{2}}(-3) + 0(-1) + \frac{1}{\sqrt{2}}(3) = 0$$

$$\|\mathbf{p}\| = \sqrt{\langle \mathbf{p}, \mathbf{p} \rangle} = \sqrt{\mathbf{p}(-1)^2 + \mathbf{p}(0)^2 + \mathbf{p}(1)^2} = \sqrt{\frac{1}{2} + 0 + \frac{1}{2}} = 1$$

From these calculations it follows that, for this particular inner product, $\mathbf{p}$ and $\mathbf{q}$ are orthogonal and $\mathbf{p}$ is a unit vector in P^2.

7.4.3 Results for Inner Product Spaces

Using the concepts of length, distance, and orthogonality, it is possible to generalize most of the previous results in this chapter to vectors in an inner product space V.

Generalization of Projections

For example, given $\mathbf{v}, \mathbf{d} \in V$ with $\mathbf{d} \neq \mathbf{0}$, the orthogonal projection of $\mathbf{v}$ onto $\mathbf{d}$ is

$$\mathbf{x} = \frac{\langle \mathbf{v}, \mathbf{d} \rangle}{\langle \mathbf{d}, \mathbf{d} \rangle} \mathbf{d}$$

and the component of $\mathbf{v}$ orthogonal to $\mathbf{d}$ is

$$\mathbf{y} = \mathbf{v} - \mathbf{x} = \mathbf{v} - \frac{\langle \mathbf{v}, \mathbf{d} \rangle}{\langle \mathbf{d}, \mathbf{d} \rangle} \mathbf{d}$$

Likewise, every vector $\mathbf{v}$ in V has a unique projection onto a subspace W of V, and that projection is the closest point in W to $\mathbf{v}$, as measured using the inner product. Also, there is a unique orthogonal decomposition of every vector $\mathbf{v} \in V$, that is, there are unique vectors $\mathbf{x} \in W$ and $\mathbf{y} \in W^\perp$ such that

$$\mathbf{v} = \mathbf{x} + \mathbf{y} \tag{7.41}$$

In particular, if $\mathbf{w}_1, \ldots, \mathbf{w}_k$ are an orthogonal basis for a subspace W of V, then

$$\mathbf{x} = \frac{\langle \mathbf{v}, \mathbf{w}_1 \rangle}{\langle \mathbf{w}_1, \mathbf{w}_1 \rangle} \mathbf{w}_1 + \cdots + \frac{\langle \mathbf{v}, \mathbf{w}_k \rangle}{\langle \mathbf{w}_k, \mathbf{w}_k \rangle} \mathbf{w}_k \quad \text{and} \quad \mathbf{y} = \mathbf{v} - \mathbf{x} \tag{7.42}$$

The Gram-Schmidt Process

It is also possible to use an appropriately modified Gram-Schmidt process to find an orthogonal basis for a subspace W of V. To illustrate, consider the basis consisting of the following polynomials for the subspace P^2 of P^3:

$$\mathbf{q}_1(x) = 1, \quad \mathbf{q}_2(x) = x, \quad \text{and} \quad \mathbf{q}_3(x) = x^2$$

You can use this basis together with the Gram-Schmidt process (see Section 7.2.3) to find an orthogonal basis, say, $\mathbf{p}_1, \mathbf{p}_2$, and $\mathbf{p}_3$ for P^2. To do so, you must first define an inner product on P^3. After doing so, the Gram-Schmidt process becomes:

$$\mathbf{p}_1 = \mathbf{q}_1$$

$$\mathbf{p}_2 = \mathbf{q}_2 - \frac{\langle \mathbf{q}_2, \mathbf{p}_1 \rangle}{\langle \mathbf{p}_1, \mathbf{p}_1 \rangle} \mathbf{p}_1$$

$$\mathbf{p}_3 = \mathbf{q}_3 - \frac{\langle \mathbf{q}_3, \mathbf{p}_1 \rangle}{\langle \mathbf{p}_1, \mathbf{p}_1 \rangle} \mathbf{p}_1 - \frac{\langle \mathbf{q}_3, \mathbf{p}_2 \rangle}{\langle \mathbf{p}_2, \mathbf{p}_2 \rangle} \mathbf{p}_2$$

In this example, suppose the inner product in Example 7.12 is used for P^3 with $x_0 = -2$, $x_1 = -1$, $x_2 = 1$, and $x_3 = 2$. Then, following the Gram-Schmidt process, $\mathbf{p}_1 = \mathbf{q}_1 = 1$. Also, as you are asked to verify in Exercise 13, $\mathbf{p}_1 = 1$ is orthogonal to $\mathbf{q}_2 = x$, so $\langle \mathbf{q}_2, \mathbf{p}_1 \rangle = 0$ and hence

$$\mathbf{p}_2 = \mathbf{q}_2 - \frac{\langle \mathbf{q}_2, \mathbf{p}_1 \rangle}{\langle \mathbf{p}_1, \mathbf{p}_1 \rangle} \mathbf{p}_1 = \mathbf{q}_2 = x$$

It remains to find $\mathbf{q}_3$ by using the formula

$$\mathbf{p}_3 = \mathbf{q}_3 - \frac{\langle \mathbf{q}_3, \mathbf{p}_1 \rangle}{\langle \mathbf{p}_1, \mathbf{p}_1 \rangle} \mathbf{p}_1 - \frac{\langle \mathbf{q}_3, \mathbf{p}_2 \rangle}{\langle \mathbf{p}_2, \mathbf{p}_2 \rangle} \mathbf{p}_2 \qquad (7.43)$$

So, compute

$$\begin{aligned}
\langle \mathbf{q}_3, \mathbf{p}_1 \rangle &= \mathbf{q}_3(-2)\mathbf{p}_1(-2) + \mathbf{q}_3(-1)\mathbf{p}_1(-1) + \mathbf{q}_3(1)\mathbf{p}_1(1) + \mathbf{q}_3(2)\mathbf{p}_1(2) \\
&= 4(1) + 1(1) + 1(1) + 4(1) = 10
\end{aligned}$$

$$\begin{aligned}
\langle \mathbf{p}_1, \mathbf{p}_1 \rangle &= \mathbf{p}_1(-2)\mathbf{p}_1(-2) + \mathbf{p}_1(-1)\mathbf{p}_1(-1) + \mathbf{p}_1(1)\mathbf{p}_1(1) + \mathbf{p}_1(2)\mathbf{p}_1(2) \\
&= 1(1) + 1(1) + 1(1) + 1(1) = 4
\end{aligned}$$

$$\begin{aligned}
\langle \mathbf{q}_3, \mathbf{p}_2 \rangle &= \mathbf{q}_3(-2)\mathbf{p}_2(-2) + \mathbf{q}_3(-1)\mathbf{p}_2(-1) + \mathbf{q}_3(1)\mathbf{p}_2(1) + \mathbf{q}_3(2)\mathbf{p}_2(2) \\
&= 4(-2) + 1(-1) + 1(1) + 4(2) = 0
\end{aligned}$$

$$\begin{aligned}
\langle \mathbf{p}_2, \mathbf{p}_2 \rangle &= \mathbf{p}_2(-2)\mathbf{p}_2(-2) + \mathbf{p}_2(-1)\mathbf{p}_2(-1) + \mathbf{p}_2(1)\mathbf{p}_2(1) + \mathbf{p}_2(2)\mathbf{p}_2(2) \\
&= (-2)(-2) + (-1)(-1) + 1(1) + 2(2) = 10
\end{aligned}$$

Making the foregoing substitutions in (7.43) yields

$$\mathbf{p}_3 = \mathbf{q}_3 - \frac{\langle \mathbf{q}_3, \mathbf{p}_1 \rangle}{\langle \mathbf{p}_1, \mathbf{p}_1 \rangle} \mathbf{p}_1 - \frac{\langle \mathbf{q}_3, \mathbf{p}_2 \rangle}{\langle \mathbf{p}_2, \mathbf{p}_2 \rangle} \mathbf{p}_2 = \mathbf{q}_3 - \frac{10}{4}\mathbf{p}_1 - \frac{0}{10}\mathbf{p}_2 = x^2 - \frac{5}{2}$$

In summary, an orthogonal basis for P^2 consists of the three polynomials $1, x,$ and $x^2 - (5/2)$.

The Cauchy Schwarz and Triangle Inequalities

Two inequalities involving an inner product are presented next. The first one provides the basis for defining the angle between two vectors, as explained subsequently.

THEOREM 7.9

(The Cauchy-Schwarz Inequality)

For all vectors $\mathbf{u}, \mathbf{v}$ in an inner product space V,

$$|\langle \mathbf{u}, \mathbf{v} \rangle| \leq \|\mathbf{u}\| \|\mathbf{v}\| \qquad (7.44)$$

Proof.

If $\mathbf{u} = \mathbf{0}$, then both sides of (7.44) are 0 and so the inequality is true. Consider, therefore, the case in which $\mathbf{u} \neq \mathbf{0}$ and let W be the subspace spanned by $\mathbf{u}$. Then

$$\| \operatorname{proj}_W \mathbf{v} \| = \left\| \frac{\langle \mathbf{v}, \mathbf{u} \rangle}{\langle \mathbf{u}, \mathbf{u} \rangle} \mathbf{u} \right\| = \frac{|\langle \mathbf{v}, \mathbf{u} \rangle|}{|\langle \mathbf{u}, \mathbf{u} \rangle|} \|\mathbf{u}\| = \frac{|\langle \mathbf{v}, \mathbf{u} \rangle|}{\|\mathbf{u}\|^2} \|\mathbf{u}\| = \frac{|\langle \mathbf{v}, \mathbf{u} \rangle|}{\|\mathbf{u}\|} \tag{7.45}$$

Now, from Figure 7.10, the Pythagorean theorem yields

$$\| \operatorname{proj}_W \mathbf{v} \|^2 = \|\mathbf{v}\|^2 - \|\mathbf{v} - \operatorname{proj}_W \mathbf{v}\|^2 \leq \|\mathbf{v}\|^2$$

So, from (7.45),

$$\frac{|\langle \mathbf{v}, \mathbf{u} \rangle|}{\|\mathbf{u}\|} = \| \operatorname{proj}_W \mathbf{v} \| \leq \|\mathbf{v}\|$$

Multiplying the foregoing inequality through by $\|\mathbf{u}\|$ provides the desired result that $|\langle \mathbf{u}, \mathbf{v} \rangle| = |\langle \mathbf{v}, \mathbf{u} \rangle| \leq \|\mathbf{u}\| \|\mathbf{v}\|$. ∎

The Cauchy-Schwarz inequality makes it possible to define the angle between any two nonzero vectors $\mathbf{u}$ and $\mathbf{v}$. In particular, assuming that $\mathbf{u} \neq \mathbf{0}$ and that $\mathbf{v} \neq \mathbf{0}$, you can divide both sides of the Cauchy-Schwarz inequality by the positive number $\|\mathbf{u}\| \|\mathbf{v}\|$ to obtain

$$\frac{|\langle \mathbf{u}, \mathbf{v} \rangle|}{\|\mathbf{u}\| \|\mathbf{v}\|} \leq 1, \quad \text{or equivalently,} \quad -1 \leq \frac{\langle \mathbf{u}, \mathbf{v} \rangle}{\|\mathbf{u}\| \|\mathbf{v}\|} \leq 1$$

Because $\langle \mathbf{u}, \mathbf{v} \rangle / (\|\mathbf{u}\| \|\mathbf{v}\|)$ is between -1 and 1, you can find an angle whose cosine is this number and define the resulting value as the angle θ between $\mathbf{u}$ and $\mathbf{v}$. That is,

$$\theta = \text{angle between } \mathbf{u} \text{ and } \mathbf{v} = \arccos\left(\frac{\langle \mathbf{u}, \mathbf{v} \rangle}{\|\mathbf{u}\| \|\mathbf{v}\|} \right)$$

The Cauchy-Schwarz inequality is also used to prove other results, such as the following inequality.

THEOREM 7.10

(The Triangle Inequality)

For all vectors $\mathbf{u}$ and $\mathbf{v}$ in an inner product space V,

$$\|\mathbf{u} + \mathbf{v}\| \leq \|\mathbf{u}\| + \|\mathbf{v}\| \tag{7.46}$$

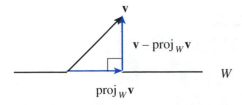

Figure 7.10 *The Pythagorean Theorem Applied to Projections.*

Proof.

Choosing vectors **u** and **v**, you have

$$\|\mathbf{u} + \mathbf{v}\|^2 = \langle \mathbf{u} + \mathbf{v}, \mathbf{u} + \mathbf{v} \rangle \qquad \text{(definition of norm)}$$

$$= \langle \mathbf{u}, \mathbf{u} \rangle + 2\langle \mathbf{u}, \mathbf{v} \rangle + \langle \mathbf{v}, \mathbf{v} \rangle \qquad \text{(laws of an inner product)}$$

$$= \|\mathbf{u}\|^2 + 2\langle \mathbf{u}, \mathbf{v} \rangle + \|\mathbf{v}\|^2 \qquad \text{(definition of norm)}$$

$$\leq \|\mathbf{u}\|^2 + 2\|\mathbf{u}\|\|\mathbf{v}\| + \|\mathbf{v}\|^2 \qquad \text{(Cauchy-Schwarz inequality)}$$

$$= (\|\mathbf{u}\| + \|\mathbf{v}\|)^2 \qquad \text{(algebra)}$$

The result now follows by taking the positive square root of both sides of the foregoing inequality. ■

7.4.4 An Application of Inner Products

In the linear regression model presented in Section 7.3.2, data for the dependent variable y and the independent variable x are used to determine values for the parameters m and b that make the following line best fit the given data:

$$y = mx + b$$

In doing so, the sum of squared errors is used to measure how well a particular line fits the data. Specifically, the measure of goodness of the line corresponding to given values of m and b, relative to the given data $(x_1, y_1), \ldots, (x_k, y_k)$, is:

$$\text{Sum of Squared Errors} = (p_1 - y_1)^2 + \cdots + (p_k - y_k)^2 \tag{7.47}$$

where, for each $i = 1, \ldots, k$,

$$p_i = mx_i + b \quad \text{(predicted value at } x_i\text{)}$$

Letting $\mathbf{y} = (y_1, \ldots, y_k)$ and $\mathbf{p} = (p_1, \ldots, p_k)$, the measure in (7.47) is $\|\mathbf{p} - \mathbf{y}\|^2$.

In some situations, the values of **y** are not equally reliable. For example, the measurements may be done at different points in time or taken under different circumstances. As another example, the values of **y** might be obtained from samples of different sizes. In such cases, it is more appropriate to weight the errors in (7.47), giving a larger weight to the more reliable measurements. This is accomplished by choosing a weight, w_i, for each data point. Then (7.47) becomes the following *weighted* squared error:

$$\text{Weighted Squared Error} = w_1^2(p_1 - y_1)^2 + \cdots + w_k^2(p_k - y_k)^2 \tag{7.48}$$

You can write (7.48) in the vector notation of an inner product space, as follows:

$$\text{Weighted Squared Error} = \|\mathbf{p} - \mathbf{y}\|^2 = \langle \mathbf{p} - \mathbf{y}, \mathbf{p} - \mathbf{y} \rangle$$

where

$$\langle \mathbf{u}, \mathbf{v} \rangle = w_1^2 u_1 v_1 + \cdots + w_k^2 u_k v_k \tag{7.49}$$

It is left to Exercise 6 to verify that (7.49) defines an inner product.

To solve the weighted least-squares problem, that is, to find values for m and b that minimize (7.48), you can solve a modified least-squares problem, as described in Section 7.3.3. There, $\mathbf{p}$ is written as

$$\mathbf{p} = A\mathbf{z} \tag{7.50}$$

where

$$A = \begin{bmatrix} x_1 & 1 \\ \vdots & \vdots \\ x_k & 1 \end{bmatrix} \quad \text{and} \quad \mathbf{z} = \begin{bmatrix} m \\ b \end{bmatrix}$$

As stated in the solution to Problem 7.4 in Section 7.3.3, to find the vector $\mathbf{z}$ that will

$$\text{minimize } \| A\mathbf{z} - \mathbf{y} \|^2 \tag{7.51}$$

you must solve the normal equations

$$(A'A)\mathbf{z} = A'\mathbf{y} \tag{7.52}$$

To solve the weighted least-squares problem, let W be the diagonal matrix whose diagonal elements are $w_1, \ldots, w_k$. Then, in matrix-vector notation, the problem of minimizing the weighted squared error in (7.48) becomes that of minimizing

$$\begin{aligned}
\text{Weighted Squared Error} &= w_1^2(p_1 - y_1)^2 + \cdots + w_k^2(p_k - y_k)^2 \\
&= (w_1 p_1 - w_1 y_1)^2 + \cdots + (w_k p_k - w_k y_k)^2 \\
&= \| W\mathbf{p} - W\mathbf{y} \|^2
\end{aligned}$$

Finally, replacing $\mathbf{p}$ with $A\mathbf{z}$ from (7.50), to solve the weighted least-squares problem, you want to find a vector $\mathbf{z}$ that will

$$\text{minimize } \| (WA)\mathbf{z} - W\mathbf{y} \|^2 \tag{7.53}$$

Comparing (7.53) with (7.51), you can see that A in (7.51) corresponds to WA in (7.53) and that $\mathbf{y}$ in (7.51) corresponds to $W\mathbf{y}$ in (7.53). Making these substitutions in (7.52) reduces solving the weighted least-squares problem to solving the following normal equations for $\mathbf{z}$:

$$[(WA)'(WA)]\mathbf{z} = (WA)'W\mathbf{y} \tag{7.54}$$

To illustrate, suppose you have a least-squares problem in which you want to find the line $y = mx + b$ that best fits the data $(1, 4)$, $(2, 3)$, $(3, 5)$, $(4, 3)$, $(5, 6)$. In this case, you have

$$A = \begin{bmatrix} 1 & 1 \\ 2 & 1 \\ 3 & 1 \\ 4 & 1 \\ 5 & 1 \end{bmatrix} \quad \text{and} \quad \mathbf{y} = \begin{bmatrix} 4 \\ 3 \\ 5 \\ 3 \\ 6 \end{bmatrix}$$

By computing

$$A^t A = \begin{bmatrix} 55 & 15 \\ 15 & 5 \end{bmatrix} \quad \text{and} \quad A^t \mathbf{y} = \begin{bmatrix} 67 \\ 21 \end{bmatrix}$$

the solution to the normal equations in (7.52) is

$$\mathbf{z} = \begin{bmatrix} m \\ b \end{bmatrix} = (A^t A)^{-1} A^t \mathbf{y} = \begin{bmatrix} 55 & 15 \\ 15 & 5 \end{bmatrix}^{-1} \begin{bmatrix} 67 \\ 21 \end{bmatrix} = \begin{bmatrix} 0.4 \\ 3.0 \end{bmatrix}$$

Thus, the line that best fits the given data, using the measure of sum of squared errors, is $y = 0.4x + 3$.

In contrast, suppose the same problem is solved as a weighted least-squares problem, in which the first and last measurements are given one quarter as much weight as the other data points, so $w_1 = 1$, $w_2 = 4$, $w_3 = 4$, $w_4 = 4$, and $w_5 = 1$. Thus,

$$W = \begin{bmatrix} 1 & 0 & 0 & 0 & 0 \\ 0 & 4 & 0 & 0 & 0 \\ 0 & 0 & 4 & 0 & 0 \\ 0 & 0 & 0 & 4 & 0 \\ 0 & 0 & 0 & 0 & 1 \end{bmatrix}, \quad WA = \begin{bmatrix} 1 & 1 \\ 8 & 4 \\ 12 & 4 \\ 16 & 4 \\ 5 & 1 \end{bmatrix}, \quad \text{and} \quad W\mathbf{y} = \begin{bmatrix} 4 \\ 12 \\ 20 \\ 12 \\ 6 \end{bmatrix}$$

By computing

$$(WA)^t (WA) = \begin{bmatrix} 490 & 150 \\ 150 & 50 \end{bmatrix} \quad \text{and} \quad (WA)^t W\mathbf{y} = \begin{bmatrix} 562 \\ 186 \end{bmatrix}$$

the solution to the normal equations in (7.54) is

$$\mathbf{z} = \begin{bmatrix} m \\ b \end{bmatrix} = [(WA)^t (WA)]^{-1} (WA)^t W\mathbf{y} = \begin{bmatrix} 490 & 150 \\ 150 & 50 \end{bmatrix}^{-1} \begin{bmatrix} 562 \\ 186 \end{bmatrix} = \begin{bmatrix} 0.10 \\ 3.42 \end{bmatrix}$$

Thus, the line that best fits the given data, using the measure of weighted squared error, is $y = 0.1x + 3.42$. You can see that these weights result in a line different from the one $y = 0.4x + 3$ obtained without the weights.

In this section, you have seen how abstraction is used to generalize the dot product of two n-vectors to an inner product for two vectors in a vector space. One application in which an inner product arises is the weighted least-squares problem.

Exercises for Section 7.4

1. Let a and b be positive real numbers. Show that the following operation on 2-vectors $\mathbf{u} = (u_1, u_2)$ and $\mathbf{v} = (v_1, v_2)$ defines an inner product on R^2 by verifying the four axioms in Definition 7.3:

$$\langle \mathbf{u}, \mathbf{v} \rangle = au_1 v_1 + bu_2 v_2$$

2. Let $a_1, \ldots, a_n$ be positive real numbers. Show that the following operation on n-vectors $\mathbf{u} = (u_1, \ldots, u_n)$ and $\mathbf{v} = (v_1, \ldots, v_n)$ defines an inner product on R^n by verifying the four axioms in Definition 7.3:

$$\langle \mathbf{u}, \mathbf{v} \rangle = a_1 u_1 v_1 + \cdots + a_n u_n v_n$$

3. For the vector space P^n of all polynomials of degree $\leq n$, let $x_0, x_1, \ldots, x_n$ be $n + 1$ distinct real numbers. Show that axioms (1), (2), and (3) in Definition 7.3 hold for the following operation on the polynomials $\mathbf{p}, \mathbf{q} \in P^n$:

$$\langle \mathbf{p}, \mathbf{q} \rangle = \mathbf{p}(x_0)\mathbf{q}(x_0) + \cdots + \mathbf{p}(x_n)\mathbf{q}(x_n)$$

4. Let $B = \{\mathbf{v}_1, \ldots, \mathbf{v}_n\}$ be a basis for a vector space V and $[\mathbf{v}]_B$ be the components of a vector $\mathbf{v} \in V$ relative to B. Show that the following operation on vectors $\mathbf{u}, \mathbf{v} \in V$ defines an inner product:

$$\langle \mathbf{u}, \mathbf{v} \rangle = ([\mathbf{u}]_B)^t [\mathbf{v}]_B$$

5. Let A be an $(n \times n)$ invertible matrix. Show that the following operation defines an inner product on R^n:

$$\langle \mathbf{u}, \mathbf{v} \rangle = (A\mathbf{u})^t (A\mathbf{v})$$

6. Let $w_1, \ldots, w_n$ be nonzero real numbers. Use the results in the previous exercise to show that the following operation on n-vectors $\mathbf{u} = (u_1, \ldots, u_n)$ and $\mathbf{v} = (v_1, \ldots, v_n)$ defines an inner product on R^n:

$$\langle \mathbf{u}, \mathbf{v} \rangle = w_1^2 u_1 v_1 + \cdots + w_n^2 u_n v_n$$

7. Use the inner product and numerical values of x_0, x_1, and x_2 from Example 7.13 in Section 7.4.2 to compute $\langle \mathbf{p}, \mathbf{q} \rangle$ for $\mathbf{p}(x) = 1 + x^2$ and $\mathbf{q}(x) = 2 - 3x$.

8. Use the inner product and numerical values of x_0, x_1, and x_2 from Example 7.13 in Section 7.4.2 to compute $\langle \mathbf{p}, \mathbf{q} \rangle$ for $\mathbf{p}(x) = 1 + x + x^2$ and $\mathbf{q}(x) = (1 - x)^2$.

9. Use the inner product and numerical values of x_0, x_1, and x_2 from Example 7.13 in Section 7.4.2 to compute the length of the polynomials $\mathbf{p}$ and $\mathbf{q}$ in Exercise 7. Then normalize $\mathbf{p}$ and $\mathbf{q}$.

10. Use the inner product and numerical values of x_0, x_1, and x_2 from Example 7.13 in Section 7.4.2 to compute the length of the polynomials $\mathbf{p}$ and $\mathbf{q}$ in Exercise 8. Then normalize $\mathbf{p}$ and $\mathbf{q}$.

11. Find the orthogonal projection of $\mathbf{p}$ onto $\mathbf{q}$ for the polynomials in Exercise 7.

12. Find the orthogonal projection of $\mathbf{p}$ onto $\mathbf{q}$ for the polynomials in Exercise 8.

13. Consider the inner product for P^3 given in Example 7.12 in Section 7.4.2, with $x_0 = -2$, $x_1 = -1, x_2 = 1$, and $x_3 = 2$. Show that for this inner product, the polynomial $\mathbf{p}(x) = 1$ is orthogonal to the polynomial $\mathbf{q}(x) = x$.

14. Compute the distance from the polynomial $\mathbf{p}$ to the polynomial $\mathbf{q}$ in Example 7.13 in Section 7.4.2.

15. Consider the inner product for P^3 in Example 7.12 in Section 7.4.2, with $x_0 = -3$, $x_1 = -1$, $x_2 = 1$, and $x_3 = 3$ and the following basis for the subspace P^2:

$$\mathbf{q}_1(x) = 1, \quad \mathbf{q}_2(x) = x, \quad \text{and} \quad \mathbf{q}_3(x) = x^2$$

   (a) Compute the orthogonal projection of $\mathbf{q}_3$ onto the subspace spanned by $\mathbf{q}_1$ and $\mathbf{q}_2$.
   (b) Use the result in part (a) to find an orthogonal basis for P^2.

16. Consider the inner product for P^4 in Example 7.12 in Section 7.4.2, with $x_0 = -2$, $x_1 = -1$, $x_2 = 0$, $x_3 = 1$, and $x_4 = 2$ and the following basis for the subspace P^2:

$$\mathbf{q}_1(x) = 1, \quad \mathbf{q}_2(x) = x, \quad \text{and} \quad \mathbf{q}_3(x) = x^2$$

   (a) Compute the orthogonal projection of $\mathbf{q}_3$ onto the subspace spanned by $\mathbf{q}_1$ and $\mathbf{q}_2$.
   (b) Use the result in part (a) to find an orthogonal basis for P^2.

17. Use the Cauchy-Schwarz inequality to prove that for any real numbers $a \geq 0$ and $b \geq 0$, the following relationship holds between the geometric mean, $\sqrt{ab}$, and the arithmetic mean, $(a + b)/2$:

$$\sqrt{ab} \leq \frac{a+b}{2}$$

   (Hint: Consider the vectors $\mathbf{u} = (\sqrt{a}, \sqrt{b})$ and $\mathbf{v} = (\sqrt{b}, \sqrt{a})$.)

18. Use the Cauchy-Schwarz inequality to establish that for any real numbers a and b,

$$(a + b)^2 \leq 2(a^2 + b^2)$$

   (Hint: Consider the vectors $\mathbf{u} = (a, b)$ and $\mathbf{v} = (1, 1)$.)

19. Use the triangle inequality to prove that for any vectors $\mathbf{u}$, $\mathbf{v}$, and $\mathbf{w}$ in an inner product space V, $d(\mathbf{u}, \mathbf{w}) \leq d(\mathbf{u}, \mathbf{v}) + d(\mathbf{v}, \mathbf{w})$, where

$$d(\mathbf{u}, \mathbf{v}) = \|\mathbf{u} - \mathbf{v}\|$$

20. Prove that if $\mathbf{u}$ and $\mathbf{v}$ are vectors in an inner product space V, then for all real numbers t, $\langle \mathbf{u}, t\mathbf{v} \rangle = t \langle \mathbf{u}, \mathbf{v} \rangle$.

7.5 Problem Solving with Orthogonality

Your knowledge of orthogonality is unified in this section to solve the problem of Philadelphia Paints presented at the beginning of the chapter.

7.5.1 The Problem of Philadelphia Paints

Philadelphia Paints sells a variety of paints. One of their newer products is a latex paint that is produced on a weekly basis. In reviewing available production costs for this paint, the CEO (Chief Executive Officer) noticed that the cost-per-gallon decreases as the number of gallons produced increases, but only up to a certain point. After this point, the cost begins to rise. She has asked you, the Manager of the Production Department, to use the cost data in Table 7.2 to determine the optimal number of gallons to produce in each batch.

Thousands of gallons produced	1	2	3	4	4.5	5
Cost-per-gallon (in $)	6.00	5.00	5.50	4.75	5.50	4.00

Thousands of gallons produced	5	6.5	7.5	8	9	10
Cost-per-gallon (in $)	5.25	4.25	5.25	4.50	5.75	5.50

Table 7.2 *Production-Cost Data for the Problem of Philadelphia Paints.*

7.5.2 Problem Formulation

The data in Table 7.2 are plotted in Figure 7.11(a) and confirm the observations of the CEO. According to the discussion in Section 7.3.1, to determine the least-cost production amount, you must specify a functional relationship between the cost-per-gallon and the number of gallons produced. To that end, define the following dependent variable, y, and independent variable, x:

y = the cost-per-gallon in dollars

x = the number of gallons produced

Looking at Figure 7.11(a), you can see that a function is needed that first decreases as x increases and then, at some point, begins to increase. On the basis of the visual plot of the points, you might notice the general pattern of a parabola, represented by the dotted curve in Figure 7.11(b). Thus, it is reasonable to postulate the following quadratic relationship between y and x:

$$y = ax^2 + bx + c \tag{7.55}$$

Your objective is to determine values for the parameters a, b, and c that make the quadratic equation in (7.55) best fit the data in Table 7.2. After doing so, you then need to find the value of x that, when substituted in the right side of (7.55), provides the smallest value of y. You will now see how the material in this chapter is used to accomplish these tasks.

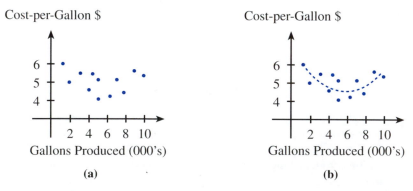

Figure 7.11 *The Data for the Problem of Philadelphia Paints.*

7.5.3 Solution Procedure

The approach of minimizing the sum of squared errors, as developed in Section 7.3.2, is used to determine the values for the parameters a, b, and c that make the quadratic equation in (7.55) best fit the data in Table 7.2.

The Measure of Sum of Squared Errors

You can use the sum of squared errors to measure how good a particular choice for a, b, and c is. Specifically, given values for a, b, and c, substitute each of the $k = 12$ values of x_i from the data in the first row of Table 7.2 in (7.55) to obtain the following predicted cost-per-gallon when x_i gallons are produced:

$$p_i = ax_i^2 + bx_i + c, \quad \text{for } i = 1, \ldots, k \tag{7.56}$$

You can then compare these predicted costs, denoted by the k-vector $\mathbf{p}$, with the actual costs-per-gallon given in the second row of Table 7.2, denoted by the k-vector $\mathbf{y}$. In particular, the following is a measure of how well the quadratic equation fits the k data values:

$$\text{Sum of Squared Errors} = \sum_{i=1}^{k} (p_i - y_i)^2 = \|\mathbf{p} - \mathbf{y}\|^2 \tag{7.57}$$

Noting from (7.56) that the values of $\mathbf{p}$ depend on a, b, and c, your objective is to find values for a, b, and c that minimize (7.57).

Using matrix-vector notation simplifies the solution procedure. From (7.56), you have that

$$\mathbf{p} = \begin{bmatrix} ax_1^2 + bx_1 + c \\ \vdots \\ ax_k^2 + bx_k + c \end{bmatrix} = \begin{bmatrix} x_1^2 & x_1 & 1 \\ \vdots & \vdots & \vdots \\ x_k^2 & x_k & 1 \end{bmatrix} \begin{bmatrix} a \\ b \\ c \end{bmatrix} \tag{7.58}$$

Let A be the $(k \times 3)$ matrix on the right side of (7.58) whose elements you can compute from the data. Also, let $\mathbf{z}$ be the 3-vector whose three components are the three unknowns a, b, and c. Then, from (7.58), the predicted values of y at the k data points $x_1, \ldots, x_k$ are the components of the k-vector $\mathbf{p} = A\mathbf{z}$. So (7.57) becomes:

$$\text{Sum of Squared Errors} = \|\mathbf{p} - \mathbf{y}\|^2 = \|A\mathbf{z} - \mathbf{y}\|^2 \tag{7.59}$$

In summary, given the $(k \times 3)$ matrix A and the k-vector $\mathbf{y}$, you want to find values for the 3-vector $\mathbf{z}$ that minimize (7.59).

Obtaining the Solution

Recognizing that you have a general least-squares problem (see Problem 7.4 in Section 7.3.2), you know that the solution is obtained by solving the following normal equations for $\mathbf{z}$:

$$(A^t A)\mathbf{z} = A^t \mathbf{y} \tag{7.60}$$

For the problem of Philadelphia Paints,

$$
A = \begin{bmatrix}
1.00 & 1.0 & 1 \\
4.00 & 2.0 & 1 \\
9.00 & 3.0 & 1 \\
16.00 & 4.0 & 1 \\
20.25 & 4.5 & 1 \\
25.00 & 5.0 & 1 \\
25.00 & 5.0 & 1 \\
42.25 & 6.5 & 1 \\
56.25 & 7.5 & 1 \\
64.00 & 8.0 & 1 \\
81.00 & 9.0 & 1 \\
100.00 & 10.0 & 1
\end{bmatrix}
\quad \text{and} \quad
y = \begin{bmatrix}
6.00 \\
5.00 \\
5.50 \\
4.75 \\
5.50 \\
4.00 \\
5.25 \\
4.25 \\
5.25 \\
4.50 \\
5.75 \\
5.50
\end{bmatrix}
$$

Using appropriate technology, you can compute

$$
A'A = \begin{bmatrix}
27620 & 3379 & 444 \\
3379 & 444 & 66 \\
444 & 66 & 12
\end{bmatrix}
\quad \text{and} \quad
A'y = \begin{bmatrix}
2272.8 \\
332.2 \\
61.2
\end{bmatrix}
$$

You can now solve the system $(A'A)z = A'y$ for z by finding $(A'A)^{-1}$. This is because the columns of A are linearly independent so, by Theorem 7.8 in Section 7.3.3, $A'A$ is invertible. Thus,

$$
z = (A'A)^{-1}A'y = \begin{bmatrix}
0.0017 & -0.0185 & 0.0392 \\
-0.0185 & 0.2164 & -0.4983 \\
0.0392 & -0.4983 & 1.3531
\end{bmatrix}
\begin{bmatrix}
2272.8 \\
332.2 \\
61.2
\end{bmatrix}
$$

$$
= \begin{bmatrix}
0.0512 \\
-0.5920 \\
6.4421
\end{bmatrix}
= \begin{bmatrix}
a \\
b \\
c
\end{bmatrix}
$$

In other words, these results indicate that the following quadratic equation best fits the data according to the measure of least-squared error:

$$
y = 0.0512x^2 - 0.592x + 6.4421 \tag{7.61}
$$

This quadratic equation is plotted together with the data in Figure 7.12.

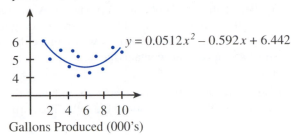

Figure 7.12 *The Quadratic Equation that Best Fits the Data for the Problem of Philadelphia Paints.*

Using the Results

Having found the quadratic equation in (7.61), you can determine the optimal production quantity, namely, the number of gallons that minimizes

$$0.0512x^2 - 0.592x + 6.4421$$

From calculus, you know that the value for x that minimizes the quadratic equation

$$q(x) = ax^2 + bx + c$$

is obtained by setting the derivative $q'(x) = 2ax + b = 0$, so

$$x = -\frac{b}{2a}$$

You can also prove this without using calculus. For the problem of Philadelphia Paints, in which $a = 0.0512$ and $b = -0.592$, it is optimal to produce

$$x = -\frac{b}{2a} = -\frac{-0.592}{2(0.0512)} \approx 5.781 \text{ thousand gallons}$$

In other words, you should recommend that the company produce 5781 gallons each week because this amount incurs the least cost-per-gallon, based on the data in Table 7.2.

In this chapter, you have seen the role of orthogonality in problem solving. All of the problems solved in this book so far have been relatively small in size. To solve large real-world problems in reasonable amounts of computer time, special numerical methods are needed. This topic is the focus of the last chapter.

Exercises for Section 7.5

PROJECT 7.1: The Production Problem of Philadelphia Paints

Use a graphing calculator, MATLAB, or other software package, as indicated by your instructor, to perform the operations in Exercise 1 through Exercise 4 for solving the production problem of Philadelphia Paints, described in Section 7.4.1. Whenever possible, use matrix and vector operations available on your system. Write the answer and the sequence of operations you performed to obtain the answer.

1. Create a 12-vector **x** containing the data in the first row of Table 7.2 pertaining to the number of thousands of gallons produced. Also create a 12-vector **y** containing the data in the second row of Table 7.2 corresponding to the cost-per-gallon. Then use **x** to construct the (12×3) matrix A whose first column contains the values $x_1^2, \ldots, x_{12}^2$, whose second column contains the values $x_1, \ldots, x_{12}$, and whose third column contains all 1's.

2. Find the values of the parameters a, b, and c for the quadratic equation $ax^2 + bx + c$ that best fits the data by solving the following normal equations for the column vector **z** of parameters:

$$(A^t A)\mathbf{z} = A^t \mathbf{y}$$

Also compute and display the optimal production amount of $x = -b/(2a)$ thousand gallons.

3. Draw a single graph that contains both the data and the quadratic equation found in the previous exercise.

4. On presenting your results at a meeting, a member of the Accounting Department suggested that the data associated with producing 1, 2, 9, and 10 thousand gallons of paint should not be counted as heavily as the other data. Repeat the previous two exercises to determine the optimal amount to produce by minimizing the weighted squared error (see Section 7.4.4). Weight the data associated with producing 1, 2, 9, and 10 thousand gallons of paint one quarter as much as the other data.

PROJECT 7.2: The Radioactive-Decay Problem of Nuclear Materials

Scientists at Nuclear Materials know that the number of grams, g, of a particular radioactive material left after t days is given approximately by the following formula:

$$g = g_0 e^{-\alpha t} \tag{7.62}$$

They have collected the following data:

Time in days (t)	10	30	50	80
Grams left (g)	0.95	0.82	0.72	0.61

Use a graphing calculator, MATLAB, or other software package, as indicated by your instructor, to determine the values of the parameters g_0 and α that best fit the given data by performing Exercise 5 through Exercise 8. Whenever possible, use matrix and vector operations available on your system. Write the answer and the sequence of operations you performed to obtain the answer.

5. Create vectors **t** and **g** that contain the data corresponding to the times and the amount of radioactive material recorded.

6. The relation in (7.62) is not linear. However, you can create a linear regression problem by taking the natural logarithm of both sides of (7.62) to obtain:

$$\ln(g) = \ln(g_0) - \alpha t \tag{7.63}$$

Think of $\ln(g)$ as a new variable, y, $\ln(g_0)$ as the value of b, and $-\alpha$ as the value of m.

Then (7.63) becomes

$$y = b + mt \tag{7.64}$$

To find the values of m and b in (7.64), create a vector **y** whose components are the natural logarithm of the corresponding components of **g** in Exercise 5. Then use **t** from Exercise 5 to construct the (4×2) matrix A whose first column contains the values $t_1, \ldots, t_4$ and whose second column contains all 1's.

7. Find the values of the parameters m and b for the line in (7.64) that best fits the data by solving the following normal equations for the column vector **z** of parameters:

$$(A^t A)\mathbf{z} = A^t \mathbf{y}$$

Then use these values of m and b to compute the values of $g_0 = e^b$ and $\alpha = -m$ that make (7.62) best fit the original data.

8. Use the values of g_0 and α from the previous exercise to answer the following questions.
   (a) Estimate the amount of radioactive material at time 0.
   (b) What is the *half-life* of this material, that is, how long does it take for g grams of material to decay to one-half of g?

Chapter Summary

In this chapter, you have seen various generalizations and uses for orthogonality.

Generalizations of Orthogonality

Sequential generalization is applied to the concept of a pair of orthogonal n-vectors, that is, two n-vectors whose dot product is 0. For example, a set of n-vectors, $\{\mathbf{v}_1, \ldots, \mathbf{v}_k\}$, is orthogonal if each distinct pair of vectors is orthogonal.

Orthogonal sets of vectors have certain desirable properties. For example, a set of nonzero orthogonal n-vectors is linearly independent. Thus, a set of n such vectors forms an orthogonal basis for R^n. One computational advantage of working with an orthogonal basis $B = \{\mathbf{v}_1, \ldots, \mathbf{v}_k\}$ for a subspace W of R^n is that you do not need a numerical method to find the components of a vector $\mathbf{v} \in W$. Rather, you can use the following closed-form expression to find component i of **v** relative to this basis:

$$([\mathbf{v}]_B)_i = \frac{\mathbf{v} \cdot \mathbf{v}_i}{\mathbf{v}_i \cdot \mathbf{v}_i}$$

Many such computations are simplified even further when working with an orthonormal basis, which is an orthogonal basis consisting of unit vectors.

A further generalization arises when you consider the case when an n-vector **v** is orthogonal to every vector in a subspace W of R^n. The orthogonal complement $W^\perp$ of W is the subspace consisting of all n-vectors that are orthogonal to every vector in W.

A final generalization is the axiomatic system of an inner product space V. Specifically, to each pair of vectors **u** and **v** in a vector space V, you associate a real number $\langle \mathbf{u}, \mathbf{v} \rangle$ that must satisfy the following axioms for all vectors $\mathbf{u}, \mathbf{v}, \mathbf{w} \in V$ and for all real numbers t:

1. $\langle \mathbf{u}, \mathbf{v} \rangle = \langle \mathbf{v}, \mathbf{u} \rangle$.
2. $\langle \mathbf{u} + \mathbf{v}, \mathbf{w} \rangle = \langle \mathbf{u}, \mathbf{w} \rangle + \langle \mathbf{v}, \mathbf{w} \rangle$.

3. $\langle t\mathbf{u}, \mathbf{v} \rangle = t\langle \mathbf{u}, \mathbf{v} \rangle$.

4. $\langle \mathbf{u}, \mathbf{u} \rangle \geq 0$; and also, $\langle \mathbf{u}, \mathbf{u} \rangle = 0$ if and only if $\mathbf{u} = \mathbf{0}$.

You can generalize the concepts of orthogonality, length, and distance to vectors in an inner product space. Specifically, for $\mathbf{u}, \mathbf{v} \in V$,

(a) $\mathbf{u}$ and $\mathbf{v}$ are orthogonal if and only if $\langle \mathbf{u}, \mathbf{v} \rangle = 0$.

(b) $\|\mathbf{v}\| = \sqrt{\langle \mathbf{v}, \mathbf{v} \rangle}$.

(c) The distance from $\mathbf{u}$ to $\mathbf{v} = \|\mathbf{u} - \mathbf{v}\|$.

All other results in this chapter pertaining to n-vectors extend to an inner product space. You also saw the Cauchy-Schwarz and the triangle inequalities:

Cauchy-Schwarz Inequality	Triangle Inequality
$|\langle \mathbf{u}, \mathbf{v} \rangle| \leq \|\mathbf{u}\|\|\mathbf{v}\|$	$\|\mathbf{u} + \mathbf{v}\| \leq \|\mathbf{u}\| + \|\mathbf{v}\|$

Applications of Orthogonality

Orthogonality is used to find the projection of one n-vector onto another n-vector. Specifically, the orthogonal projection of $\mathbf{v}$ onto a nonzero n-vector $\mathbf{d}$ is the following n-vector $\mathbf{p}$ that points in the same direction as $\mathbf{d}$:

$$\mathbf{p} = \frac{\mathbf{v} \cdot \mathbf{d}}{\mathbf{d} \cdot \mathbf{d}}\mathbf{d} \quad \text{(orthogonal projection of } \mathbf{p} \text{ onto } \mathbf{d}\text{)}$$

The component of $\mathbf{v}$ orthogonal to $\mathbf{d}$ is the following n-vector $\mathbf{q}$ that is perpendicular to $\mathbf{d}$:

$$\mathbf{q} = \mathbf{v} - \mathbf{p} = \mathbf{v} - \frac{\mathbf{v} \cdot \mathbf{d}}{\mathbf{d} \cdot \mathbf{d}}\mathbf{d} \quad \text{(component of } \mathbf{v} \text{ orthogonal to } \mathbf{d}\text{)}$$

One generalization is the projection of an n-vector onto a line through the origin in the direction $\mathbf{d}$, which is the same as the projection of $\mathbf{v}$ onto $\mathbf{d}$. A further generalization is the projection of an n-vector $\mathbf{v}$ onto a subspace W of R^n. In particular, you can decompose any n-vector $\mathbf{v}$ uniquely into the sum of a vector $\mathbf{p} \in W$ and a vector $\mathbf{q} \in W^\perp$. The vector $\mathbf{p}$ is the orthogonal projection of $\mathbf{v}$ onto W, namely, $\text{proj}_W \mathbf{v}$. You can use any orthogonal basis $\mathbf{w}_1, \ldots, \mathbf{w}_k$ for W to find $\text{proj}_W \mathbf{v}$, as follows:

$$\text{proj}_W \mathbf{v} = \frac{\mathbf{v} \cdot \mathbf{w}_1}{\mathbf{w}_1 \cdot \mathbf{w}_1}\mathbf{w}_1 + \cdots + \frac{\mathbf{v} \cdot \mathbf{w}_k}{\mathbf{w}_k \cdot \mathbf{w}_k}\mathbf{w}_k$$

The Gram-Schmidt process is a numerical method for creating an orthogonal basis of a subspace W of R^n, starting with any basis for W.

Projections are also used to solve a regression model in which you want to predict the value of a dependent variable, y, using the value of an independent variable, x. To do so, you must postulate a functional relationship between y and x. In a simple linear regression model, you assume that y depends on x according to the following relation:

$$y = mx + b$$

You then use data—in the form of k pairs of known values for x and y, say, $(x_1, y_1), \ldots, (x_k, y_k)$—to estimate values for the parameters m and b. Specifically, you want to find values for m and b that make the line $y = mx + b$ best fit the data by minimizing the following sum of squared errors:

$$\text{Sum of Squared Errors} = \sum_{i=1}^{k}(mx_i + b - y_i)^2$$

In matrix-vector notation, letting

$$A = \begin{bmatrix} x_1 & 1 \\ \vdots & \vdots \\ x_k & 1 \end{bmatrix}, \quad \mathbf{z} = \begin{bmatrix} m \\ b \end{bmatrix}, \quad \text{and} \quad \mathbf{y} = \begin{bmatrix} y_1 \\ \vdots \\ y_k \end{bmatrix}$$

you want to find values for the column vector **z** so as to

minimize $\|A\mathbf{z} - \mathbf{y}\|^2$

Through the use of projections, you learned that the solution to this problem is obtained by solving the following normal equations for **z**:

$$(A^t A)\mathbf{z} = A^t \mathbf{y}$$

Chapter 8

Numerical Methods in Linear Algebra

Y ou now know various ways in which linear algebra is used to solve certain problems. Computers perform the necessary computations when solving real-world problems because extensive time and effort are required. In this chapter, you will see the care that is needed to ensure that the computer obtains numerically accurate solutions to large problems efficiently. These issues must be addressed if you want to solve problems such as the following one of Waste Management.

The Flow Problem of Waste Management

Waste Management operates a liquid recycling operation that processes a large number of waste liquids into 75 marketable products. Each waste liquid yields a few of the sellable products at a known rate in gallons per hour. Each day the Manager of Operations solves an initial system of 75 linear equations to determine the rates at which 75 initially-chosen waste liquids are processed so as to achieve a desired rate of output for each of the 75 sellable products. Some hours later, one of the chosen waste liquids runs out and another one must be selected in its place. To find a replacement liquid that achieves the same rates of flow for the output products requires solving a new system of 75 linear equations in 75 unknowns on a trial-and-error basis. You have been hired as a consultant to design an on-line system for solving these large systems of linear equations in real time on the manager's desktop computer.

8.1 General Computational Concerns

When developing methods for solving a problem by computer, the following general issues arise and require special attention:

1. *Computational efficiency*, which enables one to obtain solutions to large real-world problems in reasonable amounts of computer time.

2. *Efficient use of computer memory*, which is a limited resource that can easily be exceeded—for example, when storing large matrices.

3. *Numerical accuracy*, so that the solutions obtained are precise and reliable.

4. *User interface*, so that the user is able to enter the data for the problem easily, make modifications and corrections, and obtain output displayed in an organized and understandable format.

Each of these items is described now in greater detail.

8.1.1 Computational Efficiency

As you know, there are various ways to solve a particular problem. For example, for a given $(n \times n)$ matrix A and a column vector $\mathbf{b} \in R^{n \times 1}$, consider methods for solving the following system of linear equations for the column vector $\mathbf{x} \in R^{n \times 1}$:

$$A\mathbf{x} = \mathbf{b} \tag{8.1}$$

You can use any of the following approaches:

1. Gauss-Jordan elimination (see Section 2.3.1).

2. Gaussian elimination (see Section 2.3.2).

3. Inverting A and then computing $\mathbf{x} = A^{-1}\mathbf{b}$ (see Section 3.2.1).

4. Cramer's rule (see Section 3.3.3).

Which of these methods is most efficient?

Measuring Computational Efficiency for Solving One Specific Problem

To answer this question, it is first necessary to understand how to measure the efficiency of an algorithm applied to one particular problem. One approach is to count the number of each type of operation—addition, subtraction, multiplication, division, and so on—needed to obtain the solution. Thus, for example, if one algorithm requires 100 multiplications and 40 additions to solve a specific problem and a second algorithm requires 50 multiplications and 20 additions to solve the same problem, then you can conclude that the second algorithm is more efficient for solving that one problem.

However, suppose the first algorithm requires 100 multiplications and 40 additions and the second algorithm requires 50 multiplications and 250 additions. Now it is not clear which algorithm is more efficient because multiplications are different from additions.

One approach to resolving this issue is to determine the relative amount of work needed to perform a multiplication in comparison to an addition. For example, suppose a multiplication requires 4 times as much time to perform as an addition. Then, calling the amount of time needed to perform an addition 1 *time unit*, the first algorithm requires the following number of time units to solve the problem:

$$\begin{pmatrix} \text{time units for} \\ \text{100 multiplications} \end{pmatrix} + \begin{pmatrix} \text{time units for} \\ \text{40 additions} \end{pmatrix} = 4(100) + 40 = 440$$

The second algorithm requires the following number of time units to solve the same problem:

$$\begin{pmatrix} \text{time units for} \\ \text{50 multiplications} \end{pmatrix} + \begin{pmatrix} \text{time units for} \\ \text{250 additions} \end{pmatrix} = 4(50) + 250 = 450$$

Comparing these total time units leads to the conclusion that the first algorithm is more efficient for solving this specific problem.

To use the foregoing approach to compare the efficiency of two algorithms requires knowing how many time units each operation takes. This information is difficult to obtain. Moreover, these values vary from one computer to another. Nevertheless, once determined, you can then compute a single number—hereafter referred to as the **running time**—that represents the total number of time units an algorithm needs to solve a specific problem.

Measuring Computational Efficiency as a Function of Problem Data

It is important to realize that the running time of an algorithm depends on the specific data for the problem being solved. For example, the running time of an algorithm for solving the (2×2) system of linear equations

$$2x_1 - 3x_2 = 7$$
$$4x_1 + 5x_2 = 3$$

is different from the running time needed by the same algorithm to solve the (75×75) systems in the problem of Waste Management presented at the beginning of the chapter.

In general, one is interested not in the running time needed by an algorithm to solve one specific problem, but rather, in the running time needed to solve a *general* problem. To determine this running time, consider the known data for a general problem. For example, the data for the general problem of solving the system of linear equations in (8.1) are:

$$n, \quad A, \quad \mathbf{b} \quad \text{(data for solving an } (n \times n) \text{ system of linear equations)}$$

Each affects the running time, so it would be desirable to express the running time of an algorithm in terms of n, A, and $\mathbf{b}$. Such an expression is usually complex and difficult to obtain. A compromise measure of efficiency is now presented that avoids having to find expressions for the running time in terms of all the general data. This new approach also avoids the need to determine the relative number of time units required by each operation.

A Practical Measure of Efficiency

The approach is to focus on those aspects of the problem and algorithm that have the greatest impact on the running time. For example, consider the problem of solving an $(n \times n)$ system of linear equations, $A\mathbf{x} = \mathbf{b}$. Careful thought, or experimentation, should lead you to conclude that, among n, A, and $\mathbf{b}$, n has the greatest impact on the running time of an algorithm for solving the system. For example, the running time needed to solve a (100×100) system is substantially greater than the running time needed to solve a (3×3) system. In contrast, the running times needed to solve a (100×100) system $A\mathbf{x} = \mathbf{b}$ and a (100×100) system $A'\mathbf{x} = \mathbf{b}'$ do not differ as greatly.

On the basis of the foregoing discussion, rather than attempt to find an expression, $R(n, A, \mathbf{b})$, for the running time of an algorithm for solving a general system of linear equations, a more reasonable approach is to find an approximation, $T(n)$, that depends only on the value of n. Keep in mind that $T(n)$ is an approximation to the true running time, $R(n, A, \mathbf{b})$.

A similar approach is used to avoid having to determine the relative number of time units needed to perform each operation. The idea is to identify one group of operations that involves the most amount of effort. For example, an algorithm for solving a system of linear equations uses additions, subtractions, multiplications, and divisions. Regardless of the computer, multiplications and divisions require substantially more computational effort to perform than additions and subtractions. Furthermore, the effort involved in performing a multiplcation is of the same order of magnitude as that involved in performing a division, so you can combine these two operations in one group, hereafter referred to as *multiplications*.

In view of the foregoing discussion, the following leads to a practical measure of the efficiency of an algorithm for solving an $(n \times n)$ system of linear equations:

$$T(n) = \begin{cases} \text{the number of multiplications needed by a particular} \\ \text{algorithm to solve a general } (n \times n) \text{ system of linear} \\ \text{equations whose matrix is invertible} \end{cases} \qquad (8.2)$$

Using the measure in (8.2) and carefully counting the number of multiplications needed by an efficient implementataion of each algorithm to solve a system of linear equations $A\mathbf{x} = \mathbf{b}$, in which A is invertible, results in the values of $T(n)$ in Table 8.1. The last group of columns in Table 8.1 shows the values of $T(n)$ for various values of n.

By an *efficient implementation* is meant one that requires as few multiplications as possible. For example, consider applying Gauss-Jordan elimination to the following augmented matrix:

$$\begin{bmatrix} 2 & 4 & | & 6 \\ 3 & 1 & | & 5 \end{bmatrix}$$

To obtain a leading 1 in row 1, you divide that row by the pivot element 2, which involves three divisions. However, there is no need to divide the pivot element, 2, by itself because you know that the result is 1. Thus, in an efficient implementation of Gauss-Jordan elimination, in this example, you need perform only two divisions to obtain a leading 1 in the first row.

One conclusion from Table 8.1 is that Gauss-Jordan and Gaussian elimination are more efficient, according to the measure $T(n)$ in (8.2), than the other two methods. Comparing the last two methods in Table 8.1, and as seen in Figure 8.1, Cramer's rule is more efficient than finding the inverse for $n = 2$, 3, and 5 but less efficient for larger values of n. This is because, for large values of n, the term in $T(n)$ containing the highest power of n makes a more dominant contribution to the value of $T(n)$ than do the terms of lower power. In other words, for large values of n, the polynomial

$$T(n) = a_k n^k + a_{k-1} n^{k-1} + \cdots + a_1 n + a_0$$

Method	$T(n)$	$T(n)$ for $n =$				
		2	3	5	10	20
Gauss-Jordan elimination	$\frac{1}{3}n^3 + n^2 - \frac{1}{3}n$	6	17	65	430	3060
Gaussian elimination	$\frac{1}{3}n^3 + n^2 - \frac{1}{3}n$	6	17	65	430	3060
Compute A^{-1} and then $\mathbf{x} = A^{-1}\mathbf{b}$	$2n^3 + n^2$	20	63	275	2100	16400
Cramer's rule	$\frac{1}{3}n^4 + \frac{1}{3}n^3 + \frac{2}{3}n - 1$	11	43	269	3739	56279

Table 8.1 *Measures of Efficiency for Various Algorithms that Solve a System of Linear Equations.*

can be approximated by $a_k n^k$, that is,

$$T(n) \approx a_k n^k \quad \text{(for large values of } n)$$

Thus, for large values of n, the running time for solving a system of linear equations by finding A^{-1} is approximately $2n^3$ and that of Cramer's rule is approximately $n^4/3$. Even the constant in the dominant term becomes unimportant as n gets large. That is, for large values of n, $n^4/3 > 2n^3$. This is why Cramer's rule is less efficient than using the inverse to solve the system when $n \geq 10$, as seen in Table 8.1 and Figure 8.1.

In summary, if the running times of two algorithms are polynomials in a parameter n

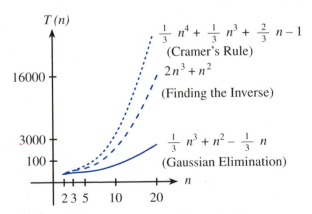

Figure 8.1 *A Comparison of the Polynomials in Table 8.1.*

(n usually indicates the size of the problem), then, for large values of n, the more efficient algorithm is the one whose largest-degree term is smallest. Thus, for large values of n, Table 8.1 indicates that Gauss-Jordan and Gaussian elimination are more efficient than finding the inverse which, in turn, is more efficient than using Cramer's rule.

8.1.2 Efficient Use of Computer Memory

Computers can store information—such as the elements of an $(m \times n)$ matrix—in a variety of formats and locations. The specific choice can affect significantly the amount of memory and time needed to solve a problem. For example, information stored in **primary memory**, which is located in a special part of the computer, can be retrieved quickly for processing but is quite limited in size and lost after the problem is solved. In contrast, computers can store information in **secondary memory**, located on a hard or floppy disk or on a magnetic tape. Secondary memory is relatively slow to access but is virtually unlimited in size, for most practical purposes. Furthermore, you can save data stored in secondary memory for later use, even after the problem is solved.

Using Sparsity to Save Storage Space

When solving a problem that involves a (10×10) matrix A, you need not be concerned with how the $10 \times 10 = 100$ elements of A are stored. However, when A is (500×500), you should be concerned because the $500 \times 500 = 250,000$ elements of A might not fit in primary memory. In many cases, you can avoid having to store a large matrix in secondary memory by reducing significantly the amount of memory needed to store the matrix. This is possible when the matrix exhibits a property called **sparsity**, which means that the vast majority—perhaps 95% or more—of the elements of the matrix are 0. Most large matrices that arise in applications are sparse. In such cases, you can conserve significant amounts of memory by storing only the nonzero elements, together with their row and column numbers, as illustrated in the following example.

EXAMPLE 8.1 Storing a Sparse Matrix

Instead of storing all 25 elements in a sparse (5×5) matrix A, you need only store the nonzero elements, as follows:

A Sparse Matrix A	**Storing the Nonzero Elements**		
	Row	Column	Value

$$\begin{bmatrix} 0 & 0 & 0 & 0 & 3 \\ 0 & 0 & 9 & 0 & 0 \\ 0 & 0 & 0 & 7 & 0 \\ 0 & 5 & 0 & 0 & 0 \\ 0 & 0 & 0 & 0 & 0 \end{bmatrix}$$

Row	Column	Value
1	5	3
2	3	9
3	4	7
4	2	5

Other examples of sparse matrices are the elementary matrices that arise in performing Gauss-Jordan and Gaussian elimination. As you are asked to show in Exercise 12 through Exercise 14, it is possible to develop a sparse representation based on how these elementary matrices differ from the identity matrix.

There are different formats—called **data structures**—for storing information in a computer. For example, one data structure for storing all 25 elements of the (5 × 5) matrix *A* in Example 8.1 is a *two-dimensional array*, which is a rectangular table of information stored in consecutive memory locations in the computer. An alternative data structure, called a *linked list*, is used to store the four nonzero elements in the format shown in Example 8.1. The choice of data structures affects both the amount of storage and time needed to solve a problem. Each data structure has advantages and disadvantages that should be taken into account when determining which alternative is best. Further details on data structures are outside the scope of this book. Many books and computer science courses deal with data structures.

Using Sparsity to Improve Computational Efficiency

As you have just discovered, one advantage of sparse vectors and matrices is the ability to store only the nonzero elements by using an appropriate data structure. Another benefit arises in performing computations with sparse vectors and matrices. For example, if **x** and **y** are *n*-vectors, then you know that the dot product is

$$\mathbf{x} \cdot \mathbf{y} = \sum_{i=1}^{n} x_i y_i \tag{8.3}$$

When **x** and **y** are sparse, many of the terms $x_i y_i$ in (8.3) are 0. You can therefore perform the computation in (8.3) more efficiently by multiplying only those components of **x** and **y** that are both nonzero, as illustrated in the following example.

EXAMPLE 8.2 Computing the Dot Product of Two Sparse Vectors Efficiently
To compute the dot product of

$$\mathbf{x} = (-2, 0, 0, 0, 3, 0, 0, -1, 4, 0, 0) \quad \text{and}$$
$$\mathbf{y} = (0, 0, -3, 0, 2, 0, 0, 0, -1, 0, 5)$$

you need only multiply the following components of **x** and **y** that are both nonzero:

$$\mathbf{x} \cdot \mathbf{y} = x_5 y_5 + x_9 y_9 = 3(2) + 4(-1) = 2$$

8.1.3 Numerical Accuracy

Computers work with the decimal representations of real numbers rather than their fractional equivalents. Due to limited memory, a computer can store only a certain number of digits for each real number. For example, a computer with a *precision* of six digits would store $2/3 = 0.666666\ldots$ as either 0.666666 or as 0.666667, depending on whether the computer *truncates* or *rounds*. In either case, the remaining digits of the real number are lost, resulting in what is called **truncation error**. If not controlled, over time, these truncation errors can accumulate to a point where the results obtained from the computer are incorrect, as shown in the following example.

EXAMPLE 8.3 Inaccuracies Caused by Truncation Errors

You can verify that $x_1 = 2$ and $x_2 = -1.2$ is the solution to the following system of linear equations:

$$100x_1 + 90x_2 = 92$$
$$11x_1 + 10x_2 = 10$$

However, consider what happens when you perform the following elementary row operations, truncating to two decimals of accuracy. Divide the second equation through by 11 to obtain

$$100x_1 + 90x_2 = 92$$
$$x_1 + 0.90x_2 = 0.90$$

Subtract 100 times the foregoing second equation from the first equation:

$$0x_1 + 0x_2 = 2$$
$$x_1 + 0.90x_2 = 0.90$$

Looking at the equation $0x_1 + 0x_2 = 2$, you would now conclude, incorrectly, that the system has no solution.

Another source of truncation error arises when adding numbers of greatly different magnitudes, as shown in the following example.

EXAMPLE 8.4 Inaccuracies Caused by Adding Numbers of Greatly Different Magnitudes

Consider adding the following list of numbers:

$$\begin{array}{r} 5.967 \\ 1325.000 \\ -1324.000 \\ \hline 6.967 \end{array} \quad \text{(total)}$$

Look at what happens when a computer adds the first two numbers (which differ greatly in magnitude):

$$\begin{array}{r} 5.967 \\ 1325.000 \\ \hline 1330.967 \end{array}$$

If the computer stores only four significant digits, then the foregoing value of 1330.967 is truncated to 1330. When this number is then added to the final number in the original list, -1324, the computer obtains the incorrect result of $1330 - 1324 = 6$, instead of 6.967.

In view of Example 8.3 and Example 8.4, you should realize the need to take measures, if possible, to avoid the problems caused by truncation errors. One technique for doing so is presented now.

Partial Pivoting

Computer scientists have determined that one way to reduce truncation errors is to avoid the division of relatively large numbers by relatively small numbers. Informally, the reason is that division by small numbers magnifies truncation errors.

Divisions are a necessity when solving a system of linear equations. For example, with Gauss-Jordan elimination, division is used to create a leading 1 in each row of the augmented matrix (see Section 2.3.1). To illustrate, consider applying Gauss-Jordan elimination to the following augmented matrix:

$$\left[\begin{array}{ccc|c} 2 & -7 & -2 & -8 \\ -2 & 2 & 4 & 6 \\ -4 & 8 & -8 & 4 \end{array}\right] \tag{8.4}$$

Since the pivot element, 2, in row 1 and column 1 is not zero, you can create a leading 1 in row 1 by dividing row 1 through by 2. Although it is possible to do so, a better approach for controlling truncation errors is to divide by as large a value as possible. To that end, consider what happens if, before performing the division, you were to interchange rows 1 and 3:

$$\left[\begin{array}{ccc|c} -4 & 8 & -8 & 4 \\ -2 & 2 & 4 & 6 \\ 2 & -7 & -2 & -8 \end{array}\right] \tag{8.5}$$

Now when you create a leading 1 in row 1, you divide by the pivot element -4, which is larger in absolute value than the previous pivot element, 2.

In other words, you can reduce truncation errors in Gauss-Jordan elimination by using the following rule.

The Rule of Partial Pivoting

The rule of **partial pivoting** for creating a leading 1 in row i and column k of the current augmented matrix C is to interchange row i with that row $j \geq i$ in which C_{jk} has the largest absolute value. In this way, the number with the largest possible absolute value becomes the pivot element that is then used as a divisor.

Partial pivoting is illustrated in the following example.

EXAMPLE 8.5 Partial Pivoting in Gauss-Jordan Elimination

Consider applying Gauss-Jordan elimination to the following augmented matrix from (8.4):

$$\left[\begin{array}{ccc|c} 2 & -7 & -2 & -8 \\ -2 & 2 & 4 & 6 \\ -4 & 8 & -8 & 4 \end{array}\right] \tag{8.6}$$

When trying to create a leading 1 in row 1, look in column 1 and identify row 3 as the row containing the element of largest absolute value, -4. Partial pivoting results in interchanging row 1 and row 3 to obtain:

$$\left[\begin{array}{ccc|c} -4 & 8 & -8 & 4 \\ -2 & 2 & 4 & 6 \\ 2 & -7 & -2 & -8 \end{array}\right] \tag{8.7}$$

Now divide row 1 of the matrix in (8.7) by the pivot element -4 to obtain a leading 1 in row 1:

$$\left[\begin{array}{ccc|c} 1 & -2 & 2 & -1 \\ -2 & 2 & 4 & 6 \\ 2 & -7 & -2 & -8 \end{array}\right] \tag{8.8}$$

For the matrix in (8.8), add 2 times row 1 to row 2 and add -2 times row 1 to row 3:

$$\left[\begin{array}{ccc|c} 1 & -2 & 2 & -1 \\ 0 & -2 & 8 & 4 \\ 0 & -3 & -6 & -6 \end{array}\right] \tag{8.9}$$

Now cross off row 1 in (8.9) and identify column 2 as the first nonzero column from the left. Look in column 2 and identify row 3 as the row below row 2 containing the element of largest absolute value, -3. Use partial pivoting to interchange row 2 and row 3, thus obtaining:

$$\left[\begin{array}{ccc|c} 1 & -2 & 2 & -1 \\ 0 & -3 & -6 & -6 \\ 0 & -2 & 8 & 4 \end{array}\right] \tag{8.10}$$

Now divide row 2 in (8.10) by the pivot element -3 to obtain a leading 1 in row 2:

$$\left[\begin{array}{ccc|c} 1 & -2 & 2 & -1 \\ 0 & 1 & 2 & 2 \\ 0 & -2 & 8 & 4 \end{array}\right]$$

The remaining steps of Gauss-Jordan elimination do not involve partial pivoting and are therefore omitted.

8.1.4 User Interface

To solve a problem that involves linear algebra, it is necessary to enter the data for the problem, regardless of the type of technology used (such as a graphing calculator, MATLAB, Mathematica, and so on). When you create a computer program for someone other than yourself to use, take special care in designing the interface used to obtain the data. One reason is that the person who is entering the data might not have the technical knowledge associated with solving the problem and may not know much about linear algebra. Such a person might not understand the following message displayed on the computer screen:

Enter a vector for the right side of the system of linear equations.

Computer programming languages, such as Visual Basic, are used to design a user-friendly interface that provides instructions, help menus, and labeled boxes on the computer screen in which the user types the needed data.

Another concern arises when substantial amounts of data are involved. In such cases, make every effort to ensure that the user has to enter the least amount of information possible. For example, suppose a problem requires the data in a large $(n \times n)$ matrix. When the matrix is sparse, you should have the user enter only the nonzero elements, as shown for the sparse matrix in Example 8.1 in Section 8.1.1.

As another example of minimizing the inputting effort, suppose you are writing a program to solve a problem that requires entering the mileages between all pairs of 100 cities. You can reduce the amount of information that must be entered from $100 \times 100 = 10000$ values to 5000 values by noting that the distance from city i to city j is the same as the distance from city j to city i. Thus, the user should be required to enter this value only once. Further inputting efficiencies are achieved by noting that the distance from city i to itself is 0, so the user should not be required to enter these values.

In this section, you have been exposed to the concerns of computational efficiency, use of memory, numerical accuracy, and creating the user interface that must be taken into account when using computers to solve large real-world problems. Developing efficient and accurate numerical methods for solving systems of linear equations and for finding eigenvalues and eigenvectors of matrices are presented in the remaining sections of this chapter.

Exercises for Section 8.1

1. Suppose that n and k are positive integers. Indicate the number of multiplications needed to compute n^{2k} using each of the following approaches. Express your answer in terms of n and k. Which method is most efficient?
   (a) Compute $n^{2k} = n \cdot n \cdots n$.
   (b) Compute $m = n^2$ and then $n^{2k} = (n^2)^k = m^k = m \cdot m \cdots m$.
   (c) Compute $m = n^k = n \cdot n \cdots n$ and then $n^{2k} = (n^k)^2 = m^2$.

2. Let n and k be positive integers and recall that $n! = n(n-1)\cdots 1$. For each of the following approaches, indicate the number of multiplications needed to compute

$$\binom{n}{k} = \frac{n!}{(n-k)!\ k!}$$

Express your answer in terms of n and k. Which method is most efficient?

(a) $\displaystyle\binom{n}{k} = \frac{n(n-1)\cdots 1}{[(n-k)(n-k-1)\cdots 1][k(k-1)\cdots 1]}$

(b) $\displaystyle\binom{n}{k} = \frac{n(n-1)\cdots(n-k+1)}{k(k-1)\cdots 1}$

(c) $\displaystyle\binom{n}{k} = \frac{n(n-1)\cdots(k+1)}{(n-k)(n-k-1)\cdots 1}$

3. Suppose that A is an $(m \times n)$ matrix and B is an $(n \times p)$ matrix. How many multiplications are required to compute the $(m \times p)$ matrix AB? Express your answer in terms of m, n, and p.

4. Suppose that A and B are $(n \times n)$ matrices and that $\mathbf{x} \in R^{n \times 1}$ is a column vector. Which computation requires fewer multiplications, $(AB)\mathbf{x}$ or $A(B\mathbf{x})$?

5. Consider two algorithms for solving a problem whose size is described by the value of a positive integer n.
   (a) Suppose that the running time of the first algorithm is $T_1(n) = 4n^2$ and that of the second algorithm is $T_2(n) = n^3$. Determine analytically the value of n for which the first algorithm is more efficient than the second algorithm.
   (b) Verify your result in part (a) by filling in the following table:

	n							
	1	2	3	4	5	6	7	8
$T_1(n) = 4n^2$								
$T_2(n) = n^3$								

6. Suppose the running time of an algorithm for solving a problem whose size is described by the value of a positive integer n is $T(n) = n^3/2 + 4n^2$.
   (a) Determine analytically the value of n for which the cubic term exceeds the quadratic term.
   (b) Verify your result in part (a) by filling in the following table:

	n								
	1	2	3	4	5	6	7	8	9
$n^3/2$									
$4n^2$									

7. Consider storing the vector $\mathbf{x} = (0, 0, 4, 0, -2, 0, 0, 6, 0)$ in the memory of a computer.
   (a) How many values are stored if you store all elements of $\mathbf{x}$?
   (b) Show how to store $\mathbf{x}$ in sparse form, that is, store only the nonzero elements and their positions. How many values, including the positions, are stored?

8. Consider storing in the memory of a computer a sparse n-vector $\mathbf{x}$ that has k nonzero elements.
   (a) How many values are stored if you store all elements of $\mathbf{x}$? Express your answer in terms of n and k.
   (b) How many values are stored if you store only the nonzero elements of $\mathbf{x}$ and their positions? Express your answer in terms of n and k.
   (c) In view of your answers to parts (a) and (b), when is it more efficient to store an n-vector $\mathbf{x}$ in sparse form? Express your answer in terms of n and k.

9. Show how to store only the nonzero elements of the following matrix:

$$A = \begin{bmatrix} 0 & 3 & 0 & 0 \\ 6 & 0 & 0 & 0 \\ 0 & 0 & 0 & 1 \\ 5 & 0 & 0 & 0 \end{bmatrix}$$

10. Show how to store only the nonzero elements of the following matrix:

$$A = \begin{bmatrix} 0 & 0 & 0 & -1 & 0 \\ -3 & 0 & 0 & 0 & 0 \\ 0 & 0 & 0 & 0 & -2 \\ 0 & -1 & 0 & 0 & 0 \\ 0 & 0 & -4 & 0 & 0 \end{bmatrix}$$

11. Show how to store an $(n \times n)$ diagonal matrix in sparse form.

12. Show how to store in sparse form an elementary matrix that multiplies row i of the identity matrix by c. Do not store the 1's on the diagonal.

13. Show how to store in sparse form an elementary matrix that interchanges row i of the identity matrix with row j. Do not store the 1's on the diagonal.

14. Show how to store in sparse form an elementary matrix that adds c times row i of the identity matrix to row j. Do not store the 1's on the diagonal.

15. Show how to store in sparse form a matrix E that differs from the $(n \times n)$ identity matrix only in column k. Do not store the 1's on the diagonal.

16. In Example 8.4, it was shown that truncation error leads to an incorrect answer on a computer with a precision of four digits when the first two numbers in the list 5.967, 1325.000, -1324.000 are added and the resulting total is then added to the last number. Show that the correct result is obtained when the last two numbers in this list are added first and the resulting total is then added to the first number in the list.

17. According to the rule of partial pivoting, which row of the following matrix should you exchange with row 1 when using Gauss-Jordan elimination to obtain a leading one in row 1:

$$\begin{bmatrix} 2 & -1 & 3 & 0 \\ 1 & 3 & 0 & -2 \\ -4 & -2 & 1 & 6 \\ 3 & 0 & -5 & 7 \end{bmatrix}$$

18. With the rule of partial pivoting, which row of the following matrix should you exchange with row 3 when using Gauss-Jordan elimination to obtain a leading one in row 3:

$$\begin{bmatrix} 1 & 0 & -9 & 0 & 5 & -3 \\ 0 & 1 & 8 & -2 & 0 & 2 \\ 0 & 0 & -2 & 6 & -1 & 0 \\ 0 & 0 & 0 & 7 & 4 & 5 \\ 0 & 0 & 3 & 1 & -2 & 6 \\ 0 & 0 & 1 & -6 & 3 & 7 \end{bmatrix}$$

8.2 Matrix Factorizations

You know from Section 2.3 that the various methods for solving a system of linear equations are composed of the following two phases:

1. *Preparation phase*, in which operations are performed to prepare the system of equations for finding a subsequent solution.

2. *Solution phase*, in which additional operations are performed to provide a solution to the system of equations.

In this section, several new approaches are presented that take into account some of the computational concerns discussed in Section 8.1. You will see that it often pays to spend more time and effort in the preparation phase to gain subsequent efficiencies in the solution phase. The key to these approaches is given in the following definition.

DEFINITION 8.1

A **factorization** of a matrix A is a collection of two or more matrices whose product is A.

As an example, suppose that A is an invertible $(n \times n)$ matrix and that I is the $(n \times n)$ identity matrix. The matrices A and A^{-1} constitute a factorization of I because I is the product of A and A^{-1}, that is,

$$I = AA^{-1}$$

Other factorizations of matrices and their roles in solving problems are now described.

8.2.1 Using a Factorization of the Inverse of a Matrix to Solve a System of Linear Equations

Throughout this section, suppose that A is an $(n \times n)$ matrix, $\mathbf{b} \in R^{n \times 1}$ is a column vector, and that you want to solve the system of linear equations

$$A\mathbf{x} = \mathbf{b} \tag{8.11}$$

One way to do so, as described in Section 3.2, is to find the inverse of A in the preparation phase and then to compute $\mathbf{x} = A^{-1}\mathbf{b}$ in the solution phase. You will now see how a factorization arises when elementary row operations are used to find A^{-1}.

Expressing the Inverse as a Product of Elementary Matrices

As shown in the proof of Theorem 3.4 in Section 3.2.3, one way to obtain the inverse of an invertible matrix A is to find elementary $(n \times n)$ matrices $E_1, \ldots, E_p$ such that

$$(E_p \cdots E_1)A = I$$

for then

$$A^{-1} = E_p \cdots E_1 \tag{8.12}$$

The matrices $E_p, \ldots, E_1$ on the right side of (8.12) constitute a factorization of A^{-1} because A^{-1} is the product of $E_p, \ldots, E_1$.

To illustrate, consider the matrix

$$A = \begin{bmatrix} 0 & 1 \\ 2 & 4 \end{bmatrix}$$

The following steps of Gauss-Jordan elimination with partial pivoting (see Section 8.1.3) yield the desired factorization of A^{-1}.

Step 1. Partial pivoting requires exchanging row 1 and row 2 of A, so

$$E_1 = \begin{bmatrix} 0 & 1 \\ 1 & 0 \end{bmatrix} \quad \text{and thus} \quad E_1 A = \begin{bmatrix} 2 & 4 \\ 0 & 1 \end{bmatrix}$$

Step 2. Multiply row 1 of $E_1 A$ by 1/2, so

$$E_2 = \begin{bmatrix} 1/2 & 0 \\ 0 & 1 \end{bmatrix} \quad \text{and thus} \quad E_2 E_1 A = \begin{bmatrix} 1 & 2 \\ 0 & 1 \end{bmatrix}$$

Step 3. Having created a leading 1 in row 1, turn now to row 2. Because there is already a leading 1 in row 2 of $E_2 E_1 A$, you need only add -2 times row 2 to row 1, so

$$E_3 = \begin{bmatrix} 1 & -2 \\ 0 & 1 \end{bmatrix} \quad \text{and thus} \quad E_3 E_2 E_1 A = \begin{bmatrix} 1 & 0 \\ 0 & 1 \end{bmatrix}$$

The foregoing factorization of A^{-1}, namely, $E_3 E_2 E_1$, is summarized in the following example.

EXAMPLE 8.6 A Factorization of the Inverse of a Matrix

A factorization of the inverse of the matrix

$$A = \begin{bmatrix} 0 & 1 \\ 2 & 4 \end{bmatrix}$$

is

$$E_1 = \begin{bmatrix} 0 & 1 \\ 1 & 0 \end{bmatrix}, \quad E_2 = \begin{bmatrix} 1/2 & 0 \\ 0 & 1 \end{bmatrix}, \quad \text{and} \quad E_3 = \begin{bmatrix} 1 & -2 \\ 0 & 1 \end{bmatrix}$$

because

$$A^{-1} = E_3 E_2 E_1 = \begin{bmatrix} -2 & 1/2 \\ 1 & 0 \end{bmatrix}$$

Using a Factorization of the Inverse to Solve a System of Linear Equations

One way to use the factorization in (8.12) for solving the system of linear equations $A\mathbf{x} = \mathbf{b}$ is to compute $A^{-1} = E_p \cdots E_2 E_1$, which requires $(p - 1)n^3$ multiplications. With an additional n^2 multiplications, you can then calculate the solution:

$$\mathbf{x} = A^{-1}\mathbf{b} \tag{8.13}$$

You can obtain the same answer with only pn^2 multiplications. To see how, replace A^{-1} in (8.13) with $E_p \cdots E_2 E_1$ to obtain

$$\mathbf{x} = (E_p \cdots E_2 E_1)\mathbf{b} \tag{8.14}$$

Then use the associative property to perform the multiplications in (8.14) in the following more efficient order:

$$\mathbf{x} = E_p \cdots (E_2(E_1 \mathbf{b})) \tag{8.15}$$

Another reason the approach in (8.15) is better than the one in (8.14) is that, by keeping the elementary matrices separate from each other, you can maintain and exploit the sparsity (see Section 8.1.3) of these matrices, which is lost when you multiply them together, as in (8.14).

EXAMPLE 8.7 Using a Factorization of the Inverse to Solve a System of Linear Equations

Consider solving the system of linear equations $A\mathbf{x} = \mathbf{b}$, in which

$$A = \begin{bmatrix} 0 & 1 \\ 2 & 4 \end{bmatrix} \quad \text{and} \quad \mathbf{b} = \begin{bmatrix} 1 \\ -2 \end{bmatrix}$$

Using the factorization of A^{-1} from Example 8.6 and performing the computations in the order presented in (8.15) yields

$$\mathbf{x} = A^{-1}\mathbf{b} = E_3 E_2 (E_1 \mathbf{b})$$

$$= E_3 E_2 \left(\begin{bmatrix} 0 & 1 \\ 1 & 0 \end{bmatrix} \begin{bmatrix} 1 \\ -2 \end{bmatrix} \right)$$

$$= E_3 \left(\begin{bmatrix} 1/2 & 0 \\ 0 & 1 \end{bmatrix} \begin{bmatrix} -2 \\ 1 \end{bmatrix} \right)$$

$$= \left(\begin{bmatrix} 1 & -2 \\ 0 & 1 \end{bmatrix} \begin{bmatrix} -1 \\ 1 \end{bmatrix} \right)$$

$$= \begin{bmatrix} -3 \\ 1 \end{bmatrix}$$

8.2.2 Using an *LU* Factorization in Gaussian Elimination

Recall, from Section 2.3.1, that when Gaussian elimination is used to solve the system of linear equations $A\mathbf{x} = \mathbf{b}$, the augmented matrix $[A \mid \mathbf{b}]$ is first reduced to a row-echelon form in the preparation phase. Then, the efficient process of back substitution is used to obtain the desired values of $\mathbf{x}$ in the solution phase. You will now see how these steps are accomplished using a matrix factorization.

The approach is to find an **LU factorization** of A, namely:

$$A = LU \tag{8.16}$$

in which L is a lower triangular matrix with 1's along the diagonal and U is an upper triangular matrix. An example of such a factorization follows. The details of how to find this factorization are explained subsequently.

EXAMPLE 8.8 An *LU* Factorization of A

An *LU* factorization of the matrix

$$A = \begin{bmatrix} 2 & -1 & -2 \\ -2 & 4 & 4 \\ -4 & 8 & 4 \end{bmatrix}$$

consists of the following lower triangular matrix, L, and upper triangular matrix, U, because

$$A = \begin{bmatrix} 2 & -1 & -2 \\ -2 & 4 & 4 \\ -4 & 8 & 4 \end{bmatrix} = \begin{bmatrix} 1 & 0 & 0 \\ -1 & 1 & 0 \\ -2 & 2 & 1 \end{bmatrix} \begin{bmatrix} 2 & -1 & -2 \\ 0 & 3 & 2 \\ 0 & 0 & -4 \end{bmatrix} = LU$$

How to Use an *LU* Factorization to Solve a System of Linear Equations

Once an LU factorization of A is found, the triangular structure of L and U are used to solve the following system efficiently by forward and back substitution (see Section 2.3.2):

$$A\mathbf{x} = \mathbf{b} \tag{8.17}$$

To see how, write (8.17) by substituting $A = LU$ from (8.16) to obtain:

$$L(U\mathbf{x}) = \mathbf{b} \tag{8.18}$$

Think of $U\mathbf{x}$ in (8.18) as some new variables, denoted by the column vector $\mathbf{y} \in R^{n \times 1}$, that is

$$U\mathbf{x} = \mathbf{y} \tag{8.19}$$

Once you find the value for $\mathbf{y}$, you can find the value for $\mathbf{x}$ by solving (8.19) efficiently using back substitution, because U is upper triangular.

To find the needed value for $\mathbf{y}$, replace $U\mathbf{x}$ with $\mathbf{y}$ in (8.18) to obtain:

$$L\mathbf{y} = \mathbf{b} \tag{8.20}$$

Now L is lower triangular, so you can solve (8.20) for $\mathbf{y}$ efficiently by using forward substitution.

Solving a System of Linear Equations with an *LU* Factorization

In summary, to solve the system $A\mathbf{x} = \mathbf{b}$ with an LU factorization, apply the following preparation and solution phases:

1. Prepare the system $A\mathbf{x} = \mathbf{b}$ by finding an LU factorization of A. That is, find a lower triangular matrix, L, and an upper triangular matrix, U, such that $A = LU$.

2. Solve the system $A\mathbf{x} = \mathbf{b}$, as follows:

   (a) Find values for the column vector $\mathbf{y}$ by applying forward substitution to the system $L\mathbf{y} = \mathbf{b}$.

   (b) Use the vector $\mathbf{y}$ found in 2(a) to find values for the column vector $\mathbf{x}$ by applying back substitution to the system $U\mathbf{x} = \mathbf{y}$.

EXAMPLE 8.9 Solving a System of Equations with an *LU* Factorization

Consider solving the system $A\mathbf{x} = \mathbf{b}$, in which

$$A = \begin{bmatrix} 2 & -1 & -2 \\ -2 & 4 & 4 \\ -4 & 8 & 4 \end{bmatrix} \quad \text{and} \quad \mathbf{b} = \begin{bmatrix} -2 \\ 6 \\ 4 \end{bmatrix}$$

To use an *LU* factorization, it is first necessary to prepare the system by finding a lower triangular matrix, L, and an upper triangular matrix, U, such that $A = LU$. The desired matrices L and U are given in Example 8.8.

Having prepared the system by finding an *LU* factorization of A, you now solve the system $A\mathbf{x} = \mathbf{b}$ by using forward substitution to solve the system $L\mathbf{y} = \mathbf{b}$ for $\mathbf{y}$. In this example, using L from Example 8.8, you solve the following system $L\mathbf{y} = \mathbf{b}$ by forward substitution:

$$\begin{bmatrix} 1 & 0 & 0 \\ -1 & 1 & 0 \\ -2 & 2 & 1 \end{bmatrix} \begin{bmatrix} y_1 \\ y_2 \\ y_3 \end{bmatrix} = \begin{bmatrix} -2 \\ 6 \\ 4 \end{bmatrix} \quad \text{or} \quad \begin{array}{rcr} y_1 & = & -2 \\ -y_1 + y_2 & = & 6 \\ -2y_1 + 2y_2 + y_3 & = & 4 \end{array}$$

From the first equation, $y_1 = -2$. Substituting this value in the second equation and solving for y_2 yields $y_2 = 4$. Substituting $y_1 = -2$ and $y_2 = 4$ in the last equation and solving for y_3 yields $y_3 = -8$.

Having found the value for $\mathbf{y}$, you obtain the desired value for $\mathbf{x}$ by using back substitution to solve the system $U\mathbf{x} = \mathbf{y}$ for $\mathbf{x}$. In this example, using U from Example 8.8, you solve the following system $U\mathbf{x} = \mathbf{y}$ by back substitution:

$$\begin{bmatrix} 2 & -1 & -2 \\ 0 & 3 & 2 \\ 0 & 0 & -4 \end{bmatrix} \begin{bmatrix} x_1 \\ x_2 \\ x_3 \end{bmatrix} = \begin{bmatrix} -2 \\ 4 \\ -8 \end{bmatrix} \quad \text{or} \quad \begin{array}{rcr} 2x_1 - x_2 - 2x_3 & = & -2 \\ 3x_2 + 2x_3 & = & 4 \\ -4x_3 & = & -8 \end{array}$$

From the last equation, $x_3 = 2$. Substituting this value in the second equation and solving for x_2 yields $x_2 = 0$. Substituting $x_2 = 0$ and $x_3 = 2$ in the first equation and solving for x_1 yields $x_1 = 1$.

Additional efficiency is realized with the foregoing computational approach when it is necessary to solve the system $A\mathbf{x} = \mathbf{b}$ many times, each time with the same A matrix but a different column vector $\mathbf{b}$. This is because the matrices L and U are computed only once in the preparation phase. Then, for each right-hand-side vector $\mathbf{b}$, you need only use forward substitution to solve the system $L\mathbf{y} = \mathbf{b}$ for $\mathbf{y}$ and and then back substitution to solve the system $U\mathbf{x} = \mathbf{y}$ for $\mathbf{x}$.

Finding an *LU* Factorization of A

The approach to finding the matrices L and U in an *LU* factorization of A is based on Gaussian elimination. Specifically, let $E_1, \ldots, E_p$ be the elementary matrices in Gaussian elimination that reduce the augmented matrix $[A \mid \mathbf{b}]$ to a row-echelon form. Then, the following matrix, being in row-echelon form, is the desired upper triangular matrix U:

$$U = (E_p \cdots E_1)A \tag{8.21}$$

To find the matrix L, observe that each elementary matrix, E_i, in (8.21) is invertible and therefore so is $E_p \cdots E_1$. Then

$$L = (E_p \cdots E_1)^{-1} \tag{8.22}$$

satisfies

$$LU = (E_p \cdots E_1)^{-1}[(E_p \cdots E_1)A] = L(L^{-1}A) = A$$

The matrix L in (8.22) will be lower triangular when each matrix E_i in (8.21) is lower triangular. This is because the product of lower triangular matrices is lower triangular, as you are asked to show in Exercise 13 and Exercise 14. Each matrix E_i in (8.21) will be lower triangular provided that no row interchanges are performed during Gaussian elimination.

Although you can find U and L by performing the computations in (8.21) and (8.22), a more efficient approach is available. To see how, consider the following two steps of Gaussian elimination that reduce the matrix A in Example 8.8 to a row-echelon form:

$$\underbrace{\begin{bmatrix} 2 & -1 & -2 \\ -2 & 4 & 4 \\ -4 & 8 & 4 \end{bmatrix}}_{A=A_1} \longrightarrow \underbrace{\begin{bmatrix} 2 & -1 & -2 \\ 0 & 3 & -2 \\ 0 & 6 & 0 \end{bmatrix}}_{A_2} \longrightarrow \underbrace{\begin{bmatrix} 2 & -1 & -2 \\ 0 & 3 & -2 \\ 0 & 0 & 4 \end{bmatrix}}_{U} \tag{8.23}$$

You can use the matrices, A_1, A_2, and so on, obtained in finding U, to create lower triangular matrices L_1, L_2, and so on, the last of which is L. In this example, A_1 and A_2 in (8.23) are used to create L_1 and L_2, with $L_2 = L$. To see how this is done, start with $L_0 = I$, the (3×3) identity matrix. The elements in the first column of I below the 1 on the diagonal are replaced with the corresponding elements in the first column of A_1 divided by the pivot element, 2, in this case. That is,

$$L_0 = \begin{bmatrix} 1 & 0 & 0 \\ 0 & 1 & 0 \\ 0 & 0 & 1 \end{bmatrix} \quad \text{leads to} \quad L_1 = \begin{bmatrix} 1 & 0 & 0 \\ -1 & 1 & 0 \\ -2 & 0 & 1 \end{bmatrix} \tag{8.24}$$

Now A_2 in (8.23) and L_1 in (8.24) are used to create $L_2 = L$. Specifically, the elements in the second column of L_2 below the 1 on the diagonal are replaced with the corresponding elements in the second column of A_2 divided by the pivot element, 3, in this case. The result is:

$$L_1 = \begin{bmatrix} 1 & 0 & 0 \\ -1 & 1 & 0 \\ -2 & 0 & 1 \end{bmatrix} \quad \text{leads to} \quad L_2 = \begin{bmatrix} 1 & 0 & 0 \\ -1 & 1 & 0 \\ -2 & 2 & 1 \end{bmatrix} = L \tag{8.25}$$

How to Find an *LU* Factorization

In summary, to obtain an LU factorization of an $(n \times n)$ matrix A, proceed as follows:

1. Use Gaussian elimination to reduce each column i of the matrix A to row-echelon form, U, obtaining the intermediate matrices $A = A_1, A_2, \ldots, A_{n-1} = U$.

2. Use $A_1, \ldots, A_{n-1}$ from (1) to create lower triangular matrices $L_1, \ldots, L_{n-1}$, the last of which is L. Specifically, start with $L_0 = I$. For each $i = 1, \ldots, n-1$, create L_i by replacing column i of L_{i-1} below the 1 on the diagonal with the corresponding elements in column i of A_i divided by the pivot element in row i and column i of A_i.

Keep in mind that L is lower triangular only when no row interchanges are used during Gaussian elimination. If row interchanges are performed, either to avoid a 0 on the diagonal of U or for the purposes of implementing partial pivoting so that the elements on the diagonal of U are as large as possible, then L is a *permuted* lower triangular matrix. In this case, solving the system $L\mathbf{y} = \mathbf{b}$ for $\mathbf{y}$ is more complicated than using simple forward substitution. The details, however, are omitted.

8.2.3 A *QR* Factorization in the Gram-Schmidt Process

Another factorization of an $(m \times n)$ matrix arises when an orthonormal basis for a subspace of R^n is constructed using the Gram-Schmidt process, which you should review from Section 7.2.3 together with the material on orthogonality in Section 7.1.

Finding a *QR* Factorization

Recall that the following Gram-Schmidt process uses a given basis of n-vectors $\mathbf{x}_1, \ldots, \mathbf{x}_k$ for a subspace W of R^n to produce an orthogonal basis $\mathbf{v}_1, \ldots, \mathbf{v}_k$ of W, that is, a basis in which the vectors $\mathbf{v}_i$ and $\mathbf{v}_j$ are orthogonal for all $i \neq j$:

The Gram-Schmidt Process

Given a basis $\{\mathbf{x}_1, \ldots, \mathbf{x}_k\}$ for a subspace W of R^n, an orthogonal basis $\{\mathbf{v}_1, \ldots, \mathbf{v}_k\}$ for W is obtained by computing:

$$\mathbf{v}_1 = \mathbf{x}_1$$

$$\mathbf{v}_2 = \mathbf{x}_2 - \frac{\mathbf{x}_2 \cdot \mathbf{v}_1}{\mathbf{v}_1 \cdot \mathbf{v}_1}\mathbf{v}_1$$

$$\mathbf{v}_3 = \mathbf{x}_3 - \frac{\mathbf{x}_3 \cdot \mathbf{v}_1}{\mathbf{v}_1 \cdot \mathbf{v}_1}\mathbf{v}_1 - \frac{\mathbf{x}_3 \cdot \mathbf{v}_2}{\mathbf{v}_2 \cdot \mathbf{v}_2}\mathbf{v}_2$$

$$\vdots$$

$$\mathbf{v}_k = \mathbf{x}_k - \frac{\mathbf{x}_k \cdot \mathbf{v}_1}{\mathbf{v}_1 \cdot \mathbf{v}_1}\mathbf{v}_1 - \frac{\mathbf{x}_k \cdot \mathbf{v}_2}{\mathbf{v}_2 \cdot \mathbf{v}_2}\mathbf{v}_2 - \cdots - \frac{\mathbf{x}_k \cdot \mathbf{v}_{k-1}}{\mathbf{v}_{k-1} \cdot \mathbf{v}_{k-1}}\mathbf{v}_{k-1}$$

If the basis vectors, say, $\mathbf{x}_1, \ldots, \mathbf{x}_n$, are linearly independent columns of an $(m \times n)$ matrix

A, then, with a small amount of additional work, you can produce an $(m \times n)$ matrix Q, whose columns form an orthonormal basis for the column space of A, together with an upper triangular $(n \times n)$ matrix R, such that $A = QR$. The matrices Q and R constitute a **QR factorization** of A.

To see how the matrices Q and R are created, start with the orthogonal vectors $\mathbf{v}_1, \ldots, \mathbf{v}_n$ produced by the Gram-Schmidt process. Column j of Q is the normalized column vector $\mathbf{v}_j$, that is

$$Q_{*j} = \frac{1}{\|\mathbf{v}_j\|} \mathbf{v}_j \qquad (8.26)$$

The columns of Q then constitute an orthonormal basis for the column space of A.

You can use Q to establish the existence of a matrix R such that $A = QR$. To do so, note that because Q is a basis for the column space of A, each of the n columns of A is a linear combination of the n columns of Q. That is, for each column j of A, there is a column vector $\mathbf{r}_j \in R^{n \times 1}$ such that

$$A_{*j} = Q\mathbf{r}_j, \quad \text{for } j = 1, \ldots, n \qquad (8.27)$$

Letting R be the $(n \times n)$ matrix in which $R_{*j} = \mathbf{r}_j$, you then have from (8.27) that

$$A = QR \qquad (8.28)$$

The proof that R is invertible and upper triangular is omitted.

To find the elements of R using (8.27) requires expressing each column of A as a linear combination of the columns of Q. In the spirit of using mathematics to solve a problem more efficiently, you will now see that you can use Q to compute R more simply, as follows:

$$R = Q^t A \qquad (8.29)$$

To see that (8.29) is valid, multiply both sides of (8.28) on the left by Q^t, so

$$Q^t A = Q^t(QR) = (Q^t Q)R \qquad (8.30)$$

The columns of Q are orthonormal, so $Q^t Q = I$ because

$$(Q^t Q)_{ij} = (Q^t)_{i*} \cdot Q_{*j} \qquad \text{(definition of matrix multiplication)}$$

$$= Q_{*i} \cdot Q_{*j} \qquad \text{(definition of transpose)}$$

$$= \begin{cases} 1, & \text{if } i = j \\ 0, & \text{if } i \neq j \end{cases} \qquad \text{(the columns of } Q \text{ are orthonormal)}$$

It now follows that the right side of (8.30) reduces to R and so $R = Q^t A$.

The process of constructing the QR factorization of a matrix A using (8.26) and (8.29) is illustrated in the following example.

EXAMPLE 8.10 Finding a *QR* Factorization of a Matrix

To find a QR factorization of the matrix

$$
A = \begin{bmatrix} 1 & 1 & 1 \\ 0 & -1 & 3 \\ -1 & 3 & -1 \end{bmatrix}
$$

first apply the Gram-Schmidt process to the columns of A to obtain the following orthogonal basis for the column space of A:

$$
\mathbf{v}_1 = \begin{bmatrix} 1 \\ 0 \\ -1 \end{bmatrix}, \quad \mathbf{v}_2 = \begin{bmatrix} 2 \\ -1 \\ 2 \end{bmatrix}, \quad \text{and} \quad \mathbf{v}_3 = \begin{bmatrix} 2/3 \\ 8/3 \\ 2/3 \end{bmatrix}
$$

From (8.26), column j of Q is $\mathbf{v}_j$ divided by the norm of $\mathbf{v}_j$, so

$$
Q = \begin{bmatrix} 0.7071 & 0.6667 & 0.2357 \\ 0.0000 & -0.3333 & 0.9428 \\ -0.7071 & 0.6667 & 0.2357 \end{bmatrix}
$$

Now compute the upper triangular matrix R using (8.29), as follows:

$$
R = Q^t A = \begin{bmatrix} 0.7071 & 0.0000 & -0.7071 \\ 0.6667 & -0.3333 & 0.6667 \\ 0.2357 & 0.9428 & 0.2357 \end{bmatrix} \begin{bmatrix} 1 & 1 & 1 \\ 0 & -1 & 3 \\ -1 & 3 & -1 \end{bmatrix}
$$

$$
= \begin{bmatrix} 1.4142 & -1.4142 & 1.4142 \\ 0.0000 & 3.0000 & -1.0000 \\ 0.0000 & 0.0000 & 2.8284 \end{bmatrix}
$$

Computational Implementation

The foregoing approach to finding a QR factorization of a matrix A requires using the Gram-Schmidt process to find an orthogonal basis $\mathbf{v}_1, \ldots, \mathbf{v}_n$ for the column space of A. The resulting vectors are then normalized to form the columns of Q. The truncation errors arising in the Gram-Schmidt process can sometimes accumulate to a point where for large i and j with $i \neq j$, $\mathbf{v}_i \cdot \mathbf{v}_j$ is not sufficiently close to 0. This problem of numerical accuracy can often be overcome by modifying the computations in the Gram-Schmidt process. The details can be found in *Fundamentals of Matrix Computations*, by David S. Watkins, (John Wiley & Sons, New York, 1991, pages 175-176).

One of the best computational approaches for finding Q is similar to that used in finding a factorization in Gaussian elimination described in Section 8.2.2. In this case, orthogonal matrices $Q_1, \ldots, Q_p$ are found so that

$$(Q_p \cdots Q_1)A = R \tag{8.31}$$

is upper triangular. Comparing (8.31) with (8.29), it follows that

$$Q^t = Q_p \cdots Q_1 \tag{8.32}$$

Equivalently, by transposing both sides of (8.32),

$$Q = Q_1^t \cdots Q_p^t \tag{8.33}$$

The details of computing $Q_1, \ldots, Q_p$ can be found in *Fundamentals of Matrix Computations*, by David S. Watkins, (John Wiley & Sons, New York, 1991, pages 143-144).

8.2.4 A Factorization of a Diagonalizable Matrix

Recall, from Definition 5.5 in Section 5.4.3, that an $(n \times n)$ matrix A is *similar* to a given $(n \times n)$ matrix D if and only if there is an invertible $(n \times n)$ matrix P such that

$$A = PDP^{-1} \tag{8.34}$$

When D is a diagonal matrix, the factorization in (8.34) is particularly useful for computing powers of A. The reason, as shown in Section 6.4.2, is that for any positive integer k,

$$A^k = (PDP^{-1})^k = PD^k P^{-1} \tag{8.35}$$

Moreover, D^k is easy to compute when D is diagonal because, as shown in Section 6.4.1,

$$D^k = \begin{bmatrix} (D_{11})^k & 0 & \cdots & 0 \\ 0 & (D_{22})^k & \cdots & 0 \\ \vdots & \vdots & \ddots & \vdots \\ 0 & 0 & \cdots & (D_{nn})^k \end{bmatrix}$$

The need to compute powers of A arises, for example, in the study of dynamical systems (see Chapter 6). Finding a factorization such as the one in (8.34), in which D is a diagonal matrix, allows you to determine the long-term behavior of a dynamical system by using (8.35) to compute large powers of A efficiently.

An $(n \times n)$ matrix A with the property that there is an $(n \times n)$ diagonal matrix D and an invertible matrix P such that (8.34) holds true is said to be *diagonalizable* (see Section 6.4.2). Such a matrix is given in the following example.

EXAMPLE 8.11 A Factorization of a Diagonalizable Matrix
The matrix

$$A = \begin{bmatrix} -6 & 4 & -10 \\ -5 & 3 & -10 \\ 2 & -2 & 3 \end{bmatrix}$$

is diagonalizable because you can verify that, for the following matrices P and D, P is invertible and $A = PDP^{-1}$:

$$P = \begin{bmatrix} 1 & 2 & -2 \\ 1 & 2 & 0 \\ 0 & -1 & 1 \end{bmatrix} \quad \text{and} \quad D = \begin{bmatrix} -2 & 0 & 0 \\ 0 & 3 & 0 \\ 0 & 0 & -1 \end{bmatrix}$$

It now follows that for any positive integer k:

$$A^k = PD^k P^{-1} = \begin{bmatrix} 1 & 2 & -2 \\ 1 & 2 & 0 \\ 0 & -1 & 1 \end{bmatrix} \begin{bmatrix} (-2)^k & 0 & 0 \\ 0 & 3^k & 0 \\ 0 & 0 & (-1)^k \end{bmatrix} \begin{bmatrix} 1 & 0 & 2 \\ -1/2 & 1/2 & -1 \\ -1/2 & 1/2 & 0 \end{bmatrix}$$

From Theorem 6.3 in Section 6.4.3, you can conclude that an $(n \times n)$ matrix A is diagonalizable if and only if A has n linearly independent eigenvectors. In this case, the columns of P are the eigenvectors and the diagonal elements of D are the corresponding eigenvalues. Thus, to find P and D requires finding the eigenvectors and eigenvalues of A. A numerical method for doing so is presented in Section 8.3.

Exercises for Section 8.2

1. Do the following matrices constitute a factorization of the identity matrix? Explain.

$$A = \begin{bmatrix} -1 & -3 \\ 1 & 4 \end{bmatrix} \quad \text{and} \quad B = \begin{bmatrix} -4 & -6 \\ 1 & 2 \end{bmatrix}$$

2. Show that the following matrices constitute a factorization of the matrix B in the previous exercise:

$$A = \begin{bmatrix} -4 & -3 \\ 1 & 1 \end{bmatrix} \quad \text{and} \quad C = \begin{bmatrix} 1 & 0 \\ 0 & 2 \end{bmatrix}$$

3. Use Gauss-Jordan elimination with partial pivoting to produce the elementary matrices that constitute a factorization of the inverse of the following matrix (show your work):

$$A = \begin{bmatrix} -2 & -3 \\ 4 & 8 \end{bmatrix}$$

4. Use Gauss-Jordan elimination with partial pivoting to produce the elementary matrices that constitute a factorization of the inverse of the following matrix (show your work):

$$A = \begin{bmatrix} -2 & -3 \\ 3 & 6 \end{bmatrix}$$

5. Use the factorization in Exercise 3 to solve the following system of linear equations:

$$-2x_1 - 3x_2 = -4$$
$$4x_1 + 8x_2 = 8$$

6. Use the factorization in Exercise 4 to solve the following system of linear equations:

$$-2x_1 - 3x_2 = -4$$
$$3x_1 + 6x_2 = 9$$

7. Use the following LU factorization of A to solve the system $Ax = b$ for the given column vector b by performing the indicated operations:

$$A = \begin{bmatrix} 3 & 6 \\ -3 & -2 \end{bmatrix}, \quad L = \begin{bmatrix} 1 & 0 \\ -1 & 1 \end{bmatrix}, \quad U = \begin{bmatrix} 3 & 6 \\ 0 & 4 \end{bmatrix}, \quad b = \begin{bmatrix} 3 \\ -7 \end{bmatrix}$$

   (a) Write the system $Ly = b$ and then use forward substitution to solve for y.
   (b) Use the vector y obtained in part (a) to write the system $Ux = y$ and then apply back substitution to solve for x. Verify that your solution satisfies the original system of linear equations $Ax = b$.

8. Use the following LU factorization of A to solve the system $Ax = b$ for the given column vector b by performing the indicated operations:

$$A = \begin{bmatrix} 2 & 1 & -1 \\ 2 & 3 & -3 \\ -2 & -2 & 5 \end{bmatrix}, L = \begin{bmatrix} 1 & 0 & 0 \\ 1 & 1 & 0 \\ -1 & -1/2 & 1 \end{bmatrix}, U = \begin{bmatrix} 2 & 1 & -1 \\ 0 & 2 & -2 \\ 0 & 0 & 3 \end{bmatrix}, b = \begin{bmatrix} -2 \\ 2 \\ 3 \end{bmatrix}$$

   (a) Write the system $Ly = b$ and then use forward substitution to solve for y.
   (b) Use the vector y obtained in part (a) to write the system $Ux = y$ and then apply back substitution to solve for x. Verify that your solution satisfies the original system of linear equations $Ax = b$.

9. Use technology to perform Gaussian elimination to obtain an LU factorization of the following matrix A.

$$A = \begin{bmatrix} -1 & -3 & -1 \\ 2 & 8 & -2 \\ 0 & 4 & -4 \end{bmatrix}$$

10. Use technology to perform Gaussian elimination to obtain an LU factorization of the following matrix A.

$$A = \begin{bmatrix} -2 & -3 & 1 \\ 0 & 3 & -9 \\ 3 & 6 & -3 \end{bmatrix}$$

11. Use the *LU* factorization in Exercise 9 to solve the system $A\mathbf{x} = \mathbf{b}$, in which

$$\mathbf{b} = \begin{bmatrix} -2 \\ 4 \\ -4 \end{bmatrix}$$

by performing the following steps.
(a) Write the system $L\mathbf{y} = \mathbf{b}$ and then use forward substitution to solve for $\mathbf{y}$.
(b) Use the vector $\mathbf{y}$ obtained in part (a) to write the system $U\mathbf{x} = \mathbf{y}$ and then use back substitution to solve for $\mathbf{x}$. Verify that your solution satisfies the original system of linear equations $A\mathbf{x} = \mathbf{b}$.

12. Use the factorization in Exercise 10 to solve the system $A\mathbf{x} = \mathbf{b}$, in which

$$\mathbf{b} = \begin{bmatrix} 1 \\ 3 \\ -6 \end{bmatrix}$$

by performing the following steps.
(a) Write the system $L\mathbf{y} = \mathbf{b}$ and then use forward substitution to solve for $\mathbf{y}$.
(b) Use the vector $\mathbf{y}$ obtained in part (a) to write the system $U\mathbf{x} = \mathbf{y}$ and then use back substitution to solve for $\mathbf{x}$. Verify that your solution satisfies the original system of linear equations $A\mathbf{x} = \mathbf{b}$.

13. Prove that if A and B are lower triangular matrices, then AB is lower triangular.

14. Use induction to prove that for any integer $p \geq 1$, if $L_1, \ldots, L_p$ are lower triangular, then $L = L_1 \cdots L_p$ is lower triangular. (Hint: Use the previous exercise.)

15. For the columns of the given matrix A, the Gram-Schmidt process produces the following three orthogonal vectors $\mathbf{v}_1$, $\mathbf{v}_2$, and $\mathbf{v}_3$:

$$A = \begin{bmatrix} 2 & 1 & 0 \\ -2 & 1 & 0 \\ 1 & 0 & -3 \end{bmatrix}, \quad \mathbf{v}_1 = \begin{bmatrix} 2 \\ -2 \\ 1 \end{bmatrix}, \quad \mathbf{v}_2 = \begin{bmatrix} 1 \\ 1 \\ 0 \end{bmatrix}, \quad \mathbf{v}_3 = \begin{bmatrix} 2/3 \\ -2/3 \\ -8/3 \end{bmatrix}$$

(a) Use $\mathbf{v}_1$, $\mathbf{v}_2$, and $\mathbf{v}_3$ to create the matrix Q in a QR factorization of A.
(b) Use technology to compute the upper triangular matrix $R = Q^t A$ and show that $A = QR$.
(c) Use technology to solve the following systems of equations for $\mathbf{r}_j$:

$$Q\mathbf{r}_j = A_{*j}, \quad j = 1, 2, 3$$

Verify that for each j, $\mathbf{r}_j$ is equal to column j of the matrix R obtained in part (b).

16. Repeat the previous exercise for the following matrix A and the three orthogonal vectors $\mathbf{v}_1$, $\mathbf{v}_2$, and $\mathbf{v}_3$ obtained from the Gram-Schmidt process:

$$A = \begin{bmatrix} 1 & 1 & -2 \\ 0 & -2 & 4 \\ -1 & 3 & 0 \end{bmatrix}, \quad \mathbf{v}_1 = \begin{bmatrix} 1 \\ 0 \\ -1 \end{bmatrix}, \quad \mathbf{v}_2 = \begin{bmatrix} 2 \\ -2 \\ 2 \end{bmatrix}, \quad \mathbf{v}_3 = \begin{bmatrix} 1 \\ 2 \\ 1 \end{bmatrix}$$

8.3 Finding Eigenvalues and Eigenvectors

In Chapter 6, you saw how eigenvalues and eigenvectors are used to solve problems, such as determining the long-term behavior of a dynamical system. You should review that material now. In many applications, it is not possible to find the eigenvalues and eigenvectors explicitly. In such cases, the numerical methods described in this section are used to obtain approximate values. Throughout this section, A is an $(n \times n)$ matrix.

8.3.1 The Power Method

The method described next is used to find a **strictly dominant eigenvalue**, that is, an eigenvalue whose absolute value is strictly greater than the absolute value of all other eigenvalues. A corresponding eigenvector is also produced. The approach is to generate a sequence of n-vectors, $\mathbf{X} = (\mathbf{x}_0, \mathbf{x}_1, \mathbf{x}_2, \ldots)$, that *converges*—that is, gets closer and closer—to an eigenvector corresponding to the strictly dominant eigenvalue of A. This sequence is obtained by selecting an initial vector, $\mathbf{x}_0$, and then computing

$$\mathbf{x}_{k+1} = A\mathbf{x}_k, \quad \text{for } k = 0, 1, 2, \ldots \tag{8.36}$$

To illustrate, consider the following matrix A, whose eigenvalues are 1.5 and 0.25, and the following initial vector $\mathbf{x}_0$:

$$A = \begin{bmatrix} 0.75 & 1.5 \\ 0.25 & 1.0 \end{bmatrix} \quad \text{and} \quad \mathbf{x}_0 = \begin{bmatrix} -1 \\ 2 \end{bmatrix}$$

You can then compute

$$\mathbf{x}_1 = A\mathbf{x}_0 = \begin{bmatrix} 0.75 & 1.5 \\ 0.25 & 1.0 \end{bmatrix} \begin{bmatrix} -1 \\ 2 \end{bmatrix} = \begin{bmatrix} 2.25 \\ 1.75 \end{bmatrix}$$

$$\mathbf{x}_2 = A\mathbf{x}_1 = \begin{bmatrix} 0.75 & 1.5 \\ 0.25 & 1.0 \end{bmatrix} \begin{bmatrix} 2.25 \\ 1.75 \end{bmatrix} = \begin{bmatrix} 4.3125 \\ 2.3125 \end{bmatrix}$$

Continuing in this manner leads to the vectors in Table 8.2. These vectors are also plotted in the graph of Figure 8.2. The vector $\mathbf{x}_4$ in the last column of Table 8.2 is an approximate eigenvector. In fact, an exact eignvector is

$$\mathbf{v} = \begin{bmatrix} 10.12 \\ 5.06 \end{bmatrix}$$

k	0	1	2	3	4
$\mathbf{x}_k$	$\begin{bmatrix} -1 \\ 2 \end{bmatrix}$	$\begin{bmatrix} 2.25 \\ 1.75 \end{bmatrix}$	$\begin{bmatrix} 4.3125 \\ 2.3125 \end{bmatrix}$	$\begin{bmatrix} 6.7031 \\ 3.3906 \end{bmatrix}$	$\begin{bmatrix} 10.1132 \\ 5.0664 \end{bmatrix}$

Table 8.2 *Values of the Sequence* $\mathbf{x}_{k+1} = A\mathbf{x}_k$.

Having found an approximate eigenvector, $\mathbf{x}_4$, you can now compute an approximate corresponding eigenvalue, λ', using the approach described in Section 6.1.1. Specifically, λ' should satisfy $A\mathbf{x}_4 \approx \lambda'\mathbf{x}_4$. Equating the first element of

$$A\mathbf{x}_4 = \begin{bmatrix} 0.75 & 1.5 \\ 0.25 & 1.0 \end{bmatrix} \begin{bmatrix} 10.1132 \\ 5.0664 \end{bmatrix} = \begin{bmatrix} 15.1845 \\ 7.5947 \end{bmatrix} \tag{8.37}$$

to the first element of

$$\lambda'\mathbf{x}_4 = \lambda' \begin{bmatrix} 10.1132 \\ 5.0664 \end{bmatrix} = \begin{bmatrix} 10.1132\lambda' \\ 5.0664\lambda' \end{bmatrix} \tag{8.38}$$

yields

$$15.1845 = 10.1132\lambda' \tag{8.39}$$

Solving (8.39) leads to $\lambda' = 1.50145$.

Alternatively, equating the second element of $A\mathbf{x}_4$ in (8.37) to the second element of $\lambda'\mathbf{x}_4$ in (8.38) yields

$$7.5947 = 5.0664\lambda' \tag{8.40}$$

Solving (8.40) leads to $\lambda' = 1.49903$.

Either of the values 1.50145 obtained from (8.39) and 1.49903 obtained from (8.40) provides a good approximation of the strictly dominant eigenvalue of A, namely, $\lambda = 1.5$.

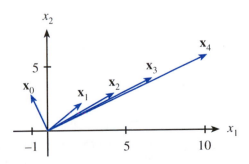

Figure 8.2 *The vectors* $\mathbf{x}_0$, $\mathbf{x}_1$, $\mathbf{x}_2$, $\mathbf{x}_3$, *and* $\mathbf{x}_4$ *from Table 8.2.*

Theoretical Justification

The reason that the foregoing method does, in fact, produce a sequence of vectors that converges to an eigenvector corresponding to a strictly dominant eigenvalue of A is now explained informally. Let $\lambda_1, \ldots, \lambda_n$ be the n eigenvalues of A, ordered so that

$$|\lambda_1| > |\lambda_2| \geq \cdots \geq |\lambda_n|$$

Thus, λ_1 is the strictly dominant eigenvalue. Also, let $\mathbf{v}_1, \ldots, \mathbf{v}_n$ be corresponding eigenvectors, so, for each i, $A\mathbf{v}_i = \lambda_i \mathbf{v}_i$. For ease of explanation, assume that these eigenvectors are linearly independent and hence form a basis for R^n.

Any initial vector, $\mathbf{x}_0$, is a linear combination of these basis vectors, so there are unique real numbers $t_1, \ldots, t_n$ such that

$$\mathbf{x}_0 = t_1 \mathbf{v}_1 + t_2 \mathbf{v}_2 + \cdots + t_n \mathbf{v}_n \tag{8.41}$$

The next vector generated by the approximation method is $\mathbf{x}_1 = A\mathbf{x}_0$. Multiplying (8.41) through by A yields

$$\mathbf{x}_1 = A\mathbf{x}_0 = t_1 A\mathbf{v}_1 + t_2 A\mathbf{v}_2 + \cdots + t_n A\mathbf{v}_n$$

Equivalently, using the fact that for each i, $A\mathbf{v}_i = \lambda_i \mathbf{v}_i$,

$$\mathbf{x}_1 = t_1 \lambda_1 \mathbf{v}_1 + t_2 \lambda_2 \mathbf{v}_2 + \cdots + t_n \lambda_n \mathbf{v}_n \tag{8.42}$$

Repeating the process of computing $\mathbf{x}_{k+1} = A\mathbf{x}_k$ and using the fact that for each i, $A\mathbf{v}_i = \lambda_i \mathbf{v}_i$, you can show by induction that for each $k = 0, 1, 2, \ldots$,

$$\mathbf{x}_k = t_1 (\lambda_1)^k \mathbf{v}_1 + t_2 (\lambda_2)^k \mathbf{v}_2 + \cdots + t_n (\lambda_n)^k \mathbf{v}_n \tag{8.43}$$

Dividing (8.43) through by $(\lambda_1)^k$ yields

$$\frac{1}{(\lambda_1)^k} \mathbf{x}_k = t_1 \mathbf{v}_1 + t_2 \left(\frac{\lambda_2}{\lambda_1} \right)^k \mathbf{v}_2 + \cdots + t_n \left(\frac{\lambda_n}{\lambda_1} \right)^k \mathbf{v}_n \tag{8.44}$$

Because λ_1 is the strictly dominant eigenvalue, each of the fractions $(\lambda_2/\lambda_1), \ldots, (\lambda_n/\lambda_1)$ on the right side of (8.44) is less than 1 in absolute value. Consequently, as k gets large, all of these terms approach 0. Thus, for large k,

$$\frac{1}{(\lambda_1)^k} \mathbf{x}_k \approx t_1 \mathbf{v}_1$$

or, equivalently,

$$\mathbf{x}_k \approx t_1 (\lambda_1)^k \mathbf{v}_1 \tag{8.45}$$

The conclusion from (8.45) is that, as long as $t_1 \neq 0$, $\mathbf{x}_k$ is approximately equal to a real number $(t_1 (\lambda_1)^k)$ times the eigenvector $\mathbf{v}_1$, for large values of k. Because any real number times an eigenvector of a matrix is also an eigenvector of that matrix, it follows from (8.45) that, as long as $t_1 \neq 0$, for large values of k, the sequence $\mathbf{X} = (\mathbf{x}_0, \mathbf{x}_1, \mathbf{x}_2, \ldots)$ approaches an eigenvector corresponding to the strictly dominant eigenvalue λ_1.

Computational Implementation

The foregoing analysis is validated in the numerical example presented at the beginning of this section. In particular, the vector $\mathbf{x}_4$ in Table 8.2 is approximately equal to a true eigenvector, $\mathbf{v}_1$, of A because

$$\mathbf{x}_4 = \begin{bmatrix} 10.1132 \\ 5.0664 \end{bmatrix} \approx \begin{bmatrix} 10.12 \\ 5.06 \end{bmatrix} = \mathbf{v}_1$$

You might notice from Table 8.2 that the components of $\mathbf{x}_k$ are getting larger as k gets larger. Mathematicians have discovered a way to keep the absolute value of the largest component of $\mathbf{x}_k$ equal to 1. To do so, for each k, let

$$c_k = \text{the element of } A\mathbf{x}_k \text{ having the largest absolute value} \qquad (8.46)$$

$$\mathbf{x}_{k+1} = \left(\frac{1}{c_k}\right) A\mathbf{x}_k \qquad (8.47)$$

The sequences defined by (8.46) and (8.47) have the following properties: (1) the largest absolute value of an element of $\mathbf{x}_k$ is 1 and (2) for almost all choices of $\mathbf{x}_0$, the sequence $\mathbf{X} = (\mathbf{x}_0, \mathbf{x}_1, \mathbf{x}_2, \ldots)$ converges to an eigenvector of A whose largest element in absolute value is 1 and the sequence $\mathbf{C} = (c_0, c_1, c_2, \ldots)$ converges to the strictly dominant eigenvalue of the matrix A.

A summary of the **power method** for finding an approximate value for the strictly dominant eigenvalue of an $(n \times n)$ matrix A and an associated eigenvector follows.

The Power Method for Approximating a Strictly Dominant Eigenvalue and an Associated Eigenvector of A

Step 1. Choose an initial vector $\mathbf{x}_0$.

Step 2. For each $k = 0, 1, 2, \ldots$, compute

(a) $A\mathbf{x}_k$.

(b) $c_k = $ the element of $A\mathbf{x}_k$ having the largest absolute value.

(c) $\mathbf{x}_{k+1} = (1/c_k)A\mathbf{x}_k$.

For almost all choices of $\mathbf{x}_0$, the sequence $\mathbf{C} = (c_0, c_1, c_2, \ldots)$ converges to the strictly dominant eigenvalue of A and the sequence $\mathbf{X} = (\mathbf{x}_0, \mathbf{x}_1, \mathbf{x}_2, \ldots)$ converges to a corresponding eigenvector.

Table 8.3 shows the results of applying the power method to the numerical example given at the beginning of this section, in which

$$A = \begin{bmatrix} 0.75 & 1.5 \\ 0.25 & 1.0 \end{bmatrix} \quad \text{and} \quad \mathbf{x}_0 = \begin{bmatrix} -1 \\ 2 \end{bmatrix}$$

k	0	1	2	3	4
$\mathbf{x}_k$	$\begin{bmatrix} -1 \\ 2 \end{bmatrix}$	$\begin{bmatrix} 1.0000 \\ 0.7778 \end{bmatrix}$	$\begin{bmatrix} 1.0000 \\ 0.5362 \end{bmatrix}$	$\begin{bmatrix} 1.0000 \\ 0.5058 \end{bmatrix}$	$\begin{bmatrix} 1.0000 \\ 0.5010 \end{bmatrix}$
$A\mathbf{x}_k$	$\begin{bmatrix} 2.25 \\ 1.75 \end{bmatrix}$	$\begin{bmatrix} 1.9167 \\ 1.0278 \end{bmatrix}$	$\begin{bmatrix} 1.5544 \\ 0.7862 \end{bmatrix}$	$\begin{bmatrix} 1.5087 \\ 0.7558 \end{bmatrix}$	$\begin{bmatrix} 1.5014 \\ 0.7510 \end{bmatrix}$
c_k	2.25	1.9167	1.5544	1.5087	1.5014

Table 8.3 *Results of Applying the Power Method.*

8.3.2 The Inverse Power Method

The power method described in Section 8.3.1 is used to find an approximate value for the strictly dominant eigenvalue of an $(n \times n)$ matrix A. In this section, you will see how the power method is used to find an approximate value for any eigenvalue of A, say, λ, provided that a sufficiently close estimate, μ, of λ is available.

The General Approach

The approach is to use a specific value for μ to create an $(n \times n)$ matrix B with a strictly dominant eigenvalue

$$\beta = \frac{1}{\lambda - \mu} \tag{8.48}$$

Once a value for β is found, you can compute the value of λ by solving (8.48) to obtain

$$\lambda = \mu + \frac{1}{\beta} \tag{8.49}$$

The desired matrix B is constructed from μ and A, as follows:

$$B = (A - \mu I)^{-1} \tag{8.50}$$

If $\lambda_1, \ldots, \lambda_n$ are the n eigenvalues of A, with corresponding eigenvectors $\mathbf{v}_1, \ldots, \mathbf{v}_n$, then the eigenvalues of B are

$$\frac{1}{\lambda_1 - \mu}, \ldots, \frac{1}{\lambda_n - \mu} \tag{8.51}$$

This is because, for each i,

$$
\begin{aligned}
B(A - \mu I) &= I \quad &\text{(multiply (8.50) by } A - \mu I) \\
BA - \mu B &= I \quad &\text{(matrix algebra)} \\
BA\mathbf{v}_i - \mu B\mathbf{v}_i &= \mathbf{v}_i \quad &\text{(multiply through by } \mathbf{v}_i)
\end{aligned}
$$

$$\lambda_i B \mathbf{v}_i - \mu B \mathbf{v}_i = \mathbf{v}_i \qquad (A\mathbf{v}_i = \lambda_i \mathbf{v}_i)$$

$$(\lambda_i - \mu) B \mathbf{v}_i = \mathbf{v}_i \qquad \text{(algebra)}$$

$$B \mathbf{v}_i = \left(\frac{1}{\lambda_i - \mu} \right) \mathbf{v}_i \quad \text{(algebra)}$$

The final equality also indicates that $\mathbf{v}_i$ is an eigenvector of B corresponding to the eigenvalue $1/(\lambda_i - \mu)$.

A Numerical Example

To illustrate the foregoing approach, recall the following matrix from Section 8.3.1, whose strictly dominant eigenvalue was determined to be 1.5:

$$A = \begin{bmatrix} 0.75 & 1.5 \\ 0.25 & 1.0 \end{bmatrix}$$

Suppose you now want to find the smaller eigenvalue, λ, of A and that you start with the initial estimate of $\mu = 0.5$. According to (8.50), you now compute

$$B = (A - \mu I)^{-1} = \begin{bmatrix} 0.25 & 1.50 \\ 0.25 & 0.50 \end{bmatrix}^{-1} = \begin{bmatrix} -2 & 6 \\ 1 & -1 \end{bmatrix}$$

You can verify that the strictly dominant eigenvalue of B is $\beta = -4$. With the values of $\mu = 0.5$ and $\beta = -4$, you can now find the corresponding value of λ from (8.49):

$$\lambda = \mu + \frac{1}{\beta} = 0.5 + \frac{1}{-4} = 0.25$$

When you cannot find the strictly dominant eigenvalue of B explicitly, the power method is used to compute an approximate value. This value, together with μ, is then substituted in (8.49) to obtain an approximate value for λ, the desired eigenvalue of A.

To understand why the choice for the value of μ is critical to the success of the foregoing approach, suppose that $\lambda_1, \ldots, \lambda_n$ are the n eigenvalues of A. Then, from (8.51), the n eigenvalues of $B = (A - \mu I)^{-1}$ are

$$\frac{1}{\lambda_1 - \mu}, \quad \frac{1}{\lambda_2 - \mu}, \quad \ldots, \quad \frac{1}{\lambda_n - \mu} \tag{8.52}$$

Now suppose that you want to estimate the value of the eigenvalue λ_1 of A. Looking at (8.52), you can see that μ must be closer to λ_1 than to any other eigenvalue of A for $1/(\lambda_1 - \mu)$ to be the strictly dominant eigenvalue of B. In other words, *you can use the power method to find an approximate value for $1/(\lambda_1 - \mu)$ only if the value of μ is closer to λ_1 than to any other eigenvalue of A.*

Computational Implementation

As you have just seen, to find an approximate value for an eigenvalue λ of A, you choose a value for μ sufficiently close to λ and then apply the power method to the matrix $(A - \mu I)^{-1}$.

That is, you choose an initial n-vector $\mathbf{x}_0$ and then compute, for each $k = 0, 1, \ldots$:

$$\mathbf{x}_{k+1} = \frac{1}{c_k}(A - \mu I)^{-1}\mathbf{x}_k \tag{8.53}$$

where c_k is the element of $(A - \mu I)^{-1}\mathbf{x}_k$ having the largest absolute value.

You can perform the computations in (8.53) more efficiently by solving the following system of linear equations for $\mathbf{y}_k$ first:

$$(A - \mu I)\mathbf{y}_k = \mathbf{x}_k \tag{8.54}$$

You then use the element c_k of $\mathbf{y}_k$ having the largest absolute value to compute $\mathbf{x}_{k+1}$:

$$\mathbf{x}_{k+1} = \frac{1}{c_k}\mathbf{y}_k \tag{8.55}$$

The reason that solving the system of linear equations in (8.54) is more efficient than the computation in (8.53) is that you can use a factorization of $A - \mu I$, such as the one described in Section 8.2.2 for Gaussian elimination, to solve the systems in (8.54) repeatedly, each time using the new value of $\mathbf{x}_k$ as the right-hand-side vector. A summary of the **inverse power method** for finding an approximate eigenvalue λ of an $(n \times n)$ matrix A follows.

The Inverse Power Method for Approximating an Eigenvalue λ and an Associated Eigenvector of A

Step 1. Select a value for μ sufficiently close to λ.

Step 2. Compute a factorization of $A - \mu I$ that you can use for solving a system of linear equations $(A - \mu I)\mathbf{y} = \mathbf{b}$.

Step 3. Choose an initial vector $\mathbf{x}_0$.

Step 4. For each $k = 0, 1, 2, \ldots$,

   (a) Use the factorization from Step 2 to solve the system of linear equations $(A - \mu I)\mathbf{y}_k = \mathbf{x}_k$ for $\mathbf{y}_k$.

   (b) Let c_k be the element of $\mathbf{y}_k$ having the largest absolute value.

   (c) Compute $\mathbf{x}_{k+1} = (1/c_k)\mathbf{y}_k$.

For almost all choices of $\mathbf{x}_0$, the sequence $\mu + (1/c_k)$ converges to the eigenvalue λ of A and the sequence $\mathbf{X} = (\mathbf{x}_0, \mathbf{x}_1, \mathbf{x}_2, \ldots)$ converges to a corresponding eigenvector.

Table 8.4 shows the results of applying the inverse power method to find the eigenvalue $\lambda = 0.25$ of the following matrix A, using $\mu = 0.5$ and starting from the given initial point:

$$A = \begin{bmatrix} 0.75 & 1.5 \\ 0.25 & 1.0 \end{bmatrix} \quad \text{and} \quad \mathbf{x}_0 = \begin{bmatrix} -1 \\ 2 \end{bmatrix}$$

From the last column in Table 8.4, you can see that

$$\mathbf{x}_4 = \begin{bmatrix} 1.0000 \\ -0.3355 \end{bmatrix}$$

k	0	1	2	3	4
$\mathbf{x}_k$	$\begin{bmatrix} -1 \\ 2 \end{bmatrix}$	$\begin{bmatrix} 1.0000 \\ -0.2143 \end{bmatrix}$	$\begin{bmatrix} 1.0000 \\ -0.3696 \end{bmatrix}$	$\begin{bmatrix} 1.0000 \\ -0.3247 \end{bmatrix}$	$\begin{bmatrix} 1.0000 \\ -0.3355 \end{bmatrix}$
$\mathbf{y}_k$	$\begin{bmatrix} 14 \\ -3 \end{bmatrix}$	$\begin{bmatrix} -3.2858 \\ 1.2143 \end{bmatrix}$	$\begin{bmatrix} -4.2176 \\ 1.3696 \end{bmatrix}$	$\begin{bmatrix} -3.9482 \\ 1.3247 \end{bmatrix}$	$\begin{bmatrix} -4.0131 \\ 1.3355 \end{bmatrix}$
c_k	14	-3.2858	-4.2176	-3.9482	-4.0131
$\mu + \dfrac{1}{c_k}$	0.5714	0.1957	0.2629	0.2467	0.2508

Table 8.4 *Results of the Inverse Power Method.*

is an approximate eigenvector corresponding to the approximate eigenvalue 0.2508, obtained from the last row in Table 8.4.

In this section, you have seen how the power method and the inverse power method are used to approximate eigenvalues and eigenvectors of a matrix. In Section 8.4, a new approach to solving large, sparse systems of linear equations is presented.

Exercises for Section 8.3

1. Perform the indicated computations for the following matrix:

$$A = \begin{bmatrix} 1 & -1 \\ -2 & 0 \end{bmatrix}$$

(a) Use the technique in Section 6.1 to find the two eigenvalues of A by solving the equation $\det(A - \lambda I) = 0$ by hand.

(b) Find an approximate value for an eigenvector corresponding to the strictly dominant eigenvalue of A by computing by hand the sequence $\mathbf{x}_{k+1} = A\mathbf{x}_k$ and filling in the following table:

k	0	1	2	3	4	5
$\mathbf{x}_k$	$\begin{bmatrix} -1 \\ 3 \end{bmatrix}$					

(c) Use the vector $\mathbf{x}_5$ from part (b) to find an approximate value for the strictly dominant eigenvalue of A.

2. Repeat the previous exercise for the following matrix:

$$A = \begin{bmatrix} 4 & 2 \\ -3 & -3 \end{bmatrix}$$

3. Use technology indicated by your instructor to apply the power method to the matrix A in Exercise 1 and fill in the following table:

k	0	1	2	3	4	5
$\mathbf{x}_k$	$\begin{bmatrix} -1 \\ 3 \end{bmatrix}$					
$A\mathbf{x}_k$						
c_k						

4. Repeat the previous exercise using the matrix A in Exercise 2.

5. Consider finding an eigenvector corresponding to the strictly dominant eigenvalue of the following matrix:

$$A = \begin{bmatrix} 9 & 20 & 7 \\ -2 & -8 & -2 \\ -13 & -20 & -11 \end{bmatrix}$$

Use technology indicated by your instructor to apply the power method to A and fill in the following table:

k	0	1	2	3	4
$\mathbf{x}_k$	$\begin{bmatrix} 4 \\ 0 \\ -1 \end{bmatrix}$				
$A\mathbf{x}_k$					
c_k					

6. Repeat the previous exercise using the following matrix:

$$A = \begin{bmatrix} 5.0 & 2.0 & 0.0 \\ -3.0 & -2.0 & 0.0 \\ 14.4 & 4.8 & -2.0 \end{bmatrix}$$

7. In the theoretical justification of the power method, it is assumed that the initial vector x_0 is written as the following linear combination of n eigenvectors $v_1, \ldots, v_n$ of the matrix A (in which v_1 is an eigenvector corresponding to the strictly dominant eigenvalue λ_1):

$$x_0 = t_1 v_1 + t_2 v_2 + \cdots + t_n v_n$$

What problem arises in the theoretical analysis if $t_1 = 0$? What does this mean for the success of the power method?

8. In the theoretical justification of the power method, it is assumed that $\lambda_1, \lambda_2, \ldots, \lambda_n$ are the n eigenvalues of A and that λ_1 is the strictly dominant eigenvalue. What problem arises in the theoretical analysis if $\lambda_2 = \lambda_1$? Explain.

9. Find an approximate value for the non-dominant eigenvalue and corresponding eigenvector of the matrix A in Exercise 1 by performing the following steps.
   (a) For the value $\mu = -2$, compute by hand the matrix $B = (A - \mu I)^{-1}$.
   (b) Find the two eigenvalues of B explicitly by solving the equation $\det(B - \beta I) = 0$ for β.
   (c) Verify your result in part (b) by using the formulas in (8.51) in Section 8.3.2.
   (d) Use technology indicated by your instructor to perform the inverse power method and fill in the following table (in which $y_k = Bx_k$ and c_k is the element of y_k having the largest absolute value):

k	0	1	2	3	4
x_k	$\begin{bmatrix} -1 \\ 3 \end{bmatrix}$				
y_k					
c_k					
$\mu + \dfrac{1}{c_k}$					

   (e) For what values of μ will the sequences generated by the inverse power method converge to the desired values? Explain.

10. Repeat the previous exercise to find an approximate value for the non-dominant eigenvalue and corresponding eigenvector of the matrix A in Exercise 2 using $\mu = -1$.

11. Perform the subsequent computations using technology indicated by your instructor to find approximate values for the three eigenvalues and corresponding eigenvectors of the following matrix:

$$A = \begin{bmatrix} 4.3571 & 1.3810 & -2.0000 \\ -1.4286 & 0.8095 & 1.0000 \\ 3.6508 & 1.9683 & -1.6667 \end{bmatrix}$$

(a) Apply the power method to find an approximate value for the strictly dominant eigenvalue and corresponding eigenvector of A. Indicate the initial vector you used and the number of iterations you performed.

(b) Use the inverse power method to find approximate values for the remaining two eigenvalues and eigenvectors of A. Try different values of μ, as necessary. Report the two values of μ and the initial vectors you used to get the remaining two eigenvalues and eigenvectors of A.

12. Repeat the previous exercise to find approximate values for the three eigenvalues and corresponding eigenvectors of the following matrix:

$$A = \begin{bmatrix} -1 & 2 & 1 \\ 0 & -1 & -4 \\ 2 & -3 & -1 \end{bmatrix}$$

8.4 Iterative Methods

Given an $(n \times n)$ matrix A and a column vector $\mathbf{b} \in R^{n \times 1}$, a new method is presented in this section for solving the system

$$A\mathbf{x} = \mathbf{b} \tag{8.56}$$

These **iterative methods** generate a sequence of column vectors:

$$\mathbf{x}_0, \mathbf{x}_1, \mathbf{x}_2, \ldots$$

Under suitable circumstances, the sequence converges to a solution of (8.56). More specifically, the methods start with an initial estimate, $\mathbf{x}_0$, of the solution to (8.56). The elements of $\mathbf{x}_0$ are then used to compute $\mathbf{x}_1$, whose elements are then used to compute $\mathbf{x}_2$, and so on.

The advantage of iterative methods is that they often require less storage space when the matrix A is large and sparse—that is, has mostly zero elements—than the factorization methods in Section 8.2. Iterative methods are fast when the initial guess is close to the actual solution. Furthermore, you can stop an iterative method when the current vector in the sequence, though not an exact solution, satisfies the system of equations to an acceptable degree of accuracy.

To use the iterative methods proposed in this section, it is assumed that the system of linear equations is written in the following form, with each diagonal element $a_{ii} \neq 0$:

$$a_{11}x_1 + a_{12}x_2 + a_{13}x_3 + \cdots + a_{1n}x_n = b_1$$
$$a_{21}x_1 + a_{22}x_2 + a_{23}x_3 + \cdots + a_{2n}x_n = b_2$$
$$\vdots$$
$$a_{n1}x_1 + a_{n2}x_2 + a_{n3}x_3 + \cdots + a_{nn}x_n = b_n$$

(8.57)

The preparation phase of the proposed methods is to solve each equation i in (8.57) for x_i in terms of the remaining variables. Doing so results in the following:

$$x_1 = \frac{1}{a_{11}}(b_1 - a_{12}x_2 - a_{13}x_3 - \cdots - a_{1n}x_n)$$

$$x_2 = \frac{1}{a_{22}}(b_2 - a_{21}x_1 - a_{23}x_3 - \cdots - a_{2n}x_n)$$

(8.58)

$$\vdots$$

$$x_n = \frac{1}{a_{nn}}(b_n - a_{n1}x_1 - a_{n2}x_2 - \cdots - a_{nn-1}x_{n-1})$$

The two methods described next differ in the subsequent solution phase.

8.4.1 Jacobi's Method

The solution phase of **Jacobi's method** uses the elements of the current vector, $\mathbf{x}_k$, to compute the elements of the next vector, $\mathbf{x}_{k+1}$. Specifically, if

$$\mathbf{x}_k = \begin{bmatrix} x_1 \\ x_2 \\ \vdots \\ x_n \end{bmatrix}$$

(8.59)

then the elements of $\mathbf{x}_{k+1}$ are obtained by substituting the values in (8.59) on the right side of (8.58) and performing the appropriate computations. The resulting values are the elements of $\mathbf{x}_{k+1}$. This process is then repeated, substituting the known elements of $\mathbf{x}_{k+1}$ on the right side of (8.58) to compute the unknown elements of $\mathbf{x}_{k+2}$, and so on. A numerical example follows.

A Numerical Example of Jacobi's Method

To illustrate Jacobi's method, consider the following system:

$$6x_1 + x_2 - x_3 = 10$$
$$2x_1 - 8x_2 + 3x_3 = 25$$
$$x_1 - 2x_2 - 9x_3 = 17$$

The preparation phase is to write the foregoing system in the form (8.58) by using each equation i to solve for x_i, as follows:

$$x_1 = (10 - x_2 + x_3)/6$$
$$x_2 = -(25 - 2x_1 - 3x_3)/8 \qquad\qquad (8.60)$$
$$x_3 = -(17 - x_1 + 2x_2)/9$$

Begin by selecting an initial estimate, $\mathbf{x}_0$, of the solution to this system. If no better estimate is available, you can use $\mathbf{x}_0 = \mathbf{0}$, as is done here.

The elements of $\mathbf{x}_0$, all 0's in this case, are then substituted on the right side of (8.60). Doing so and performing the computations yields the following elements for $\mathbf{x}_1$:

$$\mathbf{x}_1 = \begin{bmatrix} (10 - 0 + 0)/6 \\ -(25 - 2(0) - 3(0))/8 \\ -(17 - 0 + 2(0))/9 \end{bmatrix} = \begin{bmatrix} 1.6667 \\ -3.1250 \\ -1.8889 \end{bmatrix}$$

Having found the elements of $\mathbf{x}_1$, you now substitute those values on the right side of (8.60) to obtain the following elements for $\mathbf{x}_2$:

$$\mathbf{x}_2 = \begin{bmatrix} (10 - (-3.1250) + (-1.8889))/6 \\ -(25 - 2(1.6667) - 3(-1.8889))/8 \\ -(17 - 1.6667 + 2(-3.1250))/9 \end{bmatrix} = \begin{bmatrix} 1.8727 \\ -3.4167 \\ -1.0093 \end{bmatrix}$$

Continuing in this manner provides the values of the sequence shown in Table 8.5. You can see that the vector

$$\mathbf{x}_5 = \begin{bmatrix} 1.9948 \\ -2.9895 \\ -1.0082 \end{bmatrix}$$

in the last column of Table 8.5 provides a good estimate of the solution $x_1 = 2$, $x_2 = -3$, and $x_3 = -1$ to the system of linear equations.

A summary of Jacobi's method follows.

Jacobi's Method

Step 1. Prepare the system of linear equations $A\mathbf{x} = \mathbf{b}$ in the form described in (8.58), assuming that each diagonal element of A is not 0.

Step 2. Use the following solution phase to generate a sequence of column vectors:

    (a) Choose an initial estimate, $\mathbf{x}_0$, of the solution to the system of equations.

    (b) For each $k = 0, 1, 2, \ldots$, substitute the elements of $\mathbf{x}_k$ in the right side of (8.58). After performing the appropriate computations, the resulting values are the elements of $\mathbf{x}_{k+1}$.

$\mathbf{x}_0$	$\mathbf{x}_1$	$\mathbf{x}_2$	$\mathbf{x}_3$	$\mathbf{x}_4$	$\mathbf{x}_5$

$$\begin{bmatrix} 0 \\ 0 \\ 0 \end{bmatrix} \begin{bmatrix} 1.6667 \\ -3.1250 \\ -1.8889 \end{bmatrix} \begin{bmatrix} 1.8727 \\ -3.4167 \\ -1.0093 \end{bmatrix} \begin{bmatrix} 2.0679 \\ -3.0353 \\ -0.9215 \end{bmatrix} \begin{bmatrix} 2.0190 \\ -2.9536 \\ -0.9846 \end{bmatrix} \begin{bmatrix} 1.9948 \\ -2.9895 \\ -1.0082 \end{bmatrix}$$

Table 8.5 *Values of the Sequence from Jacobi's Method.*

Theoretical Justification

The sequence of column vectors generated by Jacobi's method may or may not converge to the solution of the system $A\mathbf{x} = \mathbf{b}$. One condition under which convergence is guaranteed is when A is **strictly diagonally dominant**, that is, when the absolute value of the diagonal element in row i of A exceeds the sum of the absolute values of the remaining elements in row i. The speed of convergence depends on the the amount by which the diagonal elements exceed their row sums. The matrix A used in the foregoing numerical computations is strictly diagonally dominant, as is the matrix in the following example,

EXAMPLE 8.12 A Strictly Diagonally Dominant Matrix
The following matrix is strictly diagonally dominant:

$$\begin{bmatrix} 8 & -3 & 2 \\ -1 & 6 & 4 \\ 5 & 1 & -9 \end{bmatrix} \quad \text{because} \quad \begin{array}{l} |8| > |-3| + |2| = 5 \\ |6| > |-1| + |4| = 5 \\ |-9| > \quad |5| + |1| = 6 \end{array}$$

As you are asked to show in Exercise 10, when the matrix A is not strictly diagonally dominant, it is sometimes possible to interchange the rows in such a way that the new matrix is strictly diagonally dominant.

8.4.2 The Gauss-Seidel Method

The preparation phase of the **Gauss-Seidel method** is the same as that of Jacobi, namely, to write the system in the following form:

$$x_1 = \frac{1}{a_{11}}(b_1 - a_{12}x_2 - a_{13}x_3 - a_{14}x_4 - \cdots - a_{1n}x_n) \tag{8.61}$$

$$x_2 = \frac{1}{a_{22}}(b_2 - a_{21}x_1 - a_{23}x_3 - a_{24}x_4 - \cdots - a_{2n}x_n) \tag{8.62}$$

$$x_3 = \frac{1}{a_{33}}(b_3 - a_{31}x_1 - a_{32}x_2 - a_{34}x_4 - \cdots - a_{2n}x_n) \tag{8.63}$$

$$\vdots$$

$$x_n = \frac{1}{a_{nn}}(b_n - a_{n1}x_1 - a_{n2}x_2 - a_{n3}x_3 - \cdots - a_{nn-1}x_{n-1}) \tag{8.64}$$

The solution phase of the Gauss-Seidel method also begins with an initial estimate of the solution to the system, say,

$$\mathbf{x}_0 = \begin{bmatrix} x_1 \\ x_2 \\ x_3 \\ \vdots \\ x_n \end{bmatrix} \tag{8.65}$$

The elements of $\mathbf{x}_0$ are then substituted on the right side of the first equation, (8.61). After performing the appropriate computations, you obtain the first element of $\mathbf{x}_1$, say, y_1.

Now look at (8.62). Using Jacobi's method, you substitute the values of $\mathbf{x}_0$ from (8.65) on the right side of (8.62). Using the Gauss-Seidel method, you substitute the newly-computed value, y_1, for x_1 on the right side of (8.62). The remaining values for $x_3, \ldots, x_n$ on the right side of (8.62) are obtained from $\mathbf{x}_0$ in (8.65). The result of performing the computations on the right side of (8.62) then yields the second element of $\mathbf{x}_1$, say, y_2.

You now have the first two elements of $\mathbf{x}_1$, namely, y_1 and y_2. These values are now substituted for x_1 and x_2, respectively, on the right side of (8.63). The remaining values of $x_4, \ldots, x_n$ are obtained from the corresponding elements of $\mathbf{x}_0$ in (8.65). Performing the computations on the right side of (8.63) then yields the third element of $\mathbf{x}_1$, say, y_3.

Repeating the process of substituting the newly-computed elements of $\mathbf{x}_1$ and the remaining elements of $\mathbf{x}_0$ on the right side of each subsequent equation eventually results in all n elements of $\mathbf{x}_1$. The elements of $\mathbf{x}_1$ are then used in a similar manner to compute the elements of $\mathbf{x}_2$, and so on.

A Numerical Example of the Gauss-Seidel Method

The Gauss-Seidel method is now illustrated with the following system of linear equations that was solved by Jacobi's method in Section 8.4.1:

$$6x_1 + x_2 - x_3 = 10$$
$$2x_1 - 8x_2 + 3x_3 = 25$$
$$x_1 - 2x_2 - 9x_3 = 17$$

The preparation phase is to write the foregoing system in the form:

$$x_1 = (10 - x_2 + x_3)/6 \tag{8.66}$$
$$x_2 = -(25 - 2x_1 - 3x_3)/8 \tag{8.67}$$
$$x_3 = -(17 - x_1 + 2x_2)/9 \tag{8.68}$$

Begin by selecting an initial estimate, $\mathbf{x}_0$, of the solution to this system. If no better estimate is available, you can use $\mathbf{x}_0 = \mathbf{0}$, as is done here.

The elements of $\mathbf{x}_0$, all 0's in this case, are then substituted on the right side of (8.66). Doing so and performing the appropriate computations yields the following value for the first element, y_1, of $\mathbf{x}_1$:

$$y_1 = (10 - 0 + 0)/6 = 1.6667$$

$\mathbf{x}_0$	$\mathbf{x}_1$	$\mathbf{x}_2$	$\mathbf{x}_3$	$\mathbf{x}_4$	$\mathbf{x}_5$
$\begin{bmatrix} 0 \\ 0 \\ 0 \end{bmatrix}$	$\begin{bmatrix} 1.6667 \\ -2.7083 \\ -1.1019 \end{bmatrix}$	$\begin{bmatrix} 1.9344 \\ -3.0546 \\ -0.9952 \end{bmatrix}$	$\begin{bmatrix} 2.0099 \\ -2.9957 \\ -0.9999 \end{bmatrix}$	$\begin{bmatrix} 1.9993 \\ -3.0001 \\ -1.0001 \end{bmatrix}$	$\begin{bmatrix} 2.0000 \\ -3.0000 \\ -1.0000 \end{bmatrix}$

Table 8.6 *Values of the Sequence from the Gauss-Seidel Method.*

The foregoing value of y_1 is now substituted for x_1 on the right side of (8.67). The value substituted for x_3 on the right side of (8.67) is the third element of $\mathbf{x}_0$, namely, 0. Making these two substitutions and performing the computations on the right side of (8.67) yields the following value for the second element, y_2, of $\mathbf{x}_1$:

$$y_2 = -(25 - 2(1.6667) - 3(0))/8 = -2.7083$$

Finally, substitute the foregoing values of y_1 and y_2 for x_1 and x_2, respectively, on the right side of (8.68). Doing so and performing the appropriate computations yields the following value for the third element, y_3, of $\mathbf{x}_1$:

$$y_3 = -(17 - 1.6667 + 2(-2.7083))/9 = -1.1019$$

The Gauss-Seidel method has now produced the following column vector:

$$\mathbf{x}_1 = \begin{bmatrix} y_1 \\ y_2 \\ y_3 \end{bmatrix} = \begin{bmatrix} 1.6667 \\ -2.7083 \\ -1.1019 \end{bmatrix}$$

Continuing in this manner provides the values of the sequence of vectors shown in Table 8.6. The value of $\mathbf{x}_5$ in the last column of Table 8.6 provides an estimate that agrees with the exact solution of the system, up to the four decimal places shown.

This example also shows the increased speed of convergence of the Gauss-Seidel method, as compared to Jacobi's method (see Table 8.5 in Section 8.4.1), that is the result of using the newly-computed components of $\mathbf{x}_{k+1}$ to determine the remaining components of $\mathbf{x}_{k+1}$. Although increased efficiency is usually obtained with the Gauss-Seidel method, there are numerical examples where Jacobi's method converges more rapidly.

In this section, you have seen two iterative methods for solving a system of linear equations. Your knowledge of numerical methods in linear algebra is now brought together.

Exercises for Section 8.4

1. Prepare the following system of linear equations so that you could then apply either of the iterative methods described in this section:

$$\begin{aligned} 4x_1 - x_2 + x_3 &= 10 \\ -2x_1 + 9x_2 + 5x_3 &= -2 \\ x_1 - 2x_2 - 6x_3 &= 7 \end{aligned}$$

2. For the following matrix A and vector $\mathbf{b}$, prepare the system of equations $A\mathbf{x} = \mathbf{b}$ so that you could then apply either of the iterative methods described in this section:

$$A = \begin{bmatrix} 3 & -1 & 0 \\ -3 & -6 & 1 \\ 4 & -2 & -9 \end{bmatrix} \quad \text{and} \quad \mathbf{b} = \begin{bmatrix} 7 \\ 4 \\ -26 \end{bmatrix}$$

3. Use technology indicated by your instructor to perform 5 iterations of Jacobi's method on the system of linear equations in Exercise 1. Start with $\mathbf{x}_0 = \mathbf{0}$ and show each subsequent vector.

4. Use technology indicated by your instructor to perform 5 iterations of Jacobi's method on the system of linear equations in Exercise 2. Start with $\mathbf{x}_0 = \mathbf{0}$ and show each subsequent vector.

5. Use technology indicated by your instructor to perform 5 iterations of the Gauss-Seidel method on the system in Exercise 1. Start with $\mathbf{x}_0 = \mathbf{0}$ and show each subsequent vector.

6. Use technology indicated by your instructor to perform 5 iterations of the Gauss-Seidel method on the system in Exercise 2. Start with $\mathbf{x}_0 = \mathbf{0}$ and show each subsequent vector.

7. Which of the following matrices are strictly diagonally dominant?

(a) $\begin{bmatrix} 3 & -1 & 0 \\ -3 & -6 & 1 \\ 4 & -2 & -9 \end{bmatrix}$
(b) $\begin{bmatrix} 6 & -2 & 2 \\ 5 & 1 & -7 \\ -1 & 9 & 3 \end{bmatrix}$

8. Which of the following matrices are strictly diagonally dominant?

(a) $\begin{bmatrix} 7 & 2 & -1 & -3 \\ -3 & -9 & 1 & 2 \\ 0 & -2 & -8 & 5 \\ 4 & 1 & 0 & 6 \end{bmatrix}$
(b) $\begin{bmatrix} 9 & 3 & -2 & -3 \\ 0 & -1 & 4 & 1 \\ 3 & -7 & -2 & -1 \\ -5 & 2 & 0 & 8 \end{bmatrix}$

9. (a) Show that interchanging row 2 with row 3 of the matrix in Exercise 7(b) results in a matrix that is strictly diagonally dominant.
   (b) Interchange two rows of the matrix in Exercise 8(b) so that the resulting matrix is strictly diagonally dominant.

10. Use the previous exercise as a guide for determining when it is possible to interchange rows of an $(n \times n)$ matrix A so that the resulting matrix is strictly diagonally dominant.

11. Consider the system of linear equations $A\mathbf{x} = \mathbf{b}$, in which A is the matrix in Exercise 8(b) and the elements of $\mathbf{b}$ are $-4, -12, -15, 33$. Use the results of Exercise 9(b) to interchange appropriate rows so that the resulting matrix is strictly diagonally dominant. Then use technology indicated by your instructor to solve the system of linear equations by performing five iterations of Jacobi's method. Start with $\mathbf{x}_0 = \mathbf{0}$ and show each subsequent vector.

12. Repeat the previous exercise using the Gauss-Seidel method.

8.5 Using Numerical Methods in Problem Solving

You will now see how numerical methods are used to solve the problem of Waste Management presented at the beginning of the chapter. Surprisingly, much of the work involves mathematical analysis.

8.5.1 The Problem of Waste Management

Waste Management operates a liquid recycling operation that processes a large number of waste liquids into 75 marketable products. Each waste liquid yields a few of the sellable products at a known rate in gallons per hour. Each day the Manager of Operations solves an initial system of 75 linear equations to determine the rates at which 75 initially-chosen waste liquids are processed so as to achieve a desired rate of output for each of the 75 sellable products. Some hours later, one of the chosen waste liquids runs out and another one must be selected in its place. To find a replacement liquid that achieves the same rates of flow for the output products requires solving a new system of 75 linear equations in 75 unknowns on a trial-and-error basis. You have been hired as a consultant to design an on-line system for solving these large systems of linear equations in real time on the manager's desktop computer.

Identifying the Data

The first step in solving this problem is to identify the data and the systems of linear equations that are to be solved. The data consist of the known rates at which each waste liquid yields each of the sellable products, as shown, for example, in Table 8.7. From the column in Table 8.7 corresponding to waste liquid 1, you can see that each 1000 gallons per hour of that waste liquid yields 200 gallons per hour of product 1, none of output 2, and 100 gallons per hour of product 3. A similar interpretation applies to each waste-liquid column in Table 8.7. The last column contains the desired flow rates, in gallons per hour, for the three products.

The Initial System of Linear Equations

The Manager of Operations needs to solve an initial system of linear equations to determine the flow rates of the waste liquids to be processed. In this example, there are three output products, so the Manager must identify three waste liquids to process initially. Suppose waste liquids 1, 2, and 3 are chosen. Accordingly, you can define the following three

Output	Waste Liquid					Desired Hourly
Product	1	2	3	4	5	Flow Rate
1	200	0	100	50	150	5000
2	0	150	200	0	250	6750
3	100	250	0	150	50	2250

Table 8.7 *Gallons Per Hour of Sellable Products Obtained from Processing 1000 Gallons per Hour of Each Waste Liquid.*

variables:

x_1 = the thousands of gallons per hour of waste liquid 1 to process

x_2 = the thousands of gallons per hour of waste liquid 2 to process

x_3 = the thousands of gallons per hour of waste liquid 3 to process

You want to choose values for these variables that achieve the flow rates of the output products given in the last column of Table 8.7. For example, you want the flow rate for the first product to be 5000 gallons per hour. The flow rate for product 1 depends on the flow rates of waste liquids 1, 2, and 3, as follows:

$$\begin{pmatrix} \text{Flow rate for} \\ \text{product 1} \end{pmatrix} = \begin{pmatrix} \text{Flow rate from} \\ \text{waste liquid 1} \end{pmatrix} + \begin{pmatrix} \text{Flow rate from} \\ \text{waste liquid 2} \end{pmatrix} + \begin{pmatrix} \text{Flow rate from} \\ \text{waste liquid 3} \end{pmatrix}$$

Using the data in the three columns in row 1 of Table 8.7 corresponding to the first three waste liquids, you are now led to the following linear equation for satisfying the flow rate of product 1:

$$200x_1 + 0x_2 + 100x_3 = 5000 \quad \text{(flow rate for product 1)} \tag{8.69}$$

Using the corresponding data from row 2 of Table 8.7, you also want the values of the variables to satisfy the following equation for product 2:

$$0x_1 + 150x_2 + 200x_3 = 6750 \quad \text{(flow rate for product 2)} \tag{8.70}$$

Likewise, from the data in row 3 of Table 8.7, you want the values of the variables to satisfy the following equation for product 3:

$$100x_1 + 250x_2 + 0x_3 = 2250 \quad \text{(flow rate for product 3)} \tag{8.71}$$

Putting together (8.69), (8.70), and (8.71), the values of the variables must satisfy the following system of linear equations:

$$\begin{aligned} 200x_1 + \quad 0x_2 + 100x_3 &= 5000 \quad \text{(flow rate for product 1)} \\ 0x_1 + 150x_2 + 200x_3 &= 6750 \quad \text{(flow rate for product 2)} \\ 100x_1 + 250x_2 + \quad 0x_3 &= 2250 \quad \text{(flow rate for product 3)} \end{aligned} \tag{8.72}$$

Using technology to solve (8.72) results in the following values for the variables:

$$x_1 = 10, \quad x_2 = 5, \quad \text{and} \quad x_3 = 30$$

In other words, to achieve the flow rates of 5000 gallons per hour for product 1, 6750 gallons per hour for product 2, and 2250 gallons per hour for product 3, the Manager of Operations should process 10 thousand gallons per hour of waste liquid 1, 5 thousand gallons per hour of waste liquid 2, and 30 thousand gallons per hour of waste liquid 3.

Subsequent Systems of Linear Equations

At the foregoing rates, one of the three waste liquids will be depleted eventually—say, waste liquid 3. At that time, the Manager of Operations must identify one of the other waste liquids to use in place of liquid 3. Suppose, for example, that waste liquid 4 is selected as the replacement, so the following new variable is introduced:

$x_4 = $ the thousands of gallons per hour of waste liquid 4 to process

To maintain the flow rates of the output products listed in the last column of Table 8.7, the values of x_1, x_2, and x_4 must now satisfy the following system of three linear equations:

$$
\begin{aligned}
200x_1 + 0x_2 + 50x_4 &= 5000 \quad \text{(flow rate for product 1)} \\
0x_1 + 150x_2 + 0x_4 &= 6750 \quad \text{(flow rate for product 2)} \\
100x_1 + 250x_2 + 150x_4 &= 2250 \quad \text{(flow rate for product 3)}
\end{aligned}
\qquad (8.73)
$$

Observe that the system in (8.73) is obtained by replacing, in (8.72), x_3 and the associated data from the corresponding column of Table 8.7 with x_4 and the associated data for waste liquid 4 from Table 8.7.

Using technology to solve (8.73) results in the following values for the variables:

$$x_1 = 48, \quad x_2 = 45, \quad \text{and} \quad x_4 = -92$$

The negative value for x_4 indicates that waste liquid 4 is not a valid replacement for waste liquid 3. Another alternative is needed.

To try waste liquid 5, introduce the variable x_5 and use the corresponding data from Table 8.7 to write the following system of linear equations:

$$
\begin{aligned}
200x_1 + 0x_2 + 150x_5 &= 5000 \quad \text{(flow rate for product 1)} \\
0x_1 + 150x_2 + 250x_5 &= 6750 \quad \text{(flow rate for product 2)} \\
100x_1 + 250x_2 + 50x_5 &= 2250 \quad \text{(flow rate for product 3)}
\end{aligned}
\qquad (8.74)
$$

Using technology to solve (8.74) results in the following values for the variables:

$$x_1 = 5.472, \quad x_2 = 1.604, \quad \text{and} \quad x_5 = 26.038$$

The values of the variables are nonnegative, so waste liquid 5 is a valid replacement for waste liquid 3. The new flow rates of 5.472 thousand gallons per hour of waste liquid 1, 1.604 thousand gallons per hour of waste liquid 2, and 26.038 thousand gallons per hour of waste liquid 5 will result in 5000 gallons per hour of product 1, 6750 gallons per hour of product 2, and 2250 gallons per hour of product 3.

In summary, the solution to the problem of Waste Management must

1. Read and store the data consisting of the known flow rates for the products obtained from 1000 gallons of each waste liquid, together with the desired flow rates for the products, as in Table 8.7.

2. Solve the initial system, $A\mathbf{x} = \mathbf{b}$, of 75 linear equations to determine the flow rates of 75 selected waste liquids.

3. Repeatedly allow the user to specify which waste liquid has been depleted and which waste liquid to try in its place. On so doing, it is necessary to solve the new system obtained by replacing the column of the current matrix A corresponding to the depleted waste liquid with the column of the data corresponding to the replacement liquid.

One computational approach for solving the systems that arise in item (3) is presented next.

8.5.2 An Approach to Solving the Problem

Recall, from the problem description, that each system of linear equations involves a (75×75) matrix. Rather than solve each such system independently, a more efficient approach is now developed. Keep in mind that the specific implementation depends on whether you are using a software package, such as MATLAB, Maple, and Mathematica, or writing a computer program in a language such as C++.

For illustration purposes, suppose that, to solve the initial system

$$A\mathbf{x} = \mathbf{b} \tag{8.75}$$

you have found A^{-1}, so $\mathbf{x} = A^{-1}\mathbf{b}$. Suppose you now want to solve the following system of equations in an attempt to find a replacement liquid:

$$A'\mathbf{x} = \mathbf{b} \tag{8.76}$$

The matrix A' is obtained by replacing the column of A corresponding to the waste liquid that is depleted with the column of data corresponding to the proposed replacement liquid.

Because A' in (8.76) is the same as A in (8.75) except for one column, it is reasonable to expect that the inverse of A' is somehow related to the inverse of A. If this is so, then you can use the inverse of A, which you already have, to compute the inverse of A'. The next step is to determine how A^{-1} is related to $(A')^{-1}$.

Using A^{-1} to Find $(A')^{-1}$

You know that A^{-1} satisfies

$$A^{-1}A = I$$

To find the inverse of A', you need to find a matrix $(A')^{-1}$ such that

$$(A')^{-1}A' = I$$

To that end, begin by looking at A^{-1} times A'. To illustrate, suppose that the original matrices A and A^{-1} are as follows:

$$A = \begin{bmatrix} 0 & 1 & -2 \\ 1 & 2 & -4 \\ -1 & -2 & 3 \end{bmatrix} \quad \text{and} \quad A^{-1} = \begin{bmatrix} -2 & 1 & 0 \\ 1 & -2 & -2 \\ 0 & -1 & -1 \end{bmatrix} \tag{8.77}$$

Consider now the matrix A' obtained by replacing, say, column 2 of A with the following column vector $\mathbf{a}$:

$$\mathbf{a} = \begin{bmatrix} 3 \\ 2 \\ -1 \end{bmatrix} \quad \text{and so} \quad A' = \begin{bmatrix} 0 & 3 & -2 \\ 1 & 2 & -4 \\ -1 & -1 & 3 \end{bmatrix} \tag{8.78}$$

When A' is now multiplied by A^{-1} from (8.77), you obtain the following:

$$A^{-1}A' = \begin{bmatrix} 1 & -4 & 0 \\ 0 & 1 & 0 \\ 0 & -1 & 1 \end{bmatrix} \qquad (8.79)$$

Looking at (8.79), you can see that $A^{-1}A'$ differs from the identity matrix only in column 2, the same column of A that was replaced by **a** to obtain A'. The conclusion is that the matrix A^{-1} provides a good starting point for finding $(A')^{-1}$. In fact, all that is needed is a matrix C such that multiplying (8.79) through by C results in the identity matrix. The reason is that, if you can find C such that

$$(CA^{-1})A' = I$$

then

$$(A')^{-1} = CA^{-1} \qquad (8.80)$$

To illustrate with the foregoing numerical example, you can verify that the following matrix C satisfies $(CA^{-1})A' = I$:

$$C = \begin{bmatrix} 1 & 4 & 0 \\ 0 & 1 & 0 \\ 0 & 1 & 1 \end{bmatrix}$$

It therefore follows from (8.80) that

$$(A')^{-1} = CA^{-1} = \begin{bmatrix} 1 & 4 & 0 \\ 0 & 1 & 0 \\ 0 & 1 & 1 \end{bmatrix} \begin{bmatrix} -2 & 1 & 0 \\ 1 & -2 & -2 \\ 0 & -1 & -1 \end{bmatrix} = \begin{bmatrix} 2 & -7 & -8 \\ 1 & -2 & -2 \\ 1 & -3 & -3 \end{bmatrix}$$

In summary, suppose the original system of linear equations is $A\mathbf{x} = \mathbf{b}$ and that you have already computed A^{-1}. The approach you have just seen for solving the new system $A'\mathbf{x} = \mathbf{b}$ is to find a matrix C such that

$$(CA^{-1})A' = I$$

You can then set $(A')^{-1} = CA^{-1}$ and solve the system $A'\mathbf{x} = \mathbf{b}$ by computing

$$\mathbf{x} = (A')^{-1}\mathbf{b} = (CA^{-1})\mathbf{b}$$

Further efficiencies are now described using some of the numerical methods you have learned in this chapter.

8.5.3 Computational Implementation

The foregoing approach to solving the problem is based on using the inverse of the matrix to solve the system of linear equations. To simplify the subsequent discussion, assume that the following original system is consistent:

$$A\mathbf{x} = \mathbf{b} \qquad (8.81)$$

From the discussion in Section 8.1.1 and Section 8.2.1, you know that Gauss-Jordan elimination with partial pivoting provides a collection of elementary matrices, $E_1, \ldots, E_p$, such that

$$A^{-1} = E_p \cdots E_1 \tag{8.82}$$

The elementary matrices in (8.82) are sparse, however, when multiplied together, as in (8.82), the resulting matrix, A^{-1}, can lose all sparsity. Thus, rather than perform the multiplication in (8.82), you should store each of the elementary matrices, $E_1, \ldots, E_p$, separately in a sparse form (see Section 8.1.2 and Exercise 12 through Exercise 14 in that section).

You can use the factorization of A^{-1} in (8.82) to solve the initial system efficiently, as follows (see Section 8.2.1):

$$\mathbf{x} = A^{-1}\mathbf{b} = E_p \cdots (E_2(E_1\mathbf{b}))$$

Suppose that you now want to solve the new system

$$A'\mathbf{x} = \mathbf{b} \tag{8.83}$$

in which A' is obtained by replacing column k of A with a column vector $\mathbf{a}$. Following the approach developed in Section 8.5.2, the idea is to use the matrices $E_1, \ldots, E_p$ from (8.82) together with $\mathbf{a}$ to find a few additional elementary matrices $E_{p+1}, \ldots, E_q$ such that

$$(A')^{-1} = E_q \cdots E_{p+1}(E_p \cdots E_1) \tag{8.84}$$

You can then solve the system in (8.83) efficiently, as follows:

$$\mathbf{x} = (A')^{-1}\mathbf{b} = E_q \cdots (E_{p+1}(E_p \cdots (E_1\mathbf{b}))) \tag{8.85}$$

It remains to find the matrices $E_{p+1}, \ldots, E_q$ that satisfy (8.84). To do so, follow the approach in Section 8.5.2 of starting with $E_1, \ldots, E_p$ and computing

$$E_p \cdots E_1 A' \tag{8.86}$$

On so doing, you will discover that the matrix in (8.86) differs from the identity matrix only in column k, the same column of A that is replaced with the new column vector $\mathbf{a}$. The elementary matrices $E_{p+1}, \ldots, E_q$ in (8.84) are those matrices corresponding to the elementary row operations needed to reduce the matrix in (8.86) to I.

To illustrate, recall the following matrices A and A':

$$A = \begin{bmatrix} 0 & 1 & -2 \\ 1 & 2 & -4 \\ -1 & -2 & 3 \end{bmatrix}, \quad A' = \begin{bmatrix} 0 & 3 & -2 \\ 1 & 2 & -4 \\ -1 & -1 & 3 \end{bmatrix}, \quad \text{so } \mathbf{a} = \begin{bmatrix} 3 \\ 2 \\ -1 \end{bmatrix} \tag{8.87}$$

Suppose you have already found elementary matrices $E_1, \ldots, E_p$ such that $E_p \cdots E_1 A = I$. For notational convenience, let $E = E_p \cdots E_1$. For the foregoing numerical example, you can verify that the following matrix E satisfies $EA = I$:

$$E = E_p \ldots E_1 = \begin{bmatrix} -2 & 1 & 0 \\ 1 & -2 & -2 \\ 0 & -1 & -1 \end{bmatrix} \tag{8.88}$$

Unfortunately, column 2 of EA' differs from column 2 of I because

$$EA' = \begin{bmatrix} 1 & -4 & 0 \\ 0 & 1 & 0 \\ 0 & -1 & 1 \end{bmatrix} \tag{8.89}$$

The desired matrices $E_{p+1}, \ldots, E_q$ correspond to the elementary row operations needed to reduce column 2 of the foregoing matrix EA' to column 2 of I. The results of using Gauss-Jordan elimination to do so follow.

Step 1. A leading 1 already exists in the pivot element in row 2 and column 2 of EA', so, add 4 times row 2 of EA' to row 1, hence

$$E_{p+1} = \begin{bmatrix} 1 & 4 & 0 \\ 0 & 1 & 0 \\ 0 & 0 & 1 \end{bmatrix} \quad \text{and} \quad E_{p+1}EA' = \begin{bmatrix} 1 & 0 & 0 \\ 0 & 1 & 0 \\ 0 & -1 & 1 \end{bmatrix} \tag{8.90}$$

Step 2. Add 1 times row 2 of $E_{p+1}EA'$ to row 3, so

$$E_{p+2} = \begin{bmatrix} 1 & 0 & 0 \\ 0 & 1 & 0 \\ 0 & 1 & 1 \end{bmatrix} \quad \text{and} \quad E_{p+2}E_{p+1}EA' = I \tag{8.91}$$

The foregoing matrices E_{p+1} and E_{p+2} together with E produce the following factorization of $(A')^{-1}$:

$$(A')^{-1} = E_{p+2}E_{p+1}E \tag{8.92}$$

In general, given $E_1, \ldots, E_p$, you can find $E_{p+1}, \ldots, E_q$ efficiently by first computing column k (the column of A that is replaced with $\mathbf{a}$) of $E_p \cdots E_1 A'$, denoted by $\mathbf{a}'$, as follows:

$$\mathbf{a}' = (E_p \cdots E_1 A')_{*k} = (E_p \cdots E_1) A'_{*k} = E_p \cdots (E_1 \mathbf{a}) \tag{8.93}$$

Then, $E_{p+1}, \ldots, E_q$ are the elementary row operations of Gauss-Jordan elimination needed to reduce $\mathbf{a}'$ in (8.93) to column k of I.

Solving the New System

You can then use the factorization of $(A')^{-1}$, consisting of the initial matrices $E_1, \ldots, E_p$ and the new matrices $E_{p+1}, \ldots, E_q$, to solve the new system

$$A'\mathbf{x} = \mathbf{b}$$

Specifically,

$$\mathbf{x} = (A')^{-1}\mathbf{b} = E_q \cdots (E_{p+1}(E_p \cdots (E_1 \mathbf{b}))) \tag{8.94}$$

Letting

$$\mathbf{b}' = E_p \cdots (E_1 \mathbf{b}) \tag{8.95}$$

that you have already computed when solving the original system $A\mathbf{x} = \mathbf{b}$, the solution in (8.94) to the new system becomes

$$\mathbf{x} = E_q \cdots (E_{p+1} \mathbf{b}') \tag{8.96}$$

To illustrate recall the numerical example in which A, A', and $\mathbf{a}$ are given in (8.87) and suppose that

$$\mathbf{b} = \begin{bmatrix} 6 \\ 4 \\ -3 \end{bmatrix}$$

Using $E = E_p \cdots E_1$ from (8.88), you have from (8.95) that

$$\mathbf{b}' = E_p \cdots (E_1 \mathbf{b}) = E\mathbf{b} = \begin{bmatrix} -2 & 1 & 0 \\ 1 & -2 & -2 \\ 0 & -1 & -1 \end{bmatrix} \begin{bmatrix} 6 \\ 4 \\ -3 \end{bmatrix} = \begin{bmatrix} -8 \\ 4 \\ -1 \end{bmatrix}$$

Using E_{p+1} in (8.90) and E_{p+2} in (8.91), you now have from (8.96) that

$$\mathbf{x} = E_{p+2}(E_{p+1} \mathbf{b}') = \begin{bmatrix} 1 & 0 & 0 \\ 0 & 1 & 0 \\ 0 & 1 & 1 \end{bmatrix} \left(\begin{bmatrix} 1 & 4 & 0 \\ 0 & 1 & 0 \\ 0 & 0 & 1 \end{bmatrix} \begin{bmatrix} -8 \\ 4 \\ -1 \end{bmatrix} \right) = \begin{bmatrix} 8 \\ 4 \\ 3 \end{bmatrix}$$

To summarize, suppose the initial system is $A\mathbf{x} = \mathbf{b}$ and that the new system is $A'\mathbf{x} = \mathbf{b}$, in which A' is obtained by replacing column k of A with the column vector $\mathbf{a}$. The computational approach developed here for solving the new system, given that you have already computed elementary matrices $E_1, \ldots, E_p$ such that $A^{-1} = E_p \cdots E_1$ and used them to compute $\mathbf{b}' = E_p \cdots (E_1 \mathbf{b})$, is:

Step 1. Compute $\mathbf{a}' = E_p \cdots (E_1 \mathbf{a})$.

Step 2. Use Gauss-Jordan elimination to find the elementary matrices $E_{p+1}, \ldots, E_q$ that reduce $\mathbf{a}'$ to column k of I.

Step 3. Compute $\mathbf{x}' = E_q \cdots (E_{p+1} \mathbf{b}')$.

Additional Computational Considerations and Refinements

Additional issues need to be addressed to ensure that the foregoing solution procedure is both efficient and numerically accurate. For example, you should use partial pivoting as part of your procedure for finding the new elementary matrices $E_{p+1}, \ldots, E_q$ so as to reduce truncation errors. Unfortunately, interchanging row k of $E_p \cdots E_1 A'$ with a row k' below row k, as partial pivoting can require, destroys the property that column k' of $E_p \cdots E_1 A'$ is column k' of I. The result is that additional row operations are needed to reduce $E_p \cdots E_1 A'$ to I.

You should exploit the sparsity of the elementary matrices not only in storing them, but also in performing the computations in the foregoing steps efficiently. That is, you should take advantage of the form and sparsity of each matrix so as to perform only those multiplications in which both terms are nonzero.

Observe also that each time a new system of linear equations is solved, additional elementary matrices are computed and stored in sparse form. A point will be reached eventually at which so many matrices are stored that the foregoing approach becomes inefficient, because of both excessive storage and computations needed to perform the necessary multiplications. The approach to overcoming this problem is to choose a *threshold* number of iterations after which all of the elementary matrices generated so far are discarded. The current system is then solved by applying Gaussian elimination with partial pivoting to the matrix A corresponding to the current system of linear equations. In so doing, a new set of initial elementary matrices, $E_1, \ldots, E_p$, are created and stored in sparse form. The next system of linear equations is solved by finding a few additional elementary matrices $E_{p+1}, \ldots, E_q$. This process is repeated for each new system until the number of elementary matrices exceeds the chosen threshold, at which point all elementary matrices are discarded and the whole procedure is started anew.

Further gains in efficiency are achieved by using an LU factorization of A (see Section 8.2.2) instead of the inverse of A. Mathematical analysis is needed to determine an efficient method for using L and U to find new matrices L' and U' so that $A' = L'U'$, where A' is the matrix obtained by replacing column k of A with a new column vector, $\mathbf{a}$. The details of doing so, however, are omitted.

A final consideration is the user interface. This includes allowing the user to enter and modify all the data conveniently, to specify which waste liquids are to be in the initial system of linear equations, and to enter each waste liquid as it is depleted, together with its chosen replacement liquid. The program then attempts to solve the new system of linear equations. The user must be informed whenever the new system has no solution in which the values of all variables are positive and a new replacement liquid must be requested.

Exercises for Section 8.5

PROJECT 8.1: Solving the Problem of Waste Management Using a Factorization of the Inverse

Use a graphing calculator, MATLAB, or other software package, as indicated by your instructor, to perform the operations in Exercise 1 through Exercise 5 for solving the problem of Waste Management described in Section 8.5.1. Whenever possible, use matrix and vector operations available on your system. Write the answer and the sequence of operations you performed to obtain the answer.

1. Enter the data from Table 8.7 for the five waste liquids in a matrix D and the desired flow rates of 5000, 6750, and 2250 for the output products in a column vector $\mathbf{b}$. Then create a (3×3) matrix A consisting of the first three columns of D and solve the initial system of linear equations $A\mathbf{x} = \mathbf{b}$ by computing $\mathbf{x} = A^{-1}\mathbf{b}$ to find the flow rates for waste liquids 1, 2, and 3.

2. Suppose you now want to replace waste liquid 3 with liquid 5. Recall that the matrix A' corresponding to the associated system of linear equations is obtained by replacing column 3 of A with $\mathbf{a} = D_{*5}$. Prepare the new system $A'\mathbf{x} = \mathbf{b}$ by finding elementary matrices that reduce $\mathbf{a}' = A^{-1}\mathbf{a}$ to column 3 of the identity matrix, as follows:

(a) Compute $\mathbf{a}' = A^{-1}\mathbf{a}$.

(b) Construct the elementary matrix E_1 so that the third element of $E_1\mathbf{a}'$ is 1.

(c) Construct the elementary matrix E_2 so that the second element of $E_2(E_1\mathbf{a}')$ is 0.

(d) Construct the elementary matrix E_3 so that the first element of $E_3(E_2(E_1\mathbf{a}'))$ is 0.

3. Use E_1, E_2, and E_3 from Exercise 2 to solve the system $A'\mathbf{x} = \mathbf{b}$ by computing $\mathbf{b}' = A^{-1}\mathbf{b}$ and then $\mathbf{x} = E_3(E_2(E_1\mathbf{b}'))$. Verify your result by computing $\mathbf{x} = (A')^{-1}\mathbf{b}$.

4. Suppose you now want to replace waste liquid 2 with liquid 4. Recall that the matrix A'' corresponding to the associated system of linear equations is obtained by replacing column 2 of A' with $\mathbf{a} = D_{*4}$. Prepare the new system $A''\mathbf{x} = \mathbf{b}$ by finding elementary matrices that reduce $\mathbf{a}' = E_3(E_2(E_1(A^{-1}\mathbf{a})))$ to column 2 of the identity matrix, as follows:

(a) Compute $\mathbf{a}' = E_3(E_2(E_1(A^{-1}\mathbf{a})))$.

(b) Construct the elementary matrix E_4 so that the second element of $E_4\mathbf{a}'$ is 1.

(c) Construct the elementary matrix E_5 so that the third element of $E_5(E_4\mathbf{a}')$ is 0.

(d) Construct the elementary matrix E_6 so that the first element of $E_6(E_5(E_4\mathbf{a}'))$ is 0.

5. Use E_1, E_2, and E_3 from Exercise 2 and E_4, E_5, and E_6 from Exercise 4 to solve the system $A''\mathbf{x} = \mathbf{b}$ by computing

$$\mathbf{x} = E_6(E_5(E_4(E_3(E_2(E_1(A^{-1}\mathbf{b}))))))$$

Verify your result by computing $\mathbf{x} = (A'')^{-1}\mathbf{b}$.

PROJECT 8.2 Solving a New Problem of Waste Management Using an *LU* Factorization

Use a graphing calculator, MATLAB, or other software package, as indicated by your instructor, to perform the operations in Exercise 6 through Exercise 9 for solving the following problem of Waste Management. Whenever possible, use matrix and vector operations available on your system. Write the answer and the sequence of operations you performed to obtain the answer.

Five waste liquids are used to make five sellable products. The number of gallons per hour of each output product obtained from 1000 gallons per hour of waste liquid is given in the columns of the following table, the last of which contains the desired flow rates, in gallons per hour, for the five products:

| Output | Waste Liquid | | | | | Desired Hourly |
Product	1	2	3	4	5	Flow Rate
1	100	200	150	250	100	140000
2	200	300	100	100	200	155000
3	300	200	250	150	100	160000
4	200	0	100	50	50	55000
5	100	100	250	200	150	120000

6. Formulate a system of linear equations to determine the rates at which to process the five waste liquids so as to achieve the desired rates for the outputs.

7. Let A be the (5×5) matrix whose columns contain the flow rates of the five output products given in the corresponding columns of the foregoing table. Find an LU factorization of A.

8. Use the LU factorization in the previous exercise to solve the system of linear equations in Exercise 6. Specifically, show the result of using forward substitution to solve the system $L\mathbf{y} = \mathbf{b}$ for $\mathbf{y}$. Then show the result of using back substitution to solve the system $U\mathbf{x} = \mathbf{y}$ for $\mathbf{x}$.

9. There is a virtually unlimited supply of all the waste liquids. However, the desired flow rates for the output products are changed every hour. The specific rates for each of the next three hours are given in the first row of the following table. Use the LU factorization from Exercise 7 to solve the three associated systems of equations. For each system, show the result of using forward substitution to solve the system $L\mathbf{y} = \mathbf{b}$ for $\mathbf{y}$ and the result of using back substitution to solve the system $U\mathbf{x} = \mathbf{y}$ for $\mathbf{x}$. Provide the answer in a table similar to the following one:

	Hour		
	1	**2**	**3**
Desired Flow Rates	$\begin{bmatrix} 85000 \\ 85000 \\ 115000 \\ 50000 \\ 100000 \end{bmatrix}$	$\begin{bmatrix} 75000 \\ 85000 \\ 85000 \\ 35000 \\ 85000 \end{bmatrix}$	$\begin{bmatrix} 150000 \\ 130000 \\ 170000 \\ 70000 \\ 150000 \end{bmatrix}$
y			
x			

Chapter Summary

When developing methods for solving a problem by computer, special attention is needed in the following areas: (1) computational efficiency, (2) efficient use of computer memory, (3) numerical accuracy, and (4) the user interface. One approach for estimating the running time of an algorithm—that is, the total number of time units needed to solve a problem with general data—is to express the number of most time-consuming operations the algorithm needs to solve the problem as a function of the data that have the greatest impact on the running time. For example, the efficiency of an algorithm for solving an $(n \times n)$ system of linear equations $A\mathbf{x} = \mathbf{b}$ is approximated by the number of multiplications needed to solve such a system, expressed as a function of n.

You can save a significant amount of computer memory when working with sparse matrices and vectors—that is, matrices and vectors that have mostly zero elements—by storing only the nonzero elements. Large matrices in applications typically are sparse, as are elementary matrices that arise in Gauss-Jordan and Gaussian elimination.

Numerical accuracy is lost when computers store only a fixed number of digits in the decimal representation of real numbers. The loss of the remaining digits results in truncation errors that can produce inaccurate answers. Computer scientists have determined that truncation errors are reduced when performing divisions in which the denominator is relatively large. Therefore, when solving systems of linear equations, you should use the rule of partial pivoting to interchange appropriate rows in such a way that the absolute value of the number used as a denominator in performing subsequent row operations is as large as possible.

A proper user interface enables the user to enter the data for a problem in a direct and simple way. Always be sure that the user enters the least amount of data needed to solve the problem.

A factorization of a matrix A is a collection of matrices whose product is A. Factorizations provide efficient computational methods for solving problems in linear algebra. One important example is a factorization of the inverse of an invertible $(n \times n)$ matrix A into a product of elementary matrices, $E_1, \ldots, E_p$. These elementary matrices are obtained by performing the steps of Gauss-Jordan elimination with partial pivoting. You can store these sparse matrices and solve the system of linear equations

$$A\mathbf{x} = \mathbf{b} \tag{8.97}$$

efficiently by computing

$$\mathbf{x} = A^{-1}\mathbf{b} = E_p \cdots (E_1\mathbf{b})$$

You can use Gaussian elimination to produce an LU factorization of A, in which L is a lower triangular matrix and U is an upper triangular matrix. Then, to solve the system $A\mathbf{x} = \mathbf{b}$, first use forward substitution to solve the following system for $\mathbf{y}$:

$$L\mathbf{y} = \mathbf{b}$$

Then use back substitution to solve the following system for $\mathbf{x}$:

$$U\mathbf{x} = \mathbf{y}$$

A factorization is also used to express a matrix A as the product of a matrix Q, whose columns form an orthonormal basis for the column space of A, and an invertible upper triangular matrix R. Yet another example of a factorization arises when A is similar to a diagonal matrix D.

Numerical methods are also used to estimate eigenvalues and eigenvectors of an $(n \times n)$ matrix A. When A has a strictly dominant eigenvalue, that is, an eigenvalue whose absolute value is strictly larger than that of all other eigenvalues, the power method is used to generate a sequence of n-vectors that get closer and closer to a corresponding eigenvector. This is accomplished by choosing an initial n-vector $\mathbf{x}_0$ and then computing the following sequence of n-vectors:

$$\mathbf{x}_{k+1} = A\mathbf{x}_k, \quad \text{for } k = 0, 1, 2, \ldots$$

The inverse power method is used to find an approximate value for any eigenvalue of A, say, λ, provided that a sufficiently close estimate, μ, of λ is available. To do so, apply the power method to the matrix $B = (A - \mu I)^{-1}$ to obtain an estimate, β, of the strictly dominant eigenvalue, $1/(\lambda - \mu)$, of B. You then obtain an approximate value for λ by computing

$$\lambda \approx \mu + \frac{1}{\beta}$$

Iterative methods attempt to find a solution to the system $A\mathbf{x} = \mathbf{b}$ by generating a sequence of column vectors $\mathbf{x}_0, \mathbf{x}_1, \mathbf{x}_2, \ldots$. These methods start with an initial estimate, $\mathbf{x}_0$, of the solution to (8.56). The elements of $\mathbf{x}_0$ are then used to compute $\mathbf{x}_1$, whose elements are then used to compute $\mathbf{x}_2$, and so on. The specific way in which the elements of the current vector are used to compute the elements of the next vector differ in Jacobi's method and in the Gauss-Seidel method. Nevertheless, under suitable circumstances, the generated sequence converges to a solution to the system of equations.

Although not finite in nature, iterative methods are fast when the initial vector is close to the solution. Furthermore, you can stop an iterative method when the current vector in the sequence, though not an exact solution, satisfies the system of linear equations to an acceptable degree of accuracy. Another advantage of iterative methods is that they often require less storage space when the matrix A is large and sparse than the factorization methods.

Proof Techniques

A **proof** is a logical argument for convincing someone that a particular mathematical statement is true. In this appendix, a systematic approach is presented for learning how to read, understand, and create proofs. To that end, a collection of **proof techniques** are described. You will also learn how and when to use them. The approach presented here is adapted from *How to Read and Do Proofs*, Second Edition, Daniel Solow, copyright ©1990 by John Wiley & Sons, Inc., NY[†]. The reader will find a more detailed description in that book.

A.1 Mathematical Statements and Implications

Proofs deal with **statements**, which, in mathematics, are expressions that are either true or false. For example, each of the following are statements:

1. $0 < 1$.

2. $3 > 4$.

3. The real number x is less than 2.

4. There are real numbers that are not rational.

You can see that (1) is true and (2) is false. In contrast, the truth of (3) depends on the specific value of x, which you do not know. As such, (3) is called a **conditional statement**. Nevertheless, (3) is either true or false. Finally, (4) is true, although this fact may not be so obvious. It is for this reason that a proof is used, namely, to convince someone that a particular statement is true. Throughout the rest of this appendix, p and q represent two statements.

You can use statements to create other statements. For example, for a given statement

---

[†]Reprinted with permission from John Wiley & Sons, Inc.

p	not p
T	F
F	T

Table A.1 *The Truth Table for not p.*

p, the **negation of** p, written *not p* or $\neg p$, is a statement that is false if p is true and true if p is false. This information is summarized in Table A.1, called a **truth table**, that indicates the truth of a complex statement (*not p*, in this case) in terms of the truth of simpler statements (p, in this case).

As another example, the **conjunction** of the statements p and q is the statement *p and q* (also written $p \wedge q$) that is true if both p and q are true and is false, otherwise [see Table A.2(a)]. The **disjunction** of the statements p and q is the statement *p or q* (also written $p \vee q$) that is true in all cases except when p is false and q is false [see Table A.2(b)].

A.1.1 Implications

Proofs apply to the following special kinds of statements:

If p is true, then q is true,

or, more simply,

If p, then q. (A.1)

The statement in (A.1) is called an **implication** in which p is the **hypothesis** and q is the **conclusion**. You can write (A.1) in any of the following equivalent ways:

$$p \ implies \ q, \quad p \Rightarrow q, \quad p \rightarrow q.$$

In an implication, it is important to realize that p is a statement, q is a statement, and so is *p implies q*. The truth of *p implies q* depends on the truth of p and q themselves. Mathematicians have come to the agreement that the statement *p implies q* is true in all cases *except when p is true and q is false*, as summarized in Table A.3. Three other statements related to *p implies q* are:

p	q	p and q
T	T	T
T	F	F
F	T	F
F	F	F

p	q	p or q
T	T	T
T	F	T
F	T	T
F	F	F

(a) *Truth Table for p and q.* (b) *Truth Table for p or q.*

Table A.2 *The Truth Table for (a) the Conjunction and (b) the Disjunction of Two Statements.*

p	q	p *implies* q
T	T	T
T	F	F
F	T	T
F	F	T

Table A.3 *The Truth Table for p implies q.*

1. The **contrapositive statement**: (*not* q) *implies* (*not* p) (see the truth table in Table A.4).
2. The **converse statement**: q *implies* p (see the truth table in Table A.5).
3. The **inverse statement**: (*not* p) *implies* (*not* q) (see the truth table in Table A.6).

The use of Table A.3, as applied to the original statement p *implies* q, is shown in the next example.

EXAMPLE A.1 Using the Truth Table to Prove an Implication
In the implication

If $3 > 4$, then $0 < 1$,

the hypothesis $3 > 4$ is false and the conclusion $0 < 1$ is true. Thus, according to the third row of Table A.3, the implication itself is true.

The use of Table A.3 is not always so straightforward, as seen in the next example.

EXAMPLE A.2 The Need for a New Proof Technique
In the implication:

If x and y are positive real numbers for which

$$x^2 \geq 2 \quad \text{and} \quad y = \frac{1}{2}\left(x + \frac{2}{x}\right),$$

then $x \geq y$,

p	q	p *implies* q	*not* q	*not* p	(*not* q) *implies* (*not* p)
T	T	T	F	F	T
T	F	F	T	F	F
F	T	T	F	T	T
F	F	T	T	T	T

Table A.4 *The Truth Table for (not q) implies (not p).*

p	q	q implies p
T	T	T
T	F	T
F	T	F
F	F	T

Table A.5 *The Truth Table for q implies p.*

you do not know whether the hypothesis is true or false because you do not know the specific values of x and y. Likewise for the conclusion. Thus, you do not know which row of Table A.3 is applicable. Nevertheless, as you will now see, you can still determine that this implication is true.

A.2 The Forward-Backward Method

The implication in Example A.2 illustrates the need for some method of proving that *p implies q* is true, other than using Table A.3 directly. To understand this new approach, look again at Example A.2. Although you do not know which row of Table A.3 is applicable, you can reason as follows:

"In the event that p is false, row 3 or row 4 of Table A.3 is applicable and, in either case, *p implies q* is true, so there is no need to consider the case when p is false. On the other hand, if p is true, then either row 1 or row 2 is applicable. If I want *p implies q* to be true, then I need to be sure that row 1 is applicable, that is, that q is true."

The result of this reasoning is that, if you want to prove *p implies q* is true,

You can *assume that p is true* and your objective is to *show that q is true*.

The *forward-backward method* is a technique for using the assumption that p is true to reach the conclusion that q is true. As its name implies, the forward-backward method consists of two parts, each of which is described next.

p	q	q implies p	not p	not q	(not p) implies (not q)
T	T	T	F	F	T
T	F	T	F	T	T
F	T	F	T	F	F
F	F	T	T	T	T

Table A.6 *The Truth Table for (not p) implies (not q).*

A.2.1   The Backward Process

The objective of the **backward process** is to help you reach the conclusion that q is true by obtaining, from q, a new statement, q_1, with the property that if q_1 is true, then so is q. In the fortunate event that q_1 is p, you are done because you are assuming that p is true and so you would know that q_1 is true and thus so is q. On the other hand, if you do not know that q_1 is true, then you can apply the backward process again to q_1 to obtain a new statement, q_2, with the property that if q_2 is true, then so is q_1 (and hence q). This backward process is continued until you obtain a statement that you know is true.

One way to obtain these new statements is to look at the last statement you have in the backward process, say, q, and to ask the following **key question**:

How can I show that q is true?

The specific way in which you ask this question often determines your answer. For example, you have the following key question associated with the conclusion in Example A.2:

Conclusion	Key Question
$q : x \geq y$	How can I show that a real number (namely, x) is greater than or equal to another real number (namely, y)?

Here are other examples of statements q and their associated key questions.

Statement q	Key Questions
The lines L_1 and L_2 are parallel.	How can I show that two lines are parallel?
$A \cup B = B$ (where A and B are sets).	How can I show that two sets are equal?
The right triangles ABC and RST are congruent.	How can I show that two triangles are congruent?
	or
	How can I show that two right triangles are congruent?
$f(x) = 0$ (where x is a given real number and f is a real-valued function of one variable).	How can I show that the value of a function at a given point is 0?
	or
	How can I show that a function crosses the x-axis at a given point?
	or
	How can I show that two real numbers are equal?

The following points are worth noting from these examples:

1. The key questions contain no symbols or notation from the statement under consideration. (Avoiding symbols allows you to focus on the more important aspects of the problem.)
2. There can be several different key questions associated with a statement q. (Choosing the correct one is an art that may require a trial-and-error process.)

After asking the key question, you must provide an answer. For instance, recall the key question associated with the conclusion

$$q : x \geq y$$

in the implication in Example A.2, which is:

How can I show that a real number (namely, x) is greater than another real number (namely, y)?

One answer to this question is to show that the difference of the two real numbers is nonnegative, that is, you must show that

$$q_1 : x - y \geq 0$$

Answering the key question is done in the following two steps.

How to Answer a Key Question

1. First provide an answer that contains no symbols or notation from the specific problem under consideration.
2. Then write your answer in (1) using appropriate symbols and notation from the specific problem to obtain the new statement in the backward process.

There is often more than one answer to a key question. For the foregoing statement q, another answer is to show that the first number (x) is equal to the second number (y) plus some nonnegative amount. Choosing the right answer to the key question is an art that may require a trial-and-error process.

In summary, through asking and answering the key question, the objective of the backward process is to derive, from the last statement q, a new statement, q_1, with the property that if q_1 is true, then so is q. You continue from q_1 with the backward process until you create a statement that you know is true, at which time the proof is complete. If you have difficulties with the backward process, you can use the assumption that the hypothesis p is true, as described next.

A.2.2 The Forward Process

The **forward process** is the process of deriving, from the hypothesis p, some new statements that are necessarily true as a result of the assumption that p is true. For instance, from the following hypothesis in the implication in Example A.2:

$$p : x^2 \geq 2,$$

you can create the new statement:

$$p_1 : x^2 - 2 \geq 0.$$

Observe that p_1 is true as a result of the assumption that p is true.

As another example of the forward process, from the following hypothesis in the implication in Example A.2:

$$p : y = \frac{1}{2}\left(x + \frac{2}{x}\right),$$

you can create a new statement by multiplying both sides of the equality in p by -1, adding x, and then performing algebra to obtain:

$$p_2 : x - y = \frac{x^2 - 2}{2x}.$$

The objective of the forward process is to obtain the last statement in the backward process, at which time, the proof is complete. In this case, the last statement in the backward process is

$$q_1 : x - y \geq 0,$$

which is the motivation for obtaining p_2 in the forward process (compare p_2 and q_1). In fact, you can now conclude that q_1 is true because the numerator $x^2 - 2$ in p_2 is ≥ 0 (see p_1), and the denominator $2x$ is > 0 (because $x > 0$ from the hypothesis). It therefore follows that q_1, and hence q, is true, thus completing the proof.

In the following written proof of the implication in Example A.2, observe that there is little explicit reference to the forward and backward processes.

THEOREM A.1

If x and y are positive real numbers for which

$$x^2 \geq 2 \quad \text{and} \quad y = \frac{1}{2}\left(x + \frac{2}{x}\right),$$

then $x \geq y$.

Proof.

The conclusion is obtained by showing that $x - y \geq 0$. From the hypothesis, it follows by algebra that

$$x - y = \frac{x^2 - 2}{2x}.$$

Furthermore, from the hypothesis that x is positive and that $x^2 \geq 2$, the numerator $x^2 - 2 \geq 0$ and the denominator $2x > 0$, so $x - y \geq 0$ and hence $x \geq y$, completing the proof. ∎

A.3 The Existential Quantifier and the Construction Method

The forward-backward method is only one of various proof techniques available. In many cases, you can choose a correct technique on the basis of certain key words appearing in the hypothesis or conclusion of the implication. One such set of key words involves the **existential quantifier** *there is*, (*there are*, *there exists*, and so on). Mathematicians use the symbol $\exists$ for the words *there is* and $\ni$ for the words *such that*, as illustrated in the following statements:

1. There is a real number $x > 0$ such that $x = 2^{-x}$, or
   $\exists$ a real number $x > 0 \ni x = 2^{-x}$, or
   $\exists x > 0 \ni x = 2^{-x}$.

2. There is an element x in the set A such that x is in the set B, or
   $\exists$ an element $x \in A \ni x \in B$ (where $\in$ stands for "in" or "is in"), or
   $\exists x \in A \ni x \in B$.

3. The quadratic equation $ax^2 + bx + c = 0$ has two distinct real roots when a, b, and c are real numbers with $b^2 - 4ac > 0$ and $a \neq 0$.

From these examples, you can see that such statements have the following general form:

There is an *object* with a *certain property* such that *something happens*.

In (1), the *object* is a real number x; the *certain property* is that of being > 0; and the *something that happens* is that $x = 2^{-x}$. In (2), the *object* is an element x; the *certain property* is that of being in the set A; and the *something that happens* is that x is in the set B. Even (3) has this form when rewritten as follows:

There are real numbers y and z with $y \neq z$ such that $ay^2 + by + c = 0$ and $az^2 + bz + c = 0$.

When a statement containing the existential quantifier arises in the forward process, you can assume that there is an object with the certain property such that the something happens. You should use this object to arrive at the desired conclusion.

However, when a statement containing the existential quantifier arises in the backward process, it is your job to show that there is an object with the certain property such that the something happens. The **construction method** is a technique for accomplishing this goal. With the construction method, you must first produce the object—by guessing, by trial-and-error, by devising an algorithm whose output is the object, or by any other means. However, you must subsequently prove that the object you produced is the correct one in that the object has the certain property and that the something happens. The use of the construction method is demonstrated in the proof of the following theorem.

DEFINITION A.1

An integer a **divides** an integer b if and only if there is an integer k such that $b = ka$.

THEOREM A.2

If an integer a divides an integer b, then a divides the integer b^2.

Developing the Proof

According to the forward-backward method, you can assume that the following hypothesis is true:

p : The integer a divides the integer b.

You must use this assumption to conclude that

q : a divides the integer b^2.

The backward process applied to q gives rise to the key question: How can I show that an integer (namely, a) divides another integer (namely, b^2)? By using the definition—one of the most common methods for answering a key question—you obtain the following new statement to prove:

q_1 : There is an integer k such that $b^2 = ka$.

The appearance of the existential quantifier *there is* in q_1 suggests that you now use the construction method to produce the integer k. How you find the value of k is not clear but, in general, you should use the information in the hypothesis, which you are assuming is true.

In this case, the hypothesis p is that a divides b. Working forward from Definition A.1, this means that

p_1 : there is an integer m such that $b = ma$.

The idea is to use the known integer m in p_1 to construct the unknown integer k in q_1. This is accomplished by looking at what properties you want k to satisfy. In this case, you want $b^2 = ka$. To find such a value for k, you can work forward from p_1, as follows:

p_2 : $b^2 = (ma)^2 = m^2a^2 = (m^2a)a$.

From p_2, you can see that the desired value for k is $k = m^2a$.

It is important to note that, although the value of k has now been constructed as m^2a, you must still prove that this value of k is correct, that is, that $b^2 = ka$. This, however, is true, as seen in p_2.

Proof.

By definition, it must be shown that there is an integer k such that $b^2 = ka$. However, from the hypothesis that a divides b, there is an integer m such that $b = ma$. Letting $k = m^2a$, it follows that

$$b^2 = (ma)^2 = (m^2a)a = ka.$$

Thus a divides b^2, completing the proof.

A.4 The Universal Quantifier and the Choose Method

Several other proof techniques are available when the hypothesis or conclusion of an implication contains the **universal quantifier** *for all* (*for each*, *for any*, and so on), written ∀. The way in which this quantifier arises is illustrated in the following statements:

1. For all real numbers $x > 1$, $x^2 > x$, or
   ∀ real numbers $x > 1$, $x^2 > x$, or
   $\forall x > 1, x^2 > x$.

2. For all elements x in the set A, x is in the set B, or
   ∀ elements x in A, x is in B, or
   $\forall x \in A, x \in B$.

From these examples, you can see that such statements have the following general form:

> For all *objects* with a *certain property*, *something happens*.

In (1), the *objects* are real numbers x; the *certain property* is that of being > 1; and the *something that happens* is $x^2 > x$. In (2), the *objects* are elements x; the *certain property* is that of being in the set A; and the *something that happens* is that x is in the set B.

When a statement containing the universal quantifier arises in the backward process, you should consider using the **choose method** to prove that for every object with the certain property, the something happens. This method works as follows. Rather than prove that the something happens *for every* object with the certain property (which you cannot do because there are usually too many of them), the idea is to show that, for *one representative object with the ceratin property, the something happens*. If you are successful for this one chosen object, then you could, in theory, repeat the same proof for each and every object with the certain property, thus showing that for all objects with the certain property, the something happens.

In the example that follows, observe that the act of choosing a generic object with the certain property provides a new statement in the forward process. The statement that, for this chosen object, the something happens becomes the new statement to be proved in the backward process.

DEFINITION A.2

A set A is a **subset** of a set B if and only if for every element x in A, x is in B.

THEOREM A.3

If

$$A = \{\text{real numbers } x : -1 \le x \le 2\} \quad \text{and}$$
$$B = \{\text{real numbers } x : x^2 - x - 2 \le 0\},$$

then A is a subset of B.

Developing the Proof

You must conclude that

$q : A$ is a subset of B.

The backward process gives rise to the key question: How can I show that a set (namely, A) is a subset of another set (namely, B)? According to Definition A.2, the answer is to show that

$q_1 :$ for every element x in A, x is in B.

Recognizing the quantifier *for every* in q_1, you should now proceed with the choose method, whereby, you choose one representative object with the certain property for which you must show that the something happens. In this case, that means you should choose

$p_1 :$ an element y in A

for which you must show that

$q_2 : y$ is in B

Equivalently, from the fact that $B = \{$real numbers $x : x^2 - x - 2 \leq 0\}$, you must show that

$q_3 : y^2 - y - 2 \leq 0$

Working forward from p_1, you know that y is in A so, from the hypothesis that $A = \{x : -1 \leq x \leq 2\}$, it follows that

$p_2 : -1 \leq y \leq 2$

Thus,

$p_3 : y - 2 \leq 0$ and $y + 1 \geq 0$

so

$p_4 : y^2 - y - 2 = (y - 2)(y + 1) \leq 0$

The proof is now complete because p_4 (in the forward process) is the same as q_3 (the last statement in the backward process).

In the written proof that follows, observe that the symbol x is used for the chosen object instead of y. Also note that the word *let* indicates that the choose method is being used, as is often the case.

Proof.

To see that A is a subset of B, *let* x be an element of A, so, $-1 \leq x \leq 2$. But then $x - 2 \leq 0$ and $x + 1 \geq 0$, thus

$x^2 - x - 2 = (x - 2)(x + 1) \leq 0.$

This means that x is an element of B and so it has been shown that every element x in A is also in B, thus completing the proof.

A.5 Induction

You have just seen how the choose method is used when the last statement in the backward process contains the universal quantifier *for all* in the form:

> For all *objects* with a *certain property*, *something happens*.

However, when the objects are integers, the certain property is that of being greater than or equal to some initial integer, and the something that happens is some statement, $S(n)$, that depends on the integer n, a proof technique called **induction** is often the best method to use, even before the choose method. For example, you should consider using induction to prove each of the following statements:

1. For every integer $n \geq 1$,

$$S(n) : \sum_{k=1}^{n} k = \frac{n(n+1)}{2}.$$

2. For every integer $n \geq 1$,

$$S(n) : \frac{1}{n!} \leq \frac{1}{2^{n-1}}, \quad \text{[where } n! = 1(2) \cdots (n)\text{]}.$$

The idea of induction is to begin by proving that the statement $S(n)$ is true for $n = 1$. You could then use the fact that $S(1)$ is true to prove that $S(2)$ is true. Then you could use $S(2)$ to show that $S(3)$ is true, and so on. Because there are an infinite number of statements, you cannot actually prove all of them; however, you can accomplish the same goal by performing the following two steps:

The Two Steps of a Proof by Induction

Step 1. Prove that the statement is true for $n = 1$.

Step 2. Assume that $S(n)$ is true and prove that $S(n + 1)$ is also true.

The first step of proving that $S(1)$ is true is usually straightforward, requiring little more than writing $S(1)$ and verifying that $S(1)$ is true. The second step is more challenging. Begin by writing $S(n)$ and assuming that this statement is true. You should then write $S(n + 1)$ by replacing n everywhere with $n + 1$ and try to prove that $S(n + 1)$ is true. To do so, you want to use the assumption that $S(n)$ is true—often called the **induction hypothesis**—and this requires finding a relationship between $S(n + 1)$ and $S(n)$. That is, you must express $S(n + 1)$ in terms of $S(n)$ so that you can use the induction hypothesis that $S(n)$ is true. These two steps are illustrated in the following example.

THEOREM A.4

For every integer $n \geq 1$,

$$\frac{1}{n!} \leq \frac{1}{2^{n-1}}$$

Proof.

The first step of induction requires you to prove that $S(1)$ is true, that is,

$$S(1) : \frac{1}{1!} \leq \frac{1}{2^{1-1}}$$

You can see that $S(1)$ is true because the left side of the inequality is 1 and so is the right side.

Turning to the second step of induction, you should assume that $S(n)$ is true which, in this case, means you should assume that

$$S(n) : \frac{1}{n!} \leq \frac{1}{2^{n-1}} \quad \text{(induction hypothesis)}.$$

You must now use the assumption that $S(n)$ is true to prove that $S(n + 1)$ is true which, in this case, means you must prove that

$$S(n + 1) : \frac{1}{(n + 1)!} \leq \frac{1}{2^n}.$$

To use the induction hypothesis that $S(n)$ is true, relate $S(n + 1)$ to $S(n)$. For example, from the left side of $S(n + 1)$ and the definition of $(n + 1)!$ you have that

$$\frac{1}{(n + 1)!} = \frac{1}{n! \, (n + 1)} = \left(\frac{1}{n!}\right)\left(\frac{1}{n + 1}\right). \tag{A.2}$$

You can now apply the induction hypothesis to (A.2). Specifically, from $S(n)$ you know that

$$\frac{1}{n!} \leq \frac{1}{2^{n-1}}$$

and so (A.2) becomes

$$\frac{1}{(n + 1)!} = \left(\frac{1}{n!}\right)\left(\frac{1}{n + 1}\right) \leq \left(\frac{1}{2^{n-1}}\right)\left(\frac{1}{n + 1}\right). \tag{A.3}$$

All that remains is to note that because $n \geq 1$,

$$\frac{1}{n + 1} \leq \frac{1}{2},$$

so (A.3) becomes

$$\frac{1}{(n + 1)!} \leq \left(\frac{1}{2^{n-1}}\right)\left(\frac{1}{n + 1}\right) \leq \left(\frac{1}{2^{n-1}}\right)\left(\frac{1}{2}\right) = \frac{1}{2^n},$$

which is precisely $S(n + 1)$, thus completing the proof. ■

A.6 The Universal Quantifier and the Specialization Method

In Section A.4 and Section A.5, you learned to use the choose and induction methods when the universal quantifier *for all* arises in the backward process in the form:

For all *objects* with a *certain property*, *something happens*.

When this quantifier arises in the *forward* process, the **specialization method** is often used to obtain a new statement. For example, suppose you are doing a proof in which S and T are sets satisfying the following statement in the forward process:

p : For all elements $x \in S, x \in T$.

You can use specialization to create a new statement, p_1, by applying the general knowledge in p to *one specific element in S*. For instance, if, in the course of doing this proof, you come across a particular element y in S, then you can apply specialization to the statement p to conclude that, for this element y,

$p_1 : y \in T$.

In general, the specialization method allows you to obtain a new statement in the forward process from the following statement:

p : For all *objects* X with a *certain property*, *something happens*,

by performing these steps:

Steps for Applying Specialization

Step 1. Identify one particular object, say, Y. (You may need to use trial-and-error to identify the correct object to use for specialization.)

Step 2. Verify that the object Y satisfies the certain property in p.

Step 3. Create a new statement, p_1, in the forward process by writing that the something happens for this specific object Y.

These steps are demonstrated in the following example.

DEFINITION A.3

A real number a is a **lower bound** for a set T of real numbers if and only if for every element $x \in T, a \leq x$.

THEOREM A.5

If a is a lower bound for a set T of real numbers and $S \subseteq T$, then a is a lower bound for S.

Developing the Proof

You must show that

q : a is a lower bound for S.

The backward process leads to the key question:

How can I show that a real number (namely, a) is a lower bound for a set of real numbers (namely, S)?

From Definition A.3, you must show that

q_1 : for every element $x \in S, a \leq x$.

Recognizing the universal quantifier *for every* in the backward process, you should use the choose method to choose an element

$p_1 : y \in S$,

for which you must show that

$q_2 : a \leq y$.

You will now see how specialization is used to obtain q_2.

Turning to the forward process, you know from the hypothesis p that $S \subseteq T$ which, by Definition A.2, means that

p_2 : for every element $x \in S, x \in T$.

The universal quantifier *for every* now appears in the forward statement p_2, so you should consider applying specialization. To do so, follow the three steps in the discussion preceding this example. That is, first identify one particular object. In this case, that object is the element y in the statement p_1. Next, make sure that this special object satisfies the certain property in p_2 of being in the set S. But this is the case for y, as stated in p_1. The final step of specialization is to create, from p_2, the following new statement in the forward process:

p_3 : In particular, for $y \in S$, it follows that $y \in T$.

The statement p_3 is the something that happens in p_2 applied to the particular object y. Compare the statements p_3 and p_2.

To complete the proof, you must still obtain the last statement, q_2, in the backward process. You can use specialization again to accomplish this goal. Specifically, from the hypothesis that a is a lower bound for T, by Definition A.3, you know that

p_4 : for every element $x \in T, a \leq x$.

Recognizing the universal quantifier *for every* in the forward process, consider applying specialization. Once again, the particular object to use for specializing the general statement in p_4 is y. Before doing so, however, make sure that y satisfies the certain property in p_4 of being in T. But you do know that $y \in T$ from the statement p_3 in the forward process. The final step of specialization is to create, from p_4, the following new statement in the forward process:

p_5 : In particular, for $y \in T, a \leq y$.

The statement p_5 is the something that happens in p_4 applied to the particular object y. Compare the statements p_5 and p_4.

The proof is now complete because the forward statement p_5 is the same as the last backward statement, q_2.

In the written proof that follows, observe that no specific reference is made to the specialization or choose methods and that the symbol x is used throughout instead of y.

Proof.

To see that a is a lower bound for S, let $x \in S$. (The word *let* indicates that the choose method is used.) Because $S \subseteq T$, every element of S is an element of T. In particular, for $x \in S, x \in T$. (The words *in particular* indicate that specialization is used.) Also, from the hypothesis that a is a lower bound for T, by definition, for all $y \in T, a \leq y$. It now follows that for $x \in T, a \leq x$. (Here, the words *it now follows that* indicate that specialization is used.) It has therefore been shown that a is a lower bound for S, thus completing the proof. ■

A.7 Nested Quantifiers

You have just learned the construction, choose, induction, and specialization methods for working with statements containing quantifiers. Some statements, however, contain **nested quantifiers**, that is, more than one quantifier. For example, you might encounter a backward statement of the form:

> For all *objects X* with a *certain property P*, there is an *object Y* with a *certain property Q* such that *something happens*.

When working with statements that have nested quantifiers, proceed as follows:

> **Rule for Working with Nested Quantifiers**
> Process the quantifiers one at a time as they appear from left to right, using the construction, choose, induction, and specialization methods, as appropriate.

For instance, in the foregoing example, the first quantifier from the left is *for all*, thus, you should use the choose method to choose

p_1 : an object X with the certain property P,

for which you must show that

q_1 : there is an object Y with the certain property Q such that the something happens.

In trying to establish q_1, the appearance of the existential quantifier suggests that you now use the construction method to produce the object Y with the certain property Q and for which the something happens.

Alternatively, if a backward statement contains nested quantifiers in the form:

> There is an *object X* with a *certain property P* such that for all *objects Y* with a *certain property Q*, *something happens*,

then you should proceed as follows. Because the first quantifier from the left is *there is*, you should use the construction method to produce an object X with the certain property P. After doing so, you will also have to show that, for this object X,

q_1 : for all objects Y with a certain property Q, something happens.

In trying to establish q_1, the appearance of the universal quantifier suggests that you now use the choose method to choose

p_1 : an object Y with the certain property Q,

for which you must show that

q_2 : the something happens.

This rule of processing nested quantifiers from left to right is demonstrated with the following example.

THEOREM A.6

If f is the function defined by $f(x) = x^2$, then for every real number $y \geq 0$, there is a real number x such that $f(x) > y$.

Developing the Proof

The conclusion of this theorem contains nested quantifiers, the first of which is *for every*. Accordingly, you should first use the choose method to choose

p_1 : a real number $y \geq 0$,

for which you must show that

q_1 : there is a real number x such that $f(x) > y$.

The appearance of the existential quantifier in q_1 suggests that you now use the construction method to produce a real number x for which $f(x) = x^2 > y$. Thus, if you construct

$p_2 : x > \sqrt{y}$ (which you can do because $y \geq 0$),

it follows that

$p_3 : f(x) = x^2 > y$,

thus completing the proof.

Proof.

To see that the conclusion is true, let $y \geq 0$ be a real number. (The word *let* indicates that the choose method is used.) By taking x to be a real number with $x > \sqrt{y}$, it follows that

$f(x) = x^2 > y$.

Thus, it has been shown that there is a real number x for which $f(x) > y$, and so the proof is complete.   ■

Statements can contain any number of nested quantifiers. Just remember to process those quantifiers one at a time as they appear from left to right and to use the associated construction, choose, induction, and specialization methods.

A.8 The Contrapositive Method

Another technique for proving that

  p implies q

arises by considering the following associated *contrapositive statement*:

  $\neg q\ implies\ \neg p$,   that is,   *(not q) implies (not p)*.

Table A.4 in Appendix A.1 shows how to determine the truth of the contrapositive statement on the basis of the truth of the statements *p* and *q*. From Table A.4, you can see that *(not q) implies (not p)* is true under the same conditions as *p implies q*, that is, in all cases except when *p* is true and *q* is false.

  From this observation, you can conclude that to prove *p implies q* is true, you can just as well prove that the statement *(not q) implies (not p)* is true. The **contrapositive method**, then, is the forward-backward method applied to the the contrapositive statement. That is, to prove that

  p implies q

with the contrapositive method, you

1. Assume that the statement *not q* is true—that is, that *q* is false.

2. Show that the statement *not p* is true—that is, that *p* is false.

You accomplish this objective by working forward from *not q* and backward from *not p*. The use of the contrapositve method is demonstrated in the following example.

DEFINITION A.4

An integer *n* is **even** if and only if there is an integer *k* such that $n = 2k$. An integer *n* is **odd** if and only if there is an integer *k* such that $n = 2k + 1$.

THEOREM A.7

If *n* is an integer for which n^2 is even, then *n* is even.

Developing the Proof

According to the contrapositive method, you should assume that the conclusion is not true which, in this case, means you should assume that

  not q : *n* is not even,

or, equivalently,

  p_1 : *n* is odd.

You must work forward from p_1 to show that the hypothesis is not true which, in this case, means you should show that

    not p : n^2 is not even,

or, equivalently,

    q_1 : n^2 is odd.

The remainder of this proof is to work forward from p_1 and backward form q_1.

    Working backward from q_1, you have the key question: How can I show that an integer (namely, n^2) is odd? From Definition A.4, you obtain the following new statement to prove:

    q_2 : There is an integer k such that $n^2 = 2k + 1$.

The appearance of the existential quantifier in q_2 suggests that you now use the construction method (see Section A.3) to produce the value for the integer k. You get the value for k from the forward process.

    Turning to the forward process, you can apply Definition A.4 to the statement that n is odd in p_1 to obtain the following:

    p_2 : There is an integer m such that $n = 2m + 1$.

The idea is to use the known value of m in p_2 to construct the unknown value of k in q_2. Specifically, on squaring both sides of the equality in p_2 you obtain that

    p_3 : $n^2 = (2m + 1)^2 = 4m^2 + 4m + 1 = 2(2m^2 + 2m) + 1$.

You can see from p_3 that the desired value of k in q_2 is $k = 2m^2 + 2m$. You must still show that this value for k is correct, but this follows from p_3.

Proof.

By the contrapositive method, assume that n is not even, that is, that n is odd. It must be shown that n^2 is not even, that is, that n^2 is odd. However, because n is odd, by definition, there is an integer m such that $n = 2m + 1$. Letting $k = 2m^2 + 2m$, it follows that

$$n^2 = (2m + 1)^2 = 4m^2 + 4m + 1 = 2(2m^2 + 2m) + 1 = 2k + 1,$$

and so n^2 is odd, thus completing the proof.   ■

    As a general rule, you will find the contrapositive method to be successful under either of the following circumstances:

1. When the conclusion—or the last statement in the backward process—is one of two alternatives. (This is the case in Theorem A.7 where the conclusion that n is even is one of the two alternatives of n being odd or even.)

2. When the conclusion—or the last statement in the backward process—contains the key word *no* or *not*. For example, if x and y are real numbers and the conclusion of your theorem is to show that $x \neq y$, then you should consider the contrapositive method because, in so doing, you will assume that $x = y$, and this gives you some useful information.

A.9 The Contradiction Method

Another proof technique arises when you recall that the implication

> *p implies q*

is true in all cases except when p is true and q is false. With the **contradiction method** (also called an **indirect proof**), you establish the truth of *p implies q* by ruling out this one unfavorable case, as follows:

1. Assume that the one bad case *does* happen, that is, that the statement p is true and the statement q is false (*not q*).

2. Work forward from p and *not q* to reach a contradiction to some fact that you absolutely know is true.

The specific contradiction you obtain depends on the problem under consideration, but some examples of valid contradictions are: (1) $1 < 0$, (2) $x \neq x$ (where x is a real number), and (3) some statement s is both true and false at the same time. Unfortunately, you will generally not know what the specific contradiction is beforehand, so you cannot work backward in this technique. The contradiction method is demonstrated in the following example.

DEFINITION A.5

A real number r is **rational** if and only if there are integers a and b with $b \neq 0$ such that $r = a/b$.

THEOREM A.8

If r is a real number for which $r^2 = 2$, then r is not rational.

Developing the Proof

Proceeding by the contradiction method, you should assume that the hypothesis is true, so, assume that

> $p : r$ is a real number for which $r^2 = 2$.

You should also assume that the conclusion is false, so, assume that

> *not q* : r *is* rational.

The objective now is to work forward from the statements p and *not q* to reach some kind of contradiction. For example, working forward from the statement *not q* by using Definition A.5, you can state that

> p_1 : there are integers a and b with $b \neq 0$ such that $r = a/b$.

In this problem, the contradiction arises by noting that, in p_1, you can assume further that a and b have no common divisor, in other words,

> p_2 : there is no integer that divides both a and b.

The reason p_2 is true is that, if there were an integer k that divides both a and b, then you could cancel that value from both the numerator a and the denominator b, ultimately writing $r = a/b$ in lowest terms.

You can now reach a contradiction by showing that 2 divides both a and b, that is, that both a and b are even. This is accomplished by working forward from *not q* and p_1, as shown in the following proof.

Proof.

By contradiction, assume r *is* rational, so, there are integers a and b with $b \neq 0$ such that

$$r = \frac{a}{b}. \tag{A.4}$$

It can be assumed further that a and b are in lowest terms and therefore have no common divisor. A contradiction is reached by showing that both a and b are even and thus they have 2 as a common divisor.

From the hypothesis that $r^2 = 2$ and (A.4), you have that

$$r^2 = \left(\frac{a}{b}\right)^2 = \frac{a^2}{b^2} = 2. \tag{A.5}$$

Multiplying (A.5) through by b^2 yields that

$$a^2 = 2b^2, \tag{A.6}$$

from which you conclude that

$$a^2 \text{ is even.}$$

Now from Theorem A.7, because a^2 is even, you can say that

$$a \text{ is even.} \tag{A.7}$$

It remains to show that b is even. However, from (A.7) you know by definition that there is an integer k such that

$$a = 2k.$$

By substituting this value of $a = 2k$ in (A.6), you have that

$$2b^2 = a^2 = (2k)^2 = 4k^2. \tag{A.8}$$

On dividing (A.8) through by 2, you obtain

$$b^2 = 2k^2,$$

from which it follows that b^2 is even. But now, because b^2 is even, by Theorem A.7, b is also even. You now know that a is even [see (A.7)] and that b is even and these facts contradict the assumption that a and b have no common divisor. This contradiction completes the proof of the theorm. ■

In general, you will find that a proof by contradiction is effective when the conclusion of a theorem contains the key words *no* or *not* (as is the case in Theorem A.8).

A.10 Negations of Statements

To use either the contrapositive or contradiction method, you need to write the *negation of a statement p*, namely, *not p*. In some cases, finding the negation is straightforward. For example, if x is a real number, then the negation of the statement $x > 0$ is $x \leq 0$.

In other cases, however, there are certain rules you can follow to obtain the correct negation of the statement. For example, if the statement p contains the word *not* then, when you negate the statement, you remove the word *not*. The process of doing so is shown in the following example.

> ### EXAMPLE A.3 Negating the Word *Not*
>
> Suppose that a and b are integers. The negation of the statement
>
> $p : a$ does not divide b
>
> is
>
> *not p* : a *does* divide b.

When the statement p contains the word *and* in the form

$p : p_1$ *and* p_2 (where p_1 and p_2 are statements),

the negation contains the word *or*, as follows:

not p : (*not p_1*) *or* (*not p_2*).

The following example illustrates this rule.

> ### EXAMPLE A.4 Negating the Word *And*
>
> Suppose that x is a real number. The negation of the statement
>
> $p : -1 \leq x$ *and* $x \leq 1$,
>
> in which p_1 is "$-1 \leq x$" and p_2 is "$x \leq 1$," is
>
> *not p* : (*not* $-1 \leq x$) *or* (*not* $x \leq 1$),
>
> or equivalently,
>
> *not p* : $-1 > x$ *or* $x > 1$.

Analogously, the word *or* changes to *and*. As shown in the subsequent example, the negation of the statement

$p : p_1$ *or* p_2 (where p_1 and p_2 are statements),

is the statement

not p : (*not p_1*) *and* (*not p_2*).

EXAMPLE A.5 Negating the Word *Or*
Suppose that x is a real number. The negation of the statement

$$p : x > 4 \ \textit{or} \ x < 2,$$

in which p_1 is "$x > 4$" and p_2 is "$x < 2$," is ($not \ x > 4$) and ($not \ x < 2$), or equivalently,

$$not \ \ p : x \le 4 \ \textit{and} \ x \ge 2.$$

A more challenging situation arises when the statements contain quantifiers. As shown in the subsequent example, for the existential quantifier, the negation of the statement that

there is an *object* with a *certain property* such that *something happens*,

is the statement that

for all *objects* with the *certain property*, the *something does not happen*.

EXAMPLE A.6 Negating the Words *There Is*
Suppose that S is a set of real numbers. The negation of the statement

$$p : \text{There is an element } x \in S \text{ such that } x > 2$$

is

$$not \ \ p : \text{For all elements } x \in S, \ \ x \le 2.$$

Observe, in Example A.6, that the negation of a statement containing the existential quantifier results in a statement containing the universal quantifier. Also note that the something that happens changes but the certain property does not.

These same ideas apply when negating a statement containing the universal quantifier. Specifically, the negation of the statement that

for all *objects* with a *certain property*, *something happens*,

is the statement that

there is an *object* with the *certain property* such that the *something does not happen*,

as shown in the following example.

EXAMPLE A.7 Negating the Words *For All*
Suppose that S and T are sets. The negation of the statement that

$$p : \text{for all elements } x \in S, \ \ x \in T$$

is the statement that

$$not \ \ p : \text{there is an element } x \in S \text{ such that } x \notin T.$$

When a statement contains more than one quantifier, negate the quantifiers one at a time *from left to right* as they appear in the statement. For instance, the negation of

there is an *object* X with a *certain property* P such that for all *objects* Y with *another property* Q, *something happens*,

is the statement that

for all *objects* X with the *certain property* P, there is an *object* Y with the *property* Q such that the *something does not happen*.

This rule is demonstrated in the following example. Observe what happens as the word *not* is moved from left to right through the quantifiers.

EXAMPLE A.8 Negating the Nested Quantifiers *There Is* and *For All*
Suppose that S and T are sets of real numbers. The negation of the statement that

there is an element $x \in S$ such that for all elements $y \in T$, $y \le x$,

is the statement that

for all elements $x \in S$, it is not true that for all elements $y \in T$, $y \le x$,

or, by applying the word *not* to the next quantifier to the right,

for all elements $x \in S$, there is an element $y \in T$ such that $y > x$.

Likewise, the negation of the statement that

for all *objects* X with a *certain property* P, there is an *object* Y with *another property* Q such that *something happens*,

is the statement that

there is an *object* X with the *certain property* P such that for all *objects* Y with the *property* Q, the *something does not happen*.

This rule is demonstrated in the following example.

EXAMPLE A.9 Negating the Nested Quantifiers *For All* and *There Is*
Suppose that f is a real-valued function of one variable. The negation of the statement that

for all real numbers $y \ge 0$, there is a real number $x \ge 0$ such that $f(x) \ge y$,

is the statement that

there is a real number $y \ge 0$ such that it is not true that there is a real number $x \ge 0$ for which $f(x) \ge y$,

or, by applying the word *not* to the next quantifier to the right,

there is a real number $y \ge 0$ such that for all real numbers $x \ge 0$, $f(x) < y$.

The various rules for negating statements are summarized in Table A.7.

Statement	Negation
not p	*p*
p_1 *and* p_2	*(not* p_1*) or (not* p_2*)*
p_1 *or* p_2	*(not* p_1*) and (not* p_2*)*
There is an *object* with a *certain property* such that *something happens*.	For all *objects* with the *certain property*, the *something does not happen*.
For all *objects* with a *certain property*, *something happens*.	There is an *object* with the *certain property* such that the *something does not happen*.
Nested quantifiers	Process them from left to right.

Table A.7 *Summary of the Rules for Negating Statements.*

A.11 Either/Or Methods

Special proof techniques, referred to as the **either/or methods**, are available when the key words *either/or* arise in proving that

$$p \ implies \ q.$$

One technique is used when those key words appear in the forward process. The other technique is applicable when those key words appear in the backward process. Both techniques are described in what follows.

A.11.1 Proof by Cases

A **proof by cases** is used when a statement in the forward process contains the key words *either/or* in the form

p : either p_1 is true or p_2 is true   (where p_1 and p_2 are statements).

Because p is a statement in the forward process, you know that at least one of p_1 and p_2 is true—the only question is, which one? To cover both cases, you must do two proofs:

Case 1. First, assume that p_1 is true and prove that q is true. This is accomplished by working forward from p_1 and backward from q.

Case 2. Next, assume that p_2 is true and again prove that q is true. This is accomplished by working forward from p_2 and backward from q.

A proof by cases is demonstrated in the following example where, for two sets A and B,

$$A \cup B = \{x : x \in A \text{ or } x \in B\},$$

$$A \cap B = \{x : x \in A \text{ and } x \in B\},$$

$$A^c = \{x : x \notin A\}.$$

THEOREM A.9

If A and B are sets, then $(A \cap B)^c \subseteq A^c \cup B^c$.

Developing the Proof

The forward-backward method leads to the key question:

How can I show that a set (namely, $(A \cap B)^c$) is a subset of another set (namely, $A^c \cup B^c$)?

According to the definition, you must show that

q_1 : for every element $x \in (A \cap B)^c$, $x \in A^c \cup B^c$.

Recognizing the universal quantifier *for every* in the backward statement q_1, you should proceed by the choose method, whereby you choose

p_1 : an element $x \in (A \cap B)^c$,

for which you must show that

$q_2 : x \in A^c \cup B^c$.

Working forward from p_1 you know that

$p_2 : x \notin A \cap B$,

or, in other words, it is not true that $x \in A$ and $x \in B$. Applying the rules for negating a statement containing the word *and* (see Table A.7), you can state from p_2 that

p_3 : either $x \notin A$ or $x \notin B$.

Recognizing the key words *either/or* in the forward statement p_3, you should now proceed using a proof by cases, as follows.

Case 1: Assume that $x \notin A$. In this case, $x \in A^c$ and so $x \in A^c$ or $x \in B^c$. Thus, $x \in A^c \cup B^c$, which is precisely q_2, the last statement in the backward process.

Case 2: Assume that $x \notin B$. In this case, $x \in B^c$ and so $x \in B^c$ or $x \in A^c$. Thus, $x \in A^c \cup B^c$, which again is q_2.

Thus, in either case, $x \in A^c \cup B^c$, which shows that $(A \cap B)^c \subseteq A^c \cup B^c$ and this completes the proof.

In the written proof that follows, the words *without loss of generality* mean that only one of the two foregoing cases is presented and the other case is left for the reader to verify.

Proof.

To show that $(A \cap B)^c \subseteq A^c \cup B^c$, let $x \in (A \cap B)^c$. (The word *let* indicates that the choose method is used.) It follows that either $x \notin A$ or $x \notin B$. Assume, without loss of generality, that $x \notin A$. But then $x \in A^c$ and so $x \in A^c \cup B^c$. This means that $(A \cap B)^c \subseteq A^c \cup B^c$, completing the proof. ■

A.11.2 Proof by Elimination

The other either/or method, referred to as a **proof by elimination**, is used when the key words *either/or* arise in the backward process in the form:

$$q : \text{either } q_1 \text{ or } q_2 \text{ is true.}$$

In this situation, you must reach the conclusion that one of q_1 or q_2 is true. One way to do so is to assume that q_1 is not true. You must therefore show that q_2 is true. In other words, you use a proof by elimination to prove that

$$p \ implies \ (q_1 \ or \ q_2),$$

as follows:

1. Assume that p is true and also that q_1 is not true.

2. Use the assumptions in (1) to reach the conclusion that q_2 is true. In particular, work forward from p and *not* q_1 and backward from q_2.

A proof by elimination is illustrated in the following example.

THEOREM A.10

If x is a real number for which $x^2 + x - 6 \geq 0$, then $|x| \geq 2$.

Developing the Proof

Working backward from the conclusion, you can show that $|x| \geq 2$ by showing that

$$q_1 : \text{either } x \geq 2 \text{ or } x \leq -2.$$

Recognizing the key words *either/or* in the backward process, you should proceed with a proof by elimination. Accordingly, in addition to the hypothesis, you can assume that the statement $x \geq 2$ is not true, that is, that

$$p_2 : x < 2.$$

You must now show that

$$q_2 : x \leq -2.$$

Working forward from p_2, you have that

$$p_3 : x - 2 < 0,$$

and from the hypothesis, it follows that

$$p : x^2 + x - 6 = (x - 2)(x + 3) \geq 0.$$

On dividing both sides of p by the negative number $x - 2$ (see p_3), you have:

$$p_4 : x + 3 \leq 0,$$

from which it follows that

$$p_5 : x \leq -3,$$

and so q_2 is true, thus completing the proof.

Proof.
To show that $|x| \geq 2$, it is shown that either $x \geq 2$ or $x \leq -2$, so assume that $x < 2$. It now follows from the hypothesis that

$$x^2 + x - 6 = (x - 2)(x + 3) \geq 0.$$

Because $x < 2$, you have that $x + 3 \leq 0$, that is, that $x \leq -3$ and hence $x \leq -2$, as desired. ■

When trying to prove that

$$p \; implies \; (q_1 \; or \; q_2),$$

you can equally well use a proof by elimination in which you assume that p and *not* q_2 are true and then show that q_1 is true. Try this approach for proving Theorem A.10.

A.12 Uniqueness Methods

The final group of proof techniques presented here are the **uniqueness methods**, which are used when a statement has the following form:

There is a *unique object* with a *certain property* such that *something happens*.

When the key word *unique* (or equivalent words like *one and only one*) arise in the forward process, you use this knowledge as follows. If you encounter two objects, say, X and Y, both of which satisfy the certain property and the something happens, then, by the uniqueness property, you can conclude that $X = Y$, that is, that X and Y are the same.

In contrast, when you encounter the key word *unique* in the backward process, you can use one of two proof techniques to establish that there is a unique such object. These methods are described in what follows.

A.12.1 The Direct Uniqueness Method

To prove that

there is a *unique object* with a *certain property* such that *something happens*,

use the **direct uniqueness method**, as follows:

1. First establish that there is an object, say, X, with the certain property and for which the something happens. (You can use the construction method described in Section A.3 to do so.)

2. Assume that Y is also an object with the certain property and for which the something happens.

3. Work forward from the properties that X and Y satisfy to show that $X = Y$, that is, that X and Y are the same. You can also work backward from the statement that $X = Y$; however, the key question depends on the specific objects X and Y. For example, trying to show that two real numbers are equal is different from trying to show that two functions are equal, which is different from trying to show that two sets are equal.

On completing these three steps, you can conclude that the object X is unique because you have shown that any other object Y with the same properties as X is, in fact, the same as X. The direct uniqueness method is demonstrated in the following example to show that two sets are equal.

THEOREM A.11

If

$$C^1 = \{\text{points } (x, y) : x^2 + y^2 = 4\} \quad \text{and}$$
$$C^2 = \{\text{points } (x, y) : (x - 3)^2 + y^2 = 1\},$$

then the circles C^1 and C^2 intersect in exactly one point.

Developing the Proof

The appearance of the key word *exactly one* in the conclusion suggests using a uniqueness method. According to the direct uniqueness method, you must first construct a point (x, y) in the intersection of both circles. In this case, it is easy to see that $(2, 0)$ is such a point.

Having constructed one such point, you should now assume that

$p_1 : (a, b)$ is also a point in the intersection of C^1 and C^2,

from which it follows that

$p_2 : a^2 + b^2 = 4$

and

$p_3 : (a - 3)^2 + b^2 = a^2 - 6a + 9 + b^2 = 1.$

Working forward from p_2 and p_3, you must show that

$q_1 : (a, b) = (2, 0).$

In this case, that means showing that two pairs of real numbers are equal, that is,

$q_2 : a = 2 \quad \text{and} \quad b = 0.$

To establish q_2, replace the expression $a^2 + b^2$ in p_3 with the value of 4 from p_2 to obtain

$p_4 : 13 - 6a = 1,$

from which it follows that

$$p_5 : a = 2.$$

By replacing a in p_2 with its known value of 2, you then have that

$$p_6 : b = 0.$$

Thus, from p_5 and p_6, you know that q_2 is true and so you have established the uniqueness of the point $(2, 0)$ in the intersection of the two circles C^1 and C^2. This is because, by the direct uniqueness method, you have shown that if (a, b) is *any* point in the intersection, then $(a, b) = (2, 0)$.

Proof.
You can easily verify that the point $(2, 0)$ is in the intersection of the two circles C^1 and C^2. For the uniqueness, assume that (a, b) is also in the intersection, so

$$a^2 + b^2 = 4 \tag{A.9}$$

and

$$(a - 3)^2 + b^2 = a^2 - 6a + 9 + b^2 = 1. \tag{A.10}$$

Substituting (A.9) in (A.10) yields that $13 - 6a = 1$, or equivalently, that $a = 2$. Replacing this value of a in (A.9) yields that $b = 0$. Thus, if (a, b) is a point in the intersection of the two circles C^1 and C^2, then $(a, b) = (2, 0)$, and so the point $(2, 0)$ is unique. ∎

A.12.2 The Indirect Uniqueness Method

To prove that

there is a *unique object* with a *certain property* such that *something happens*,

you can also use the **indirect uniqueness method**, as follows:

1. First establish that there is an object, say, X, with the certain property and for which the something happens. (You can use the construction method described in Section A.3 to do so.)

2. Assume that Y is a *different* object with the certain property and for which the something happens.

3. You rule out the existence of the object Y by working forward from the properties that X and Y satisfy, especially the fact that $X \neq Y$, to reach a contradiction.

The indirect uniqueness method is demonstrated in the following example.

THEOREM A.12

There is a unique real number $x > 0$ such that $x^2 = 2$.

Developing the Proof

According to the indirect uniqueness method, you must first construct a real number $x > 0$ such that $x^2 = 2$. Doing so involves quite a bit of work and is omitted so as to focus on the issue of uniqueness. So suppose that you have constructed a real number x such that

$$p_1 : x > 0 \quad \text{and} \quad x^2 = 2.$$

Having constructed one such real number, to use the indirect uniqueness method, you should now assume that y is also a real number for which

$$p_2 : y > 0 \quad \text{and} \quad y^2 = 2,$$

and also that

$$p_3 : y \neq x.$$

Now work forward from p_1, p_2, and p_3 to reach a contradiction which, in this case, is that $x < 0$ (contradicting p_1). You obtain this contradiction by combining p_1 and p_2 to obtain

$$p_4 : x^2 = y^2,$$

from which it follows that

$$p_5 : x^2 - y^2 = (x - y)(x + y) = 0.$$

From p_3, you know that $y \neq x$ and so $x - y \neq 0$. You can therefore divide both sides of p_5 by $x - y$ to obtain that

$$p_6 : x + y = 0,$$

or equivalently, that

$$p_7 : x = -y.$$

But from p_2, you know that $y > 0$ so $-y < 0$, hence, p_7 provides the contradiction that $x = -y < 0$.

Proof.

The fact that there is a real number $x > 0$ such that $x^2 = 2$ is omitted, so assume that $x > 0$ is a real number such that $x^2 = 2$. To prove the uniqueness of x, suppose that $y > 0$ is also a real number for which $y^2 = 2$ and further that $y \neq x$. It then follows that $x^2 = y^2$, so

$$x^2 - y^2 = (x - y)(x + y) = 0. \tag{A.11}$$

Because $y \neq x$, you can divide (A.11) through by $x - y$ to conclude that $x + y = 0$, that is, that $x = -y < 0$. But this contradicts the fact that $x > 0$, hence showing that x is the unique positive real number for which $x^2 = 2$, thus completing the proof. ■

You have now learned various techniques for proving that the implication p *implies* q is true. You have also learned to select a proof technique on the basis of certain key words that appear in the hypothesis and conclusion of the implication. When no key words are present, try the forward-backward method. Many additional details of these techniques are presented in the book *How to Read and Do Proofs* by D. Solow (John Wiley & Sons, 1990).

Mathematical Thinking Processes

I n Appendix A, you learned the thinking processes involved in doing proofs. In this appendix, a summary of various other mathematical thinking processes you have seen throughout the book is presented. A description of each one is given together with a brief explanation of how and when to use the technique. Advantages and disadvantages are also discussed.

B.1 Closed-form and Numerical-method Solutions

A mathematical problem consists of some known *data*—that is, information—that you can use to obtain the solution. For example, if you want to find the distance between two points in the plane, then the data are the coordinates of the two points. The solution itself takes one of two forms:

1. A *closed-form solution*, which is a solution obtained from the data by a simple rule or formula.

2. A *numerical-method solution*, which is a solution obtained from the data by performing a sequence of computations, usually involving repetition.

Mathematics is also used to make an existing solution procedure more efficient. For example, if a numerical method for finding the solution to a particular problem requires too much computational effort, then you might try using mathematics to develop an algorithm that provides an approximate solution in a reasonable amount of time.

B.2 Unification

In the spirit of using mathematics to increase efficiency, *unification* is the process of combining two or more problems (ideas, concepts, theories, and so on) into a single class of problems whose solution procedure can be used to solve any specific problem—that is, *spe-*

cial case—in that class. For example, consider the following two formulas for computing the distance between the two points (x_1, x_2) and (y_1, y_2) in the plane:

$$|x_1 - y_1| + |x_2 - y_2| \tag{B.1}$$

$$\sqrt{(x_1 - y_1)^2 + (x_2 - y_2)^2} \tag{B.2}$$

If you can recognize the similarity between the two foregoing formulas, then, you can unify them into the following single formula, in which p is a positive integer:

$$(|x_1 - y_1|^p + |x_2 - y_2|^p)^{1/p} \quad \text{(unification)} \tag{B.3}$$

Observe that (B.1) is a special case of (B.3) obtained by substituting $p = 1$ in (B.3). Likewise, (B.2) is a special case obtained by substituting $p = 2$.

Other examples of unification you have seen in this book include:

1. The unification of the various different representations of a plane into the single expression $a_1 x_1 + \cdots + a_n x_n = b$ (see Section 1.3.3).

2. The representation of many individual systems of linear equations with the unified notation $A\mathbf{x} = \mathbf{b}$.

3. The unification of n-vectors and matrices into the single framework of a vector space (see Section 4.1.1).

A summary of how to apply unification and the associated advantages and disadvantages follows.

How to Apply Unification

Step 1. Identify two or more problems whose data and solution you know.

Step 2. By recognizing similarities between the problems in Step 1, create a single common problem with its data that includes all of the specific problems as special cases.

Step 3. By using the solution procedures for the special cases in Step 1 as a guide, develop a solution procedure for the common problem identified in Step 2.

Step 4. Verify that applying the solution procedure for the common problem to each of the specific problems in Step 1 results in the solution to those special cases.

Step 5. Apply the solution procedure for the common problem to any special case, including any new ones you encounter.

Advantages of Unification

1. Unifying many individual problems into one common problem allows you to develop a single solution procedure that you can then use to solve all problems in that class, even new ones you have not yet encountered.

2. Unifying problems into one common problem makes it easier to understand and to study the properties of all problems in that class.

Disadvantage of Unification

1. When you study the common problem, you lose sight of the individual problems in the class. Thus, for example, when trying to solve the common problem, you do not use any of the special properties that pertain to one specific problem in the class.

B.3 Generalization

Generalization is the process of creating, from an original mathematical concept (problem, definition, theorem, and so on) a more general concept (problem, definition, theorem, and so on), that includes not only the original one, but many other new, and perhaps different, ones as special cases. Each special case is obtained from the generalization by an appropriate substitution of symbols. For example, recall the following formula in (B.3) for computing the distance between two points in a plane:

$$(|x_1 - y_1|^p + |x_2 - y_2|^p)^{1/p} \tag{B.4}$$

One generalization of (B.4) is to create a distance function, d, whose specific form is not specified. That is, for points (x_1, x_2) and (y_1, y_2), let

$$d((x_1, x_2), (y_1, y_2)) = \text{the distance from } (x_1, x_2) \text{ to } (y_1, y_2) \tag{B.5}$$

Then (B.4) is a special case of (B.5) in which d has the specific form in (B.4). Note, however, that (B.5) allows for other forms for computing the distance and thus, is more general than (B.4).

It is often the case that one generalization leads to another, thus giving rise to a *sequential generalization*. For example, a further generalization of (B.5) arises by working with two n-vectors, **x** and **y**, rather than two points in the plane. Thus, you now have

$$d(\mathbf{x}, \mathbf{y}) = \text{the distance from the } n\text{-vector } \mathbf{x} \text{ to the } n\text{-vector } \mathbf{y} \tag{B.6}$$

Additional examples of sequential generalization that you have seen in this book include:

1. The problem of solving a single linear equation in a single unknown, to the problem of solving a system of n linear equations in n unknowns, to the problem of solving a system of m linear equations in n unknowns (see Section 2.3).

2. The concept of a real number extended first to an n-vector and then to a matrix.

3. The problem of projecting an n-vector **v** onto another n-vector, to the problem of projecting **v** onto a line through the origin, to the problem of projecting **v** onto a subspace of R^n (see Section 7.2).

A summary of how to apply generalization and the associated advantages and disadvantages follows.

How to Apply Generalization

Step 1. Identify a specific problem whose data and solution you know.

Step 2. Create a more general problem with its larger set of data.

Step 3. Verify that the specific problem is a special case of the general problem. You do this by using the data for the specific problem in Step 1 to create appropriate data for the general problem in Step 2 that, when substituted in the general problem, results in the specific problem.

Step 4. Develop a solution procedure for the general problem identified in Step 2.

Step 5. Verify that applying the solution procedure for the general problem in Step 2 to the specific problem in Step 1 results in the corresponding solution to that special case.

Step 6. Apply the solution procedure for the general problem to any special case of interest.

Advantages of Generalization

1. Generalization results in a class of problems and a single solution procedure that you can use to solve any problem in that class, including all the special cases, and even new problems you have not yet encountered.

2. Generalization allows you to work with concepts that you cannot visualize, such as n-vectors, when $n > 3$.

Disadvantages of Generalization

1. You must develop a solution for the general problem that may be quite different from the solutions for the special cases and thus requires significant effort to create.

2. The solution to the general problem may be less efficient computationally than solving the special cases directly. This is because the general problem has more data than the specific problem.

B.4 Abstraction and Axiomatic Systems

Abstraction is the process of thinking in terms of general objects rather than specific items. Doing so allows you to create a problem-solving framework that is even more general than that obtained by applying unification and generalization.

The first step of abstraction is to create an *abstract system*, consisting of a set of objects together with one or more ways to perform operations on those objects. For example, recall the foregoing example of finding the distance between two n-vectors, as given by the following distance function, d, in (B.6):

$$d(\mathbf{x}, \mathbf{y}) = \text{the distance from the } n\text{-vector } \mathbf{x} \text{ to the } n\text{-vector } \mathbf{y} \qquad (\text{B.7})$$

To apply abstraction, think of $\mathbf{x}$ and $\mathbf{y}$ as general objects, x and y, belonging to a set, say, S.

Then (B.7) becomes:

$$d(x, y) = \text{the distance from the object } x \text{ to the object } y \tag{B.8}$$

The pair (S, d) constitutes an abstract system in which the function d is an operation on pairs of elements in the set S.

Abstraction greatly expands your ability to measure distance, in this case. For example, with (B.8), you can now consider measuring the distance between any of the following:

1. The distance between two points in n dimensions (rather than just in the plane).

2. The distance between two sets of real numbers.

3. The distance between two functions.

4. The distance between two computer programs.

Of course, in each of these cases, you would need to develop an appropriate method for defining and computing the distance between two specific objects. Observe that you cannot use the formulas in (B.1), (B.2), and (B.3) because they apply only to points in the plane, and not to general objects.

The distance between two objects would also have to satisfy certain properties, such as being nonnegative. Here you see one of the disadvantages of abstraction, namely, that you lose properties of the special cases that give rise to the abstract system in the first place.

One way to overcome this loss is to include, with the abstract system, desirable properties, in the form of *axioms*, that are assumed to hold. Which axioms you choose to include depend on what properties of the special cases you want to study, the kind of results you eventually want to obtain about the abstract system, and more. In any event, the abstract system together with the selected axioms constitute an *axiomatic system*. For example, for the foregoing abstract system (S, d), you might create an axiomatic system that includes the following axioms to ensure that d has the desirable properties of a distance function:

1. For all $x, y \in S, d(x, y) \geq 0$.

2. For all $x, y \in S, d(x, y) = 0$ if and only if $x = y$.

3. For all $x, y \in S, d(x, y) = d(y, x)$.

4. For all $x, y, z \in S, d(x, z) \leq d(x, y) + d(y, z)$.

Other examples of axiomatic systems you have seen in this book include:

1. A vector space (see Section 4.1).

2. An inner product space (see Section 7.4.1).

A summary of how to use abstraction to develop an axiomatic system and the associated advantages and disadvantages follows.

How to Develop an Axiomatic System

Step 1. Identify one or more special cases that you want to unify and study.

Step 2. Create an abstract system by replacing specific items with general objects belonging to a set and identifying operations of interest on the general objects.

Step 3. Identify the axioms you want the abstract system to satisfy.

Step 4. By an appropriate interpretation of the general objects and the associated operations, verify that each of the special cases satisfies the axioms listed in Step 3.

Step 5. Derive results—in the form of theorems and proofs—pertaining to the axiomatic system that you can then apply to any special case, including new ones not identified in Step 1.

Advantage of an Axiomatic System

1. An axiomatic system allows you to expand greatly the corresponding problem to include not only the special cases, but many other related problems obtained by interpreting the meaning of the general objects appropriately in the context of the special cases.

Disadvantages of an Axiomatic System

1. The process of abstraction causes you to lose sight of the specific items to the point where the general objects may have little meaning you can relate to.

2. You can become so abstract that it is difficult to obtain meaningful results about the abstract system.

B.5 Mathematical Skills

A number of mathematical skills have been presented throughout the book. A summary of these skills is given now.

B.5.1 Identifying Similarities and Differences

To unify two problems into a single framework, such as an axiomatic system, you must first identify common features of the problems. Any differentiating features you encounter must eventually be eliminated so that only common properties remain. More generally, the technique of *identifying similarities and differences* is the process of comparing and contrasting two or more mathematical concepts (problems, definitions, theorems, and so on) with the objective of gaining a deeper understanding of their relationship. For example, identifying differences between the integers, the rational numbers, and the real numbers has led to significant mathematical developments and axiomatic systems.

B.5.2 Visualization and Translating to Symbolic Form

Visualization is the process of creating an image that captures the essential features of a mathematical concept so that you can think about and work with that concept more easily. For example, the images of *n*-vectors as a point and as an arrow provide visual ways to think about vectors when solving problems.

However, the solution to a problem must eventually be written in a *symbolic form*.

You should therefore use the technique of *translating visual images to symbolic form* to convert a visual image of a mathematical concept to a formal written form in terms of the vocabulary and syntax of the language of mathematics. The advantages of a symbolic form include:

1. The ability to work with and to perform symbolic operations on the objects.

2. The ability to develop solution procedures—in the form algorithms that computers can perform—for solving a problem.

B.5.3 Creating Mathematical Definitions

A *definition* is a name given to a collection of objects that satisfy a desirable property (for example, linearly independent vectors). The art of creating a correct definition involves looking at many examples of objects satisfying the desirable property and other examples of objects that do not satisfy the property. The definition is then created by performing the following steps:

Steps for Creating a Mathematical Definition

Step 1. Identify a common property shared by all desirable objects you are considering.

Step 2. Choose a representative name for the common desirable property.

Step 3. Translate the visual image of this property to a symbolic form that then becomes the definition.

Step 4. Verify that the resulting definition works in two ways: (1) all desirable objects should satisfy the stated property and (2) no other object should satisfy the stated property.

The thinking processes presented in this appendix arise not only in linear algebra, but also in all subjects involving advanced mathematics. These techniques provide systematic problem-solving tools that you can use over and over again.

Solutions to Selected Exercises

SOLUTIONS TO SELECTED EXERCISES IN CHAPTER 1

Solutions to Exercises in Section 1.1

1. (a) $\mathbf{x} = (-3, 4)$ miles. The dimension of $\mathbf{x}$ is two.

   (b) $\mathbf{y} = (5 \text{ miles}, 12 \text{ minutes})$. This is because the distance from the library to city hall is $\sqrt{(-3)^2 + (4)^2} = 5$, so

   $$\mathbf{y} = (5, (5/25)60) = (5, 12)$$

   The dimension of $\mathbf{y}$ is two.

3. (a) $\mathbf{x} = (5, 5, 5)$ inches. The dimension of $\mathbf{x}$ is three.

   (b) Let $\mathbf{x} = (x_1, \ldots, x_n)$, in which

   $$x_i = \begin{cases} 1, & \text{if the result on flip } i \text{ is heads} \\ 0, & \text{if the result on flip } i \text{ is tails} \end{cases}$$

5. (a) $\mathbf{p} = (3/6, 2/6, 1/6) = (1/2, 1/3, 1/6)$.

   (b) The n-vector $\mathbf{x} = (x_1, \ldots, x_n)$ is a *probability vector* if and only if (1) $x_1 + \cdots + x_n = 1$ and (2) for each $i = 1, \ldots, n$, $x_i \geq 0$.

7. (a) The figure is:

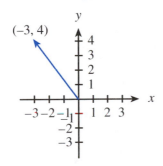

   (b) The figure is:

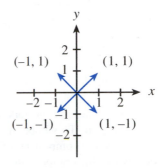

9. (a) The vectors $1 + 2i$ and $2 - i$ are perpendicular:

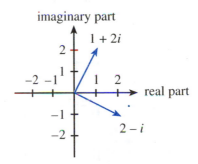

   (b) The vector is $(2, 4)$:

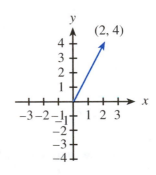

11. (a) $(1, 0) = (0, 1)$ is false.
    (b) $(0, 1, 0) \neq \mathbf{k}$ is true since $\mathbf{k} = (0, 0, 1)$.
    (c) $(1 - 1, 2 + (-2)) = \mathbf{0}$ is true because both vectors are equivalent to $(0, 0)$.

13. (a) $(-1, 0) \leq (0, 0)$ is true because $-1 \leq 0$ and $0 \leq 0$. In contrast, $(-1, 0)$ is not less than $(0, 0)$ because 0 is not less than 0.
    (b) The negation of the statement $\mathbf{u} \leq \mathbf{v}$ is "there is an integer i with $1 \leq i \leq n$ such that $u_i > v_i$."
    (c) For $\mathbf{u} = (1, 2, -1)$ and $\mathbf{v} = (3, -1, 0)$, it is not true that $\mathbf{u} \leq \mathbf{v}$ because $u_2 = 2 > -1 = v_2$. It is not true that $\mathbf{u} \geq \mathbf{v}$ because $u_1 = 1 < 3 = v_1$.
    (d) It is not true that for n-vectors $\mathbf{u}$ and $\mathbf{v}$, $\mathbf{u} \leq \mathbf{v}$ or $\mathbf{v} \leq \mathbf{u}$. This is because it can happen that some components of $\mathbf{u}$ are $<$ the corresponding components of $\mathbf{v}$ while other components of $\mathbf{u}$ are $>$ the corresponding components of $\mathbf{v}$.
    (e) The following shaded region is $\{(v_1, v_2) : v_1 \leq u_1 \text{ and } v_2 \leq u_2\}$:

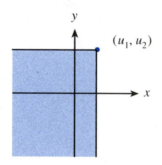

15. (a) $(2667.25 \text{cm}^2, 9134.52 \text{cm}^3)$ is the vector for the surface area and the volume of the cube. This is because the dimension of the box in cm is given by the vector $\mathbf{x} = (54/2.54, 66/2.54, 42/2.54)$, so

    $$\text{surface area} = 2(x_1 x_2 + x_1 x_3 + x_2 x_3)$$
    $$= 2667.25 \text{cm}^2$$
    $$\text{volume} = x_1 x_2 x_3 = 9134.52 \text{cm}^3$$

    (b) The components of the vector $(-1, 0, 3)$ when the origin is moved to $(2, 4, 1)$ are $(-1 - 2, 0 - 4, 3 - 1) = (-3, -4, 2)$.

Solutions to Exercises in Section 1.2

1. When $\mathbf{u} = (1, 0)$ and $\mathbf{v} = (0, 2)$,
   (a) $\|\mathbf{u}\| = \sqrt{1^2 + 0^2} = 1$ and $\|\mathbf{v}\| = \sqrt{0^2 + 2^2} = 2$.
   (b) $d(\mathbf{u}, \mathbf{v}) = \sqrt{(1 - 0)^2 + (0 - 2)^2} = \sqrt{5}$.
   (c) $-\mathbf{u} = (-1, 0)$.
   (d) $2\mathbf{u} - \mathbf{v} = (2, 0) - (0, 2) = (2, -2)$.
   (e) $\mathbf{u} \cdot \mathbf{v} = (1, 0) \cdot (0, 2) = 1(0) + 0(2) = 0$. The angle θ between $\mathbf{u}$ and $\mathbf{v}$ is 90 degrees or $\pi/2$ radians. This is because

   $$\theta = \arccos\left(\frac{\mathbf{u} \cdot \mathbf{v}}{\|\mathbf{u}\| \|\mathbf{v}\|}\right)$$
   $$= \arccos(0)$$
   $$= 90 \text{ degrees} = \frac{\pi}{2} \text{ radians}$$

3. When $\mathbf{u} = (1, -1)$ and $\mathbf{v} = (-1, 0)$,
   (a) $\mathbf{u} + 2\mathbf{v} = (1, -1) + (-2, 0) = (-1, -1)$.
   (b) $\|2\mathbf{u}\| = \|(2, -2)\| = \sqrt{2^2 + (-2)^2} = \sqrt{8} = 2\sqrt{2}$.
   (c) $\mathbf{u} \cdot \mathbf{v} = (1, -1) \cdot (-1, 0) = 1(-1) + (-1)(0) = -1$. The angle θ between $\mathbf{u}$ and $\mathbf{v}$ is 135 degrees or $3\pi/4$ radians. This is because $\|\mathbf{u}\| = \sqrt{2}$ and $\|\mathbf{v}\| = 1$, so

   $$\theta = \arccos\left(\frac{\mathbf{u} \cdot \mathbf{v}}{\|\mathbf{u}\| \|\mathbf{v}\|}\right)$$
   $$= \arccos\left(\frac{-1}{\sqrt{2}}\right)$$
   $$= 135 \text{ degrees} = \frac{3\pi}{4} \text{ radians}$$

5. When $\mathbf{u} = (1, 0, -1)$ and $\mathbf{v} = (0, 2, 0)$,
   (a) $\|\mathbf{u}\| = \sqrt{1^2 + 0^2 + (-1)^2} = \sqrt{2}$ and $\|\mathbf{v}\| = \sqrt{0^2 + 2^2 + 0^2} = 2$.
   (b) $d(\mathbf{u}, \mathbf{v}) = \sqrt{6} = \sqrt{(1 - 0)^2 + (0 - 2)^2 + (-1 - 0)^2}$.
   (c) $-3\mathbf{v} = (0, -6, 0)$.
   (d) $2\mathbf{u} - \mathbf{v} = (2, 0, -2) - (0, 2, 0) = (2, -2, -2)$.
   (e) $\mathbf{u} \cdot \mathbf{v} = (1, 0, -1) \cdot (0, 2, 0) = 1(0) + 0(2) + (-1)(0) = 0$. The angle θ between $\mathbf{u}$ and $\mathbf{v}$ is 90 degrees or $\pi/2$ radians. This

is because

$$\theta = \arccos\left(\frac{\mathbf{u} \cdot \mathbf{v}}{\|\mathbf{u}\|\|\mathbf{v}\|}\right)$$

$$= \arccos(0)$$

$$= 90 \text{ degrees} = \frac{\pi}{2} \text{ radians}$$

(f) $\mathbf{u} \times \mathbf{v} = (2, 0, 2)$. The angle θ between $\mathbf{u}$ and $\mathbf{v}$ is 90 degrees. This is because $\|\mathbf{u} \times \mathbf{v}\| = 2\sqrt{2}, \|\mathbf{u}\| = \sqrt{2},$ and $\|\mathbf{v}\| = 2$ [see part(a)], so

$$\theta = \arcsin\left(\frac{\|\mathbf{u} \times \mathbf{v}\|}{\|\mathbf{u}\|\|\mathbf{v}\|}\right)$$

$$= \arcsin\left(\frac{2\sqrt{2}}{2\sqrt{2}}\right) = \arcsin(1)$$

$$= 90 \text{ degrees} = \frac{\pi}{2}$$

7. When $\mathbf{u} = (-1, 1, 0)$ and $\mathbf{v} = (0, 2, 0)$,

(a) $\mathbf{u} + 2\mathbf{v} = (-1, 1, 0) + (0, 4, 0) = (-1, 5, 0)$.

(b) $\|\mathbf{u}\| = \sqrt{(-1)^2 + 1^2 + 0^2} = \sqrt{2}$.

(c) $\mathbf{u} \cdot \mathbf{v} = (-1, 1, 0) \cdot (0, 2, 0) = (-1)(0) + 1(2) + 0(0) = 2$. The angle θ between $\mathbf{u}$ and $\mathbf{v}$ is 45 degrees. This is because

$$\theta = \arccos\left(\frac{\mathbf{u} \cdot \mathbf{v}}{\|\mathbf{u}\|\|\mathbf{v}\|}\right)$$

$$= \arccos\left(\frac{2}{2\sqrt{2}}\right) = 45 \text{ degrees}$$

(d) $\mathbf{u} \times \mathbf{v} = (0, 0, -2)$. The angle θ between $\mathbf{u}$ and $\mathbf{v}$ is 45 degrees. This is because $\|\mathbf{u} \times \mathbf{v}\| = 2, \|\mathbf{u}\| = \sqrt{2}$ [(see part (b)], and $\|\mathbf{v}\| = 2$, so

$$\theta = \arcsin\left(\frac{\|\mathbf{u} \times \mathbf{v}\|}{\|\mathbf{u}\|\|\mathbf{v}\|}\right)$$

$$= \arcsin\left(\frac{2}{2\sqrt{2}}\right) = \arcsin\left(\frac{1}{\sqrt{2}}\right)$$

$$= 45 \text{ degrees}$$

9. (a) The following half line through the origin in the direction $\mathbf{d}$ is $\{t\mathbf{d} : t \geq 0\}$:

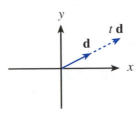

(b) The following region bounded by the half lines containing $\mathbf{u}$ and $\mathbf{v}$ represents $\{s\mathbf{u} + t\mathbf{v} : s, t \geq 0\}$:

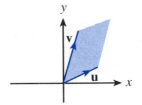

11. A similarity is that both a closed-form and a numerical-method solution use the given data to obtain a solution to a problem. A difference is that a closed-form solution is expressed as a simple formula in terms of the data. In contrast, a numerical-method solution requires performing the steps of an algorithm.

13. (a) Let the following vector be the original cost, in yen, of the three foods:

$$\mathbf{x} = (183.75, 168.00, 367.50)$$

Then, at the exchange rate of 105 yen per $, the costs of the these foods in $ is:

$$\frac{1}{105}\mathbf{x} = (1.75, 1.60, 3.50)$$

(b) Let the components of the following vector $\mathbf{x}$ be the speed, in miles per hour, for each of the three segments traveled and the components of $\mathbf{y}$ be the number of minutes driven on each segment:

$$\mathbf{x} = (50, 60, 55)$$

$$\mathbf{y} = (240, 180, 300)$$

Then, by converting the driving time to hours, you can compute the total distance in miles, as follows:

$$\mathbf{x} \cdot \frac{1}{60}\mathbf{y} = (50, 60, 55) \cdot (4, 3, 5) = 655$$

15. (a) Given the list of measurements in inches, $\mathbf{d} = (d_1, \ldots, d_n)$, the list of those measurements in centimeters is 2.54$\mathbf{d}$.
    (b) You have that 2.54$\mathbf{d}$ = 2.54(2, 3, 5) = (5.08, 7.62, 12.70).

17. The units of $\mathbf{l} = \mathbf{r} \times \mathbf{p}$ are kg-m^2 / second which, from the formula for the cross product, are the units of $\mathbf{r}$ times those of $\mathbf{p}$, namely:

$$m \times \frac{\text{kg-m}}{\text{second}} = \frac{\text{kg-m}^2}{\text{second}}$$

19. Area = $\|(1, 0, -1) \times (0, 1, 1)\|$ = $\|(1, -1, 1)\|$ = $\sqrt{3}$ = 1.7321.

21. (a) dim($\mathbf{u}$) is a unary operation that results in a number.
    (b) sum($\mathbf{u}$) is a unary operation that results in a number.
    (c) times($\mathbf{u}$, $\mathbf{v}$) is a binary operation that results in an n-vector.

23. In each part, it is necessary to show that two n-vectors are equal. By definition, this means that it is necessary to show that each component i of one vector is equal to the corresponding component i of the second vector.
    (a) $[(\mathbf{u} + \mathbf{v}) + \mathbf{w}]_i = (u_i + v_i) + w_i = u_i + (v_i + w_i) = [\mathbf{u} + (\mathbf{v} + \mathbf{w})]_i$.
    (b) $(\mathbf{u} + \mathbf{0})_i = u_i + 0 = u_i = \mathbf{u}_i$. Similarly, $(\mathbf{0} + \mathbf{u})_i = u_i$.
    (c) $[\mathbf{u} + (-\mathbf{u})]_i = u_i + (-u_i) = 0 = \mathbf{0}_i$.
    (d) $[s(\mathbf{u} + \mathbf{v})]_i = s(u_i + v_i) = (su_i) + (sv_i) = (s\mathbf{u})_i + (s\mathbf{v})_i$.
    (e) $[(s+t)\mathbf{u}]_i = (s+t)u_i = (su_i) + (tu_i) = (s\mathbf{u})_i + (t\mathbf{u})_i$.
    (f) $(1\mathbf{u})_i = 1u_i = u_i = \mathbf{u}_i$.
    (g) $(0\mathbf{u})_i = 0u_i = 0 = \mathbf{0}_i$.

25. Let $\mathbf{u} = (u_1, u_2, u_3)$, $\mathbf{v} = (v_1, v_2, v_3)$, and $\mathbf{w} = (w_1, w_2, w_3)$ throughout.

(a) $\mathbf{u} \times (\mathbf{v} + \mathbf{w})$
$$= (u_1, u_2, u_3) \times (v_1 + w_1, v_2 + w_2, v_3 + w_3)$$
$$= (u_2(v_3 + w_3) - u_3(v_2 + w_2),$$
$$u_3(v_1 + w_1) - u_1(v_3 + w_3),$$
$$u_1(v_2 + w_2) - u_2(v_1 + w_1))$$
$$= (u_2 v_3 - u_3 v_2, u_3 v_1 - u_1 v_3,$$
$$u_1 v_2 - u_2 v_1) + (u_2 w_3 - u_3 w_2,$$
$$u_3 w_1 - u_1 w_3, u_1 w_2 - u_2 w_1)$$
$$= (\mathbf{u} \times \mathbf{v}) + (\mathbf{u} \times \mathbf{w})$$

(b) $(\mathbf{u} + \mathbf{v}) \times \mathbf{w}$
$$= [(u_1, u_2, u_3) + (v_1, v_2, v_3)] \times$$
$$(w_1, w_2, w_3)$$
$$= (u_1 + v_1, u_2 + v_2, u_3 + v_3) \times$$
$$(w_1, w_2, w_3)$$
$$= [(u_2 + v_2)w_3 - (u_3 + v_3)w_2,$$
$$(u_3 + v_3)w_1 - (u_1 + v_1)w_3,$$
$$(u_1 + v_1)w_2 - (u_2 + v_2)w_1]$$
$$= (u_2 w_3 - u_3 w_2, u_3 w_1 - u_1 w_3,$$
$$u_1 w_2 - u_2 w_1) + (v_2 w_3 - v_3 w_2,$$
$$v_3 w_1 - v_1 w_3, v_1 w_2 - v_2 w_1)$$
$$= (\mathbf{u} \times \mathbf{w}) + (\mathbf{v} \times \mathbf{w})$$

(c) $t(\mathbf{u} \times \mathbf{v})$
$$= t(u_2 v_3 - u_3 v_2, u_3 v_1 - u_1 v_3,$$
$$u_1 v_2 - u_2 v_1)$$
$$= ((tu_2)v_3 - (tu_3)v_2,$$
$$(tu_3)v_1 - (tu_1)v_3,$$
$$(tu_1)v_2 - (tu_2)v_1)$$
$$= (t\mathbf{u}) \times \mathbf{v}$$
Similarly, $t(\mathbf{u} \times \mathbf{v}) = \mathbf{u} \times (t\mathbf{v})$.

(d) $\mathbf{u} \times \mathbf{0} = (u_2(0) - u_3(0), u_3(0) - u_1(0), u_1(0) - u_2(0)) = (0, 0, 0) = \mathbf{0}$. Likewise, $\mathbf{0} \times \mathbf{u} = \mathbf{0}$.

(e) $\mathbf{u} \times \mathbf{u} = (u_2 u_3 - u_3 u_2, u_3 u_1 - u_1 u_3, u_1 u_2 - u_2 u_1) = (0, 0, 0) = \mathbf{0}$.

Solutions to Exercises in Section 1.3

1. The slope of 40 means that in each additional second, the object increases its distance from the measuring device by 40 feet. The intercept 5000 means that at time $t = 0$, the object is 5000 feet from the measuring device.

3. The slope of -7 means that each additional second reduces the speed of the car by 7 feet per second. The intercept 88 means that at time $t = 0$, the speed of the car is 88 feet per second.

5. (a) Slope-intercept form: $y = -(3/2)x + 3$.
   (b) Point-normal form: $3(x-2)+2(y-0) = 0$, that is, $3x + 2y - 6 = 0$.
   (c) Parametric-equation form: $(2, 0) + t(-2, 3)$, that is, $(2 - 2t, 3t)$.

7. (a) Inclination-intercept form: $z = (1, 2) \cdot (x, y) - 8$, or, $z = x + 2y - 8$.
   (b) Point-normal form: $(1, 2, -1) \cdot [(x, y, z) - (2, 3, 0)] = 0$, or, $x + 2y - z - 8 = 0$.

9. (a) Substitute $n = 2$, $x_1 = x$, $x_2 = y$, $a_1 = 2$, $a_2 = 3$, and $b = 3$.
   (b) Substitute $n = 2$, $a_1 = 1$, $a_2 = -2$, $b = -3$.

11. (a) Data: A point (x_0, y_0) on the line and the slope m.
    Closed-form solution:
    $y = mx + (y_0 - mx_0) = m(x - x_0) + y_0$.
    (b) Data: (x_1, y_1) and (x_2, y_2) on the line.
    Closed-form solution: $y = [(y_2 - y_1)/(x_2 - x_1)]x + (y_1 x_2 - y_2 x_1)/(x_2 - x_1)$.

13. (a) Data: The slope m and the intercept b.
    Closed-form solution: $(0, b) + t(1, m)$.
    (b) Data: A point (x_0, y_0) on the line and a normal vector, (a, b).
    Closed-form solution:
    $(x_0, y_0) + t(-b, a)$, or $(x_0 - tb, y_0 + ta)$.

15. (a) The point-normal equation of the plane is $(\mathbf{u} \times \mathbf{v}) \cdot (\mathbf{x} - \mathbf{0}) = 0$, that is $(\mathbf{u} \times \mathbf{v}) \cdot \mathbf{x} = 0$.
    (b) For the three points, $(0, 0, 0)$, $(0, 2, 0)$, and $(1, 1, 1)$, you have $\mathbf{u} = (0, 2, 0)$ and $\mathbf{v} = (1, 1, 1)$. So, $\mathbf{u} \times \mathbf{v} = (2, 0, -2)$ and the equation of the plane is $(2, 0, -2) \cdot (x_1, x_2, x_3) = 0$, that is, $2x_1 - 2x_3 = 0$.

17. (a) The two ways to measure the distances are shown in the following figure:

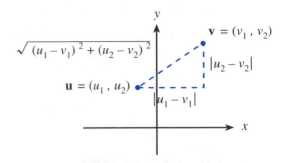

   (b) One unification of the two formulas:

   (i) $|u_1 - v_1| + |u_2 - v_2|$

   (ii) $\sqrt{(u_1 - v_1)^2 + (u_2 - v_2)^2}$

   is the following expression, in which p is a positive integer:

   $$\left(|u_1 - v_1|^p + |u_2 - v_2|^p\right)^{\frac{1}{p}}$$

   When $p = 1$, the foregoing expression yields the special case in (i) and when $p = 2$, the foregoing expression yields the special case in (ii).

19. (a) It is not possible to evaluate the expression times($2\mathbf{u}$) because times is a binary operator that requires two n-vectors and $2\mathbf{u}$ represents only one n-vector.
    (b) sum(times($\mathbf{u}, \mathbf{v}$)) $= u_1 v_1 + \cdots + u_n v_n$.
    (c) It is not possible to evaluate times($\mathbf{u}$, sum($\mathbf{v}$)) because times is a binary operator that requires two n-vectors and sum($\mathbf{v}$) is a real number, not an n-vector.
    (d) dim(sum($\mathbf{v}$)) $= 1$.

21. (a) A syntax error arises in the expression $|u - w\mathbf{v}|$ because you cannot subtract the n-vector $w\mathbf{v}$ from the real number u. A syntax error also arises in $t|\mathbf{v}|$ because you cannot evaluate the absolute value of the n-vector $\mathbf{v}$.
    (b) The generalization is $|u - \mathbf{v} \cdot \mathbf{w}| \leq t \|\mathbf{v}\|$.

23. For the given vector $\mathbf{u} = (u_1, u_2)$, any vector (x, y) on the same (opposite) side of the line as $\mathbf{u}$ has a dot product with $\mathbf{u}$ that is ≥ 0 (≤ 0). Thus, the two half spaces are:

$$\{(x, y) : (u_1, u_2) \cdot (x, y) \geq 0\} =$$
$$\{(x, y) : u_1 x + u_2 y \geq 0\}$$

$$\{(x, y) : (u_1, u_2) \cdot (x, y) \leq 0\} =$$
$$\{(x, y) : u_1 x + u_2 y \leq 0\}$$

SOLUTIONS TO SELECTED EXERCISES IN CHAPTER 2

Solutions to Exercises in Section 2.1

1. (a) Let

G = the annual output of goods (in $)

S = the annual output of services (in $)

Then the two balance equations are:

$$G = 0.7G + 0.4S \quad \text{(goods)}$$
$$S = 0.3G + 0.6S \quad \text{(services)}$$

(b) From the balance equation for goods, $0.3G = 0.4S$. Thus, any equilibrium point is defined by $G = (4/3)S$.

(c) The equilibrium point is $S = 30$ and $G = 40$ billion dollars because, from (b), $G = (4/3)S = (4/3)30 = 40$ billions dollars.

3. Let

A = annual production of Asia (in $)

E = annual production of Europe (in $)

N = annual prod. of North America (in $)

S = annual prod. of South America (in $)

The four balance equations are:

$$A = 0.4A + 0.1E + 0.2N + 0.1S$$
$$E = 0.2A + 0.6E + 0.2N + 0.1S$$
$$N = 0.3A + 0.2E + 0.4N + 0.3S$$
$$S = 0.1A + 0.1E + 0.2N + 0.5S$$

5. To formulate a linear programming problem, define the following three variables, in addition to $x_1, \ldots, x_5$ in Figure 2.1:

x_6 = thousands of gallons to ship from Oakland to Livermore

x_7 = thousands of gallons to ship from Oakland to Concord

x_8 = thousands of gallons to ship from Antioch to Sacramento

The linear programming problem is to

$$\max \ x_6 + x_7$$

$$\text{s.t.} \quad x_6 - x_1 - x_5 = 0 \quad \text{(bal. at Livermore)}$$
$$x_7 + x_1 - x_2 = 0 \quad \text{(bal. at Concord)}$$
$$x_5 - x_3 - x_4 = 0 \quad \text{(bal. at Stockton)}$$
$$x_2 + x_3 - x_8 = 0 \quad \text{(bal. at Antioch)}$$
$$x_1, \ldots, x_8 \quad \geq 0$$

7. (a) The network diagram is:

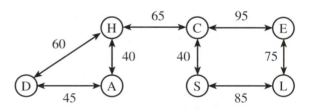

(b) For each $i, j = A, C, D, E, H, L, S$ for which there is a nonstop connection between city i and city j, define the variables x_{ij} as follows, to indicate whether to include that segment in the least-cost route from Dallas to El Paso:

$$x_{ij} = \begin{cases} 1, & \text{if the route includes the segment from city } i \text{ to city } j \\ 0, & \text{otherwise} \end{cases}$$

The mathematical model is to

$$\min \ 45x_{AD} + 40x_{AH} + 95x_{CE} + 65x_{CH} +$$
$$40x_{CS} + 45x_{DA} + 60x_{DH} + 95x_{EC} +$$
$$75x_{EL} + 40x_{HA} + 65x_{HC} + 60x_{HD} +$$
$$75x_{LE} + 85x_{LS} + 40x_{SC} + 85x_{SL}$$

s.t.

leave Dallas

$$x_{DA} + x_{DH} = 1$$

balance at Austin

$$x_{DA} + x_{HA} - x_{AD} - x_{AH} = 0$$

balance at Corpus Christi

$$x_{EC} + x_{HC} + x_{SC} - $$
$$x_{CE} - x_{CH} - x_{CS} = 0$$

balance at Houston

$$x_{AH} + x_{CH} + x_{DH} - $$
$$x_{HA} - x_{HC} - x_{HD} = 0$$

balance at Lubbock

$$x_{EL} + x_{SL} - x_{LE} - x_{LS} = 0$$

balance at San Antonio

$$x_{CS} + x_{LS} - x_{SC} - x_{SL} = 0$$

arrive at El Paso

$$x_{CE} + x_{LE} = 1$$

9. Define the following variables pertaining to the amount of each food to include in the diet:

M = number of gallons of milk

C = number of pounds of cheese

A = number of pounds of apples

The linear programming problem is to

min $2M + 3.50C + 0.90A$

s. t.

(pro.) $40M + 20C + 10A \geq 80$

(Vit. A) $5M + 40C + 30A \geq 60$

(Vit. B) $20M + 30C + 40A \geq 50$

(Vit. C) $30M + 50C + 60A \geq 30$

M , C , $A \geq 0$

Solutions to Exercises in Section 2.2

1. Substitute $n = 2$, $x_1 = x$, $x_2 = y$, $a_{11} = p$, $a_{12} = q$, $b_1 = t$, $a_{21} = r$, $a_{22} = s$, $b_2 = u$.

3. No, it is not correct to conclude that the system of linear equations is inconsistent because only one set of values for the variables does not satisfy the system. Perhaps other values will satisfy the system.

5. (a) The system has infinitely many solutions, as shown in the following figure. The general solution is $x = 2 + 2y$ and y is a free variable.

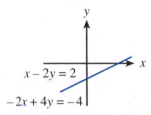

(b) The unique solution to the system of linear equations is $x = 2$ and $y = 0$, as shown in the following figure:

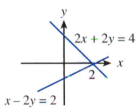

(c) The system has no solution, as shown in the following figure:

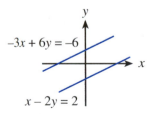

7. (a) If $ad - bc = 0$ and either $de - bf \neq 0$ or $ce - af \neq 0$, then the system has no solution.

(b) If $ad - bc = 0$ and $de - bf = 0$ and $ce - af = 0$, then the system has infinitely many solutions.

9. (a) Solving for x yields

$$x = \frac{d}{a} - \frac{b}{a}y - \frac{c}{a}z$$

Assigning an arbitrary value of s to y and t to z results in the following general solution to the system:

$$x = \frac{d}{a} - \frac{b}{a}s - \frac{c}{a}t, \quad y = s, \quad z = t$$

(b) Solving for y yields

$$y = \frac{d}{b} - \frac{a}{b}x - \frac{c}{b}z$$

Assigning an arbitrary value of s to x and t to z results in the following general solution to the system:

$$x = s, \quad y = \frac{d}{b} - \frac{a}{b}s - \frac{c}{b}t, \quad z = t$$

(c) Solving for z yields

$$z = \frac{d}{c} - \frac{a}{c}x - \frac{b}{c}y$$

Assigning an arbitrary value of s to x and t to y results in the following general solution to the system:

$$x = s, \quad y = t, \quad z = \frac{d}{c} - \frac{a}{c}s - \frac{b}{c}t$$

Solutions to Exercises in Section 2.3

1. (a) The augmented matrix is:

$$\left[\begin{array}{cccc|c} 2 & 0 & 1 & -3 & 4 \\ 0 & -1 & 2 & 1 & -2 \end{array}\right]$$

(b) The augmented matrix is:

$$\left[\begin{array}{ccc|c} 2 & 3 & -1 & -1 \\ -3 & -1 & 4 & 0 \\ 6 & 7 & 5 & 1 \end{array}\right]$$

3. (a)

$$\begin{aligned} x_1 + 2x_2 - 4x_3 &= 3 \\ x_2 - 2x_3 &= 2 \\ -x_1 - 2x_2 + 5x_3 &= -2 \end{aligned}$$

(b)

$$\begin{aligned} 3x_1 - x_2 - 4x_3 + 3x_4 &= 0 \\ 2x_1 + x_2 + 3x_3 - 2x_4 &= -1 \\ -x_1 + x_2 - x_3 + x_4 &= 2 \end{aligned}$$

5. (a) The results are:

System of Equations

$$\begin{aligned} x \quad\quad - 2z &= 3 \\ 3x + y - z &= 4 \\ 2y + 3z &= 0 \end{aligned}$$

Augmented Matrix

$$\left[\begin{array}{ccc|c} 1 & 0 & -2 & 3 \\ 3 & 1 & -1 & 4 \\ 0 & 2 & 3 & 0 \end{array}\right]$$

(b) The results are:

System of Equations

$$\begin{aligned} x \quad\quad - 2z &= 3 \\ y + 5z &= -5 \\ 2y + 3z &= 0 \end{aligned}$$

Augmented Matrix

$$\left[\begin{array}{ccc|c} 1 & 0 & -2 & 3 \\ 0 & 1 & 5 & -5 \\ 0 & 2 & 3 & 0 \end{array}\right]$$

(c) The results are:

System of Equations

$$\begin{aligned} x \quad\quad - 2z &= 3 \\ 2y + 3z &= 0 \\ y + 5z &= -5 \end{aligned}$$

Augmented Matrix

$$\left[\begin{array}{ccc|c} 1 & 0 & -2 & 3 \\ 0 & 2 & 3 & 0 \\ 0 & 1 & 5 & -5 \end{array}\right]$$

7. (a) The general solution is $x = 4 - 2t$ and $y = t$.

(b) The matrix is not in reduced row-echelon form since column 3 does not contain all zeroes except for the leading 1 in row 3.

(c) The matrix is not in reduced row-echelon form since there is no leading 1 in row 3.

(d) The general solution to the system of linear equations is

$$x_1 = 1 + s - t, \quad x_2 = s, \quad x_3 = 2t,$$
$$x_4 = t, \quad x_5 = -2$$

9. (a) The general solution to the system of equations is

$$x_1 = -13 + s + 10t, \quad x_2 = s,$$
$$x_3 = -5 + 2t, \quad x_4 = t$$

This solution is obtained by applying back substitution to the following system:

$$x_1 - x_2 - 3x_3 - 4x_4 = 2$$
$$x_3 - 2x_4 = -5$$

Solving the last equation for the leading variable x_3 yields

$$x_3 = -5 + 2x_4$$

Solving the first equation for the leading variable x_1 and substituting $-5 + 2x_4$ for x_3 yields

$$x_1 = 2 + x_2 + 3(-5 + 2x_4) + 4x_4$$
$$= -13 + x_2 + 10x_4$$

Assigning an arbitrary value s to the free variable x_2 and a value t to the free variable x_4 provides the desired general solution.

(b) The augmented matrix is not in row-echelon form because the leading value in row 2 is not to the right of the leading value in row 1.

(c) The general solution to the system of equations is

$$x_1 = 4, \quad x_2 = 1, \quad x_3 = 3$$

This solution is obtained by applying back substitution to the following system:

$$x_1 - 3x_2 = 1$$
$$x_2 - x_3 = -2$$
$$x_3 = 3$$

The last equation is solved for the leading variable x_3, so,

$$x_3 = 3$$

Solving the second equation for the leading variable x_2 and substituting $x_3 = 3$ yields

$$x_2 = -2 + x_3 = -2 + 3 = 1$$

Solving the first equation for the leading variable x_1 and substituting $x_2 = 1$ and $x_3 = 3$ yields

$$x_1 = 1 + 3x_2 = 1 + 3(1) = 4$$

(d) The general solution to the system of equations is

$$x_1 = 3, \quad x_2 = -4, \quad x_3 = 2$$

This solution is obtained by applying back substitution to the following system:

$$x_1 - 2x_2 - 3x_3 = 5$$
$$3x_2 + 6x_3 = 0$$
$$2x_3 = 4$$

The last equation is solved for the leading variable x_3, so,

$$x_3 = 2$$

Solving the second equation for the leading variable x_2 and substituting $x_3 = 3$ yields

$$x_2 = -2x_3 = -2(2) = -4$$

Solving the first equation for the leading variable x_1 and substituting $x_2 = -4$ and

$x_3 = 2$ yields

$$x_1 = 5 + 2x_2 + 3x_3 = 3$$

11. (a) The solution

$$x = \frac{d}{ad - bc} \quad \text{and} \quad y = \frac{-c}{ad - bc}$$

is obtained by applying the following steps of Gauss-Jordan elimination to the augmented matrix:

$$\left[\begin{array}{cc|c} a & b & 1 \\ c & d & 0 \end{array}\right]$$

Step 1. Multiply row 1 of the foregoing matrix by $1/a$:

$$\left[\begin{array}{cc|c} 1 & b/a & 1/a \\ c & d & 0 \end{array}\right]$$

Step 2. Add $-c$ times row 1 of the foregoing matrix to row 2:

$$\left[\begin{array}{cc|c} 1 & b/a & 1/a \\ 0 & (ad - bc)/a & -c/a \end{array}\right]$$

Step 3. Multiply row 2 of the foregoing matrix by $a/(ad - bc)$:

$$\left[\begin{array}{cc|c} 1 & b/a & 1/a \\ 0 & 1 & -c/(ad - bc) \end{array}\right]$$

Step 4. Add $-b/a$ times row 2 of the foregoing matrix to row 1:

$$\left[\begin{array}{cc|c} 1 & 0 & d/(ad - bc) \\ 0 & 1 & -c/(ad - bc) \end{array}\right]$$

(b) The solution

$$x = \frac{-b}{ad - bc} \quad \text{and} \quad y = \frac{a}{ad - bc}$$

is obtained by applying the same sequence of elementary row operations as in the solution to part (a) to the aug-

mented matrix:

$$\left[\begin{array}{cc|c} a & b & 0 \\ c & d & 1 \end{array}\right]$$

yielding

$$\left[\begin{array}{cc|c} 1 & 0 & -b/(ad - bc) \\ 0 & 1 & a/(ad - bc) \end{array}\right]$$

(c) The same solutions as in parts (a) and (b) are obtained by performing the same sequence of elementary row operations as in part (a) to the augmented matrix

$$\left[\begin{array}{cc|cc} a & b & 1 & 0 \\ c & d & 0 & 1 \end{array}\right]$$

yielding

$$\left[\begin{array}{cc|cc} 1 & 0 & d/(ad - bc) & -b/(ad - bc) \\ 0 & 1 & -c/(ad - bc) & a/(ad - bc) \end{array}\right]$$

(d) Less computational effort is required to obtain the solutions by using the approach in part (c) than by solving the systems in parts (a) and (b).

13. (a) The solution $x = 1$, $y = 3$, and $z = 0$ is obtained by performing the following steps of Gauss-Jordan elimination to the augmented matrix:

$$\left[\begin{array}{ccc|c} 0 & -2 & 1 & -6 \\ 1 & 0 & -3 & 1 \\ -1 & 1 & 2 & 2 \end{array}\right]$$

Step 1. Interchange row 1 of the foregoing matrix with row 2:

$$\left[\begin{array}{ccc|c} 1 & 0 & -3 & 1 \\ 0 & -2 & 1 & -6 \\ -1 & 1 & 2 & 2 \end{array}\right]$$

Step 2. Add 1 times row 1 of the foregoing matrix to row 3:

$$\begin{bmatrix} 1 & 0 & -3 & | & 1 \\ 0 & -2 & 1 & | & -6 \\ 0 & 1 & -1 & | & 3 \end{bmatrix}$$

Step 3. Multiply row 2 of the foregoing matrix by $-1/2$:

$$\begin{bmatrix} 1 & 0 & -3 & | & 1 \\ 0 & 1 & -1/2 & | & 3 \\ 0 & 1 & -1 & | & 3 \end{bmatrix}$$

Step 4. Add -1 times row 2 of the foregoing matrix to row 3:

$$\begin{bmatrix} 1 & 0 & -3 & | & 1 \\ 0 & 1 & -1/2 & | & 3 \\ 0 & 0 & -1/2 & | & 0 \end{bmatrix}$$

Step 5. Multiply row 3 of the foregoing matrix by -2:

$$\begin{bmatrix} 1 & 0 & -3 & | & 1 \\ 0 & 1 & -1/2 & | & 3 \\ 0 & 0 & 1 & | & 0 \end{bmatrix}$$

Step 6. Add 3 times row 3 of the foregoing matrix to row 1 and add $1/2$ of row 3 to row 2:

$$\begin{bmatrix} 1 & 0 & 0 & | & 1 \\ 0 & 1 & 0 & | & 3 \\ 0 & 0 & 1 & | & 0 \end{bmatrix}$$

(b) The sequence of instructions depends on the specific technology used.

15. The solution $x = 1$, $y = 3$, and $z = 0$ is obtained by performing the following steps of Gaussian elimination to the augmented matrix:

$$\begin{bmatrix} 0 & -2 & 1 & | & -6 \\ 1 & 0 & -3 & | & 1 \\ -1 & 1 & 2 & | & 2 \end{bmatrix}$$

The first four steps are the same as shown in the solution to Exercise 13(a), which result in the following augmented matrix:

$$\begin{bmatrix} 1 & 0 & -3 & | & 1 \\ 0 & 1 & -1/2 & | & 3 \\ 0 & 0 & -1/2 & | & 0 \end{bmatrix}$$

Now apply back substitution to the associated system of linear equations:

$$\begin{aligned} x \quad - \quad 3z &= 1 \\ y - (1/2)z &= 3 \\ -(1/2)z &= 0 \end{aligned}$$

Solving the last equation for the leading variable z yields

$$z = 0$$

Solving the second equation for the leading variable y and substituting $z = 0$ yields

$$y = 3 + (1/2)z = 3 + 0 = 3$$

Solving the first equation for the leading variable x and substituting $y = 3$ and $z = 0$ yields

$$x = 1 + 3z = 1 + 3(0) = 1$$

17. Reduced row-echelon form is a special case of row-echelon form in which the leading value in each row is required to be 1 and the entries above as well as below the leading 1 must be 0.

19. You can stop Gauss-Jordan elimination when a row consisting of all zeros, except for a nonzero value in the last column, is encountered because this situation indicates that the original system has no solution. This is because the linear equation corresponding to this row in the augmented matrix is:

$$0x_1 + \cdots + 0x_n = b \quad \text{(where } b \neq 0\text{)}$$

The foregoing equation, and hence the original system, has no solution.

1. For the Leontief economic model, you have:

$$A = \begin{bmatrix} 0.5 & -0.2 & -0.2 \\ -0.1 & 0.6 & -0.1 \\ -0.4 & -0.4 & 0.3 \end{bmatrix},$$

$$\mathbf{x} = \begin{bmatrix} G \\ J \\ U \end{bmatrix}, \quad \mathbf{b} = \begin{bmatrix} 0 \\ 0 \\ 0 \end{bmatrix}$$

3. For the network-flow problem, you have:

$$A = \begin{bmatrix} -1 & 1 & 0 & 0 & 0 \\ 1 & 0 & 0 & 0 & 1 \\ 0 & 1 & 1 & 0 & 0 \\ 0 & 0 & -1 & -1 & 1 \end{bmatrix}$$

$$\mathbf{x} = \begin{bmatrix} x_1 \\ x_2 \\ x_3 \\ x_4 \\ x_5 \end{bmatrix}, \quad \mathbf{b} = \begin{bmatrix} 30 \\ 50 \\ 60 \\ 0 \end{bmatrix}$$

5. (a) $A\mathbf{x} = \mathbf{0}$, where A is an $(m \times n)$ matrix, $\mathbf{x} \in R^{n \times 1}$ is a column vector, and $\mathbf{0} \in R^{m \times 1}$ is a column vector.
   (b) $A\mathbf{x} = \mathbf{0}$, where A is an $(n \times n)$ matrix, $\mathbf{x} \in R^{n \times 1}$ is a column vector, and $\mathbf{0} \in R^{n \times 1}$ is a column vector.

7. (a) $A - B$

$$= \begin{bmatrix} -1-1 & 2-(-4) & 0-(-1) \\ 0-1 & -3-0 & 1-3 \end{bmatrix}$$

$$= \begin{bmatrix} -2 & 6 & 1 \\ -1 & -3 & -2 \end{bmatrix}$$

(b) $2A + B$

$$= \begin{bmatrix} -2+1 & 4+(-4) & 0+(-1) \\ 0+1 & -6+0 & 2+3 \end{bmatrix}$$

$$= \begin{bmatrix} -1 & 0 & -1 \\ 1 & -6 & 5 \end{bmatrix}$$

(c) AC

$$= \begin{bmatrix} -1(1)+2(0)+0(-2) \\ 0(1)-3(0)+1(-2) \end{bmatrix} = \begin{bmatrix} -1 \\ -2 \end{bmatrix}$$

(d) It is not possible to compute CB because C has 1 column and B has 2 rows.
(e) It is not possible to compute $C'B$ because C' has 3 columns B has 2 rows.

9. (a) You have that

$$C' = \begin{bmatrix} 1 & 0 \\ 0 & 2 \\ 0 & -1 \end{bmatrix}$$

(b) $A + 2B$

$$= \begin{bmatrix} 2+2 & 1+6 \\ 0+(-2) & -1+4 \end{bmatrix}$$

$$= \begin{bmatrix} 4 & 7 \\ -2 & 3 \end{bmatrix}$$

(c) AB

$$= \begin{bmatrix} 2(1)+1(-1) & 2(3)+1(2) \\ 0(1)+(-1)(-1) & 0(3)+(-1)(2) \end{bmatrix}$$

$$= \begin{bmatrix} 1 & 8 \\ 1 & -2 \end{bmatrix}$$

(d) AC

$$= \begin{bmatrix} 2(1)+1(0) & 2(0)+1(2) \\ 0(1)+(-1)(0) & 0(0)+(-1)(2) \end{bmatrix}$$
$$\begin{matrix} 2(0)+1(-1) \\ 0(0)+(-1)(-1) \end{matrix}$$

$$= \begin{bmatrix} 2 & 2 & -1 \\ 0 & -2 & 1 \end{bmatrix}$$

(e) It is not possible to compute CA because C has 3 columns and A has 2 rows.

11. (a) This elementary matrix E interchanges row 2 and row 3 of A because

EA

$$= \begin{bmatrix} 1 & 0 & 0 \\ 0 & 0 & 1 \\ 0 & 1 & 0 \end{bmatrix} \begin{bmatrix} -1 & 0 & -3 \\ 4 & -2 & 6 \\ 3 & 1 & 5 \end{bmatrix}$$

$$= \begin{bmatrix} -1 & 0 & -3 \\ 3 & 1 & 5 \\ 4 & -2 & 6 \end{bmatrix}$$

(b) This elementary matrix E multiplies row 1 of A by -1 because

EA

$$= \begin{bmatrix} -1 & 0 & 0 \\ 0 & 1 & 0 \\ 0 & 0 & 1 \end{bmatrix} \begin{bmatrix} -1 & 0 & -3 \\ 4 & -2 & 6 \\ 3 & 1 & 5 \end{bmatrix}$$

$$= \begin{bmatrix} 1 & 0 & 3 \\ 4 & -2 & 6 \\ 3 & 1 & 5 \end{bmatrix}$$

(c) This elementary matrix E adds 1/2 of row 2 of A to row 3 because

EA

$$= \begin{bmatrix} 1 & 0 & 0 \\ 0 & 1 & 0 \\ 0 & 1/2 & 1 \end{bmatrix} \begin{bmatrix} -1 & 0 & -3 \\ 4 & -2 & 6 \\ 3 & 1 & 5 \end{bmatrix}$$

$$= \begin{bmatrix} -1 & 0 & -3 \\ 4 & -2 & 6 \\ 5 & 0 & 8 \end{bmatrix}$$

13. (a) The dimensions of $\mathbf{x}$ must be $(n \times 1)$. The resulting dimensions of $A\mathbf{x}$ are $(m \times 1)$.

(b) You have that

$$A\mathbf{x} = \begin{bmatrix} -2 \\ -6 \\ 10 \end{bmatrix}$$

This is obtained from the rows of A, as follows:

$$(A\mathbf{x})_1 = A_{1*} \cdot \mathbf{x} = \begin{bmatrix} 1 & -1 & 0 \end{bmatrix} \cdot \begin{bmatrix} -1 \\ 1 \\ 2 \end{bmatrix}$$
$$= 1(-1)+(-1)(1)+0(2)$$
$$= -2$$

$$(A\mathbf{x})_2 = A_{2*} \cdot \mathbf{x} = \begin{bmatrix} 2 & 0 & -2 \end{bmatrix} \cdot \begin{bmatrix} -1 \\ 1 \\ 2 \end{bmatrix}$$
$$= 2(-1)+0(1)+(-2)(2)$$
$$= -6$$

$$(A\mathbf{x})_3 = A_{3*} \cdot \mathbf{x} = \begin{bmatrix} -3 & 1 & 3 \end{bmatrix} \cdot \begin{bmatrix} -1 \\ 1 \\ 2 \end{bmatrix}$$
$$= (-3)(-1)+1(1)+3(2)$$
$$= 10$$

(c) Using the columns of A, you have that

$$Ax = A_{*1}x_1 + A_{*2}x_2 + A_{*3}x_3$$

$$= \begin{bmatrix} 1 \\ 2 \\ -3 \end{bmatrix}(-1) + \begin{bmatrix} -1 \\ 0 \\ 1 \end{bmatrix}(1) + \begin{bmatrix} 0 \\ -2 \\ 3 \end{bmatrix}(2)$$

$$= \begin{bmatrix} -1 \\ -2 \\ 3 \end{bmatrix} + \begin{bmatrix} -1 \\ 0 \\ 1 \end{bmatrix} + \begin{bmatrix} 0 \\ -4 \\ 6 \end{bmatrix} = \begin{bmatrix} -2 \\ -6 \\ 10 \end{bmatrix}$$

15. (Answers can vary.)

(a) A similarity is that, in both operations, the real number is multiplied by each element of the vector and matrix. A difference is that multiplying a real number by a vector results in a vector while multiplying a real number by a matrix results in a matrix.

(b) A similarity is that, in both operations, corresponding elements of the vectors and matrices are added. A difference is that adding two vectors results in a vector while adding two matrices results in a matrix.

(c) A similarity is that, in both operations, the elements of the vectors and matrices are used to obtain the result. A difference is that the result of multiplying two vectors with the dot product is a real number while the result of multiplying two matrices is a matrix.

17. (a) The key question associated with trying to prove that $(AB)_{i*} = A_{i*}B$ is, "How can I show that two row vectors are equal?" According to the definition, you must show that corresponding elements of the two row vectors are equal. That is, you must show that for all $k = 1, \ldots, p$, $[(AB)_{i*}]_k = (A_{i*}B)_k$. Recognizing the key words *for all*, use the choose method to choose an integer k with $1 \leq k \leq p$

and show that, for this chosen value of k, $[(AB)_{i*}]_k = (A_{i*}B)_k$. But

$$[(AB)_{i*}]_k = (AB)_{ik} = A_{i*} \cdot B_{*k}$$
$$= (A_{i*}B)_k$$

Similarly, to show that $(AB)_{*j} = AB_{*j}$, choose an integer k with $1 \leq k \leq m$ and show that, for this chosen value of k, $[(AB)_{*j}]_k = (AB_{*j})_k$. But

$$[(AB)_{*j}]_k = (AB)_{kj} = A_{k*} \cdot B_{*j}$$
$$= (AB_{*j})_k$$

(b) Computing AB requires mnp multiplications. Computing $A_{i*}B$ requires only np multiplications, so the difference is $mnp - np = (m-1)np$ multiplications. Similarly, computing AB_{*j} requires only mn multiplications, so the difference is $mnp - mn = mn(p-1)$ multiplications.

19. In each part, it is necessary to show that two matrices are equal. By definition, this is accomplished by showing that all corresponding elements of the two $(m \times n)$ matrices are equal. The choose method is therefore used to choose integers i and j with $1 \leq i \leq m$ and $1 \leq j \leq n$ and showing that the elements in row i and column j of the two matrices are equal.

(a) Now transform $[(A+B)+C]_{ij}$ into $[A+(B+C)]_{ij}$, as follows:

$$[(A+B)+C]_{ij} = (A+B)_{ij} + C_{ij}$$
$$= (A_{ij} + B_{ij}) + C_{ij}$$
$$= A_{ij} + (B_{ij} + C_{ij})$$
$$= A_{ij} + (B+C)_{ij}$$
$$= [A+(B+C)]_{ij}$$

(b) The proof is completed by noting that

$$(A+0)_{ij} = A_{ij} + 0_{ij}$$
$$= 0_{ij} + A_{ij} = (0+A)_{ij}$$

and also

$$(A + 0)_{ij} = A_{ij} + 0_{ij} = A_{ij} + 0 = A_{ij}$$

(c) The proof is completed by noting that

$$(A - A)_{ij} = A_{ij} - A_{ij} = 0 = 0_{ij}$$

(d) The proof is completed by noting that

$$(0 - A)_{ij} = 0_{ij} - A_{ij} = 0 - A_{ij}$$
$$= -A_{ij} = (-A)_{ij}$$

21. In both parts, it is necessary to show that two matrices are equal. By definition, this is accomplished by showing that all corresponding elements of the two matrices are equal. The choose method is used to do so.

(a) Suppose that A is an $(m \times n)$ matrix. So A^t is an $(n \times m)$ matrix. Now let i and j be integers with $1 \le i \le n$ and $1 \le j \le m$. The proof is completed by noting that

$$[(sA)^t]_{ij} = (sA)_{ji} = sA_{ji} = s(A^t)_{ij}$$

(b) Suppose that A and B are $(m \times n)$ matrices. So A^t and B^t are $(n \times m)$ matrix. Now let i and j be integers with $1 \le i \le n$ and $1 \le j \le m$. The proof is completed by noting that

$$[(A + B)^t]_{ij} = (A + B)_{ji} = A_{ji} + B_{ji}$$
$$= (A^t)_{ij} + (B^t)_{ij}$$

SOLUTIONS TO SELECTED EXERCISES IN CHAPTER 3

Solutions to Exercises in Section 3.1

1. (a) Let

G = the annual output of the goods sector in \$
S = the annual output of the services sector in \$

Then the two required linear equations are:

$$G = 0.3G + 0.2S + 3$$
$$S = 0.1G + 0.4S + 5$$

or

$$0.7G - 0.2S = 3$$
$$-0.1G + 0.6S = 5$$

(b) The solution to the system in part (a) is for the goods sector to produce $G = 7$ billion dollars of output and for the services sector to produce $S = 9.5$ billion dollars.

3. From Kirchoff's voltage law, the branch current I must satisfy $5I + 7I = 32 + 40$, or, equivalently, $12I = 72$. Thus, $I = 6$ amps.

5. From Kirchoff's current law, you have that

$$I_3 = I_1 + I_2$$

From the voltage law, you have that

$$-4I_1 = 12 \quad \text{(left loop)}$$
$$2I_2 = 12 \quad \text{(right loop)}$$

The solution to these equations is $I_1 = -3$, $I_2 = 6$, and $I_3 = 3$.

7. From Kirchoff's current law, you have that

$$I_1 = I_2 + I_3$$

From the voltage law, you have that

$$2I_1 + 5I_2 \qquad = 25 \quad \text{(top loop)}$$
$$- 5I_2 + 6I_3 = 10 \quad \text{(bottom loop)}$$

The solution to these equations is $I_1 = 6.25$, $I_2 = 2.50$, and $I_3 = 3.75$.

9. The three equations in the three unknowns a, b, and c are:

$$ax^2 + bx + c = f(x)$$
$$ay^2 + by + c = f(y)$$
$$az^2 + bz + c = f(z)$$

or, in matrix-vector notation:

$$\begin{bmatrix} x^2 & x & 1 \\ y^2 & y & 1 \\ z^2 & z & 1 \end{bmatrix} \begin{bmatrix} a \\ b \\ c \end{bmatrix} = \begin{bmatrix} f(x) \\ f(y) \\ f(z) \end{bmatrix}$$

Solutions to Exercises in Section 3.2

1. (a) The given matrix is the desired inverse because:

$$\begin{bmatrix} 2 & 3 \\ 1 & 2 \end{bmatrix} \begin{bmatrix} 2 & -3 \\ -1 & 2 \end{bmatrix} = \begin{bmatrix} 1 & 0 \\ 0 & 1 \end{bmatrix}$$

and

$$\begin{bmatrix} 2 & -3 \\ -1 & 2 \end{bmatrix} \begin{bmatrix} 2 & 3 \\ 1 & 2 \end{bmatrix} = \begin{bmatrix} 1 & 0 \\ 0 & 1 \end{bmatrix}$$

(b) The solution to the system is $x = 2$ and $y = 0$:

$$\begin{bmatrix} x \\ y \end{bmatrix} = \begin{bmatrix} 2 & 3 \\ 1 & 2 \end{bmatrix} \begin{bmatrix} 4 \\ -2 \end{bmatrix} = \begin{bmatrix} 2 \\ 0 \end{bmatrix}$$

3. The solution to the system is:

$$\begin{bmatrix} x \\ y \end{bmatrix} = \frac{1}{ad - bc} \begin{bmatrix} d & -b \\ -c & a \end{bmatrix} \begin{bmatrix} e \\ f \end{bmatrix}$$

$$= \begin{bmatrix} (de - bf)/(ad - bc) \\ (af - ce)/(ad - bc) \end{bmatrix}$$

5. (a) This diagonal matrix has no inverse because the diagonal element in row 3 and column 3 is 0.
   (b) This matrix has no inverse because the number of rows is not equal to the number of columns.
   (c) The diagonal elements of the following inverse matrix are the reciprocals of the diagonal elements of the original matrix:

$$\begin{bmatrix} 1 & 0 & 0 & 0 \\ 0 & -1/3 & 0 & 0 \\ 0 & 0 & 1 & 0 \\ 0 & 0 & 0 & 1 \end{bmatrix}$$

7. The inverse of the following elementary matrix E is obtained by replacing the value c in E_{ji} with $-c$:

$$E$$

$$\begin{array}{c} \\ 1 \\ \vdots \\ i \\ \vdots \\ j \\ \vdots \\ n \end{array} \begin{array}{cc} \quad i \quad\quad j \\ \begin{bmatrix} 1 & \cdots & 0 & \cdots & 0 & \cdots & 0 \\ \vdots & \vdots & \vdots & \vdots & \vdots & \vdots & \vdots \\ 0 & \cdots & 1 & \cdots & 0 & \cdots & 0 \\ \vdots & \vdots & \vdots & \vdots & \vdots & \vdots & \vdots \\ 0 & \cdots & c & \cdots & 1 & \cdots & 0 \\ \vdots & \vdots & \vdots & \vdots & \vdots & \vdots & \vdots \\ 0 & \cdots & 0 & \cdots & 0 & \cdots & 1 \end{bmatrix} \end{array}$$

$$E^{-1}$$

$$\begin{array}{c} \\ 1 \\ \vdots \\ i \\ \vdots \\ j \\ \vdots \\ n \end{array} \begin{array}{cc} \quad i \quad\quad j \\ \begin{bmatrix} 1 & \cdots & 0 & \cdots & 0 & \cdots & 0 \\ \vdots & \vdots & \vdots & \vdots & \vdots & \vdots & \vdots \\ 0 & \cdots & 1 & \cdots & 0 & \cdots & 0 \\ \vdots & \vdots & \vdots & \vdots & \vdots & \vdots & \vdots \\ 0 & \cdots & -c & \cdots & 1 & \cdots & 0 \\ \vdots & \vdots & \vdots & \vdots & \vdots & \vdots & \vdots \\ 0 & \cdots & 0 & \cdots & 0 & \cdots & 1 \end{bmatrix} \end{array}$$

To prove that E^{-1} satisfies $EE^{-1} = E^{-1}E = I$ requires showing that for all integers $s, t = 1, \ldots, n$:

$$(EE^{-1})_{st} = \begin{cases} 1, & \text{if } s = t \\ 0, & \text{if } s \neq t \end{cases}$$

and

$$(E^{-1}E)_{st} = \begin{cases} 1, & \text{if } s = t \\ 0, & \text{if } s \neq t \end{cases}$$

From the structure of E and E^{-1}, you can verify by direct computation that for $s \neq t$, $(EE^{-1})_{st} = (E^{-1}E)_{st} = 0$. When $s = t$, you have the following three cases:

Case 1: $s = t = i$. In this case, $(EE^{-1})_{st}$

$$= E_{i*} \cdot (E^{-1})_{*i}$$

$$= [0, \ldots, 1, \ldots, 0, \ldots, 0] \cdot \begin{bmatrix} 0 \\ \vdots \\ 1 \\ \vdots \\ -c \\ \vdots \\ 0 \end{bmatrix} = 1$$

Case 2: $s = t = j$. In this case, $(EE^{-1})_{st}$

$$= E_{j*} \cdot (E^{-1})_{*j}$$

$$= [0, \ldots, c, \ldots, 1, \ldots, 0] \cdot \begin{bmatrix} 0 \\ \vdots \\ 0 \\ \vdots \\ 1 \\ \vdots \\ 0 \end{bmatrix} = 1$$

Case 3: $s = t \neq i, j$. In this case,

$$(EE^{-1})_{st} = E_{s*} \cdot (E^{-1})_{*s}$$
$$= I_{s*} \cdot I_{*s} = 1$$

Thus, in each case when $s = t$, $(EE^{-1})_{st} = 1$. You can show similarly that $(E^{-1}E)_{st} = 1$.

9. (Answers can vary.) The matrices $A = I$ and $B = -I$ are invertible but $A+B = I - I = 0$ is not invertible.

11. The inverse of the given matrix is contained in the final two columns of the following matrix:

$$\left[\begin{array}{cc|cc} 1 & 0 & -3 & -1 \\ 0 & 1 & -2 & -1 \end{array} \right]$$

This matrix is obtained by performing the

following steps of Gauss-Jordan elimination on the matrix:

$$\left[\begin{array}{cc|cc} -1 & 1 & 1 & 0 \\ 2 & -3 & 0 & 1 \end{array} \right]$$

Step 1. Multiply row 1 of the foregoing matrix by -1:

$$\left[\begin{array}{cc|cc} 1 & -1 & -1 & 0 \\ 2 & -3 & 0 & 1 \end{array} \right]$$

Step 2. Add -2 times row 1 of the foregoing matrix to row 2:

$$\left[\begin{array}{cc|cc} 1 & -1 & -1 & 0 \\ 0 & -1 & 2 & 1 \end{array} \right]$$

Step 3. Multiply row 2 of the foregoing matrix by -1:

$$\left[\begin{array}{cc|cc} 1 & -1 & -1 & 0 \\ 0 & 1 & -2 & -1 \end{array} \right]$$

Step 4. Add 1 times row 2 of the foregoing matrix to row 1:

$$\left[\begin{array}{cc|cc} 1 & 0 & -3 & -1 \\ 0 & 1 & -2 & -1 \end{array} \right]$$

13. The inverse of the given matrix is contained in the final three columns of the following matrix:

$$\left[\begin{array}{ccc|ccc} 1 & 0 & 0 & -1 & 3 & -7 \\ 0 & 1 & 0 & 0 & 1 & -3 \\ 0 & 0 & 1 & 0 & 0 & -1 \end{array} \right]$$

This matrix is obtained by performing the following steps of Gauss-Jordan elimination on the matrix:

$$\left[\begin{array}{ccc|ccc} -1 & 3 & -2 & 1 & 0 & 0 \\ 0 & 1 & -3 & 0 & 1 & 0 \\ 0 & 0 & -1 & 0 & 0 & 1 \end{array} \right]$$

Step 1. Multiply row 1 of the foregoing matrix by -1:

$$\left[\begin{array}{ccc|ccc} 1 & -3 & 2 & -1 & 0 & 0 \\ 0 & 1 & -3 & 0 & 1 & 0 \\ 0 & 0 & -1 & 0 & 0 & 1 \end{array}\right]$$

Step 2. Add 3 times row 2 of the foregoing matrix to row 1:

$$\left[\begin{array}{ccc|ccc} 1 & 0 & -7 & -1 & 3 & 0 \\ 0 & 1 & -3 & 0 & 1 & 0 \\ 0 & 0 & -1 & 0 & 0 & 1 \end{array}\right]$$

Step 3. Multiply row 3 of the foregoing matrix by -1:

$$\left[\begin{array}{ccc|ccc} 1 & 0 & -7 & -1 & 3 & 0 \\ 0 & 1 & -3 & 0 & 1 & 0 \\ 0 & 0 & 1 & 0 & 0 & -1 \end{array}\right]$$

Step 4. Add 7 times row 3 of the foregoing matrix to row 1 and add 3 times row 3 to row 2:

$$\left[\begin{array}{ccc|ccc} 1 & 0 & 0 & -1 & 3 & -7 \\ 0 & 1 & 0 & 0 & 1 & -3 \\ 0 & 0 & 1 & 0 & 0 & -1 \end{array}\right]$$

Solutions to Exercises in Section 3.3

1. The 24 permuatations of $\{1, 2, 3, 4\}$ are:

$\{1, 2, 3, 4\}$ $\{2, 1, 3, 4\}$ $\{3, 1, 2, 4\}$ $\{4, 1, 2, 3\}$

$\{1, 2, 4, 3\}$ $\{2, 1, 4, 3\}$ $\{3, 1, 4, 2\}$ $\{4, 1, 3, 2\}$

$\{1, 3, 2, 4\}$ $\{2, 3, 1, 4\}$ $\{3, 2, 1, 4\}$ $\{4, 2, 1, 3\}$

$\{1, 3, 4, 2\}$ $\{2, 3, 4, 1\}$ $\{3, 2, 4, 1\}$ $\{4, 2, 3, 1\}$

$\{1, 4, 2, 3\}$ $\{2, 4, 1, 3\}$ $\{3, 4, 1, 2\}$ $\{4, 3, 1, 2\}$

$\{1, 4, 3, 2\}$ $\{2, 4, 3, 1\}$ $\{3, 4, 2, 1\}$ $\{4, 3, 2, 1\}$

3. The following table shows the inversions for each permutation:

| | Inversions for | | | | |
| | Position | | | Total | Odd/ |
Permutation	1	2	3	Inver.	Even
$\{3, 1, 4, 2\}$	2	0	1	3	odd
$\{4, 2, 1, 3\}$	3	1	0	4	even

5. (a) $-A_{13}A_{21}A_{34}A_{42}$.
   (b) $A_{14}A_{22}A_{31}A_{43}$.

7. (a) Following the formula in the book,

$$\det\left[\begin{array}{cc} 1 & -2 \\ -3 & 4 \end{array}\right] = 1(4) - (-2)(-3)$$

$$= -2$$

(b) Following the formula in the book:

$$\left|\begin{array}{ccc} 1 & -2 & 0 \\ -3 & 4 & 1 \\ 0 & 3 & -1 \end{array}\right|$$

$$= 1(4)(-1) - 1(1)(3) -$$
$$(-2)(-3)(-1) + (-2)(1)(0) +$$
$$0(-3)(3) - 0(4)(0)$$

$$= -1$$

9. (a) $\det(A^{-1}) = 1/\det(A) = 1/5$.
   (b) $\det(AB) = \det(A)\det(B) = 5(-2) = -10$.
   (c) $\det(B') = \det(B) = -2$.
   (d) $\det(2A) = (2^3)\det(A) = 8(5) = 40$.

11. Using the formula for the determinant of a (3×3) matrix and then substituting $\mathbf{i} = (1, 0, 0), \mathbf{j} = (0, 1, 0)$, and $\mathbf{k} = (0, 0, 1)$, it follows that

$$\det \begin{bmatrix} \mathbf{i} & \mathbf{j} & \mathbf{k} \\ u_1 & u_2 & u_3 \\ v_1 & v_2 & v_3 \end{bmatrix}$$

$$= \mathbf{i} u_2 v_3 - \mathbf{i} u_3 v_2 - \mathbf{j} u_1 v_3 +$$
$$\mathbf{j} u_3 v_1 + \mathbf{k} u_1 v_2 - \mathbf{k} u_2 v_1$$

$$= (u_2 v_3 - u_3 v_2, \, u_3 v_1 - u_1 v_3, \, u_1 v_2 - u_2 v_1)$$

$$= \mathbf{u} \times \mathbf{v}$$

13. Applying the absolute value to both sides of (3.19) yields:

$$|\det(A)| = |\sum \pm A_{1j_1} A_{2j_2} \cdots A_{nj_n}|$$
$$\leq \sum |\pm A_{1j_1} A_{2j_2} \cdots A_{nj_n}| \quad (1)$$
$$= \sum |A_{1j_1}| |A_{2j_2}| \cdots |A_{nj_n}| \quad (2)$$
$$\leq \sum 1 \quad (3)$$
$$= n! \quad (4)$$

Each of the foregoing numbered steps is justified, as follows:

Step	Justification
(1)	$|a + b| \leq |a| + |b|$
(2)	$|ab| = |a||b|$
(3)	each $A_{ij} = -1, 0$, or 1
(4)	$n!$ terms are added

15. (a) The minors and cofactors are listed in the following table:

Entry	Minor	Cofactor
a	d	d
b	c	$-c$
c	b	$-b$
d	a	a

(b) From part (a), you have

Matrix of cofactors

$$\begin{bmatrix} d & -c \\ -b & a \end{bmatrix}$$

Adjoint

$$\begin{bmatrix} d & -b \\ -c & a \end{bmatrix}$$

17. You have that

$$\begin{vmatrix} -1 & -2 & 5 \\ 0 & 1 & -2 \\ -2 & 0 & 3 \end{vmatrix} = \begin{vmatrix} -1 & -2 & 5 \\ 0 & 1 & -2 \\ 0 & 4 & -7 \end{vmatrix} \quad (1)$$

$$= \begin{vmatrix} -1 & -2 & 5 \\ 0 & 1 & -2 \\ 0 & 0 & 1 \end{vmatrix} \quad (2)$$

$$= -1(1)(1) \quad (3)$$

$$= -1$$

Each of the foregoing numbered steps is justified, as follows:

Step	Justification
(1)	by Theorem 3.7(c), because -2 times row 1 is added to row 3
(2)	by Theorem 3.7(c), because -4 times row 2 is added to row 3.
(3)	because the matrix in (2) is upper triangular

19. The minors and cofactors are listed in the following table:

Entry	Minor	Cofactor
A_{11}	3	3
A_{12}	-4	4
A_{13}	2	2

21. The minors and cofactors are listed in the following table:

Entry	Minor	Cofactor
A_{21}	-6	6
A_{22}	7	7
A_{23}	-4	4

23. The minors and cofactors are listed in the

following table:

Entry	Minor	Cofactor
A_{31}	-1	-1
A_{32}	2	-2
A_{33}	-1	-1

25. Using the results of Exercise 19 together with the cofactor expansion along row 1, you have that

$$\begin{vmatrix} -1 & -2 & 5 \\ 0 & 1 & -2 \\ -2 & 0 & 3 \end{vmatrix}$$

$$= A_{11}C_{11} + A_{12}C_{12} + A_{13}C_{13}$$

$$= (-1)(3) + (-2)(4) + 5(2) = -1$$

27. From the results of Exercises 19, 21, and 23, you have that

Matrix of cofactors

$$\begin{bmatrix} 3 & 4 & 2 \\ 6 & 7 & 4 \\ -1 & -2 & -1 \end{bmatrix}$$

Adjoint

$$\begin{bmatrix} 3 & 6 & -1 \\ 4 & 7 & -2 \\ 2 & 4 & -1 \end{bmatrix}$$

29. From the results of Exercise 25 and Exercise 27, you have

$$A^{-1} = \frac{1}{\det(A)} \operatorname{adj}(A)$$

$$= \frac{1}{-1} \begin{bmatrix} 3 & 6 & -1 \\ 4 & 7 & -2 \\ 2 & 4 & -1 \end{bmatrix}$$

$$= \begin{bmatrix} -3 & -6 & 1 \\ -4 & -7 & 2 \\ -2 & -4 & 1 \end{bmatrix}$$

31. From the results of Exercise 25, you know that $\det(A) = -1$. You must also compute the following:

$$\det(A_1) = \begin{vmatrix} 6 & -2 & 5 \\ -4 & 1 & -2 \\ -3 & 0 & 3 \end{vmatrix} = -3$$

$$\det(A_2) = \begin{vmatrix} -1 & 6 & 5 \\ 0 & -4 & -2 \\ -2 & -3 & 3 \end{vmatrix} = 2$$

$$\det(A_3) = \begin{vmatrix} -1 & -2 & 6 \\ 0 & 1 & -4 \\ -2 & 0 & 3 \end{vmatrix} = -1$$

So, by Cramer's rule, the solution to the system is:

$$x = \frac{\det(A_1)}{\det(A)} = 3, \, y = \frac{\det(A_2)}{\det(A)} = -2,$$

$$z = \frac{\det(A_3)}{\det(A)} = 1$$

SOLUTIONS TO SELECTED EXERCISES IN CHAPTER 4

Solutions to Exercises in Section 4.1

1. Abstraction is the process of thinking in terms of general objects rather than specific items. Abstraction enables you to create a problem-solving framework for studying a large number of special cases.

3. To apply abstraction to the components of an n-vector $\mathbf{x} = (x_1, \ldots, x_n)$, think of those components as objects, rather than real numbers. Some special cases are when each component represents (1) a real number, (2) an n-vector, (3) a matrix, (4) a function, (5) a set, and so on.

5. It is necessary to show that all of the axioms

of a vector space are satisfied. To that end, note that because there is only one vector $\mathbf{v}$ and by the facts that $\mathbf{v} \oplus \mathbf{v} = \mathbf{v}$ and $t \odot \mathbf{v} = \mathbf{v}$, it follows that

(a) $\mathbf{v} \oplus \mathbf{v} = \mathbf{v} \in V$.

(b) $t \odot \mathbf{v} = \mathbf{v} \in V$.

(c) $\mathbf{v}$ is the zero element of V, that is, $\mathbf{0} = \mathbf{v}$, because $\mathbf{v} \oplus \mathbf{v} = \mathbf{v}$.

(d) The vector $-\mathbf{v} = \mathbf{v}$ because $\mathbf{v} \oplus (-\mathbf{v}) = (-\mathbf{v}) \oplus \mathbf{v} = \mathbf{v} \oplus \mathbf{v} = \mathbf{v} = \mathbf{0}$.

(e) $(\mathbf{v} \oplus \mathbf{v}) \oplus \mathbf{v} = \mathbf{v} \oplus (\mathbf{v} \oplus \mathbf{v}) = \mathbf{v} \oplus \mathbf{v} = \mathbf{v}$.

(f) $\mathbf{v} \oplus \mathbf{v} = \mathbf{v} \oplus \mathbf{v} = \mathbf{v}$.

(g) For a scalar s, $s \odot (\mathbf{v} \oplus \mathbf{v}) = s \odot \mathbf{v} = \mathbf{v} = \mathbf{v} \oplus \mathbf{v} = (s \odot \mathbf{v}) \oplus (s \odot \mathbf{v})$.

(h) For scalars s and t, $(s + t) \odot \mathbf{v} = \mathbf{v} = \mathbf{v} \oplus \mathbf{v} = (s \odot \mathbf{v}) \oplus (t \odot \mathbf{v})$.

(i) For scalars s and t, $(st) \odot \mathbf{v} = \mathbf{v} = s \odot \mathbf{v} = s \odot (t \odot \mathbf{v})$.

(j) Finally, $1 \odot \mathbf{v} = \mathbf{v}$.

7. It is necessary to show that all of the axioms of a vector space are satisfied. To that end, let $\mathbf{u} = (u_1, u_2, u_3)$, $\mathbf{v} = (v_1, v_2, v_3)$, and $\mathbf{w} = (w_1, w_2, w_3)$ be in V, so, by definition,

$$au_1 + bu_2 + cu_3 = 0$$
$$av_1 + bv_2 + cv_3 = 0$$
$$aw_1 + bw_2 + cw_3 = 0$$

Also, let s and t be scalars.

(a) $\mathbf{u} \oplus \mathbf{v} \in V$ because $\mathbf{u} \oplus \mathbf{v} = (u_1, u_2, u_3) + (v_1, v_2, v_3) = (u_1 + v_1, u_2 + v_2, u_3 + v_3)$ and so

$$a(u_1 + v_1) + b(u_2 + v_2) + c(u_3 + v_3)$$
$$= (au_1 + bu_2 + cu_3) + (av_1 + bu_2 + cu_3)$$
$$= 0 + 0 = 0$$

(b) $t \odot \mathbf{v} \in V$ because $t \odot \mathbf{v} = t\mathbf{v} = (tv_1, tv_2, tv_3)$ and so

$$a(tv_1) + b(tv_2) + c(tv_3)$$
$$= t(av_1 + bv_2 + cv_3)$$
$$= t(0) = 0$$

(c) The zero element, $\mathbf{0} = (0, 0, 0)$, is in V because $a(0) + b(0) + c(0) = 0$. Also,

$$\mathbf{v} \oplus \mathbf{0} = (v_1, v_2, v_3) + (0, 0, 0)$$
$$= (v_1, v_2, v_3)$$
$$= \mathbf{v}$$

Likewise, $\mathbf{0} + \mathbf{v} = \mathbf{v}$.

(d) For $\mathbf{v} = (v_1, v_2, v_3) \in V$, you have that $-\mathbf{v} = (-v_1, -v_2, -v_3)$, which is in V because

$$a(-v_1) + b(-v_2) + c(-v_3)$$
$$= -(av_1 + bv_2 + cv_3)$$
$$= 0$$

Also,

$$\mathbf{v} \oplus (-\mathbf{v}) = (v_1, v_2, v_3) + (-v_1, -v_2, -v_3)$$
$$= (0, 0, 0)$$
$$= \mathbf{0}$$

Likewise, $(-\mathbf{v}) \oplus \mathbf{v} = \mathbf{0}$. The remaining properties for a vector space hold because those properties hold for any vectors in R^3 and $V \subseteq R^3$.

9. It is necessary to show that all of the axioms of a vector space are satisfied. To that end, let $\mathbf{U} = (u_0, u_1, u_2, \ldots)$, $\mathbf{V} = (v_0, v_1, v_2, \ldots)$, and $\mathbf{W} = (w_0, w_1, w_2, \ldots)$ be signals and let s and t be scalars.

(a) $\mathbf{U} \oplus \mathbf{V} \in V$ because you have $\mathbf{U} \oplus \mathbf{V} = (u_0 + v_0, u_1 + v_1, \ldots)$ is a signal.

(b) $t \odot \mathbf{V} \in V$ because $t \odot \mathbf{V} = (tv_0, tv_1, \ldots)$ is a signal.

(c) The zero element is $\mathbf{0} = (0, 0, \ldots)$, which is in V because $(0, 0, \ldots)$ is a signal for which

$$\mathbf{V} \oplus \mathbf{0} = (v_0, v_1, \ldots) + (0, 0, \ldots)$$
$$= (v_0, v_1, \ldots)$$
$$= \mathbf{V}$$

Likewise, $\mathbf{0} \oplus \mathbf{V} = \mathbf{V}$.

(d) For a signal $\mathbf{V}$, $-\mathbf{V} = (-v_0, -v_1, \ldots)$ is in V because $(-v_0, -v_1, \ldots)$ is a signal for which

$$\mathbf{V} \oplus (-\mathbf{V}) = (v_0, v_1, \ldots)+$$
$$(-v_0, -v_1, \ldots)$$
$$= (0, 0, \ldots)$$
$$= \mathbf{0}$$

Likewise, $(-\mathbf{V}) \oplus \mathbf{V} = \mathbf{0}$.
The remaining properties for a vector space hold by an extension of those properties in R^n.

11. The set V of invertible (2×2) matrices is not a vector space because the sum of two matrices in V need not be in V, as would be necessary for V to be a vector space. For example, the (2×2) identity matrix I and $-I$ are both in V, yet, $I + (-I) = 0$ is not invertible and hence is not in V.

13. To unify the absolute value of a number and the norm of an n-vector, create an abstract system consisting of a set of objects, V, and a unary operation, say, $\| \ \|$, that associates to each object $v \in V$, a real number denoted by $\|v\|$ that is meant to represent the "length" of v. This operation is not closed on V because $\|v\|$ is a number and not an element in V.

15. (a) The number 1 satisfies the property that, for all real numbers a, $1(a) = a(1) = a$.
    (b) The $(n \times n)$ identity matrix I satisfies the property that, for all $(n \times n)$ matrices A, $IA = AI = A$.
    (c) The desired axiom for the abstract system $(S, \odot)$ is that there should exist an element $e \in S$ such that for all $x \in S$, $x \odot e = e \odot x = x$.

Solutions to Exercises in Section 4.2

1. Let A be an $(m \times n)$ matrix, 0 be the $(m \times n)$ zero matrix, and t be a real number. Then Theorem 4.1 applied to matrices says that
   (a) The matrix 0 is the only zero matrix.
   (b) The matrix $-A$ is the only negative of

the matrix A.
   (c) $0A = 0$ (the real number 0 times the matrix A is the zero matrix).
   (d) $t0 = 0$ (the real number t times the zero matrix is the zero matrix).
   (e) $(-1)A = -A$ (the real number -1 times A is the negative of A).
   (f) If $tA = 0$, then either $t = 0$ or $A = 0$.

3. To show $V = \{t(a, b, c) : t \text{ is a real number}\}$ is a subspace of R^3 by Theorem 4.2, let $\mathbf{u} = r(a, b, c)$ and $\mathbf{v} = s(a, b, c)$ be elements of V and let t be a real number. Then $\mathbf{u} \oplus \mathbf{v} \in V$ because

$$\mathbf{u} \oplus \mathbf{v} = r(a, b, c) + s(a, b, c)$$
$$= (r + s)(a, b, c)$$

Also, $t \odot \mathbf{u} \in V$ because

$$t \odot \mathbf{u} = t[r(a, b, c)] = (tr)(a, b, c)$$

5. To show that the set $U = \{f : R^1 \rightarrow R^1 : f \text{ is differentiable}\}$ is a subspace of all real-valued functions, according to Theorem 4.2, let f and g be two elements of U and let t be a real number. Then $f + g$ is a differentiable function because the sum of two differentiable functions is a differentiable function. Thus, $f \oplus g = f + g \in U$. Also, tf is a differentiable function because a real number times a differentiable function is a differentiable function. Thus, $t \odot f = tf \in U$.

7. The set $U = \{(-1, 0) + t(2, 1) : t \text{ is a non-negative real number}\}$ is shown in the following figure:

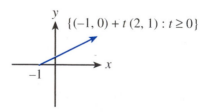

(Answers can vary.) Theorem 4.2(a) fails to hold for the vectors $(-1, 0)$ and $(-1, 0)$, both of which are in U, yet $(-1, 0) + (-1, 0) = (-2, 0)$ is not in U. Theorem 4.2(b) fails to hold for $s = 0$ and $\mathbf{u} = (-1, 0) \in U$ because $s \odot \mathbf{u} = 0(-1, 0) = (0, 0)$, which is not an element of U.

9. The set $U = \{(x, y) : x^2 + y^2 \leq 4\}$ is shown in the following figure:

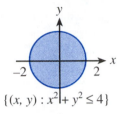

$\{(x, y) : x^2 + y^2 \leq 4\}$

(Answers can vary.) Theorem 4.2(a) fails to hold for the vectors $\mathbf{u} = (1, 0) \in U$ and $\mathbf{v} = (2, 0) \in U$ because $\mathbf{u} \oplus \mathbf{v} = (1, 0) + (2, 0) = (3, 0)$, which is not in U since $3^2 + 0^2 = 9 > 4$. Theorem 4.2(b) fails to hold for $t = 3$ and $\mathbf{u} = (1, 0) \in U$ because $t \odot \mathbf{u} = 3(1, 0) = (3, 0)$, which is not in U.

11. Recognizing the key words *for all*, choose a scalar t and a vector $\mathbf{v} \in V$. It must be shown that $(-t) \odot \mathbf{v} = t \odot (-\mathbf{v}) = -(t \odot \mathbf{v})$. It is first established that $(-t) \odot \mathbf{v} = -(t \odot \mathbf{v})$ by showing that $[(-t) \odot \mathbf{v}] \oplus (t \odot \mathbf{v}) = \mathbf{0}$. But this is true because

$$[(-t) \odot \mathbf{v}] \oplus (t \odot \mathbf{v}) = [(-t) + t] \odot \mathbf{v} \quad (1)$$
$$= 0 \odot \mathbf{v} \quad (2)$$
$$= \mathbf{0} \quad (3)$$

Each of the foregoing numbered steps is justified, as follows:

Step	Justification
(1)	by axiom 8
(2)	because $(-t) + t = 0$
(3)	by part (c) of Theorem 4.1

It is now established that $t \odot (-\mathbf{v}) = -(t \odot \mathbf{v})$ by showing that $[t \odot (-\mathbf{v})] \oplus (t \odot \mathbf{v}) = \mathbf{0}$. But this is true because

$$[t \odot (-\mathbf{v})] \oplus (t \odot \mathbf{v}) = t \odot [(-\mathbf{v}) \oplus \mathbf{v}] \quad (1)$$
$$= t \odot \mathbf{0} \quad (2)$$
$$= \mathbf{0} \quad (3)$$

Each of the foregoing numbered steps is justified, as follows:

Step	Justification
(1)	by axiom 7
(2)	by axiom 4
(3)	by part (d) of Theorem 4.1

13. The sets in (a), (c), and (d) are subspaces of the vector space of real-valued functions for the following reasons.

(a) The set $U = \{f \in V : f(0) = 0\}$ is a subspace of V by Theorem 4.2. To see that this is so, let $f, g \in U$. Then $f(0) = 0$ and $g(0) = 0$. Now $f + g \in U$ because $(f + g)(0) = f(0) + g(0) = 0 + 0 = 0$. Also, for a real number t, $tf \in U$ because $(tf)(0) = tf(0) = t(0) = 0$.

(b) The set $U = \{f \in V : f(0) = 3\}$ is not a subspace of V. This is because Theorem 4.2(b) fails to hold. Specifically, for $f \in U$ and the number $t = 0$, tf is not in U since $(tf)(0) = tf(0) = (0)3 = 0 \neq 3$.

(c) The set $U = \{f \in V : f(1) = 0\}$ is a subspace of V by Theorem 4.2. To see that this is so, let $f, g \in U$. Then $f(1) = 0$ and $g(1) = 0$. Now $f + g \in U$ because $(f + g)(1) = f(1) + g(1) = 0 + 0 = 0$. Also, for a real number t, $tf \in U$ because $(tf)(1) = tf(1) = t(0) = 0$.

(d) For a given constant c, the set $U = \{f \in V : f(x) = c\}$ is a subspace of V by Theorem 4.2. To see that this is so, let $f, g \in U$. Then there are real numbers c and d such that for all real numbers x, $f(x) = c$ and $g(x) = d$. Now $f + g \in U$ because, for all real numbers x, $(f + g)(x) = f(x) + g(x) = c + d = $ a constant. Also, for a real number t, $tf \in U$ because, for all real numbers x, $(tf)(x) = tf(x) = tc = $ a constant.

Solutions to Exercises in Section 4.3

1. The vector $(4, 2) = 2(2, 0) + (-1)(0, -2)$.

3. The polynomial $-1 - 2x + x^2 = (-1)(1 + x) + (-1)(x + x^2) + 2x^2$.

5. You want to find values for the real numbers a, b, and c that satisfy:

$$(1, -2, 0, 1) = a(1, 0, 3, -1) +$$
$$b(0, -1, 0, -2) +$$
$$c(0, 0, -1, 2)$$

Performing the multiplications and equating corresponding components, you want to find values for a, b, and c that satisfy the following system of linear equations:

$$
\begin{aligned}
a & & & & & = 1 \\
& - b & & & & = -2 \\
3a & & - c & & & = 0 \\
-a & - 2b & + 2c & & & = 1
\end{aligned}
$$

7. The vectors $(2, 0)$ and $(0, -2)$ span R^2 because you can write any 2-vector $\mathbf{v} = (v_1, v_2)$ as the following linear combination:

$$(v_1, v_2) = \frac{v_1}{2}(2, 0) - \frac{v_2}{2}(0, -2)$$

9. The four given matrices span the set of all (2×2) matrices because for any (2×2) matrix

$$E = \begin{bmatrix} a & b \\ c & d \end{bmatrix}$$

you can write E as the following linear combination:

$$
\begin{bmatrix} a & b \\ c & d \end{bmatrix} = a \begin{bmatrix} 1 & 0 \\ 0 & 0 \end{bmatrix} + (d - b) \begin{bmatrix} 0 & -1 \\ 0 & 0 \end{bmatrix} +
$$

$$
\frac{d - c}{2} \begin{bmatrix} 0 & 0 \\ -2 & 0 \end{bmatrix} + d \begin{bmatrix} 0 & 1 \\ 1 & 1 \end{bmatrix}
$$

11. The key question is, "How can I show that a set $(\text{span}\{\mathbf{v}_1, \ldots, \mathbf{v}_k\})$ is a subset of an-

other set $(\text{span}\{\mathbf{v}_1, \ldots, \mathbf{v}_{k+1}\})$?" From the definition of subset, it is necessary to show that for every element $\mathbf{v}$ in $\text{span}\{\mathbf{v}_1, \ldots, \mathbf{v}_k\}$, $\mathbf{v}$ is in $\text{span}\{\mathbf{v}_1, \ldots, \mathbf{v}_{k+1}\}$. Recognizing the key words *for all*, choose an element $\mathbf{v}$ in $\text{span}\{\mathbf{v}_1, \ldots, \mathbf{v}_k\}$ and then show that $\mathbf{v}$ is in $\text{span}\{\mathbf{v}_1, \ldots, \mathbf{v}_{k+1}\}$. To show that $\mathbf{v}$ is in $\text{span}\{\mathbf{v}_1, \ldots, \mathbf{v}_{k+1}\}$, by definition of span, you must show that $\mathbf{v}$ is a linear combination of $\mathbf{v}_1, \ldots, \mathbf{v}_{k+1}$, that is, that there are scalars $t_1, \ldots, t_{k+1}$ such that

$$\mathbf{v} = (t_1 \odot \mathbf{v}_1) \oplus \cdots \oplus (t_{k+1} \odot \mathbf{v}_{k+1})$$

These scalars are constructed from the fact that $\mathbf{v}$ is in $\text{span}\{\mathbf{v}_1, \ldots, \mathbf{v}_k\}$, which means that there are scalars $s_1, \ldots, s_k$ such that

$$\mathbf{v} = (s_1 \odot \mathbf{v}_1) \oplus \cdots \oplus (s_k \odot \mathbf{v}_k)$$

You can now construct the desired scalars as follows. Let $t_1 = s_1, \ldots, s_k = t_k$, and $t_{k+1} = 0$, which satisfy

$$\mathbf{v} = (t_1 \odot \mathbf{v}_1) \oplus \cdots \oplus (t_{k+1} \odot \mathbf{v}_{k+1})$$

13. To show that the vectors $(-1, 1)$ and $(2, 1)$ are linearly independent, use the definition and let s and t be real numbers for which

$$s(-1, 1) + t(2, 1) = (0, 0)$$

Performing the multiplications and additions and then equating corresponding components of the vectors on each side of the equality, it follows that

$$
\begin{aligned}
-s + 2t &= 0 \\
s + t &= 0
\end{aligned}
$$

The only solution to the foregoing equations is $s = t = 0$, from which it follows that the two vectors are linearly independent.

15. To show that the matrices in Exercise 9 are linearly independent, use the definition and

let s, t, u and v be real numbers for which

$$s \begin{bmatrix} 1 & 0 \\ 0 & 0 \end{bmatrix} + t \begin{bmatrix} 0 & -1 \\ 0 & 0 \end{bmatrix} +$$

$$u \begin{bmatrix} 0 & 0 \\ -2 & 0 \end{bmatrix} + v \begin{bmatrix} 0 & 1 \\ 1 & 1 \end{bmatrix} = \begin{bmatrix} 0 & 0 \\ 0 & 0 \end{bmatrix}$$

Performing the multiplications and additions and then equating corresponding components of the matrices on both sides of the equality, it follows that

$$s = 0$$
$$-t + v = 0$$
$$-2u + v = 0$$
$$v = 0$$

The only solution to the foregoing equations is $s = t = u = v = 0$, from which it follows that the four matrices are linearly independent.

17. (a) To prove that **u** and **v** are linearly independent, use the definition and let s and t be real numbers for which

$$s\mathbf{u} + t\mathbf{v} = \mathbf{0}$$

Now dot product both sides of the foregoing equality with the vector **u** to obtain

$$s(\mathbf{u} \cdot \mathbf{u}) + t(\mathbf{u} \cdot \mathbf{v}) = \mathbf{u} \cdot \mathbf{0} = 0$$

Using the assumption that $\mathbf{u} \cdot \mathbf{v} = 0$, you now have

$$s(\mathbf{u} \cdot \mathbf{u}) = 0$$

Dividing through by $\mathbf{u} \cdot \mathbf{u}$ (which is not 0 because $\mathbf{u} \neq \mathbf{0}$), you know that $s = 0$. It now follows that $t\mathbf{v} = \mathbf{0}$ and, because $\mathbf{v} \neq \mathbf{0}$, it must be that $t = 0$. Thus, $s = t = 0$ and so the vectors **u** and **v** are linearly independent.

(b) To prove that **u** and $\mathbf{u} \times \mathbf{v}$ are linearly independent, use the definition and let s

and t be real numbers for which

$$s\mathbf{u} + t(\mathbf{u} \times \mathbf{v}) = \mathbf{0}$$

Now dot product both sides of the foregoing equality with the vector **u** to obtain

$$s(\mathbf{u} \cdot \mathbf{u}) + t[\mathbf{u} \cdot (\mathbf{u} \times \mathbf{v})] = \mathbf{u} \cdot \mathbf{0} = 0$$

Using the fact that **u** and $\mathbf{u} \times \mathbf{v}$ are perpendicular, it follows that $\mathbf{u} \cdot (\mathbf{u} \times \mathbf{v}) = 0$ and so

$$s(\mathbf{u} \cdot \mathbf{u}) = 0$$

Dividing through by $\mathbf{u} \cdot \mathbf{u}$ (which is not 0 because $\mathbf{u} \neq \mathbf{0}$), you know that $s = 0$. It now follows that $t(\mathbf{u} \times \mathbf{v}) = \mathbf{0}$ and, because $\mathbf{u} \times \mathbf{v} \neq \mathbf{0}$, it must be that $t = 0$. Thus, $s = t = 0$ and so the vectors **u** and **v** are linearly independent.

19. To show that the vectors $\mathbf{v}_1, \ldots, \mathbf{v}_k$ are linearly independent, use the definition and let $t_1, \ldots, t_k$ be real numbers for which

$$\mathbf{0} = (t_1 \odot \mathbf{v}_1) \oplus \cdots \oplus (t_k \odot \mathbf{v}_k)$$

Now $t_k = 0$ for otherwise, you could solve the foregoing equation for $\mathbf{v}_k$ and conclude that $\mathbf{v}_k$ is a linear combination of the preceding vectors, which cannot happen. Thus, $t_k = 0$ and so

$$\mathbf{0} = (t_1 \odot \mathbf{v}_1) \oplus \cdots \oplus (t_{k-1} \odot \mathbf{v}_{k-1})$$

Using a similar argument, $t_{k-1} = 0$. It therefore follows that $t_1 = \cdots = t_k = 0$, from which you can conclude that $\mathbf{v}_1, \ldots, \mathbf{v}_k$ are linearly independent.

21. To show that the vectors $(1, 1)$ and $(-2, -2)$ are linearly dependent, it is necessary to determine real numbers s and t, at least one of which is not 0, such that

$$s(1, 1) + t(-2, -2) = (0, 0)$$

One such set of values is $s = 2$ and $t = 1$, so these two vectors are linearly dependent.

23. The given matrices are linearly dependent be-

cause

$$\begin{bmatrix} 6 & -6 \\ -5 & 4 \end{bmatrix} = (3)\begin{bmatrix} 1 & -2 \\ -1 & 0 \end{bmatrix} +$$

$$(-1)\begin{bmatrix} -3 & 0 \\ 2 & -4 \end{bmatrix}$$

25. To show that the two statements (i) and (ii) are equivalent, two proofs are needed. So first, to show that (i) implies (ii), assume that

> There are real numbers a, b, and c, at least one of which is 1, such that $a\mathbf{u} + b\mathbf{v} + c\mathbf{w} = \mathbf{0}$.

It must be shown that

> There are real numbers d, e, and f, not all 0, such that $d\mathbf{u} + e\mathbf{v} + f\mathbf{w} = \mathbf{0}$.

Recognizing the key words *there are*, you must construct appropriate values for d, e, and f. You can use a, b, and c to do so. Specifically, let $d = a, e = b$, and $f = c$. At least one of d, e, and f is not zero because one of a, b, and c is 1. It is also clear that $d\mathbf{u} + e\mathbf{v} + f\mathbf{w} = a\mathbf{u} + b\mathbf{v} + c\mathbf{w} = \mathbf{0}$.

It remains to show that (ii) implies (i). Thus, you can assume that

> There are real numbers d, e, and f, not all 0, such that $d\mathbf{u} + e\mathbf{v} + f\mathbf{w} = \mathbf{0}$.

It must be shown that

> There are real numbers a, b, and c, at least one of which is 1, such that $a\mathbf{u} + b\mathbf{v} + c\mathbf{w} = \mathbf{0}$.

Recognizing the key words *there are*, you must construct appropriate values for a, b, and c. The specific way in which you use d, e, and f to do so depends on which of these values is not zero. Thus, a proof by cases is used, depending on whether $d \neq 0, e \neq 0$, or $f \neq 0$. Assume, without loss of generality, that $d \neq 0$. Now define a, b, and c as follows: $a = 1, b = e/d$, and $c = f/d$. You can check that $a\mathbf{u} + b\mathbf{v} + c\mathbf{w} = \mathbf{0}$.

27. To show that the vectors $A\mathbf{v}_1, \ldots, A\mathbf{v}_k$ are linearly dependent, by definition, it is necessary to show that there are real numbers $t_1, \ldots, t_k$, not all zero, such that

$$t_1(A\mathbf{v}_1) + \cdots + t_k(A\mathbf{v}_k) = \mathbf{0}$$

These numbers are constructed from the hypothesis that $\mathbf{v}_1, \ldots, \mathbf{v}_k$ are linearly dependent. By definition, this means that there are real numbers $s_1, \ldots, s_k$, not all zero, such that

$$s_1\mathbf{v}_1 + \cdots + s_k\mathbf{v}_k = \mathbf{0}$$

Now define each t_i as follows: $t_1 = s_1, \ldots,$ $t_k = s_k$, which are not all zero. You then have that

$$\begin{aligned} t_1(A\mathbf{v}_1) + \cdots + t_k(A\mathbf{v}_k) \\ = A(t_1\mathbf{v}_1 + \cdots + t_k\mathbf{v}_k) \\ = A(s_1\mathbf{v}_1 + \cdots + s_k\mathbf{v}_k) \\ = A(\mathbf{0}) \\ = \mathbf{0} \end{aligned}$$

29. To prove that $(2, 0)$ and $(2, 1)$ form a basis for R^2, it is necessary to show that these vectors are linearly independent and span R^2. To see that the vectors are linearly independent, use the definition and let s and t be real numbers for which

$$s(2, 0) + t(2, 1) = (0, 0)$$

Performing the multiplications and additions and then equating corresponding components of the vectors on each side of the equality, it follows that

$$2s + 2t = 0$$
$$t = 0$$

The only solution to the foregoing equations is $s = t = 0$, from which it follows that the two vectors are linearly independent.

The vectors $(2, 0)$ and $(2, 1)$ span R^2 because you can write any 2-vector $\mathbf{v} = (v_1, v_2)$ as the following linear combination:

$$(v_1, v_2) = \frac{v_1 - 2v_2}{2}(2, 0) + v_2(2, 1)$$

31. To prove that $(1, 2)$ and $(-2, 4)$ form a basis for R^2, it is necessary to show that these vectors are linearly independent and span R^2. To see that the vectors are linearly independent, use the definition and let s and t be real numbers for which

$$s(1, 2) + t(-2, 4) = (0, 0)$$

Performing the multiplications and additions and then equating corresponding components of the vectors on each side of the equality, it follows that

$$s - 2t = 0$$
$$2s + 4t = 0$$

The only solution to the foregoing equations is $s = t = 0$, from which it follows that the two vectors are linearly independent.

The vectors $(1, 2)$ and $(-2, 4)$ span R^2 because you can write any 2-vector $\mathbf{v} = (v_1, v_2)$ as the following linear combination:

$$(v_1, v_2) = \frac{2v_1 + v_2}{4}(1, 2) + \frac{v_2 - 2v_1}{8}(-2, 4)$$

33. To prove that $\mathbf{v}_1, \mathbf{v}_1 \oplus \mathbf{v}_2$, and $\mathbf{v}_1 \oplus \mathbf{v}_2 \oplus \mathbf{v}_3$ form a basis for V, you must show that these vectors are linearly independent and span V. To see that these vectors are linearly independent, use the definition and let s, t, and u be real numbers for which

$$[s \odot \mathbf{v}_1] \oplus [t \odot (\mathbf{v}_1 \oplus \mathbf{v}_2)] \oplus$$
$$\left[u \odot (\mathbf{v}_1 \oplus \mathbf{v}_2 \oplus \mathbf{v}_3)\right] = \mathbf{0}$$

Removing the parentheses and collecting like terms, it then follows that

$$[(s+t+u) \odot \mathbf{v}_1] \oplus [(t+u) \odot \mathbf{v}_2] \oplus [u \odot \mathbf{v}_3] = \mathbf{0}$$

From the hypothesis that $\mathbf{v}_1$, $\mathbf{v}_2$, and $\mathbf{v}_3$ form a basis, you know that these vectors are linearly independent and so $s + t + u = 0$, $t + u = 0$, and $u = 0$. But then $s = t = u = 0$, from which it follows that $\mathbf{v}_1$, $\mathbf{v}_1 \oplus \mathbf{v}_2$, and $\mathbf{v}_1 \oplus \mathbf{v}_2 \oplus \mathbf{v}_3$ are linearly independent.

It remains to show that $\mathbf{v}_1$, $\mathbf{v}_1 \oplus \mathbf{v}_2$, and $\mathbf{v}_1 \oplus \mathbf{v}_2 \oplus \mathbf{v}_3$ span V. To that end, let $\mathbf{v} \in V$.

You must show that there are scalars s, t, and u such that

$$\mathbf{v} = [s \odot \mathbf{v}_1] \oplus [t \odot (\mathbf{v}_1 \oplus \mathbf{v}_2)] \oplus$$
$$[u \odot (\mathbf{v}_1 \oplus \mathbf{v}_2 \oplus \mathbf{v}_3)]$$

To construct these scalars, use the hypothesis that $\mathbf{v}_1$, $\mathbf{v}_2$, and $\mathbf{v}_3$ is a basis for V and hence these vectors span V. Thus, there are scalars a, b, and c such that

$$\mathbf{v} = (a \odot \mathbf{v}_1) \oplus (b \odot \mathbf{v}_2) \oplus (u \odot \mathbf{v}_3)$$

Now construct

$$s = a - b - c, \quad t = b - c, \quad u = c$$

You can verify that

$$(s \odot \mathbf{v}_1) \oplus t(\mathbf{v}_1 \oplus \mathbf{v}_2) \oplus u(\mathbf{v}_1 \oplus \mathbf{v}_2 \oplus \mathbf{v}_3)$$
$$= (a \odot \mathbf{v}_1) \oplus (b \odot \mathbf{v}_2) \oplus (c \odot \mathbf{v}_3) = \mathbf{v}$$

35. To show that $\mathbf{v}_1, \ldots, \mathbf{v}_{k-1}, \mathbf{w}$ form a basis for V by definition, it is necessary to show that these vectors (1) are linearly independent and (2) span V. To see that these vectors are linearly independent, by definition, let $t_1, \ldots, t_k$ be scalars such that

$$(t_1 \odot \mathbf{v}_1) \oplus \cdots \oplus (t_{k-1} \odot \mathbf{v}_{k-1}) \oplus (t_k \odot \mathbf{w}) = \mathbf{0}$$

It must be shown that $t_1 = \cdots = t_k = 0$. The idea is to replace $\mathbf{w}$ with an expression involving $\mathbf{v}_1, \ldots, \mathbf{v}_k$ and then to use the fact that $\mathbf{v}_1, \ldots, \mathbf{v}_k$ form a basis for V and hence are linearly independent. Specifically, because $\mathbf{v}_1, \ldots, \mathbf{v}_k$ form a basis for V and $\mathbf{w} \in V$, $\mathbf{w}$ is a linear combination of these basis vectors. That is, there are scalars $s_1, \ldots, s_k$ such that

$$\mathbf{w} = (s_1 \odot \mathbf{v}_1) \oplus \cdots \oplus (s_k \odot \mathbf{v}_k)$$

Substituting this expression for $\mathbf{w}$ in the previous equation yields that

$$(t_1 \odot \mathbf{v}_1) \oplus \cdots \oplus (t_{k-1} \odot \mathbf{v}_{k-1}) \oplus$$
$$t_k \odot (s_1 \odot \mathbf{v}_1 \oplus \cdots \oplus s_{k-1} \odot \mathbf{v}_{k-1} \oplus$$
$$s_k \odot \mathbf{v}_k) = \mathbf{0}$$

Collecting like terms leads to

$$(t_1 + t_k s_1) \odot \mathbf{v}_1 \oplus \cdots \oplus$$

$$(t_{k-1} + t_k s_{k-1}) \odot \mathbf{v}_{k-1} \oplus (t_k s_k) \odot \mathbf{v}_k = \mathbf{0}$$

Now, from the fact that $\mathbf{v}_1, \ldots, \mathbf{v}_k$ are linearly independent, you know that for all scalars $r_1, \ldots, r_k$ with

$$\mathbf{0} = (r_1 \odot \mathbf{v}_1) \oplus \cdots \oplus (r_k \odot \mathbf{v}_k)$$

it follows that $r_1 = \cdots = r_k = 0$. Specializing this statement to the particular values $r_1 = t_1 + t_k s_1, \ldots, r_{k-1} = t_{k-1} + t_k s_{k-1}$, and $r_k = t_k s_k$, you can conclude that all of these scalars are 0. In particular, $r_k = t_k s_k = 0$ from which it follows that either $t_k = 0$ or $s_k = 0$. Now it cannot happen that $s_k = 0$ because if $s_k = 0$, then

$$\mathbf{w} = (s_1 \odot \mathbf{v}_1) \oplus \cdots \oplus (s_{k-1} \odot \mathbf{v}_{k-1})$$

This contradicts the hypothesis that $\mathbf{w}$ is not in $\mathrm{span}\{\mathbf{v}_1, \ldots, \mathbf{v}_{k-1}\}$. Therefore, $t_k = 0$ and so

$$0 = r_1 = t_1 + t_k s_1 = t_1$$
$$\vdots \qquad \vdots \qquad \vdots$$
$$0 = r_{k-1} = t_{k-1} + t_k s_{k-1} = t_{k-1}$$

The fact that $t_1 = \cdots = t_k = 0$ establishes that $\mathbf{v}_1, \ldots, \mathbf{v}_{k-1}, \mathbf{w}$ are linearly independent.

It remains to show that these vectors span V. To that end, let $\mathbf{v} \in V$. It must be shown that there are scalars $t_1, \ldots, t_k$ such that

$$(t_1 \odot \mathbf{v}_1) \oplus \cdots \oplus (t_{k-1} \odot \mathbf{v}_{k-1}) \oplus (t_k \odot \mathbf{w}) = \mathbf{v}$$

The construction method is used to produce these scalars. Specifically, you can use the hypothesis that $\mathbf{v}_1, \ldots, \mathbf{v}_k$ form a basis for V to conclude that there are scalars $s_1, \ldots, s_k$ such that

$$\mathbf{v} = (s_1 \odot \mathbf{v}_1) \oplus \cdots \oplus (s_k \odot \mathbf{v}_k)$$

Also, $\mathbf{w} \in V$, so there are scalars $r_1, \ldots, r_k$ such that

$$\mathbf{w} = (r_1 \odot \mathbf{v}_1) \oplus \cdots \oplus (r_k \odot \mathbf{v}_k)$$

Note that $r_k \neq 0$, otherwise, $\mathbf{w}$ would belong to $\mathrm{span}\{\mathbf{v}_1, \ldots, \mathbf{v}_{k-1}\}$. Solving for $\mathbf{v}_k$ and substituting this expression in the preceding equation yields

$$\mathbf{v} = (s_1 \odot \mathbf{v}_1) \oplus \cdots \oplus (s_{k-1} \odot \mathbf{v}_{k-1}) \oplus$$

$$\left(s_k \odot \left[\left(\frac{1}{r_k} \odot \mathbf{w} \right) \oplus \left(-\frac{r_1}{r_k} \mathbf{v}_1 \right) \oplus \right. \right.$$

$$\left. \left. \cdots \oplus \left(-\frac{r_{k-1}}{r_k} \mathbf{v}_{k-1} \right) \right] \right)$$

Collecting like terms, you can see that the desired values of the scalars $t_1, \ldots, t_k$ are

$$t_1 = s_1 - \frac{s_k r_1}{r_k}, \ldots, t_{k-1} = s_{k-1} - \frac{s_k r_{k-1}}{r_k},$$

$$t_k = \frac{s_k}{r_k}$$

The proof is complete because you now know that $\mathbf{v}_1, \ldots, \mathbf{v}_{k-1}, \mathbf{w}$ form a basis for V.

37. (a) The components of the given matrix, relative to the given basis, are $(1, 1, 1/2, 1)$. This is because

$$\begin{bmatrix} 1 & 0 \\ 0 & 1 \end{bmatrix} = (1) \begin{bmatrix} 1 & 0 \\ 0 & 0 \end{bmatrix} + (1) \begin{bmatrix} 0 & -1 \\ 0 & 0 \end{bmatrix} +$$

$$\left(\frac{1}{2} \right) \begin{bmatrix} 0 & 0 \\ -2 & 0 \end{bmatrix} + (1) \begin{bmatrix} 0 & 1 \\ 1 & 1 \end{bmatrix}$$

(b) The components of the given matrix, relative to the basis, are $(-1, -1, -1, 0)$. This is because you can write the matrix

$$\begin{bmatrix} -1 & 1 \\ 2 & 0 \end{bmatrix}$$

as the following linear combination:

$$(-1)\begin{bmatrix}1 & 0\\0 & 0\end{bmatrix} + (-1)\begin{bmatrix}0 & -1\\0 & 0\end{bmatrix} +$$

$$(-1)\begin{bmatrix}0 & 0\\-2 & 0\end{bmatrix} + (0)\begin{bmatrix}0 & 1\\1 & 1\end{bmatrix}$$

Solutions to Exercises in Section 4.4

1. To prove that a line L through the origin in R^n is one dimensional, you would have to prove that a basis for L consists of one vector.

3. These matrices are already in row-echelon form, so the bases consists of the nonzero rows. Thus,
   (a) A basis for the row space of the given matrix is $(1, 0, -2)$ and $(0, 0, 1)$. The dimension of the row space is 2.
   (b) A basis for the row space of the given matrix is $(1, 0, -1)$ and $(0, 1, 2)$. The dimension of the row space is 2.

5. The hypothesis states that $\mathbf{v}_1, \ldots, \mathbf{v}_n$ are linearly independent. By definition, for these vectors to form a basis for V, it remains to show that these vectors span V. To that end, let $\mathbf{v} \in V$. It must be shown that there are scalars $t_1, \ldots, t_n$ such that

$$\mathbf{v} = (t_1 \odot \mathbf{v}_1) \oplus \cdots \oplus (t_n \odot \mathbf{v}_n)$$

The construction method is used to produce these scalars. Specifically, from the hypothesis that V is an n-dimensional vector space, you know that there is a basis for V consisting of n vectors, say, $\mathbf{w}_1, \ldots, \mathbf{w}_n$. Now these vectors span V, so each vector $\mathbf{v}_i$ is a linear combination of $\mathbf{w}_1, \ldots, \mathbf{w}_n$. That is, for each $i = 1, \ldots, n$, there are scalars $s_{i1}, \ldots, s_{in}$ such that

$$\mathbf{v}_i = (s_{i1} \odot \mathbf{w}_1) \oplus \cdots \oplus (s_{in} \odot \mathbf{w}_n)$$

So, you want to find values for $t_1, \ldots, t_n$ that

satisfy

$$\begin{aligned}\mathbf{v} &= (t_1 \odot \mathbf{v}_1) \oplus \cdots \oplus (t_n \odot \mathbf{v}_n)\\&= t_1 \odot (s_{11} \odot \mathbf{w}_1 \oplus \cdots \oplus s_{1n} \odot \mathbf{w}_n)\oplus\\&\quad \cdots \oplus t_n \odot (s_{n1} \odot \mathbf{w}_1 \oplus \cdots \oplus s_{nn} \odot \mathbf{w}_n)\\&= [(t_1 s_{11} + \cdots + t_n s_{n1}) \odot \mathbf{w}_1] \oplus \cdots \oplus\\&\quad [(t_1 s_{1n} + \cdots + t_n s_{nn}) \odot \mathbf{w}_n]\end{aligned}$$

You also know that, because $\mathbf{v} \in V$, there are scalars $r_1, \ldots, r_n$ such that

$$\mathbf{v} = (r_1 \odot \mathbf{w}_1) \oplus \cdots \oplus (r_n \odot \mathbf{w}_n)$$

Equating corresponding coefficients of $\mathbf{w}_i$, you want to find values for $t_1, \ldots, t_n$ such that, for each $i = 1, \ldots, n$,

$$t_1 s_{1i} + \cdots + t_n s_{ni} = r_i$$

Equivalently, letting S be the be $(n \times n)$ matrix in which $S_{ij} = s_{ij}$ and

$$\mathbf{r} = \begin{bmatrix}r_1\\\vdots\\r_n\end{bmatrix} \quad \text{and} \quad \mathbf{t} = \begin{bmatrix}t_1\\\vdots\\t_n\end{bmatrix}$$

you want to find an n-vector $\mathbf{t}$ such that

$$S\mathbf{t} = \mathbf{r}$$

The $(n \times n)$ matrix S is invertible because the vectors $\mathbf{v}_1, \ldots, \mathbf{v}_n$ are linearly independent, so you can compute the value for $\mathbf{t}$ using S^{-1}, that is, $\mathbf{t} = S^{-1}\mathbf{r}$. It has thus been shown that the linearly independent vectors $\mathbf{v}_1, \ldots, \mathbf{v}_n$ span V and hence form a basis for V, completing the proof.

7. To show that dim $V \leq k$, the construction method is used to show that V has a basis consisting of k or fewer vectors. To that end, consider the vectors $\mathbf{v}_1, \ldots, \mathbf{v}_k$. If these vectors, which span V, are linearly independent, then they form a basis for V and the proof is done. If the vectors are not linearly independent, then one of them is a linear combination of the remaining vectors. Without loss of generality, assume that $\mathbf{v}_k$ is a linear

combination of $\mathbf{v}_1, \ldots, \mathbf{v}_{k-1}$. It then follows that span$\{\mathbf{v}_1, \ldots, \mathbf{v}_{k-1}\}$ = span$\{\mathbf{v}_1, \ldots, \mathbf{v}_k\}$ = span V. If $\mathbf{v}_1, \ldots, \mathbf{v}_{k-1}$ are linearly independent, then these vectors form a basis for V. Otherwise, one of them is a linear combination of the remaining vectors. This means that one more vector can be removed from the list and the remaining vectors still span V. Eventually, a set of k or fewer linearly independent vectors that span V is obtained. These vectors provide a basis for V and show that dim $V \leq k$, completing the proof.

9. Let $S = \{\mathbf{v}_1, \mathbf{v}_2, \mathbf{v}_3, \mathbf{v}_4\}$ and $\mathbf{e}_1, \ldots, \mathbf{e}_5$ be the five standard unit vectors in R^5. You can see that for each $i = 1, 2, 3,$ and 4, $\mathbf{e}_i$ is not in span(S) by verifying that each of the following systems of linear equations has no solution for the real numbers $t_1, t_2, t_3,$ and t_4:

$$t_1\mathbf{v}_1 + t_2\mathbf{v}_2 + t_3\mathbf{v}_3 + t_4\mathbf{v}_4 = \mathbf{e}_i \quad (i = 1, 2, 3, 4)$$

In contrast, $\mathbf{e}_5$ is in span(S) because a solution to the system

$$t_1\mathbf{v}_1 + t_2\mathbf{v}_2 + t_3\mathbf{v}_3 + t_4\mathbf{v}_4 = \mathbf{e}_5$$

is $t_1 = 1/2, t_2 = 0, t_3 = 1/4,$ and $t_4 = 0.$

11. (a) The matrix in Exercise 3(a) is in row-echelon form, so the following columns 1 and 3, corresponding to the columns containing the leading 1's, form a basis for the column space of the matrix:

$$\begin{bmatrix} 1 \\ 0 \end{bmatrix} \quad \text{and} \quad \begin{bmatrix} -2 \\ 1 \end{bmatrix}$$

The dimension of the column space is 2.
(b) The matrix in Exercise 3(b) is in row-echelon form, so the following columns 1 and 2, corresponding to the columns containing the leading 1's, form a basis for the column space of the matrix:

$$\begin{bmatrix} 1 \\ 0 \\ 0 \end{bmatrix} \quad \text{and} \quad \begin{bmatrix} 0 \\ 1 \\ 0 \end{bmatrix}$$

The dimension of the column space is 2.

13. (Answers can vary.) A basis for the row space consists of rows 1 and 3 of the given matrix A, namely:

$$[-2 \quad 4 \quad 6] \quad \text{and} \quad [0 \quad 3 \quad 0]$$

The reason is that columns 1 and 3 of the following row-echelon form of A' correspond to the columns that contain the leading values:

$$\begin{bmatrix} 1 & -2 & 0 \\ 0 & 0 & 1 \\ 0 & 0 & 0 \end{bmatrix}$$

15. The rank of the matrix A in Exercise 13 is 2, which is the dimension of both the row space and column space of A. From Theorem 4.15, the dimension of the null space of A is $n - \text{rank}(A) = 3 - 2 = 1$.

17. (a) A particular solution to the system is

$$\mathbf{x}_0 = \begin{bmatrix} 0 \\ -1 \\ 0 \end{bmatrix}$$

This solution is obtained by setting the free variable, x_3, equal to zero in the system corresponding to the following reduced row-echelon form of the augmented matrix associated with the original system:

$$\left[\begin{array}{ccc|c} 1 & 0 & -3 & 0 \\ 0 & 1 & 0 & -1 \\ 0 & 0 & 0 & 0 \end{array}\right]$$

(b) A basis for the null space of A is:

$$\mathbf{v}_1 = \begin{bmatrix} 3 \\ 0 \\ 1 \end{bmatrix}$$

You obtain $\mathbf{v}_1$ by applying Gauss-Jordan elimination to $[A \mid \mathbf{0}]$, which results in the following augmented matrix:

$$\begin{bmatrix} 1 & 0 & -3 & 0 \\ 0 & 1 & 0 & 0 \\ 0 & 0 & 0 & 0 \end{bmatrix}$$

Solving for the leading variables x_1 and x_2 in terms of the free variable x_3 yields:

$$x_1 = 3x_3$$
$$x_2 = 0$$

Assigning an arbitrary value t to the free variable, the solution is

$$\begin{bmatrix} x_1 \\ x_2 \\ x_3 \end{bmatrix} = t \begin{bmatrix} 3 \\ 0 \\ 1 \end{bmatrix}$$

The vector on the right side is the desired basis for the null space of A.

(c) The general solution to the system is:

$$\mathbf{x}_0 + t_1 \mathbf{v}_1 = \begin{bmatrix} 0 \\ -1 \\ 0 \end{bmatrix} + t_1 \begin{bmatrix} 3 \\ 0 \\ 1 \end{bmatrix}$$

19. The proof is accomplished by showing that the columns of an $(m \times n)$ matrix R in row-echelon form that contain the leading values in the rows of R span the column space of R and are linearly independent.

To show that these vectors span the column space of R, let $\mathbf{v}$ be a vector in the column space of R and suppose that there are leading values, say, $a_1, \ldots, a_k$, in rows 1 through k of R. Because R is in row-echelon form, $\mathbf{v}$ has the following form:

$$\mathbf{v} = \begin{bmatrix} v_1 \\ \vdots \\ v_k \\ 0 \\ \vdots \\ 0 \end{bmatrix} = \frac{v_1}{a_1} \begin{bmatrix} a_1 \\ \vdots \\ 0 \\ 0 \\ \vdots \\ 0 \end{bmatrix} + \cdots + \frac{v_k}{a_k} \begin{bmatrix} 0 \\ \vdots \\ a_k \\ 0 \\ \vdots \\ 0 \end{bmatrix}$$

In other words, $\mathbf{v}$ is a linear combination of the columns of R containing the leading values in the rows of R. Hence, these columns span the column space of R.

It remains to show that the columns of R that contain the leading values in the rows of R are linearly independent. But these k columns are shown on the right side of the foregoing equality. Being multiples of different standard unit vectors, these columns are linearly independent and so form a basis for the column space of R.

21. Note first that because A has m rows, the dimension of the row space of A is $\leq m$. Likewise, A has n columns, so the dimension of the column space is $\leq n$. Now in the hypothesis, it is given that $n < m$, so the rank of A, which is the dimension of the row and column space of A, must be $\leq n$. It now follows that A has full rank if and only if rank of $A = n$, which is true if and only if the dimension of the column space of $A = n$, which is true if and only if the columns of A are linearly independent.

SOLUTIONS TO SELECTED EXERCISES IN CHAPTER 5

Solutions to Exercises in Section 5.1

1. (a) The domain of F is {real numbers x : $x \neq 0$}. The type of value output by the function is a nonzero real number. For the value $1/4$, $F(1/4) = 4$.
   (b) The domain of F is R^n. The type of value output by the function is a positive integer less than or equal to n. For the vector $(1, 0, -2, 1)$, $F(1, 0, -2, 1) = 3$.
   (c) The domain of the function is the set of all sets that contain a finite number of elements. The type of value output by the function is a nonnegative integer. For the set $\{0\}$, $|\{0\}| = 1$.

3. A closed-form expression for the function is $F(x, y) = (x, -y)$. A visual image of this function is the following:

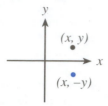

5. (a) The two coordinate functions are:

$$F_1(x, y, z) = zx + y$$
$$F_2(x, y, z) = yz + x$$

(b) The three coordinate functions are:

$$F_1(x) = 1, \quad F_2(x) = x, \quad F_3(x) = x^2$$

(c) The n coordinate functions are:

$$F_1(x_1, \ldots, x_n) = x_1, \ldots,$$
$$F_n(x_1, \ldots, x_n) = x_n$$

7. From the definitions of F and G, you have that:

(a) $(G \circ F)(1) = G(F(1)) = G(-1) = -2$. You cannot evaluate $F \circ G$ at $x = 1$ because $(F \circ G)(1) = F(G(1)) = F(0)$ but F is not defined at 0.

(b) $(G \circ F)(1) = G(F(1)) = G(0) = \sqrt{0} = 0$. Also, $(F \circ G)(1) = F(G(1)) = F(1) = 0$.

9. $G(F(x, y)) = G(x, -y) = (-x, -y)$.

Solutions to Exercises in Section 5.2

1. (a) You have that $T(\mathbf{x}) = A\mathbf{x}$, where

$$A = \begin{bmatrix} 0.5 & -0.2 & -0.2 \\ -0.1 & 0.6 & -0.1 \\ -0.4 & -0.4 & 0.3 \end{bmatrix}$$

(b) You have that $T(\mathbf{x}) = A\mathbf{x}$, where

$$A = \begin{bmatrix} -1 & 1 & 0 & 0 & 0 \\ 1 & 0 & 0 & 0 & 1 \\ 0 & 1 & 1 & 0 & 0 \\ 0 & 0 & -1 & -1 & 1 \end{bmatrix}$$

3. Only the transformation in part (a) is linear.

5. (a) According to Definition 5.4, choose vectors $\mathbf{u} = (u_1, u_2)$ and $\mathbf{v} = (v_1, v_2)$ and a scalar a and show that $T_1(\mathbf{u} + \mathbf{v}) = T_1(\mathbf{u}) + T_1(\mathbf{v})$ and $T_1(a\mathbf{u}) = aT_1(\mathbf{u})$. From the definition of T_1,

$$
\begin{aligned}
T_1(\mathbf{u} + \mathbf{v}) &= T_1(u_1 + v_1, u_2 + v_2) \\
&= u_1 + v_1 \\
&= T_1(u_1, u_2) + T_1(v_1, v_2) \\
&= T_1(\mathbf{u}) + T_1(\mathbf{v})
\end{aligned}
$$

Also,

$$
\begin{aligned}
T_1(a\mathbf{u}) &= T_1(au_1, au_2) \\
&= au_1 \\
&= aT_1(u_1, u_2) \\
&= aT_1(\mathbf{u})
\end{aligned}
$$

(b) A visual representation of T_1 is the following:

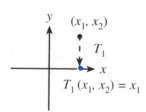

7. According to Definition 5.4, choose vectors $\mathbf{u}$ and $\mathbf{v}$ and a scalar a and show that $T(\mathbf{u} + \mathbf{v}) = T(\mathbf{u}) + T(\mathbf{v})$ and $T(a\mathbf{u}) = aT(\mathbf{u})$. From the definition of T,

$$
\begin{aligned}
T(\mathbf{u} + \mathbf{v}) &= k(\mathbf{u} + \mathbf{v}) \\
&= (k\mathbf{u}) + (k\mathbf{v}) \\
&= T(\mathbf{u}) + T(\mathbf{v})
\end{aligned}
$$

Also,

$$T(a\mathbf{u}) = k(a\mathbf{u}) = a(k\mathbf{u}) = aT(\mathbf{u})$$

9. According to Definition 5.4, choose vectors $\mathbf{u}$ and $\mathbf{v}$ and a scalar a and show that $T(\mathbf{u} + \mathbf{v}) = T(\mathbf{u}) + T(\mathbf{v})$ and $T(a\mathbf{u}) = aT(\mathbf{u})$. From the

definition of T,

$$T(\mathbf{u} + \mathbf{v}) = \mathbf{c} \cdot (\mathbf{u} + \mathbf{v})$$
$$= (\mathbf{c} \cdot \mathbf{u}) + (\mathbf{c} \cdot \mathbf{v})$$
$$= T(\mathbf{u}) + T(\mathbf{v})$$

Also,

$$T(a\mathbf{u}) = \mathbf{c} \cdot (a\mathbf{u}) = a(\mathbf{c} \cdot \mathbf{u}) = aT(\mathbf{u})$$

11. According to Definition 5.4, you choose two polynomials of degree n, say, $\mathbf{p} = p(x) = a_0 + a_1 x + a_2 x^2 + \cdots + a_n x^n$ and $\mathbf{q} = q(x) = b_0 + b_1 x + b_2 x^2 + \cdots + b_n x^n$ and a scalar c and show that $T(\mathbf{p} + \mathbf{q}) = T(\mathbf{p}) + T(\mathbf{q})$ and $T(c\mathbf{p}) = cT(\mathbf{p})$. So,

$$T(\mathbf{p} + \mathbf{q})$$
$$= T((a_0 + b_0) + (a_1 + b_1)x + (a_2 + b_2)x^2 + \cdots + (a_n + b_n)x^n)$$
$$= (a_1 + b_1) + 2(a_2 + b_2)x + \cdots + n(a_n + b_n)x^{n-1}$$
$$= (a_1 + 2a_2 x + \cdots + na_n x^{n-1}) + (b_1 + 2b_2 x + \cdots + nb_n x^{n-1})$$
$$= T(\mathbf{p}) + T(\mathbf{q})$$

Also,

$$T(c\mathbf{p}) = T(ca_0 + ca_1 x + ca_2 x^2 + \cdots + ca_n x^n)$$
$$= ca_1 + 2ca_2 x + \cdots + nca_n x^{n-1}$$
$$= c(a_1 + 2a_2 x + \cdots + na_n x^{n-1})$$
$$= cT(\mathbf{p})$$

13. You cannot generalize the statement "there are real numbers $a_1, \ldots, a_n$ such that for each n-vector $\mathbf{x} = (x_1, \ldots, x_n)$, $T(\mathbf{x}) = a_1 x_1 + \cdots + a_n x_n$" to a transformation $T : V \to W$ from a vector space V to a vector space W. The real numbers $x_1, \ldots, x_n$ that constitute the components of an n-vector $\mathbf{x}$ do not exist for a vector $\mathbf{x}$ belonging to a general vector space V.

15. According to Definition 5.4, choose vectors

$\mathbf{s} = (s_1, \ldots, s_n)$ and $\mathbf{t} = (t_1, \ldots, t_n)$ and a scalar a and show that $U(\mathbf{s} + \mathbf{t}) = U(\mathbf{s}) + U(\mathbf{t})$ and $U(a\mathbf{s}) = aU(\mathbf{s})$. So, let $\mathbf{u} = s_1 \mathbf{v}_1 + \cdots + s_n \mathbf{v}_n$ and $\mathbf{v} = t_1 \mathbf{v}_1 + \cdots + t_n \mathbf{v}_n$. Then

$$U(\mathbf{s} + \mathbf{t}) = U(s_1 + t_1, \ldots, s_n + t_n)$$
$$= [T(\mathbf{u} + \mathbf{v})]_{B'}$$
$$= [T(\mathbf{u}) + T(\mathbf{v})]_{B'}$$
$$= [T(\mathbf{u})]_{B'} + [T(\mathbf{v})]_{B'}$$
$$= U(s_1, \ldots, s_n) + U(t_1, \ldots, t_n)$$
$$= U(\mathbf{s}) + U(\mathbf{t})$$

Also,

$$U(a\mathbf{s}) = U(as_1, \ldots, as_n)$$
$$= [T(a\mathbf{u})]_{B'}$$
$$= a[T(\mathbf{u})]_{B'}$$
$$= aU(s_1, \ldots, s_n)$$
$$= aU(\mathbf{s})$$

Solutions to Exercises in Section 5.3

1. The standard matrix is

$$A = \begin{bmatrix} 1 & 0 \\ 0 & -1 \end{bmatrix}$$

This is because

$$T\left(\begin{bmatrix} 1 \\ 0 \end{bmatrix}\right) = \begin{bmatrix} 1 \\ 0 \end{bmatrix}$$

and

$$T\left(\begin{bmatrix} 0 \\ 1 \end{bmatrix}\right) = \begin{bmatrix} 0 \\ -1 \end{bmatrix}$$

3. The standard matrix is

$$A = \begin{bmatrix} -3 & 0 \\ 0 & -1 \\ 1 & 2 \end{bmatrix}$$

This is because, for the standard unit vectors

e_1 and e_2,

$$e_1 = \begin{bmatrix} 1 \\ 0 \end{bmatrix} = (-1)\begin{bmatrix} -1 \\ 0 \end{bmatrix} + (0)\begin{bmatrix} 0 \\ 2 \end{bmatrix}$$

and

$$e_2 = \begin{bmatrix} 0 \\ 1 \end{bmatrix} = (0)\begin{bmatrix} -1 \\ 0 \end{bmatrix} + (1/2)\begin{bmatrix} 0 \\ 2 \end{bmatrix}$$

and so

$$T(e_1) = (-1)T\left(\begin{bmatrix} -1 \\ 0 \end{bmatrix}\right) + (0)T\left(\begin{bmatrix} 0 \\ 2 \end{bmatrix}\right)$$

$$= (-1)\begin{bmatrix} 3 \\ 0 \\ -1 \end{bmatrix} = \begin{bmatrix} -3 \\ 0 \\ 1 \end{bmatrix}$$

$$T(e_2) = (0)T\left(\begin{bmatrix} -1 \\ 0 \end{bmatrix}\right) + \left(\frac{1}{2}\right)T\left(\begin{bmatrix} 0 \\ 2 \end{bmatrix}\right)$$

$$= \left(\frac{1}{2}\right)\begin{bmatrix} 0 \\ -2 \\ 4 \end{bmatrix} = \begin{bmatrix} 0 \\ -1 \\ 2 \end{bmatrix}$$

5. The word *unique* suggests using a uniqueness method. So the first step is to use the construction method to construct an $(m \times n)$ matrix A such that for all column vectors $\mathbf{x} \in R^{n \times 1}$, $T(\mathbf{x}) = A\mathbf{x}$. The desired matrix A is given in the problem statement, namely,

$$A = [\ T(e_1) \quad \cdots \quad T(e_n)\ ]$$

To show that this matrix is correct, the words *for all* suggest using the choose method to choose a column vector $\mathbf{x} \in R^{n \times 1}$, for which it must be shown that $T(\mathbf{x}) = A\mathbf{x}$. To do so, express $\mathbf{x} = (x_1, \ldots, x_n)$ as a linear combination of the standard unit vectors in R^n, that is,

$$\mathbf{x} = x_1 e_1 + \cdots + x_n e_n$$

Applying T to the vectors on both sides of the foregoing equality and using the properties of the linear transformation T, you have

$$\begin{aligned} T(\mathbf{x}) &= T(x_1 e_1 + \cdots + x_n e_n) \\ &= x_1 T(e_1) + \cdots + x_n T(e_n) \\ &= [T(e_1) \quad \cdots \quad T(e_n)]\mathbf{x} \\ &= A\mathbf{x} \end{aligned}$$

To show the uniqueness of the matrix A, according to the given hint, suppose that B is also an $(m \times n)$ matrix such that for all column vectors $\mathbf{x} \in R^{n \times 1}$, $T(\mathbf{x}) = B\mathbf{x}$. It must be shown that $A = B$. This is accomplished by showing that for each column $j = 1, \ldots, n$, $A_{*j} = B_{*j}$. So, choose a column j and show that $A_{*j} = B_{*j}$. This, in turn, is accomplished by specializing the statement that for all column vectors $\mathbf{x} \in R^{n \times 1}$, $T(\mathbf{x}) = B\mathbf{x}$, as follows:

$$\begin{aligned} A_{*j} &= T(e_j) \quad \text{(definition of } A_{*j}) \\ &= Be_j \quad \text{(specialize } T(\mathbf{x}) = B\mathbf{x} \text{ to} \\ &\qquad\qquad \mathbf{x} = e_j) \\ &= B_{*j} \quad \text{(by the structure of } e_j) \end{aligned}$$

It has now been shown that $A = B$, thus establishing the uniqueness of A and completing the proof.

7. The matrix A of T relative to the given bases B and B' is obtained by expressing T applied to each vector $\mathbf{v}_i$ in the basis B relative to B', resulting in:

$$A = \left[\ [T(\mathbf{v}_1)]_{B'} \quad [T(\mathbf{v}_2)]_{B'} \quad [T(\mathbf{v}_3)]_{B'}\ \right]$$

$$= \begin{bmatrix} 1 & -4 & 5 \\ 0 & -6 & 5 \end{bmatrix}$$

This is because, for

$$\mathbf{v}_1 = \begin{bmatrix} 1 \\ 0 \\ -1 \end{bmatrix}, \quad \mathbf{v}_2 = \begin{bmatrix} 0 \\ 2 \\ 0 \end{bmatrix}, \quad \mathbf{v}_3 = \begin{bmatrix} 1 \\ -2 \\ 0 \end{bmatrix}$$

you have that

$$T(\mathbf{v}_1) = \begin{bmatrix} 1 \\ 1 \end{bmatrix} = (1)\begin{bmatrix} 1 \\ 1 \end{bmatrix} + (0)\begin{bmatrix} -1 \\ 0 \end{bmatrix}$$

$$T(\mathbf{v}_2) = \begin{bmatrix} 2 \\ -4 \end{bmatrix} = (-4)\begin{bmatrix} 1 \\ 1 \end{bmatrix} + (-6)\begin{bmatrix} -1 \\ 0 \end{bmatrix}$$

$$T(\mathbf{v}_3) = \begin{bmatrix} 0 \\ 5 \end{bmatrix} = (5)\begin{bmatrix} 1 \\ 1 \end{bmatrix} + (5)\begin{bmatrix} -1 \\ 0 \end{bmatrix}$$

Solutions to Exercises in Section 5.4

1. (a) The components of $(1, 0)$, relative to the given basis matrix B, are 0 and -1, respectively. This is because

$$\begin{bmatrix} 1 \\ 0 \end{bmatrix}_B = B^{-1}\begin{bmatrix} 1 \\ 0 \end{bmatrix}$$

$$= \begin{bmatrix} 0 & 1 \\ -1 & 1 \end{bmatrix}\begin{bmatrix} 1 \\ 0 \end{bmatrix} = \begin{bmatrix} 0 \\ -1 \end{bmatrix}$$

   (b) The components of $(0, 1)$, relative to the given basis matrix B, are 1 and 1, respectively. This is because

$$\begin{bmatrix} 0 \\ 1 \end{bmatrix}_B = B^{-1}\begin{bmatrix} 0 \\ 1 \end{bmatrix}$$

$$= \begin{bmatrix} 0 & 1 \\ -1 & 1 \end{bmatrix}\begin{bmatrix} 0 \\ 1 \end{bmatrix} = \begin{bmatrix} 1 \\ 1 \end{bmatrix}$$

   (c) The components of $(2, -1)$, relative to the given basis matrix B, are -1 and -3, respectively. This is because

$$\begin{bmatrix} 2 \\ -1 \end{bmatrix}_B = B^{-1}\begin{bmatrix} 2 \\ -1 \end{bmatrix}$$

$$= \begin{bmatrix} 0 & 1 \\ -1 & 1 \end{bmatrix}\begin{bmatrix} 2 \\ -1 \end{bmatrix}$$

$$= \begin{bmatrix} -1 \\ -3 \end{bmatrix}$$

3. $T(\mathbf{x}) = AB[\mathbf{x}]_B$ because $T(\mathbf{x}) = A\mathbf{x} = A(BB^{-1}\mathbf{x}) = A(B[\mathbf{x}]_B)$.

5. (a) The components of $\mathbf{v}$, relative to the two standard unit vectors in R^2, are 5 and 3. This is because

$$\mathbf{v} = B[\mathbf{v}]_B = \begin{bmatrix} 1 & -1 \\ 1 & 0 \end{bmatrix}\begin{bmatrix} 3 \\ -2 \end{bmatrix} = \begin{bmatrix} 5 \\ 3 \end{bmatrix}$$

   (b) The components of $\mathbf{v}$, relative to the basis B', are 3 and -11. This is because

$$[\mathbf{v}]_{B'} = (B')^{-1}\mathbf{v} = \begin{bmatrix} -2 & -1 \\ 1 & 0 \end{bmatrix}^{-1}\begin{bmatrix} 5 \\ 3 \end{bmatrix}$$

$$= \begin{bmatrix} 0 & 1 \\ -1 & -2 \end{bmatrix}\begin{bmatrix} 5 \\ 3 \end{bmatrix} = \begin{bmatrix} 3 \\ -11 \end{bmatrix}$$

7. (a) You have that

$$[T]_B = \begin{bmatrix} 1 & 0 \\ -2 & 3 \end{bmatrix}$$

This is because the standard matrix associated with T is

$$A = [\,T(\mathbf{e}_1) \quad T(\mathbf{e}_2)\,] = \begin{bmatrix} 3 & 0 \\ 0 & 1 \end{bmatrix}$$

and so

$$[T]_B = B^{-1}AB$$

$$= \begin{bmatrix} 0 & 1 \\ -1 & 1 \end{bmatrix} \begin{bmatrix} 3 & 0 \\ 0 & 1 \end{bmatrix} \begin{bmatrix} 1 & -1 \\ 1 & 0 \end{bmatrix}$$

$$= \begin{bmatrix} 1 & 0 \\ -2 & 3 \end{bmatrix}$$

(b) You have that $T(-1, -2) = (-3, -2)$. This is because

$$T\left(\begin{bmatrix} -1 \\ -2 \end{bmatrix}\right) = B[T]_B B^{-1} \begin{bmatrix} -1 \\ -2 \end{bmatrix}$$

$$= \begin{bmatrix} 1 & -1 \\ 1 & 0 \end{bmatrix} \begin{bmatrix} 1 & 0 \\ -2 & 3 \end{bmatrix}$$

$$\begin{bmatrix} 0 & 1 \\ -1 & 1 \end{bmatrix} \begin{bmatrix} -1 \\ -2 \end{bmatrix}$$

$$= \begin{bmatrix} -3 \\ -2 \end{bmatrix}$$

(c) The result in part (b) is the same as that obtained by using the standard matrix A given in the solution to part (a) because

$$T\left(\begin{bmatrix} -1 \\ -2 \end{bmatrix}\right) = A \begin{bmatrix} -1 \\ -2 \end{bmatrix}$$

$$= \begin{bmatrix} 3 & 0 \\ 0 & 1 \end{bmatrix} \begin{bmatrix} -1 \\ -2 \end{bmatrix} = \begin{bmatrix} -3 \\ -2 \end{bmatrix}$$

9. The standard matrix associated with T is:

$$A = \begin{bmatrix} -3 & 0 \\ 0 & -2 \end{bmatrix}$$

This is because

$$A = B[T]_B B^{-1}$$

$$= \begin{bmatrix} 1 & -1 \\ 1 & 0 \end{bmatrix} \begin{bmatrix} -2 & 0 \\ 1 & -3 \end{bmatrix} \begin{bmatrix} 0 & 1 \\ -1 & 1 \end{bmatrix}$$

$$= \begin{bmatrix} -3 & 0 \\ 0 & -2 \end{bmatrix}$$

11. (a) Letting I be the (4×4) identity matrix, the matrix $[T]_B$ is:

$$[T]_B = \begin{bmatrix} [T(M^{11})]_B & [T(M^{12})]_B \\ [T(M^{21})]_B & [T(M^{22})]_B \end{bmatrix}$$

$$= [I_{*2}, I_{*3}, I_{*4}, I_{*1}]$$

This is because,

$$[T(M^{11})]_B = [M^{12}]_B = I_{*2} \text{ since}$$
$$M^{12} = [(0)M^{11} + (1)M^{12} +$$
$$(0)M^{21} + (0)M^{22}]_B$$
$$[T(M^{12})]_B = [M^{21}]_B = I_{*3} \text{ since}$$
$$M^{21} = [(0)M^{11} + (0)M^{12} +$$
$$(1)M^{21} + (0)M^{22}]_B$$
$$[T(M^{21})]_B = [M^{22}]_B = I_{*4} \text{ since}$$
$$M^{22} = [(0)M^{11} + (0)M^{12} +$$
$$(0)M^{21} + (1)M^{22}]_B$$
$$[T(M^{22})]_B = [M^{11}]_B = I_{*1} \text{ since}$$
$$M^{11} = [(1)M^{11} + (0)M^{12} +$$
$$(0)M^{21} + (0)M^{22}]_B$$

(b) You have that

$$T(M) = \begin{bmatrix} 4 & 1 \\ 2 & 3 \end{bmatrix}$$

This is because

$$[M]_B = \begin{bmatrix} 1 \\ 2 \\ 3 \\ 4 \end{bmatrix}$$

so

$$[T(M)_B] = [T]_B[M]_B$$

$$= \begin{bmatrix} 0 & 0 & 0 & 1 \\ 1 & 0 & 0 & 0 \\ 0 & 1 & 0 & 0 \\ 0 & 0 & 1 & 0 \end{bmatrix} \begin{bmatrix} 1 \\ 2 \\ 3 \\ 4 \end{bmatrix} = \begin{bmatrix} 4 \\ 1 \\ 2 \\ 3 \end{bmatrix}$$

Thus,

$$T(M) = 4M^{11} + 1M^{12} +$$
$$2M^{21} + 3M^{22}$$

$$= \begin{bmatrix} 4 & 1 \\ 2 & 3 \end{bmatrix}$$

(c) The result in part (b) is correct because

$$T(M) = T(1M^{11} + 2M^{12} +$$
$$3M^{21} + 4M^{22})$$
$$= T(M^{11}) + 2T(M^{12}) +$$
$$3T(M^{21}) + 4T(M^{22})$$
$$= M^{21} + 2M^{21} +$$
$$3M^{22} + 4M^{11}$$

$$= \begin{bmatrix} 4 & 1 \\ 2 & 3 \end{bmatrix}$$

13. (a) The matrices A and C are not similar using the given matrix B. This is because $A \neq BCB^{-1}$ since

$$A = \begin{bmatrix} 4 & 0 \\ 5 & 2 \end{bmatrix}$$

while

$$BCB^{-1} = \begin{bmatrix} 1 & -1 \\ 1 & 0 \end{bmatrix} \begin{bmatrix} 2 & -3 \\ 0 & 4 \end{bmatrix} \begin{bmatrix} 0 & 1 \\ -1 & 1 \end{bmatrix}$$

$$= \begin{bmatrix} 7 & -5 \\ 3 & -1 \end{bmatrix}$$

(b) The matrices A and C are similar using the given matrix B. This is because

$$A = BCB^{-1}$$

$$= \begin{bmatrix} 1 & -1 \\ 1 & 0 \end{bmatrix} \begin{bmatrix} -3 & 4 \\ 0 & -1 \end{bmatrix} \begin{bmatrix} 0 & 1 \\ -1 & 1 \end{bmatrix}$$

$$= \begin{bmatrix} -5 & 2 \\ -4 & 1 \end{bmatrix}$$

15. You have that

$$[T(\mathbf{x})]_{B'} = (B')^{-1}T(\mathbf{x})$$
$$= (B')^{-1}(A\mathbf{x})$$
$$= (B')^{-1}(A(B[\mathbf{x}]_B))$$

Solutions to Exercises in Section 5.5

1. T is not onto because, for example, there is no value for x such that $T(x) = 1$. The range of T is $\{0\}$.

3. T is not onto because, for example, there are no values for x and y such that $T(x, y) = (3x + 3y, -x - y) = (1, 2)$. The range of T is $\{(3t, -t) : t$ is a real number$\}$.

5. T is onto because, for any values of $a, b,$ and c, the following values for $x, y,$ and z satisfy $T(x, y, z) = (a, b, c)$:

$$x = \frac{6a + 2b + c}{12}, \quad y = \frac{b + 2c}{3}, \quad z = \frac{c}{2}$$

7. T is not one-to-one because, for example, $T(1) = T(2) = 0$, yet $1 \neq 2$.

9. T is one-to-one, for if $T(x_1, y_1) = T(x_2, y_2)$,

it follows that $(y_1, x_1) = (y_2, x_2)$ and thus $y_1 = y_2$ and $x_1 = x_2$, so $(x_1, y_1) = (x_2, y_2)$.

11. T is one-to-one because, if $T(x_1, y_1, z_1) = T(x_2, y_2, z_2)$, then $(2x_1 - y_1 + z_1, 3y_1 - 4z_1, 2z_1) = (2x_2 - y_2 + z_2, 3y_2 - 4z_2, 2z_2)$. Equating corresponding components and performing appropriate operations leads to the conclusion that $(x_1, y_1, z_1) = (x_2, y_2, z_2)$.

13. (a) T is onto because, for an arbitrary polynomial of degree 1 or less, say, $a + bx$, the polynomial $ax + (b/2)x^2$ satisfies $T(ax + (b/2)x^2) = a + 2(b/2)x = a + bx$.

    (b) T is not one-to-one because $T(1 + x + x^2) = T(2 + x + x^2) = 1 + 2x$, yet $1 + x + x^2 \neq 2 + x + x^2$.

    (c) One polynomial, $\mathbf{p}$, for which $T(\mathbf{p}) = 3x + 4$ is $\mathbf{p} = p(x) = 4x + (3/2)x^2$. This polynomial is not unique because $\mathbf{q} = q(x) = 1 + 4x + (3/2)x^2$, for example, also satisfies $T(\mathbf{q}) = 3x + 4$.

15. To show that F is one-to-one, by definition, you must show that for all signals $\mathbf{X} = (x_0, x_1, x_2, \ldots)$ and $\mathbf{Y} = (y_0, y_1, y_2, \ldots)$ in null(T) with $F(\mathbf{X}) = F(\mathbf{Y})$, it follows that $\mathbf{X} = \mathbf{Y}$. Recognizing the key words *for all*, use the choose method to choose two signals $\mathbf{X} = (x_0, x_1, x_2, \ldots)$, $\mathbf{Y} = (y_0, y_1, y_2, \ldots)$ in null(T) with $F(\mathbf{X}) = F(\mathbf{Y})$. You must show that $\mathbf{X} = \mathbf{Y}$. However, because $F(\mathbf{X}) = F(\mathbf{Y})$, you know that

$$(x_0, x_1, \ldots, x_{n-1}) = (y_0, y_1, \ldots, y_{n-1})$$

It remains to show that $x_n = y_n$, $x_{n+1} = y_{n+1}, \ldots$. To see that $x_n = y_n$, use the fact that $\mathbf{X}, \mathbf{Y} \in$ null(T), so, $x_n + a_1 x_{n-1} + \cdots + a_n x_0 = 0$ and $y_n + a_1 y_{n-1} + \cdots + a_n y_0 = 0$. You now have that

$$x_n = -a_1 x_{n-1} - \cdots - a_n x_0 \quad (1)$$
$$= -a_1 y_{n-1} - \cdots - a_n y_0 \quad (2)$$
$$= y_n \quad (3)$$

The foregoing numbered steps are obtained as follows:

Step	Justification
(1)	solve for x_n
(2)	$x_{n-1} = y_{n-1}, \ldots, x_0 = y_0$
(3)	solve for y_n

A similar idea shows that $x_{n+1} = y_{n+1}, \ldots$.

To see that F is onto, by definition, you must show that for any vector

$$\mathbf{y} = (y_0, y_1, \ldots, y_{n-1})$$

there is a signal $\mathbf{X} = (x_0, x_1, x_2, \ldots)$ in null(T) such that $F(\mathbf{X}) = \mathbf{y}$. Recognizing the key words *for all*, choose a vector $\mathbf{y} = (y_0, y_1, \ldots, y_{n-1})$. You must show that there is a signal $\mathbf{X} = (x_0, x_1, x_2, \ldots)$ in null(T) such that $F(\mathbf{X}) = \mathbf{y}$. The key words *there is* mean you must now construct the desired signal $\mathbf{X} = (x_0, x_1, x_2, \ldots)$ in null(T). To do so, set $x_0 = y_0$, $x_1 = y_1, \ldots, x_{n-1} = y_{n-1}$. It remains to construct the components $x_n, x_{n+1}, \ldots$. To construct x_n, for example, set

$$x_n = -a_1 x_{n-1} - \cdots - a_n x_0$$

You can then compute

$$x_{n+1} = -a_1 x_n - \cdots - a_n x_1$$

More generally, for each $k = 0, 1, \ldots$, set

$$x_{n+k} = -a_1 x_{n+k-1} - \cdots - a_n x_k$$

Constructing $\mathbf{X} = (x_0, x_1, x_2, \ldots)$ in this way ensures that $\mathbf{X} \in$ null(T).

17. To show that range(T) is a subspace of W, by Theorem 4.2 in Section 4.2.1, you need only show that (1) for all vectors $\mathbf{u}, \mathbf{v} \in$ range(T), $\mathbf{u} + \mathbf{v} \in$ range(T) and (2) for all real numbers a and vectors $\mathbf{u} \in$ range(T), $a\mathbf{u} \in$ range(T). The choose method is used for both parts. Specifically, for (1), choose vectors $\mathbf{u}, \mathbf{v} \in$ range(T). By definition of the range, this means that there are vectors $\mathbf{x}, \mathbf{y} \in V$ such $T(\mathbf{x}) = \mathbf{u}$ and $T(\mathbf{y}) = \mathbf{v}$. But then,

$$T(\mathbf{x} + \mathbf{y}) = T(\mathbf{x}) + T(\mathbf{y}) = \mathbf{u} + \mathbf{v}$$

This means that $\mathbf{u} + \mathbf{v} \in$ range(T).

To see (2), choose a real number a and a vector $\mathbf{u} \in \text{range}(T)$. By definition of the range, this means that there is a vector $\mathbf{x} \in V$ such that $T(\mathbf{x}) = \mathbf{u}$. But then,

$$T(a\mathbf{x}) = aT(\mathbf{x}) = a\mathbf{u}$$

This means that $a\mathbf{u} \in \text{range}(T)$. This completes the proof.

19. To show that $\text{null}(T)$ is a subspace of V, by Theorem 4.2 in Section 4.2.1, you need only show that (1) for all vectors $\mathbf{u}, \mathbf{v} \in \text{null}(T)$, $\mathbf{u} + \mathbf{v} \in \text{null}(T)$ and (2) for all real numbers a and vectors $\mathbf{u} \in \text{null}(T)$, $a\mathbf{u} \in \text{null}(T)$. The choose method is used for both parts. Specifically, for (1), choose vectors $\mathbf{u}, \mathbf{v} \in \text{null}(T)$. By definition of the null space, this means that $T(\mathbf{u}) = \mathbf{0}$ and $T(\mathbf{v}) = \mathbf{0}$. But then,

$$T(\mathbf{u} + \mathbf{v}) = T(\mathbf{u}) + T(\mathbf{v}) = \mathbf{0} + \mathbf{0} = \mathbf{0}$$

This means that $\mathbf{u} + \mathbf{v} \in \text{null}(T)$.

To see (2), choose a real number a and a vector $\mathbf{u} \in \text{null}(T)$. By definition of the null space, this means that $T(\mathbf{u}) = \mathbf{0}$. But then,

$$T(a\mathbf{u}) = aT(\mathbf{u}) = a\mathbf{0} = \mathbf{0}$$

This means that $a\mathbf{u} \in \text{null}(T)$. This completes the proof.

SOLUTIONS TO SELECTED EXERCISES IN CHAPTER 6

Solutions to Exercises in Section 6.1

1. No, $\mathbf{x}$ is not an eigenvector of A because there is no value of λ such that $A\mathbf{x} = \lambda\mathbf{x}$ since for any real number λ,

$$A\mathbf{x} = \begin{bmatrix} 2 & 1 \\ 3 & 4 \end{bmatrix}\begin{bmatrix} -1 \\ 3 \end{bmatrix} = \begin{bmatrix} 1 \\ 9 \end{bmatrix} \neq \lambda\begin{bmatrix} -1 \\ 3 \end{bmatrix}$$

3. The eigenvalue is $\lambda = 2$. This value is obtained by equating one component of $A\mathbf{x}$ to $\lambda\mathbf{x}$, solving to obtain $\lambda = 2$, and then verify-

ing that $A\mathbf{x} = 2\mathbf{x}$:

$$A\mathbf{x} = \begin{bmatrix} 4 & 0 & -2 \\ 1 & 3 & -2 \\ 1 & 0 & 1 \end{bmatrix}\begin{bmatrix} 1 \\ 1 \\ 1 \end{bmatrix}$$

$$= \begin{bmatrix} 2 \\ 2 \\ 2 \end{bmatrix} = 2\begin{bmatrix} 1 \\ 1 \\ 1 \end{bmatrix} = \lambda\mathbf{x}$$

5. No, $\lambda = 2$ is not an eigenvalue of A because the only vector $\mathbf{x}$ that satisfies $A\mathbf{x} = 2\mathbf{x}$ is $\mathbf{x} = \mathbf{0}$.

7. (Answers can vary.) One eigenvector associated with the eigenvalue $\lambda = 0$ is

$$\mathbf{x} = \begin{bmatrix} 1 \\ 1 \\ 0 \end{bmatrix}$$

This eigenvector is obtained by finding $\mathbf{x} \neq \mathbf{0}$ such that $A\mathbf{x} = \mathbf{0}$. To do so requires solving the following system of linear equations:

$$\begin{aligned} x_1 - x_2 + 2x_3 &= 0 \\ 2x_1 - 2x_2 + 4x_3 &= 0 \\ 3x_3 &= 0 \end{aligned}$$

This system has an infinite number of nonzero solutions, one of which is $x_1 = 1$, $x_2 = 1$, and $x_3 = 0$.

9. (a) The characteristic equation is $\lambda^2 - 3\lambda = 0$ because

$$\det(A - \lambda I) = \begin{vmatrix} 1 - \lambda & -1 \\ -2 & 2 - \lambda \end{vmatrix}$$

$$= (1 - \lambda)(2 - \lambda) - 2$$

$$= \lambda^2 - 3\lambda = 0$$

(b) The eigenvalues are $\lambda_1 = 3$ and $\lambda_2 = 0$, which are obtained by solving the characteristic equation $\lambda^2 - 3\lambda = 0$. Two corresponding eigenvectors are:

$$\mathbf{v}_1 = \begin{bmatrix} 1 \\ -2 \end{bmatrix} \quad \text{and} \quad \mathbf{v}_2 = \begin{bmatrix} 1 \\ 1 \end{bmatrix}$$

(c) A basis for the eigenspace associated with the eigenvalue $\lambda_1 = 3$ is

$$\mathbf{v}_1 = \begin{bmatrix} -1/2 \\ 1 \end{bmatrix}$$

This basis is obtained by reducing the augmented matrix associated with the system $(A - \lambda_1 I)\mathbf{x} = 0$ to the following reduced row-echelon form:

$$\begin{bmatrix} 1 & 1/2 & \bigm| & 0 \\ 0 & 0 & \bigm| & 0 \end{bmatrix}$$

The general solution to this system is: $x_1 = -(1/2)x_2$, with x_2 free, so

$$\begin{bmatrix} x_1 \\ x_2 \end{bmatrix} = x_2 \begin{bmatrix} -1/2 \\ 1 \end{bmatrix}$$

A basis for the eigenspace associated with the eigenvalue $\lambda_2 = 0$ is

$$\mathbf{v}_1 = \begin{bmatrix} 1 \\ 1 \end{bmatrix}$$

This basis is obtained by reducing the augmented matrix associated with the system $(A - \lambda_2 I)\mathbf{x} = 0$ to the following reduced row-echelon form:

$$\begin{bmatrix} 1 & -1 & \bigm| & 0 \\ 0 & 0 & \bigm| & 0 \end{bmatrix}$$

The general solution to this system is: $x_1 = x_2$, with x_2 free, so

$$\begin{bmatrix} x_1 \\ x_2 \end{bmatrix} = x_2 \begin{bmatrix} 1 \\ 1 \end{bmatrix}$$

The two eigenspaces are shown in the following figure:

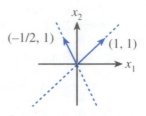

11. The characteristic equation is $\lambda^2 - 4\lambda + 5 = 0$ because

$$\det(A - \lambda I) = \det \begin{bmatrix} 1 - \lambda & 2 \\ -1 & 3 - \lambda \end{bmatrix}$$
$$= (1 - \lambda)(3 - \lambda) + 2$$
$$= \lambda^2 - 4\lambda + 5 = 0$$

13. No, $\lambda = 0$ is not always an eigenvalue value of A because, even though $A\mathbf{0} = (0)\mathbf{0}$, the vector $\mathbf{x} = \mathbf{0}$ is not an eigenvector. By definition, an eigenvector cannot be the zero vector.

15. The eigenvalues of A are the same as those of A^t. This is because λ is an eigenvalue of A if and only if $\det(A - \lambda I) = 0$ which, from Theorem 3.8(b) in Section 3.3.2, is true if and only if $\det[(A - \lambda I)^t] = 0$, which is true if and only if $\det(A^t - \lambda I) = 0$, which is true if and only if λ is an eigenvalue of A^t.

Solutions to Exercises in Section 6.2

1. (a) The state of the system on each day k is described by the following vector:

$$\mathbf{x}_k = \begin{bmatrix} \text{number of cars downtown} \\ \text{number of cars at the airport} \end{bmatrix}$$

(b) The matrix A in the linear difference equation $\mathbf{x}_{k+1} = A\mathbf{x}_k$ is:

$$A = \begin{bmatrix} 0.60 & 0.05 \\ 0.40 & 0.95 \end{bmatrix}$$

(c) The objective is to determine the first integer k such that the first component of $\mathbf{x}_{k+1}$ is < 20. Yes, this is a Markov chain because all elements of A are nonnegative and each column of A sums to 1.

3. (a) The state of the system in each month k is described by the following vector:

$$\mathbf{x}_k = \begin{bmatrix} \text{number who drink coke} \\ \text{number who drink pepsi} \\ \text{number who drink other brands} \end{bmatrix}$$

(b) The matrix A in the linear difference equation $\mathbf{x}_{k+1} = A\mathbf{x}_k$ is

$$A = \begin{bmatrix} 0.90 & 0.05 & 0.09 \\ 0.05 & 0.85 & 0.11 \\ 0.05 & 0.10 & 0.80 \end{bmatrix}$$

(c) The objective of the problem is to determine the components of $\mathbf{x}_k$, as k gets large. Yes, this is a Markov chain because all elements of A are nonnegative and each column of A sums to 1.

5. (a) The state of the system in each month k is described by the following vector, where the inventory is determined after meeting the demand for that month:

$$\mathbf{x}_k = \begin{bmatrix} \text{tons of steel produced} \\ \text{tons of steel in inventory} \\ \text{tons of steel demanded} \end{bmatrix}$$

(b) The matrix A in the linear difference equation $\mathbf{x}_{k+1} = A\mathbf{x}_k$ is

$$A = \begin{bmatrix} 0.5 & 0 & 0.5 \\ 1.0 & 1 & -1.0 \\ 0.0 & 0 & 1.0 \end{bmatrix}$$

(c) The objective of the problem is to determine the components of $\mathbf{x}_{24}$. No, this is not a Markov chain because one element of A is less than 0. Also, not all columns of A sum to 1.

Solutions to Exercises in Section 6.3

1. The given matrix A is not regular because no power of A is strictly positive. That is, for all integers $k = 0, 1, \ldots, (A^k)_{12} = 0$.

3. In terms of the eigenvalues and eigenvectors of A, Theorem 6.1 says that $\lambda = 1$ is an eigenvalue of A with corresponding eigenvector $\mathbf{x}$, where $\mathbf{x}$ is the solution to the following system, in which $\mathbf{e} = [1, 1, \ldots, 1]$:

$$(A - I)\mathbf{x} = \mathbf{0}$$
$$\mathbf{e}\mathbf{x} = 1$$

5. The steady-state vector to the dynamical system in which

$$A = \begin{bmatrix} 0.60 & 0.05 \\ 0.40 & 0.95 \end{bmatrix} \quad \text{is} \quad \mathbf{x} = \begin{bmatrix} 1/9 \\ 8/9 \end{bmatrix}$$

This vector is obtained by solving the following system of linear equations, (see the solution to Exercise 3), in which $\mathbf{e} = [1, 1]$:

$$(A - I)\mathbf{x} = \mathbf{0}$$
$$\mathbf{e}\mathbf{x} = 1$$

that is

$$\begin{bmatrix} -0.40 & 0.05 \\ 0.40 & -0.05 \\ 1.00 & 1.00 \end{bmatrix} \begin{bmatrix} x_1 \\ x_2 \end{bmatrix} = \begin{bmatrix} 0 \\ 0 \\ 1 \end{bmatrix}$$

7. Suppose that $\mathbf{x}_0$ is an eigenvector of A corresponding to an eigenvalue, λ, of A. According to the induction method, you must show that the statement is true for $k = 0$, that is, that $\mathbf{x}_0 = \lambda^0 \mathbf{x}_0$. But this is true because $\lambda^0 = 1$.

The next step of induction is to assume the statement is true for k and prove that the statement is true for $k + 1$. Thus, you should assume that $\mathbf{x}_k = \lambda^k \mathbf{x}_0$. You must show that $\mathbf{x}_{k+1} = \lambda^{k+1} \mathbf{x}_0$. But this is true because

$$\begin{aligned} \mathbf{x}_{k+1} &= A\mathbf{x}_k && \text{(linear difference eqn.)} \\ &= A(\lambda^k \mathbf{x}_0) && \text{(induction hypothesis)} \\ &= \lambda^k A\mathbf{x}_0 && \text{(algebra)} \\ &= \lambda^k (\lambda \mathbf{x}_0) && (\mathbf{x}_0 \text{ is an eigenvector of } A) \\ &= \lambda^{k+1} \mathbf{x}_0 && \text{(algebra)} \end{aligned}$$

9. Following the analysis in Section 6.3.2 and

using the result in Exercise 7, you know that

$$\mathbf{x}_k = \lambda^k \mathbf{x}_0, \quad \text{for } k = 0, 1, 2, \dots$$

The analysis now depends on which of the following cases occurs: $-1 < \lambda < 0$, $\lambda = -1$, or $\lambda < -1$.

Analysis for the Case $-1 < \lambda < 0$. If $-1 < \lambda < 0$, then, as $k \to \infty$, λ^k gets closer to 0. Thus, $\mathbf{x}_k = \lambda^k \mathbf{x}_0$ gets closer to the zero vector, as $k \to \infty$. In other words, if the system is started in the state described by an eigenvector $\mathbf{x}_0$ corresponding to an eigenvalue λ with $-1 < \lambda < 0$, then the state of the system approaches the zero vector as $k \to \infty$.

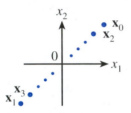

Analysis for the Case $\lambda = -1$. If $\lambda = -1$, then, for each k, $\lambda^k = \pm 1$. Thus, for each k, $\mathbf{x}_k = \lambda^k \mathbf{x}_0 = \pm \mathbf{x}_0$. In other words, if the system is started in the initial state described by an eigenvector $\mathbf{x}_0$ corresponding to the eigenvalue $\lambda = -1$, then the system alternates between $\mathbf{x}_0$ and $-\mathbf{x}_0$ forever.

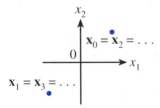

Analysis for the Case $\lambda < -1$. If $\lambda < -1$, then $|\lambda|^k \to \infty$ as $k \to \infty$. Consequently, the nonzero components of $|\lambda|^k \mathbf{x}_0$ approach $\pm \infty$. In other words, if the system is started in the state described by an eigenvector $\mathbf{x}_0$ cor-

responding to an eigenvalue $\lambda < -1$, then the system *diverges*. That is, one or more components of the sequence $\mathbf{X} = (\mathbf{x}_0, \mathbf{x}_1, \mathbf{x}_2, \dots)$ approaches ∞ in absolute value as $k \to \infty$.

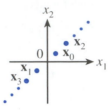

11. (a) The origin is a saddle point. This is because one eigenvalue of A, namely, $\lambda_1 = 1.1$, is greater than 1 and the other eigenvalue, $\lambda_2 = 0.5$, is less than 1. Thus, the analysis of the sequence $\mathbf{x}_k$, as k approaches ∞, is the same as that for the saddle-point discussion in Section 6.3.4.

(b) The origin is a repellor. This is because both eigenvalues of A, namely, $\lambda_1 = 1.1$ and $\lambda_2 = 2.5$, are greater than 1. Thus, the analysis of the sequence $\mathbf{x}_k$, as k approaches ∞, is the same as that for the repellor discussion in Section 6.3.4.

(c) The origin is an attractor. This is because both eigenvalues of A, namely, $\lambda_1 = 0.1$ and $\lambda_2 = 0.5$, are less than 1. Thus, the analysis of the sequence $\mathbf{x}_k$, as k approaches ∞, is the same as that for the attractor discussion in Section 6.3.4.

13. (Answers can vary.) The only eigenvalue of the following matrix is 1 and there is only one linearly independent eigenvector, for example, $(1, 0)$:

$$\begin{bmatrix} 1 & 2 \\ 0 & 1 \end{bmatrix}$$

15. It is not possible to construct such a matrix because, according to Theorem 6.2, any two eigenvectors corresponding to two distinct eigenvalues will, necessarily, be linearly independent and hence form a basis for R^2.

Solutions to Exercises in Section 6.4

1. No, A might still be diagonalizable. This is because, although $A \neq PDP^{-1}$ for the given choices of P and D, other choices of P and D might satisfy $A = PDP^{-1}$.

3. Yes, you can conclude that A is diagonalizable because

$$A = \begin{bmatrix} 8 & 5 \\ -10 & -7 \end{bmatrix}$$

$$= \begin{bmatrix} 1 & -1 \\ -2 & 1 \end{bmatrix} \begin{bmatrix} -2 & 0 \\ 0 & 3 \end{bmatrix} \begin{bmatrix} -1 & -1 \\ -2 & -1 \end{bmatrix}$$

$$= PDP^{-1}$$

5. For the given matrices, $A = PDP^{-1}$ because

$$AP = \begin{bmatrix} -2 & 6 & -2 \\ -2 & 6 & 0 \\ 0 & -3 & 1 \end{bmatrix} = PD$$

7. According to the induction method, you must show that the statement is true for $k = 1$, that is, that

$$D^1 \mathbf{x} = \begin{bmatrix} (D_{11})^1 x_1 \\ \vdots \\ (D_{nn})^1 x_n \end{bmatrix} = \begin{bmatrix} D_{11} x_1 \\ \vdots \\ D_{nn} x_n \end{bmatrix}$$

But this is true because $D^1 = D$ and D is a diagonal matrix.

   The next step of induction is to assume the statement is true for k and prove that the statement is true for $k + 1$. Thus, you should assume that

$$D^k \mathbf{x} = \begin{bmatrix} (D_{11})^k x_1 \\ \vdots \\ (D_{nn})^k x_n \end{bmatrix}$$

You must show that

$$D^{k+1} \mathbf{x} = \begin{bmatrix} (D_{11})^{k+1} x_1 \\ \vdots \\ (D_{nn})^{k+1} x_n \end{bmatrix}$$

But this is true because

$$D^{k+1} \mathbf{x} = D(D^k \mathbf{x}) \tag{1}$$

$$= D \begin{bmatrix} (D_{11})^k x_1 \\ \vdots \\ (D_{nn})^k x_n \end{bmatrix} \tag{2}$$

$$= \begin{bmatrix} D_{11}(D_{11})^k x_1 \\ \vdots \\ D_{nn}(D_{nn})^k x_n \end{bmatrix} \tag{3}$$

$$= \begin{bmatrix} (D_{11})^{k+1} x_1 \\ \vdots \\ (D_{nn})^{k+1} x_n \end{bmatrix} \tag{4}$$

Each of the foregoing numbered steps is justified, as follows:

Step	Justification
(1)	matrix algebra
(2)	induction hypothesis
(3)	D is a diagonal matrix
(4)	algebra

9. The desired matrices D and P in the diagonalization of A are:

$$D = \begin{bmatrix} -3 & 0 \\ 0 & 2 \end{bmatrix}, \quad P = \begin{bmatrix} -2 & 1 \\ 3 & 1 \end{bmatrix}$$

This is because, from the results in Section 6.1.1, the eigenvalues of A are $\lambda_1 = -3$ and

$\lambda_2 = 2$ and corresponding eigenvectors are:

$$\mathbf{v}_1 = \begin{bmatrix} -2 \\ 3 \end{bmatrix} \quad \text{and} \quad \mathbf{v}_2 = \begin{bmatrix} 1 \\ 1 \end{bmatrix}$$

11. The desired matrices D and P in the diagonalization of A are:

$$\overset{D}{\begin{bmatrix} 2 & 0 & 0 \\ 0 & -3 & 0 \\ 0 & 0 & 4 \end{bmatrix}} \quad \overset{P}{\begin{bmatrix} -10 & 0 & 0 \\ 2 & 1 & 0 \\ 5 & 0 & 1 \end{bmatrix}}$$

This is becasue the eigenvalues of A are the diagonal elements, $A_{11} = 2$, $A_{22} = -3$, and $A_{33} = 4$, of A (since A is lower triangular). Also, the columns of P are the three corresponding eigenvectors, $\mathbf{v}_1$, $\mathbf{v}_2$, and $\mathbf{v}_3$, given in the problem.

13. (a) A basis for the eigenspace corresponding to the eigenvalue 4 is the standard unit vector $\mathbf{e}_1$ in R^3. You find this basis by applying Gauss-Jordan elimination to the augmented matrix associated with the system of linear equations $(A - 4I)\mathbf{x} = \mathbf{0}$ to obtain

$$\begin{bmatrix} 0 & 1 & 0 & | & 0 \\ 0 & 0 & 1 & | & 0 \\ 0 & 0 & 0 & | & 0 \end{bmatrix}$$

The general solution to the system associated with the foregoing augmented matrix is $x_2 = 0$, $x_3 = 0$, and x_1 is free. A basis vector for this set of solutions is $\mathbf{e}_1$.

(b) A basis for the eigenspace corresponding to the eigenvalue 6 is

$$\mathbf{v} = \begin{bmatrix} 1 \\ 1 \\ 0 \end{bmatrix}$$

You find this basis by applying Gauss-Jordan elimination to the augmented matrix associated with the system of linear

equations $(A - 6I)\mathbf{x} = \mathbf{0}$ to obtain

$$\begin{bmatrix} 1 & -1 & 0 & | & 0 \\ 0 & 0 & 1 & | & 0 \\ 0 & 0 & 0 & | & 0 \end{bmatrix}$$

The general solution to the system associated with the foregoing augmented matrix is $x_1 = x_2$, $x_3 = 0$, and x_2 is free. A basis vector for this set of solutions is the foregoing vector $\mathbf{v}$.

(c) From the results in parts (a) and (b), you can conclude that it is not possible to find three linearly independent eigenvectors. Hence, by Theorem 6.3, you cannot diagonalize A.

15. When A is a diagonal matrix, the computation is reduced because, you have by induction that for $k = 1, 2, \ldots$,

$$A^k = \begin{bmatrix} (A_{11})^k & \cdots & 0 \\ 0 & \ddots & 0 \\ 0 & \cdots & (A_{nn})^k \end{bmatrix}$$

so

$$\sum_{k=1}^{N} A^k = \begin{bmatrix} \sum_{k=1}^{N}(A_{11})^k & \cdots & 0 \\ 0 & \ddots & 0 \\ 0 & \cdots & \sum_{k=1}^{N}(A_{nn})^k \end{bmatrix}$$

SOLUTIONS TO SELECTED EXERCISES IN CHAPTER 7

Solutions to Exercises in Section 7.1

1. For $\mathbf{u} = (-1, 2, 2)$ and $\mathbf{v} = (0, -3, 3)$, you have
   (a) $\mathbf{u} \cdot \mathbf{v} = (-1)(0) + 2(-3) + 2(3) = 0$, thus $\mathbf{u}$ and $\mathbf{v}$ are orthogonal.
   (b) Distance from $\mathbf{u}$ to $\mathbf{v}$ is $3\sqrt{3}$, that is,

$$\sqrt{(-1 - 0)^2 + (2 - (-3))^2 + (2 - 3)^2}$$

   (c) $(-1/3, 2/3, 2/3)$ is the normalized vector because

$$\frac{1}{\|\mathbf{u}\|}\mathbf{u} = \frac{1}{3}(-1, 2, 2)$$

3. Yes, the vectors $\mathbf{v}_1$, $\mathbf{v}_2$, and $\mathbf{v}_3$ form an orthogonal set because

$$\mathbf{v}_1 \cdot \mathbf{v}_2 = (-2)(2) + 1(4) + 0(0) = 0$$
$$\mathbf{v}_1 \cdot \mathbf{v}_3 = (-2)(0) + 1(0) + 0(-3) = 0$$
$$\mathbf{v}_2 \cdot \mathbf{v}_3 = 2(0) + 4(0) + 0(-3) = 0$$

5. Yes, these vectors form a basis for R^3 because they form an orthogonal set (see the solution to Exercise 3) so, by Theorem 7.2, these three vectors are linearly independent.

7. The components of the vector $(1, 2, -9)$, relative to the basis in Exercise 1, are $t_1 = 0$, $t_2 = 1/2$, and $t_3 = 3$. This is because, by Theorem 7.3,

$$t_1 = \frac{\mathbf{v} \cdot \mathbf{v}_1}{\mathbf{v}_1 \cdot \mathbf{v}_1} = \frac{(1, 2, -9) \cdot (-2, 1, 0)}{(-2, 1, 0) \cdot (-2, 1, 0)} = 0$$

$$t_2 = \frac{\mathbf{v} \cdot \mathbf{v}_2}{\mathbf{v}_2 \cdot \mathbf{v}_2} = \frac{(1, 2, -9) \cdot (2, 4, 0)}{(2, 4, 0) \cdot (2, 4, 0)} = \frac{1}{2}$$

$$t_3 = \frac{\mathbf{v} \cdot \mathbf{v}_3}{\mathbf{v}_3 \cdot \mathbf{v}_3} = \frac{(1, 2, -9) \cdot (0, 0, -3)}{(0, 0, -3) \cdot (0, 0, -3)} = 3$$

9. (a) To show that the vector $(-2, 1)$ is orthogonal to the subspace W, by definition, you must show that $(-2, 1)$ is orthogonal to every vector in W. To that end, note that the subspace W corresponding to the line $y = 2x$ is described by $W = \{(t, 2t) : t \text{ is a real number}\}$. Thus, for a vector $\mathbf{w} = (t, 2t)$ in W, you have that $\mathbf{w}$ is orthogonal to $(-2, 1)$ because $(-2, 1) \cdot \mathbf{w} = (-2, 1) \cdot (t, 2t) = -2t + 2t = 0$.

   (b) $W^\perp = \{(-2t, t) : t \text{ is a real number}\}$, which is the line perpendicular to W. This is obtained using the vector $(-2, 1)$ from part (a). The associated figure is:

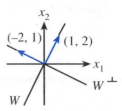

11. The key words *if and only if* indicate that two proofs are required. So first it is shown that if $\mathbf{w}_1, \ldots, \mathbf{w}_k$ is a basis for a subspace W of R^n and $\mathbf{v} \in W^\perp$, then for each $i = 1, \ldots, k$, $\mathbf{v}$ is orthogonal to $\mathbf{w}_i$. Recgonizing the key words *for all*, choose i with $1 \le i \le k$, for which it must be shown that $\mathbf{v}$ is orthogonal to $\mathbf{w}_i$. However, because $\mathbf{v} \in W^\perp$, by definition, you know that for all $\mathbf{w} \in W$, $\mathbf{v} \cdot \mathbf{w} = 0$. Specializing this statement to the particular value $\mathbf{w} = \mathbf{w}_i$, you have that $\mathbf{v} \cdot \mathbf{w}_i = 0$ and so $\mathbf{v}$ is orthogonal to $\mathbf{w}_i$.

   For the converse, assume that $\mathbf{w}_1, \ldots, \mathbf{w}_k$ form a basis for a subspace W of R^n and that for each $i = 1, \ldots, k$, $\mathbf{v}$ is orthogonal to $\mathbf{w}_i$. You must show that $\mathbf{v} \in W^\perp$. By definition of $W^\perp$, this means you must show that for all $\mathbf{w} \in W$, $\mathbf{v} \cdot \mathbf{w} = 0$. Recognizing the key words *for all*, choose a vector $\mathbf{w} \in W$ for which you must show that $\mathbf{v} \cdot \mathbf{w} = 0$. Because $\mathbf{w} \in W$ and $\mathbf{w}_1, \ldots, \mathbf{w}_k$ is a basis for W, there are real numbers $t_1, \ldots, t_k$ such that

$$\mathbf{w} = t_1\mathbf{w}_1 + \cdots + t_k\mathbf{w}_k$$

Taking the dot product of both sides of the foregoing equation with $\mathbf{v}$ and using the assumption that each $\mathbf{v} \cdot \mathbf{w}_i = 0$, you have that

$$\mathbf{v} \cdot \mathbf{w} = t_1(\mathbf{v} \cdot \mathbf{w}_1) + \cdots + t_k(\mathbf{v} \cdot \mathbf{w}_k)$$
$$= t_1(0) + \cdots t_k(0) = 0$$

This means that $\mathbf{v} \in W^\perp$ and so the proof is complete.

13. Assume that $\mathbf{v}$ belongs to a subspace W of R^n and also to $W^\perp$. You must show that $\mathbf{v} = \mathbf{0}$. However, because $\mathbf{v} \in W^\perp$, you know that for all $\mathbf{w} \in W$, $\mathbf{v} \cdot \mathbf{w} = 0$. Specializing this statement to the particular value $\mathbf{w} = \mathbf{v}$ (which is

in W), you have that $\mathbf{v} \cdot \mathbf{v} = 0$. The only way this can happen is if $\mathbf{v} = \mathbf{0}$, thus completing the proof.

Solutions to Exercises in Section 7.2

1. For these vectors $\mathbf{x}$ and $\mathbf{y}$, you have the following projection, $\mathbf{p}$, of $\mathbf{x}$ onto $\mathbf{y}$ and the component, $\mathbf{q}$, of $\mathbf{x}$ orthognal to $\mathbf{y}$:

$$\mathbf{p} = \frac{\mathbf{x} \cdot \mathbf{y}}{\mathbf{y} \cdot \mathbf{y}} \mathbf{y}$$

$$= \frac{(-1, 3) \cdot (-4, 2)}{(-4, 2) \cdot (-4, 2)} (-4, 2)$$

$$= \frac{10}{20} (-4, 2) = (-2, 1)$$

$$\mathbf{q} = \mathbf{x} - \mathbf{p} = (-1, 3) - (-2, 1) = (1, 2)$$

The associated figure is:

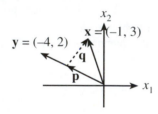

3. The projection is $\mathbf{p} = (2, 0, -3)$, which is obtained from Theorem 7.5, as follows:

$$\mathbf{p} = \frac{\mathbf{v} \cdot \mathbf{v}_1}{\mathbf{v}_1 \cdot \mathbf{v}_1} \mathbf{v}_1 + \frac{\mathbf{v} \cdot \mathbf{v}_2}{\mathbf{v}_2 \cdot \mathbf{v}_2} \mathbf{v}_2$$

$$= \frac{(2, 1, -3) \cdot (1, 0 - 1)}{(1, 0, -1) \cdot (1, 0, -1)} (1, 0, -1) +$$

$$\frac{(2, 1, -3) \cdot (1, 0, 1)}{(1, 0, 1) \cdot (1, 0, 1)} (1, 0, 1)$$

$$= \frac{5}{2} (1, 0, -1) + \frac{-1}{2} (1, 0, 1)$$

$$= (2, 0, -3)$$

5. The orthogonal basis produced by the Gram-Schmidt process is:

$$\mathbf{v}_1 = \mathbf{x}_1 = (-3, 1)$$

$$\mathbf{v}_2 = \mathbf{x}_2 - \frac{\mathbf{x}_2 \cdot \mathbf{v}_1}{\mathbf{v}_1 \cdot \mathbf{v}_1} \mathbf{v}_1$$

$$= (0, -1) - \frac{(0, -1) \cdot (-3, 1)}{(-3, 1) \cdot (-3, 1)} (-3, 1)$$

$$= (-3/10, -9/10)$$

The associated figure is:

$\mathbf{v}_1 = \mathbf{x}_1 = (-3, 1)$

$\mathbf{v}_2 = (-3/10, -9/10)$ $\mathbf{x}_2 = (0, -1)$

7. The orthogonal basis produced by the Gram-Schmidt process is:

$$\mathbf{v}_1 = \mathbf{x}_1 = (1, 0, -3, 0)$$

$$\mathbf{v}_2 = \mathbf{x}_2 - \frac{\mathbf{x}_2 \cdot \mathbf{v}_1}{\mathbf{v}_1 \cdot \mathbf{v}_1} \mathbf{v}_1$$

$$= (0, -4, 0, 2) -$$

$$\frac{(0, -4, 0, 2) \cdot (1, 0, -3, 0)}{(1, 0, -3, 0) \cdot (1, 0, -3, 0)} (1, 0, -3, 0)$$

$$= (0, -4, 0, 2) - \frac{0}{10} (1, 0, -3, 0)$$

$$= (0, -4, 0, 2)$$

9. This proof requires a uniqueness method to show that the projection of $\mathbf{v}$ onto any vector $\mathbf{d}$ on the line is always the same. To that end, let $\mathbf{d}_1$ and $\mathbf{d}_2$ be two vectors on the line L. You

must show that the orthogonal projection, $\mathbf{p}_1$, of $\mathbf{v}$ onto $\mathbf{d}_1$ is equal to the orthogonal projection, $\mathbf{p}_2$, of $\mathbf{v}$ onto $\mathbf{d}_2$. However, because $\mathbf{d}_1$ and $\mathbf{d}_2$ are both on the line L, there is a real number t such that $\mathbf{d}_1 = t\mathbf{d}_2$. Thus, you have that

$$\mathbf{p}_1 = \frac{\mathbf{v} \cdot \mathbf{d}_1}{\mathbf{d}_1 \cdot \mathbf{d}_1}\mathbf{d}_1 \qquad \text{(def. of proj.)}$$

$$= \frac{\mathbf{v} \cdot (t\mathbf{d}_2)}{(t\mathbf{d}_2) \cdot (t\mathbf{d}_2)}(t\mathbf{d}_2) \qquad (\mathbf{d}_1 = t\mathbf{d}_2)$$

$$= \frac{\mathbf{v} \cdot \mathbf{d}_2}{\mathbf{d}_2 \cdot \mathbf{d}_2}\mathbf{d}_2 \qquad \text{(algebra)}$$

$$= \mathbf{p}_2 \qquad \text{(def. of proj.)}$$

11. (a) To find the orthogonal projection, $\mathbf{p}$, of a vector $\mathbf{v}$ onto the subspace of a line L through the origin in the direction of the n-vector $\mathbf{d}$, note that $\mathbf{d}$ is a basis for L. So,

$$\mathbf{p} = \frac{\mathbf{v} \cdot \mathbf{d}}{\mathbf{d} \cdot \mathbf{d}}\mathbf{d}$$

However, because $\mathbf{v}$ is on the line L, there is a real number t such that $\mathbf{v} = t\mathbf{d}$. Substituting $\mathbf{v} = t\mathbf{d}$ in the foregoing equation yields

$$\mathbf{p} = \frac{(t\mathbf{d}) \cdot \mathbf{d}}{\mathbf{d} \cdot \mathbf{d}}\mathbf{d} = t\mathbf{d} = \mathbf{v}$$

It has now been shown that the projection onto a line L of a vector $\mathbf{v}$ on L is $\mathbf{v}$. Thus the proof is complete.

(b) The associated figure is:

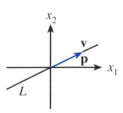

(c) A generalization is the statement that if

$\mathbf{v}$ is a vector belonging to a subspace W of R^n, then the orthogonal projection of $\mathbf{v}$ onto W is $\mathbf{v}$.

Solutions to Exercises in Section 7.3

1. The dependent variable, y, is the number of feet the object is from the measuring device. The independent variable, x, is the number of seconds that have passed.

2. The units of the slope, m, are feet per second and indicate the additional feet the object is from the device in one second. The intercept, b, is the number of feet the object is from the device at time 0.

3. The design matrix is:

$$A = \begin{bmatrix} 0 & 1 \\ 2 & 1 \\ 4 & 1 \\ 7 & 1 \\ 10 & 1 \\ 12 & 1 \end{bmatrix}$$

4. The slope, m, and intercept, b, are found by solving the following normal equations:

$$313m + 35b = 187950$$
$$35m + 6b = 31450$$

These equations are obtained from the system

$$(A^tA)\mathbf{z} = A^t\mathbf{y}$$

in which A is the matrix in the solution to the previous exercise and

$$\mathbf{z} = \begin{bmatrix} m \\ b \end{bmatrix} \quad \text{and} \quad \mathbf{y} = \begin{bmatrix} 5000 \\ 5075 \\ 5175 \\ 5300 \\ 5400 \\ 5500 \end{bmatrix}$$

5. The solution to the normal equations results

in $m = 41.3$ and $b = 5001$, so the equation of the regression line is $y = 41.3x + 5001$. After 30 seconds, the object is $y = 41.3(30) + 5001 = 6240$ feet from the measuring device.

11. The dependent variable, y, is the cost per gallon, in dollars, to produce the paint. The independent variable, x, is the number of thousands of gallons of paint produced in a batch.

12. Note that the graph of the given data is shown in Figure 7.11 in Section 7.5. From this graph, you can see that one relation between the dependent variable, y, and the independent variable, x, that conforms to the data is a quadratic relationship of the form:

$$y = ax^2 + bx + c$$

The parameters are a, b, and c.

13. Once the values of the parameters a, b, and c are obtained, the quadratic function $y = ax^2 + bx + c$ is used to determine y by substituting any appropriate value for x, the number of thousands of gallons to produce in a batch. The resulting value of y, namely, $ax^2 + bx + c$, is an estimate of the associated production cost per gallon. You can therefore solve the problem by finding the value of x that minimizes the value of y.

19. The normal equations are:

$$\left(\sum_{i=1}^{k} x_i^2\right) m + \left(\sum_{i=1}^{k} x_i\right) b = \sum_{i=1}^{k} x_i y_i$$

$$\left(\sum_{i=1}^{k} x_i\right) m + \qquad (k)b = \sum_{i=1}^{k} y_i$$

This is because,

$$A^t A = \begin{bmatrix} \sum\limits_{i=1}^{k} x_i^2 & \sum\limits_{i=1}^{k} x_i \\ \\ \sum\limits_{i=1}^{k} x_i & k \end{bmatrix}, \quad A^t \mathbf{y} = \begin{bmatrix} \sum\limits_{i=1}^{k} x_i y_i \\ \\ \sum\limits_{i=1}^{k} y_i \end{bmatrix}$$

The desired expression is obtained by substituting the foregoing expressions for $A^t A$ and $A^t \mathbf{y}$ in the following normal equations and

performing the multiplication:

$$(A^t A) \begin{bmatrix} m \\ b \end{bmatrix} = A^t \mathbf{y}$$

21. (a) The $(k \times 3)$ design matrix, A, and the three-dimensional parameter vector, $\mathbf{z}$, are defined as follows:

$$A = \begin{bmatrix} 1 & u_1 & v_1 \\ \vdots & \vdots & \vdots \\ 1 & u_k & v_k \end{bmatrix} \quad \text{and} \quad \mathbf{z} = \begin{bmatrix} \beta_0 \\ \beta_1 \\ \beta_2 \end{bmatrix}$$

(b) The parameter vector, $\mathbf{z}$, is found by solving the normal equations:

$$(A^t A)\mathbf{z} = A^t \mathbf{y}$$

where the elements of $\mathbf{y}$ are obtained from the data:

$$\mathbf{y} = \begin{bmatrix} y_1 \\ \vdots \\ y_k \end{bmatrix}$$

Solutions to Exercises in Section 7.4

1. To verify the four conditions in Definition 7.3, let $\mathbf{u} = (u_1, u_2)$, $\mathbf{v} = (v_1, v_2)$, and $\mathbf{w} = (w_1, w_2)$ and let t be a real number.
   (a) $\langle \mathbf{u}, \mathbf{v} \rangle = au_1 v_1 + bu_2 v_2 = av_1 u_1 + bv_2 u_2 = \langle \mathbf{v}, \mathbf{u} \rangle$.
   (b) $\langle \mathbf{u} + \mathbf{v}, \mathbf{w} \rangle$

   $$= a(u_1 + v_1)w_1 + b(u_2 + v_2)w_2$$
   $$= (au_1 w_1 + bu_2 w_2) +$$
   $$(av_1 w_1 + bv_2 w_2)$$
   $$= \langle \mathbf{u}, \mathbf{w} \rangle + \langle \mathbf{v}, \mathbf{w} \rangle$$

   (c) $\langle t\mathbf{u}, \mathbf{v} \rangle = a(tu_1)v_1 + b(tu_2)v_2 = t(au_1 v_1 + bu_2 v_2) = t \langle \mathbf{u}, \mathbf{v} \rangle$.
   (d) $\langle \mathbf{u}, \mathbf{u} \rangle = au_1 u_1 + bu_2 u_2 \geq 0$. Also, if $\mathbf{u} = \mathbf{0}$, then $u_1 = 0$ and $u_2 = 0$, so $\langle \mathbf{u}, \mathbf{u} \rangle = au_1 u_1 + bu_2 u_2 = 0$. Finally, if $\langle \mathbf{u}, \mathbf{u} \rangle = au_1 u_1 + bu_2 u_2 = 0$, then the nonnegative numbers $au_1 u_1$ and

bu_2u_2 must be such that $au_1u_1 = 0$ and $bu_2u_2 = 0$. But this means that $u_1 = 0$ and $u_2 = 0$ and hence $\mathbf{u} = \mathbf{0}$.

3. Let $\mathbf{p} = p(x)$, $\mathbf{q} = q(x)$, and $\mathbf{r} = r(x)$ be three polynomials in P^n and let t be a real number.

   (a) $\langle \mathbf{p}, \mathbf{q} \rangle$

$$= p(x_0)q(x_0) + \cdots + p(x_n)q(x_n)$$
$$= q(x_0)p(x_0) + \cdots + q(x_n)p(x_n)$$
$$= \langle \mathbf{q}, \mathbf{p} \rangle$$

   (b) $\langle \mathbf{p} + \mathbf{q}, \mathbf{r} \rangle$

$$= [p(x_0) + q(x_0)]r(x_0) + \cdots +$$
$$[p(x_n) + q(x_n)]r(x_n)$$
$$= [p(x_0)r(x_0) + \cdots + p(x_n)r(x_n)] +$$
$$[q(x_0)r(x_0) + \cdots + q(x_n)r(x_n)]$$
$$= \langle \mathbf{p}, \mathbf{r} \rangle + \langle \mathbf{q}, \mathbf{r} \rangle$$

   (c) $\langle t\mathbf{p}, \mathbf{q} \rangle$

$$= [tp(x_0)]q(x_0) + \cdots + [tp(x_n)]q(x_n)$$
$$= t[p(x_0)q(x_0) + \cdots + p(x_n)q(x_n)]$$
$$= t\langle \mathbf{p}, \mathbf{q} \rangle$$

5. To verify the four conditions in Definition 7.3, let $\mathbf{u}$, $\mathbf{v}$, and $\mathbf{w}$ be n-vectors and let t be a real number.

   (a) $\langle \mathbf{u}, \mathbf{v} \rangle = (A\mathbf{u})^t(A\mathbf{v}) = (A\mathbf{v})^t(A\mathbf{u}) = \langle \mathbf{v}, \mathbf{u} \rangle$.

   (b) $\langle \mathbf{u} + \mathbf{v}, \mathbf{w} \rangle$

$$= [A(\mathbf{u} + \mathbf{v})]^t(A\mathbf{w})$$
$$= [(A\mathbf{u})^t + (A\mathbf{v})^t](A\mathbf{w})$$
$$= (A\mathbf{u})^t(A\mathbf{w}) + (A\mathbf{v})^t(A\mathbf{w})$$
$$= \langle \mathbf{u}, \mathbf{w} \rangle + \langle \mathbf{v}, \mathbf{w} \rangle$$

   (c) $\langle t\mathbf{u}, \mathbf{v} \rangle = [A(t\mathbf{u})]^t(A\mathbf{v}) = t[(A\mathbf{u})^t(A\mathbf{v})] = t\langle \mathbf{u}, \mathbf{v} \rangle$.

   (d) $\langle \mathbf{u}, \mathbf{u} \rangle = (A\mathbf{u})^t(A\mathbf{u}) \geq 0$ (because the dot product of a vector with itself is ≥ 0.) Also, if $\langle \mathbf{u}, \mathbf{u} \rangle = (A\mathbf{u})^t(A\mathbf{u}) = 0$, then it is equivalent to say that the dot product of

the vector $A\mathbf{u}$ with itself is 0, so $A\mathbf{u} = \mathbf{0}$. Because A is invertible, it follows, after multiplying each side of the foregoing equality by A^{-1}, that $\mathbf{u} = \mathbf{0}$. Finally, if $\mathbf{u} = \mathbf{0}$, then $\langle \mathbf{u}, \mathbf{u} \rangle = (A\mathbf{0})^t(A\mathbf{0}) = 0$.

7. For $\mathbf{p} = p(x) = 1 + x^2$, $\mathbf{q} = q(x) = 2 - 3x$, and $x_0 = -1$, $x_1 = 0$, and $x_2 = 1$, you have that $\langle \mathbf{p}, \mathbf{q} \rangle = p(x_0)q(x_0) + p(x_1)q(x_1) + p(x_2)q(x_2) = 2(5) + 1(2) + 2(-1) = 10$.

9. For $\mathbf{p} = p(x) = 1 + x^2$, $\mathbf{q} = q(x) = 2 - 3x$, and $x_0 = -1$, $x_1 = 0$, and $x_2 = 1$, you have that

$$\|\mathbf{p}\| = \sqrt{\langle \mathbf{p}, \mathbf{p} \rangle}$$

$$= \sqrt{p(x_0)^2 + p(x_1)^2 + p(x_2)^2}$$

$$= \sqrt{2^2 + 1^2 + 2^2} = 3$$

$$\|\mathbf{q}\| = \sqrt{\langle \mathbf{q}, \mathbf{q} \rangle}$$

$$= \sqrt{q(x_0)^2 + q(x_1)^2 + q(x_2)^2}$$

$$= \sqrt{5^2 + 2^2 + (-1)^2} = \sqrt{30}$$

The normalized polynomials are:

$$\frac{1}{\|\mathbf{p}\|}\mathbf{p} = \frac{1}{3}(1 + x^2), \quad \frac{1}{\|\mathbf{q}\|}\mathbf{q} = \frac{1}{\sqrt{30}}(2 - 3x)$$

11. For $\mathbf{p} = p(x) = 1 + x^2$, $\mathbf{q} = q(x) = 2 - 3x$, and $x_0 = -1$, $x_1 = 0$, and $x_2 = 1$ you have, using the solutions to Exercise 7 and Exercise 9, that the orthogonal projection of $\mathbf{p}$ onto $\mathbf{q}$ is

$$\frac{\langle \mathbf{p}, \mathbf{q} \rangle}{\langle \mathbf{q}, \mathbf{q} \rangle}\mathbf{q} = \frac{10}{30}(2 - 3x) = \frac{1}{3}(2 - 3x)$$

13. To show that the polynomials $\mathbf{p} = p(x) = 1$ and $\mathbf{q} = q(x) = x$ are orthogonal, you must show that $\langle \mathbf{p}, \mathbf{q} \rangle = 0$. Using $x_0 = -2$, $x_1 = -1$, $x_2 = 1$, and $x_3 = 2$ in the definition of

$\langle \mathbf{p}, \mathbf{q} \rangle$, you have that

$$\langle \mathbf{p}, \mathbf{q} \rangle = p(x_0)q(x_0) + p(x_1)q(x_1) +$$
$$p(x_2)q(x_2) + p(x_3)q(x_3)$$
$$= 1(-2) + 1(-1) + 1(1) + 1(2)$$
$$= 0$$

15. (a) The orthogonal projection, $\mathbf{p}_3$, of $\mathbf{q}_3$ onto the subspace spanned by $\mathbf{q}_1$ and $\mathbf{q}_2$ is

$$\mathbf{p}_3 = x^2 - 5x$$

This answer is obtained by calculating the following inner products:

$\langle \mathbf{q}_3, \mathbf{q}_1 \rangle$	$\langle \mathbf{q}_1, \mathbf{q}_1 \rangle$
$\mathbf{q}_3(-3)\mathbf{q}_1(-3)$	$\mathbf{q}_1(-3)\mathbf{q}_1(-3)$
$+\, \mathbf{q}_3(-1)\mathbf{q}_1(-1)$	$+\, \mathbf{q}_1(-1)\mathbf{q}_1(-1)$
$+\, \mathbf{q}_3(1)\mathbf{q}_1(1)$	$+\, \mathbf{q}_1(1)\mathbf{q}_1(1)$
$+\, \mathbf{q}_3(3)\mathbf{q}_1(3)$	$+\, \mathbf{q}_1(3)\mathbf{q}1(3)$
$= 20$	$= 4$

$\langle \mathbf{q}_3, \mathbf{q}_2 \rangle$	$\langle \mathbf{q}_2, \mathbf{q}_2 \rangle$
$\mathbf{q}_3(-3)\mathbf{q}_2(-3)$	$\mathbf{q}_2(-3)\mathbf{q}_2(-3)$
$+\, \mathbf{q}_3(-1)\mathbf{q}_2(-1)$	$+\, \mathbf{q}_2(-1)\mathbf{q}_2(-1)$
$+\, \mathbf{q}_3(1)\mathbf{q}_2(1)$	$+\, \mathbf{q}_2(1)\mathbf{q}_2(1)$
$+\, \mathbf{q}_3(3)\mathbf{q}_2(3)$	$+\, \mathbf{q}_2(3)\mathbf{q}2(3)$
$= 0$	$= 20$

So, the projection, $\mathbf{p}_3$, is:

$$\mathbf{p}_3 = \mathbf{q}_3 - \frac{\langle \mathbf{q}_3, \mathbf{q}_1 \rangle}{\langle \mathbf{q}_1, \mathbf{q}_1 \rangle}\mathbf{q}_1 - \frac{\langle \mathbf{q}_3, \mathbf{q}_2 \rangle}{\langle \mathbf{q}_2, \mathbf{q}_2 \rangle}\mathbf{q}_2$$

$$= \mathbf{q}_3 - \frac{20}{4}\mathbf{q}_1 - \frac{0}{20}\mathbf{q}_2 = x^2 - 5x$$

(b) Applying the Gram-Schmidt process using the result in part (a) and computing $\langle \mathbf{q}_1, \mathbf{q}_2 \rangle = 0$, the orthogonal basis for P^2

is:

$$\mathbf{p}_1 = \mathbf{q}_1 = 1$$

$$\mathbf{p}_2 = \mathbf{q}_2 - \frac{\langle \mathbf{q}_2, \mathbf{p}_1 \rangle}{\langle \mathbf{p}_1, \mathbf{p}_1 \rangle}\mathbf{p}_1 = \mathbf{q}_2 = x$$

$$\mathbf{p}_3 = \mathbf{q}_3 - \frac{\langle \mathbf{q}_3, \mathbf{p}_1 \rangle}{\langle \mathbf{p}_1, \mathbf{p}_1 \rangle}\mathbf{p}_1 - \frac{\langle \mathbf{q}_3, \mathbf{p}_2 \rangle}{\langle \mathbf{p}_2, \mathbf{p}_2 \rangle}\mathbf{p}_2$$

$$= x^2 - 5x$$

17. Using the hint, for the vectors $\mathbf{u} = (\sqrt{a}, \sqrt{b})$ and $\mathbf{v} = (\sqrt{b}, \sqrt{a})$:

$$|\mathbf{u} \cdot \mathbf{v}| = |(\sqrt{a}, \sqrt{b}) \cdot (\sqrt{b}, \sqrt{a})| = 2\sqrt{ab}$$

$$\|\mathbf{u}\| = \sqrt{\left(\sqrt{a}\right)^2 + \left(\sqrt{b}\right)^2} = \sqrt{a + b}$$

$$\|\mathbf{v}\| = \sqrt{\left(\sqrt{b}\right)^2 + \left(\sqrt{a}\right)^2} = \sqrt{a + b}$$

Thus, applying the Cauchy-Schwarz inequality to $\mathbf{u}$ and $\mathbf{v}$, it follows that

$$2\sqrt{ab} = |\mathbf{u} \cdot \mathbf{v}|$$
$$\leq \|\mathbf{u}\|\|\mathbf{v}\|$$
$$= \sqrt{a + b}\,\sqrt{a + b}$$
$$= a + b$$

The desired result follows by dividing the foregoing inequality by 2.

19. Recognizing the key words *for any*, choose vectors $\mathbf{u}$, $\mathbf{v}$, and $\mathbf{w}$ in V, for which it must be shown that

$$d(\mathbf{u}, \mathbf{w}) \leq d(\mathbf{u}, \mathbf{v}) + d(\mathbf{v}, \mathbf{w})$$

Using the definition of the distance function d and specializing the triangle inequality to that of the inner product space, V, you have that

$$d(\mathbf{u}, \mathbf{w}) = \|\mathbf{u} - \mathbf{w}\| \qquad (1)$$

$$= \|\mathbf{u} - \mathbf{v} + \mathbf{v} - \mathbf{w}\| \qquad (2)$$

$$\leq \|\mathbf{u} - \mathbf{v}\| + \|\mathbf{v} - \mathbf{w}\| \qquad (3)$$

$$= d(\mathbf{u}, \mathbf{v}) + d(\mathbf{v}, \mathbf{w}) \qquad (4)$$

Each of the foregoing numbered steps is justified, as follows:

Step	Justification
(1)	definition of d
(2)	add and subtract $\mathbf{v}$
(3)	specialize the triangle inequality
(4)	definition of d

SOLUTIONS TO SELECTED EXERCISES IN CHAPTER 8

Solutions to Exercises in Section 8.1

1. The approaches in (b) and (c) each require k multiplications and are therefore more efficient than the approach in (a), which requires $2k - 1$ multiplications.

3. It requires mnp multiplications to compute AB. This is because multiplying each row of A by each column of B requires n multiplications and you must multiply each of the m rows of A by each of the p columns of B.

5. (a) The first algorithm is more efficient than the second algorithm when $n > 4$. This value is determined by setting the running time of the first algorithm, $4n^2$, less than the running time of the second algorithm, n^3, and solving for n.

   (b) The results in part (a) are verified in the following table:

	n			
	1	2	3	4
$4n^2$	4	16	36	64
n^3	1	8	27	64

	n			
	5	6	7	8
$4n^2$	100	144	196	256
n^3	125	216	343	512

7. (a) Storing all elements of the 9-vector $\mathbf{x}$ requires storing 9 values.

   (b) Storing $\mathbf{x}$ in the following sparse form requires storing six values: 4 in position 3, -2 in position 5, and 6 in position 8.

9. The sparse form for storing the nonzero elements of A is: 3 in row 1 and column 2, 6 in row 2 and column 1, 1 in row 3 and column 4, 5 in row 4 and column 1.

11. The sparse form for storing an $(n \times n)$ diagonal matrix D requires storing the n pairs of values i and D_{ii}, for $i = 1, \ldots, n$.

13. The sparse form for storing an elementary matrix that interchanges row i of the identity matrix with row j requires storing only the values of i and j.

15. The sparse form for storing a matrix E that differs from the $(n \times n)$ identity matrix only in column k requires storing k and the n pairs of values i and E_{ik}, for $i = 1, \ldots, n$.

17. Partial pivoting requires interchanging row 1 and row 3 of A because the value -4 in row 3 of column 1 has the largest absolute value of any element in column 1 below row 1.

Solutions to Exercises in Section 8.2

1. No, A and B do not constitute a factorization of I because

$$AB = \begin{bmatrix} -1 & -3 \\ 1 & 4 \end{bmatrix} \begin{bmatrix} -4 & -6 \\ 1 & 2 \end{bmatrix}$$

$$= \begin{bmatrix} 1 & 0 \\ 0 & 2 \end{bmatrix} \neq \begin{bmatrix} 1 & 0 \\ 0 & 1 \end{bmatrix}$$

3. The desired factorization is

$$A^{-1} = E_4 E_3 E_2 E_1 = \begin{bmatrix} -2 & -3/4 \\ 1 & 1/2 \end{bmatrix}$$

in which

$$E_1 = \begin{bmatrix} 0 & 1 \\ 1 & 0 \end{bmatrix}, \quad E_2 = \begin{bmatrix} 1/4 & 0 \\ 0 & 1 \end{bmatrix},$$

$$E_3 = \begin{bmatrix} 1 & 0 \\ 2 & 1 \end{bmatrix}, \quad E_4 = \begin{bmatrix} 1 & -2 \\ 0 & 1 \end{bmatrix}$$

These E matrices are obtained as follows:

Step 1. Partial pivoting requires exchanging row 1 and row 2 of A, so

$$E_1 = \begin{bmatrix} 0 & 1 \\ 1 & 0 \end{bmatrix}, \text{ thus } E_1 A = \begin{bmatrix} 4 & 8 \\ -2 & -3 \end{bmatrix}$$

Step 2. Multiply row 1 of $E_1 A$ by 1/4, so

$$E_2 = \begin{bmatrix} \frac{1}{4} & 0 \\ 0 & 1 \end{bmatrix}, \text{ thus } E_2 E_1 A = \begin{bmatrix} 1 & 2 \\ -2 & -3 \end{bmatrix}$$

Step 3. Add 2 times row 1 of $E_2 E_1 A$ to row 2, so

$$E_3 = \begin{bmatrix} 1 & 0 \\ 2 & 1 \end{bmatrix}, \text{ thus } E_3 E_2 E_1 A = \begin{bmatrix} 1 & 2 \\ 0 & 1 \end{bmatrix}$$

Step 4. It remains only to add -2 times row 2 of $E_3 E_2 E_1 A$ to row 1, so

$$E_4 = \begin{bmatrix} 1 & -2 \\ 0 & 1 \end{bmatrix}$$

and thus

$$E_4 E_3 E_2 E_1 A = \begin{bmatrix} 1 & 0 \\ 0 & 1 \end{bmatrix}$$

5. The solution is $x_1 = 2$ and $x_2 = 0$ because, from the factorization in the solution to Ex-

ercise 3:

$$\mathbf{x} = E_4(E_3(E_2(E_1 \mathbf{b})))$$

$$= E_4 \left(E_3 \left(E_2 \left(\begin{bmatrix} 0 & 1 \\ 1 & 0 \end{bmatrix} \begin{bmatrix} -4 \\ 8 \end{bmatrix} \right) \right) \right)$$

$$= E_4 \left(E_3 \left(\begin{bmatrix} 1/4 & 0 \\ 0 & 1 \end{bmatrix} \begin{bmatrix} 8 \\ -4 \end{bmatrix} \right) \right)$$

$$= E_4 \left(\begin{bmatrix} 1 & 0 \\ 2 & 1 \end{bmatrix} \begin{bmatrix} 2 \\ -4 \end{bmatrix} \right)$$

$$= \begin{bmatrix} 1 & -2 \\ 0 & 1 \end{bmatrix} \begin{bmatrix} 2 \\ 0 \end{bmatrix} = \begin{bmatrix} 2 \\ 0 \end{bmatrix}$$

7. The solution to the system is $x_1 = 3$ and $x_2 = -1$. This solution is obtained as follows:

(a) The system $L\mathbf{y} = \mathbf{b}$ is

$$\begin{bmatrix} 1 & 0 \\ -1 & 1 \end{bmatrix} \begin{bmatrix} y_1 \\ y_2 \end{bmatrix} = \begin{bmatrix} 3 \\ -7 \end{bmatrix}$$

or

$$\begin{aligned} y_1 &= 3 \\ -y_1 + y_2 &= -7 \end{aligned}$$

Applying forward substitution leads to $y_1 = 3$ and $y_2 = -4$.

(b) The system $U\mathbf{x} = \mathbf{y}$ is

$$\begin{bmatrix} 3 & 6 \\ 0 & 4 \end{bmatrix} \begin{bmatrix} x_1 \\ x_2 \end{bmatrix} = \begin{bmatrix} 3 \\ -4 \end{bmatrix}$$

or

$$\begin{aligned} 3x_1 + 6x_2 &= 3 \\ 4x_2 &= -4 \end{aligned}$$

Applying back substitution leads to $x_2 = -1$ and $x_1 = 3$.

9. The *LU* factorization of *A* is

$$
\begin{bmatrix} 1 & 0 & 0 \\ -2 & 1 & 0 \\ 0 & 2 & 1 \end{bmatrix}
\underbrace{}_{L}
\begin{bmatrix} -1 & -3 & -1 \\ 0 & 2 & -4 \\ 0 & 0 & 4 \end{bmatrix}
\underbrace{}_{U}
$$

This answer is obtained by applying Gaussian elimination, as follows:

$$
\underbrace{\begin{bmatrix} -1 & -3 & -1 \\ 2 & 8 & -2 \\ 0 & 4 & -4 \end{bmatrix}}_{A=A_1}
\longrightarrow
\underbrace{\begin{bmatrix} -1 & -1 & -2 \\ 0 & 2 & -4 \\ 0 & 4 & -4 \end{bmatrix}}_{A_2}
$$

$$
\longrightarrow
\underbrace{\begin{bmatrix} -1 & -3 & -1 \\ 0 & 2 & -4 \\ 0 & 0 & 4 \end{bmatrix}}_{U}
$$

Using column 1 of A_1 and column 2 of A_2, you have:

$$
\underbrace{\begin{bmatrix} 1 & 0 & 0 \\ 0 & 1 & 0 \\ 0 & 0 & 1 \end{bmatrix}}_{L_0=I}
\longrightarrow
\underbrace{\begin{bmatrix} 1 & 0 & 0 \\ -2 & 1 & 0 \\ 0 & 0 & 1 \end{bmatrix}}_{L_1}
$$

$$
\longrightarrow
\underbrace{\begin{bmatrix} 1 & 0 & 0 \\ -2 & 1 & 0 \\ 0 & 2 & 1 \end{bmatrix}}_{L}
$$

11. The solution to the system is $x_1 = 9$, $x_2 = -2$, and $x_3 = -1$. This solution is obtained as follows.

(a) The system $L\mathbf{y} = \mathbf{b}$ is:

$$
\begin{bmatrix} 1 & 0 & 0 \\ -2 & 1 & 0 \\ 0 & 2 & 1 \end{bmatrix}
\begin{bmatrix} y_1 \\ y_2 \\ y_3 \end{bmatrix}
=
\begin{bmatrix} -2 \\ 4 \\ -4 \end{bmatrix}
$$

or

$$
\begin{aligned}
y_1 &= -2 \\
-2y_1 + y_2 &= 4 \\
2y_2 + y_3 &= -4
\end{aligned}
$$

From the first equation, $y_1 = -2$. Substituting this value in the second equation and solving for y_2 yields $y_2 = 0$. Substituting $y_1 = -2$ and $y_2 = 0$ in the last equation and solving for y_3 yields $y_3 = -4$.

(b) The system $U\mathbf{x} = \mathbf{y}$ is:

$$
\begin{bmatrix} -1 & -3 & -1 \\ 0 & 2 & -4 \\ 0 & 0 & 4 \end{bmatrix}
\begin{bmatrix} x_1 \\ x_2 \\ x_3 \end{bmatrix}
=
\begin{bmatrix} -2 \\ 0 \\ -4 \end{bmatrix}
$$

or

$$
\begin{aligned}
-x_1 - 3x_2 - x_3 &= -2 \\
2x_2 - 4x_3 &= 0 \\
4x_3 &= -4
\end{aligned}
$$

From the last equation, $x_3 = -1$. Substituting this value in the second equation and solving for x_2 yields $x_2 = -2$. Substituting $x_2 = -2$ and $x_3 = -1$ in the first equation and solving for x_1 yields $x_1 = 9$.

13. From the hypothesis that A and B are lower triangular, you know that for all $i < j$, $A_{ij} = B_{ij} = 0$. Likewise, to show that AB is lower triangular, it must be shown that for all $i < j$, $(AB)_{ij} = 0$. The key words *for all* suggests using the choose method to choose i and j with $i < j$ and showing that $(AB)_{ij} = 0$. So,

$$
(AB)_{ij} = A_{i*} \cdot B_{*j} = \sum_{k=1}^{n} A_{ik} B_{kj}
$$

$$
= \sum_{k<j} A_{ik} B_{kj} + \sum_{k\geq j} A_{ik} B_{kj}
$$

$$
= \sum_{k<j} A_{ik}(0) + \sum_{k\geq j} (0) B_{kj} = 0
$$

Thus, AB is lower triangular, completing the proof.

15. (a) Utilizing the basis vectors, you have that

$$Q = \left[\frac{1}{\|v_1\|}v_1, \quad \frac{1}{\|v_2\|}v_2, \quad \frac{1}{\|v_3\|}v_3 \right]$$

$$= \begin{bmatrix} 0.6667 & 0.7071 & 0.2357 \\ -0.6667 & 0.7071 & -0.2357 \\ 0.3333 & 0.0000 & -0.9428 \end{bmatrix}$$

(b) You have that $Q^t A =$

$$R = \begin{bmatrix} 3.0000 & 0.0000 & -1.0000 \\ 0.0000 & 1.4142 & 0.0000 \\ 0.0000 & 0.0000 & 2.8284 \end{bmatrix}$$

Note that $A = QR$.

(c) Utilizing the matrix Q defined in part (a) and solving the system $Qr_j = A_{*j}$ yields the desired result. That is, the solution to each of the systems $Qr_j = A_{*j}$ is column j of R.

Solutions to Exercises in Section 8.3

1. (a) The eigenvalues of A, namely, $\lambda_1 = -1$ and $\lambda_2 = 2$, are the roots of the following characteristic equation:

$$\det(A - \lambda I) = \det \begin{bmatrix} 1 - \lambda & -1 \\ -2 & -\lambda \end{bmatrix}$$

$$= (1 - \lambda)(-\lambda) - 2 = 0$$

(b) An approximate eigenvector is given in the last column of the following table:

k	0	1	2
x_k	$\begin{bmatrix} -1 \\ 3 \end{bmatrix}$	$\begin{bmatrix} -4 \\ 2 \end{bmatrix}$	$\begin{bmatrix} -6 \\ 8 \end{bmatrix}$

k	3	4	5
x_k	$\begin{bmatrix} -14 \\ 12 \end{bmatrix}$	$\begin{bmatrix} -26 \\ 28 \end{bmatrix}$	$\begin{bmatrix} -54 \\ 52 \end{bmatrix}$

(c) Using x_5 from part (b), approximate values for the strictly dominant eigenvalue are 1.9630 and 2.0769. This is obtained by equating each of the components of Ax_5 to λx_5 and solving for λ. That is,

$$Ax_5 = \begin{bmatrix} 1 & -1 \\ -2 & 0 \end{bmatrix} \begin{bmatrix} -54 \\ 52 \end{bmatrix} = \lambda \begin{bmatrix} -54 \\ 52 \end{bmatrix}$$

or

$$\begin{matrix} -106 = -54\lambda \\ 108 = 52\lambda \end{matrix} \quad \text{so} \quad \begin{matrix} \lambda_1 = 1.9630 \\ \lambda_2 = 2.0769 \end{matrix}$$

3. The results of for the power method are:

k	0	1	2
x_k	$\begin{bmatrix} -1 \\ 3 \end{bmatrix}$	$\begin{bmatrix} 1.0 \\ -0.5 \end{bmatrix}$	$\begin{bmatrix} -0.75 \\ 1.00 \end{bmatrix}$
Ax_k	$\begin{bmatrix} -4 \\ 2 \end{bmatrix}$	$\begin{bmatrix} 1.5 \\ -2.0 \end{bmatrix}$	$\begin{bmatrix} -1.75 \\ 1.50 \end{bmatrix}$
c_k	-4	-2.0	-1.75

k	3	4	5
x_k	$\begin{bmatrix} 1.0000 \\ -0.8571 \end{bmatrix}$	$\begin{bmatrix} -0.9286 \\ 1.0000 \end{bmatrix}$	$\begin{bmatrix} 1.0000 \\ -0.9630 \end{bmatrix}$
Ax_k	$\begin{bmatrix} 1.8571 \\ -2.0000 \end{bmatrix}$	$\begin{bmatrix} -1.9286 \\ 1.8571 \end{bmatrix}$	$\begin{bmatrix} 1.9630 \\ -2.0000 \end{bmatrix}$
c_k	-2.0000	-1.9286	-2.0000

5. The results of the power method are:

k	0	1	2
$\mathbf{x}_k$	$\begin{bmatrix} 4 \\ 0 \\ -1 \end{bmatrix}$	$\begin{bmatrix} -0.7073 \\ 0.1463 \\ 1.0000 \end{bmatrix}$	$\begin{bmatrix} -0.7526 \\ 0.3711 \\ 1.0000 \end{bmatrix}$
$A\mathbf{x}_k$	$\begin{bmatrix} 29 \\ -6 \\ -41 \end{bmatrix}$	$\begin{bmatrix} 3.5610 \\ -1.7561 \\ -4.7317 \end{bmatrix}$	$\begin{bmatrix} 7.6495 \\ -3.4639 \\ -8.6392 \end{bmatrix}$
c_k	-41	-4.7317	-8.6392

k	3	4
$\mathbf{x}_k$	$\begin{bmatrix} -0.8854 \\ 0.4010 \\ 1.0000 \end{bmatrix}$	$\begin{bmatrix} -0.9390 \\ 0.4577 \\ 1.0000 \end{bmatrix}$
$A\mathbf{x}_k$	$\begin{bmatrix} 7.0501 \\ -3.4368 \\ -7.5084 \end{bmatrix}$	$\begin{bmatrix} 7.7038 \\ -3.7839 \\ -7.9479 \end{bmatrix}$
c_k	-7.5084	-7.9479

7. When $t_1 = 0$, the theoretical analysis results in the conclusion that, as k approaches infinity, $\mathbf{x}_k$ approaches $\mathbf{0}$, which is not an eigenvector. The reason the sequence approaches $\mathbf{0}$ is because, when $t_1 = 0$, (8.45) becomes $\mathbf{x}_k \approx t_1(\lambda_1)^k \mathbf{v}_1 = \mathbf{0}$.

9. (a) You have that

$$B = (A - \mu I)^{-1} = \begin{bmatrix} 3 & -1 \\ -2 & 2 \end{bmatrix}^{-1}$$

$$= \begin{bmatrix} 1/2 & 1/4 \\ 1/2 & 3/4 \end{bmatrix}$$

(b) The eigenvalues of B, namely, $\beta_1 = 1$ and $\beta_2 = 1/4$, are the roots of the following characteristic equation:

$$\det(B - \beta I)$$

$$= \det \begin{bmatrix} \dfrac{1}{2} - \beta & \dfrac{1}{4} \\ \dfrac{1}{2} & \dfrac{3}{4} - \beta \end{bmatrix}$$

$$= \left(\frac{1}{2} - \beta \right)\left(\frac{3}{4} - \beta \right) - \frac{1}{8}$$

$$= 0$$

(c) From the solution to Exercise 1, the two eigenvalues of A are $\lambda_1 = -1$ and $\lambda_2 = 2$, so, from (8.51):

$$\beta_1 = \frac{1}{\lambda_1 - \mu} = \frac{1}{-1 - (-2)} = 1$$

$$\beta_2 = \frac{1}{\lambda_2 - \mu} = \frac{1}{2 - (-2)} = \frac{1}{4}$$

(d) The results of applying the inverse power method are:

k	0	1	2
$\mathbf{x}_k$	$\begin{bmatrix} -1 \\ 3 \end{bmatrix}$	$\begin{bmatrix} 0.1429 \\ 1.0000 \end{bmatrix}$	$\begin{bmatrix} 0.3913 \\ 1.0000 \end{bmatrix}$
$\mathbf{y}_k$	$\begin{bmatrix} 0.25 \\ 1.75 \end{bmatrix}$	$\begin{bmatrix} 0.3214 \\ 0.8214 \end{bmatrix}$	$\begin{bmatrix} 0.4457 \\ 0.9457 \end{bmatrix}$
c_k	1.75	0.8214	0.9457
$\mu + \dfrac{1}{c_k}$	-1.4286	-0.7826	-0.9425

k	3	4
x_k	$\begin{bmatrix} 0.4713 \\ 1.0000 \end{bmatrix}$	$\begin{bmatrix} 0.4927 \\ 1.0000 \end{bmatrix}$
y_k	$\begin{bmatrix} 0.4856 \\ 0.9856 \end{bmatrix}$	$\begin{bmatrix} 0.4964 \\ 0.9964 \end{bmatrix}$
c_k	0.9856	0.9964
$\mu + \dfrac{1}{c_k}$	-0.9854	-0.9963

(e) The sequence generated by the inverse power method converges for all values of $\mu < 1/2$, other than $\mu = -1$. This is because when $\mu < 1/2$, μ is closer to the eigenvalue $\lambda_1 = -1$ of A than to $\lambda_2 = 2$.

11. (Answers can vary.) Starting with the vector x_0, each of whose components is 1, after 10 iterations, the power method and the inverse power method produce the following approximate eigenvalues and eigenvectors:

Method	Approx. λ	True λ	Approx. ev (x_{10})
Power	2.3369	$\dfrac{7}{3}$	$\begin{bmatrix} 1.0000 \\ -0.4901 \\ 0.6723 \end{bmatrix}$
Inverse ($\mu = 1$)	1.4998	$\dfrac{3}{2}$	$\begin{bmatrix} -0.2489 \\ 1.0000 \\ 0.3347 \end{bmatrix}$
Inverse ($\mu = -1$)	-0.3335	$-\dfrac{1}{3}$	$\begin{bmatrix} 0.5001 \\ -0.2502 \\ 1.0000 \end{bmatrix}$

Solutions to Exercises in Section 8.4

1. Solving each equation i for x_i results in

$$x_1 = 10/4 + (1/4)x_2 - (1/4)x_3$$
$$x_2 = -2/9 + (2/9)x_1 - (5/9)x_3$$
$$x_3 = -7/6 + (1/6)x_1 - (2/6)x_2$$

3. The sequence of vectors obtained from Jacobi's method is:

k	1	2	3
x_k	$\begin{bmatrix} 2.5000 \\ -0.2222 \\ -1.1667 \end{bmatrix}$	$\begin{bmatrix} 2.7361 \\ 0.9815 \\ -0.6759 \end{bmatrix}$	$\begin{bmatrix} 2.9144 \\ 0.7613 \\ -1.0378 \end{bmatrix}$

k	4	5
x_k	$\begin{bmatrix} 2.9498 \\ 1.0020 \\ -0.9347 \end{bmatrix}$	$\begin{bmatrix} 2.9842 \\ 0.9526 \\ -1.0090 \end{bmatrix}$

5. The sequence of vectors obtained from the Gauss-Seidel method is:

k	1	2	3
x_k	$\begin{bmatrix} 2.5000 \\ -0.3333 \\ -0.8611 \end{bmatrix}$	$\begin{bmatrix} 2.7986 \\ 0.8781 \\ -0.9929 \end{bmatrix}$	$\begin{bmatrix} 2.9678 \\ 0.9889 \\ -1.0017 \end{bmatrix}$

k	4	5
x_k	$\begin{bmatrix} 2.9976 \\ 1.0004 \\ -1.0005 \end{bmatrix}$	$\begin{bmatrix} 3.0002 \\ 1.0003 \\ -1.0001 \end{bmatrix}$

7. The matrix in (a) is strictly diagonally dominant because, for row 1, $|3| > |-1|+|0| = 1$, for row 2, $|-6| > |-3| + |1| = 4$, and for

row 3, $|-9| > |4| + |-2| = 6$. The matrix in (b) is not strictly diagonally dominant because, for row 2, $|1| \leq |5| + |-7| = 12$. Also, for row 3, $|3| \leq |-1| + |9| = 10$.

9. (a) The following matrix, obtained by interchanging rows 2 and 3 of the matrix in Exercise 7(b), is strictly diagonally dominant:

$$\begin{bmatrix} 6 & -2 & 2 \\ -1 & 9 & 3 \\ 5 & 1 & -7 \end{bmatrix}$$

because

$$|6| > |-2| + |2| = 4$$
$$|9| > |-1| + |3| = 4$$
$$|-7| > \qquad |5| + |1| = 6$$

(b) The following matrix, obtained by interchanging rows 2 and 3 of the matrix in Exercise 8(b), is strictly diagonally dominant:

$$\begin{bmatrix} 9 & 3 & -2 & -3 \\ 3 & -7 & -2 & -1 \\ 0 & -1 & 4 & 1 \\ -5 & 2 & 0 & 8 \end{bmatrix}$$

because

$$|9| > |3| + |-2| + |-3| = 8$$
$$|-7| > |3| + |-2| + |-1| = 6$$
$$|4| > |0| + |-1| + |1| = 2$$
$$|8| > |-5| + |2| + |0| = 7$$

11. Using the matrix in the solution to Exercise 9(b), the sequence of vectors obtained from Jacobi's method is given in the following table (the vector $\mathbf{x}_5$ is an approximate solution to the system of equations):

k	1	2	3
$\mathbf{x}_k$	$\begin{bmatrix} -0.4444 \\ 1.7143 \\ -3.7500 \\ 4.1250 \end{bmatrix}$	$\begin{bmatrix} -0.4742 \\ 2.0060 \\ -4.3527 \\ 3.4187 \end{bmatrix}$	$\begin{bmatrix} -0.9408 \\ 2.2663 \\ -4.1032 \\ 3.3271 \end{bmatrix}$

k	4	5
$\mathbf{x}_k$	$\begin{bmatrix} -1.0026 \\ 2.0081 \\ -4.0152 \\ 2.9704 \end{bmatrix}$	$\begin{bmatrix} -1.0159 \\ 2.0074 \\ -3.9906 \\ 2.9963 \end{bmatrix}$

Index